SEAWATER INTRUSION IN COASTAL AQUIFERS –
CONCEPTS, METHODS AND PRACTICES

Theory and Applications of Transport in Porous Media

Series Editor:
Jacob Bear, *Technion – Israel Institute of Technology, Haifa, Israel*

Volume 14

The titles published in this series are listed at the end of this volume.

Seawater Intrusion in Coastal Aquifers – Concepts, Methods and Practices

Edited by

Jacob Bear

Technion-Israel Institute of Technology,
Haifa, Israel

Alexander H.-D. Cheng

University of Delaware,
Newark, Delaware, U.S.A.

Shaul Sorek

Ben-Gurion University,
Sede Boker, Israel

Driss Ouazar

Ecole Mohammadia d'Ingenieur,
Rabat, Morocco

and

Ismael Herrera

National Autonomous University of Mexico,
Mexico City, Mexico

KLUWER ACADEMIC PUBLISHERS
DORDRECHT / BOSTON / LONDON

A C.I.P. Catalogue record for this book is available from the Library of Congress.

ISBN 978-90-481-5172-1

Published by Kluwer Academic Publishers,
P.O. Box 17, 3300 AA Dordrecht, The Netherlands.

Sold and distributed in North, Central and South America
by Kluwer Academic Publishers,
101 Philip Drive, Norwell, MA 02061, U.S.A.

In all other countries, sold and distributed
by Kluwer Academic Publishers,
P.O. Box 322, 3300 AH Dordrecht, The Netherlands.

Printed on acid-free paper

Contents

Preface   xi

List of Contributors   xiii

1   Introduction   1
    J. Bear & A.H.-D. Cheng
    1.1   Freshwater Resources 1
    1.2   Historical Development 2
    1.3   Organization of the Book 5

2   Geophysical Investigations   9
    M.T. Stewart
    2.1   Introduction 9
    2.2   Surface Geophysical Methods 13
    2.3   Borehole Methods 42
    2.4   Integrated Geophysical Surveys 47
    2.5   Summary 50

3   Geochemical Investigations   51
    B.F. Jones, A. Vengosh, E. Rosenthal & Y. Yechieli
    3.1   Introduction 51
    3.2   World-Wide Phenomena 52
    3.3   Chemical Modifications 57
    3.4   Mixing 60
    3.5   Water-Rock Interaction 60
    3.6   Intrusion of Fossil Seawater 68
    3.7   Criteria to Distinguish Saltwater Intrusions 69

4 Exploitation, Restoration and Management   73
   J.C. van Dam
   4.1   Introduction 73
   4.2   Optimal Exploitation of Fresh Groundwater in Coastal
         Aquifer Systems 81
   4.3   Measures to Restore Disturbed Fresh Groundwater
         Systems in Coastal Aquifers 105
   4.4   Groundwater Management 117

5 Conceptual and Mathematical Modeling   127
   J. Bear
   5.1   Introduction 127
   5.2   Three-Dimensional Sharp Interface Model (3DSIM) . 131
   5.3   Two-Dimensional Sharp Interface Model (2DSIM) . . 145
   5.4   Transition Zone Model (3DTZM) 151
   5.5   Conclusion 161

6 Analytical Solutions   163
   A.H.-D. Cheng & D. Ouazar
   6.1   Introduction 163
   6.2   Ghyben-Herzberg Solution 164
   6.3   Glover Solution 167
   6.4   Fetter Oceanic Island Solution 169
   6.5   Strack Pumping Well Solution 170
   6.6   Superposition Solution 177
   6.7   Bear and Dagan Upconing Solution 184
   6.8   Stochastic Solutions 187

7 Steady Interface in Stratified Aquifers of Random
   G. Dagan & D.G. Zeitoun
   Permeability Distribution   193
   7.1   Introduction 193
   7.2   Mathematical Formulation 195
   7.3   Solution of Planar Problem 199
   7.4   Upconing of Interface Beneath a Pumping Well . . . 205
   7.5   Summary and Conclusions 210

8 USGS SHARP Model   213
   H.I. Essaid
   8.1   Introduction 213
   8.2   Finite Difference Approximation of Freshwater and
          Saltwater Flow Equations 218
   8.3   General Model Features and Use 222
   8.4   Model Evaluation 225
   8.5   Soquel-Aptos Basin, California, a Multilayered Coastal
          Aquifer System 238
   8.6   Conclusions 246

9 USGS SUTRA Code—History, Practical Use, and
   Application in Hawaii   249
   C.I. Voss
   9.1   Introduction 249
   9.2   History of SUTRA 250
   9.3   Uses of SUTRA 253
   9.4   SUTRA Equations and Algorithms 256
   9.5   Application of SUTRA to Seawater Intrusion 262
   9.6   Setting Up a Coastal Seawater Intrusion Simulation 270
   9.7   Application of SUTRA to Major Coastal Aquifer of
          Southern Oahu, Hawaii 275

10 Three-Dimensional Model of Coupled Density-Dependent Flow
   and Miscible Salt Transport   315
   G. Gambolati, M. Putti & C. Paniconi
   10.1   Introduction 315
   10.2   Mathematical Model 317
   10.3   Numerical Discretization 321
   10.4   Linearization 329
   10.5   Projection Solvers for Linear Systems 341
   10.6   Applications 344

11 Modified Eulerian Lagrangian Method for Density Dependent
   Miscible Transport   363
   S. Sorek, V. Borisov & A. Yakirevich
   11.1   Introduction 363
   11.2   Mathematical Statement 364
   11.3   Formulation of Saltwater Intrusion Problem 373
   11.4   Numerical Examples 383

11.5   Conclusion 395

12 Survey of Computer Codes and Case Histories   399
S. Sorek & G.F. Pinder
12.1   General 399
12.2   3DFEMFAT/2DFEMFAT 401
12.3   CODESA-3D 404
12.4   DSTRAM 407
12.5   FAST-C(2D/3D) 408
12.6   FEFLOW 413
12.7   HST3D 419
12.8   MEL2DSLT 424
12.9   MLAEM/VD 428
12.10 MOCDENSE 431
12.11 MOC-DENSITY/MOCDENS3D 434
12.12 SALTFRES 439
12.13 SALTHERM/3D 443
12.14 SHARP and SWIP 446
12.15 T3DVAP.F/MOR3D.F 453
12.16 TVD-2D/TVD-3D 456
12.17 Summary 458

13 Seawater Intrusion in the United States   463
L.F. Konikow & T.E. Reilly
13.1   Introduction 463
13.2   Early History 464
13.3   Case Studies 466
13.4   Alternative Modeling Approaches 493
13.5   Closing Remarks 505

14 Impact of Sea Level Rise in the Netherlands   507
G.H.P. Oude Essink
14.1   Introduction 507
14.2   The Model 509
14.3   Propagation of Sea Level Rise 520
14.4   Saltwater Intrusion 522
14.5   Freshwater Lenses in Sand-Dune Areas 525
14.6   Seepage in Low-Lying Polders 527
14.7   Compensating Measures 528
14.8   Conclusions and Recommendations 529

15 Movement of Brackish Groundwater Near a Deep-Well
   Infiltration System in the Netherlands   531
   A. Stakelbeek
   15.1   Introduction 531
   15.2   Design of the Deep-Well Infiltration System 532
   15.3   Measurements of Movement of Brackish
          Groundwater During Operation 536
   15.4   Conclusions and Recommendations 540

16 A Semi-Empirical Approach to Intrusion Monitoring in Israeli
   Coastal Aquifer   543
   A.J. Melloul & D.G. Zeitoun
   16.1   Introduction 543
   16.2   Hydrogeological Background and Saltwater
          Monitoring of Study Area 545
   16.3   Semi-Empirical Procedure to Assess Saltwater
          Intrusion 550
   16.4   Actual Implementation to Israel Coastal Aquifer . . 556
   16.5   Conclusions 557

17 Nile Delta Aquifer in Egypt   559
   M. Sherif
   17.1   Introduction 559
   17.2   Nile Delta Aquifer 560
   17.3   Origin of Brackish Water 568
   17.4   Intrusion Mechanism and Boundary Conditions . . . 570
   17.5   Modeling Seawater Intrusion in Nile Delta Aquifer . 573
   17.6   Effect of Pumping 579
   17.7   Effect of Climate Change 583
   17.8   Concluding Remarks 590

   Bibliography   591

Preface

Coastal aquifers serve as major sources for freshwater supply in many countries around the world, especially in arid and semi-arid zones. Many coastal areas are also heavily urbanized, a fact that makes the need for freshwater even more acute. Coastal aquifers are highly sensitive to disturbances. Inappropriate management of a coastal aquifer may lead to its destruction as a source for freshwater much earlier than other aquifers which are not connected to the sea. The reason is the threat of *seawater intrusion*.

In many coastal aquifers, intrusion of seawater has become one of the major constraints imposed on groundwater utilization. As seawater intrusion progresses, existing pumping wells, especially those close to the coast, become saline and have to be abandoned. Also, the area above the intruding seawater wedge is lost as a source of natural replenishment to the aquifer.

Despite the importance of this subject, so far there does not exist a book that integrates our present knowledge of seawater intrusion, its occurrences, physical mechanism, chemistry, exploration by geophysical and geochemical techniques, conceptual and mathematical modeling, analytical and numerical solution methods, engineering measures of combating seawater intrusion, management strategies, and experience learned from case studies. By presenting this fairly comprehensive volume on the state-of-the-art of knowledge and experience on saltwater intrusion, we hoped to transfer this body of knowledge to the geologists, hydrologists, hydraulic engineers, water resources planners, managers, and governmental policy makers, who are engaged in the sustainable development of coastal fresh groundwater resources.

The editors would like to take this opportunity to express our gratitude toward Dr. Leonard Konikow of U.S. Geological Survey, who provided valuable suggestions and coordinated a few of the contributions. The assistance provided by the publishing editor, Petra D. van Steenbergen, at Kluwer, is also appreciated. The ultimate credit of the book goes to the authors of the chapters, who have contributed their time and effort, and generously shared their expert knowledge with the community.

JACOB BEAR (Haifa, Israel)
ALEXANDER CHENG (Delaware, USA)
SHAUL SOREK (Sede Boker, Israel)
DRISS OUAZAR (Rabat, Morocco)
ISMAEL HERRERA (Mexico City, Mexico)

September, 1998

List of Contributors

Prof. Jacob Bear
Department of Civil Engineering, Technion—Israel Institute of Technology, Haifa 32000, Israel.

Dr. Viacheslav Borisov
Water Resources Research Center, J. Blaustein Institute for Desert Research, Ben-Gurion University of the Negev, Sde Boker Campus 84990, Israel.

Prof. Alexander H.-D. Cheng
Department of Civil & Environmental Engineering, University of Delaware, Newark, Delaware 19716, USA.

Prof. Gedeon Dagan
Department of Fluid Mechanics and Heat Transfer, Faculty of Engineering, Tel-Aviv University, P.O. Box 39040, Ramat-Aviv, Tel-Aviv 69978, Israel.

Dr. Hedeff Essaid
Water Resources Division, U.S. Geological Survey, 345 Middlefield Rd., Menlo Park, California 94025, USA.

Prof. Giuseppe Gambolati
Dipartimento di Metodi e Modelli Matematici per le Scienze Applicate (DMMMSA), Universitá degli Studi di Padova, Via Belzoni 7, 35131 Padova, Italy.

Dr. Blair F. Jones
Water Resources Division, U.S. Geological Survey, 431 National Center, Reston, Virginia 20192, USA.

Dr. Leonard F. Konikow
Water Resources Division, U.S. Geological Survey, 431 National Center, Reston, Virginia 20192, USA.

Dr. Abraham J. Melloul
Hydrological Service, Ministry of Agriculture, Water Commission,
P.O. Box 6381, Jerusalem 91063, Israel.

Prof. Driss Ouazar
Laboratoire d'Analyse de Systémees Hydrauliques, Ecole Mohamma-
dia d'Ingénieurs, Université Mohammed V, B.P. 765, Agdal Rabat,
Morocco.

Dr.Ir. Gualbert H.P. Oude Essink
Department of Geophysics, Faculty of Earth Sciences, University of
Utrecht, P.O. Box 80021, 3508 TA Utrecht, The Netherlands. For-
merly, Department of Water Management, Environmental and San-
itary Engineering, Faculty of Civil Engineering, Delft University of
Technology, P.O. Box 5048, 2600 GA Delft, The Netherlands.

Dr. Claudio Paniconi
Environment Group, Center for Advanced Studies, Research and De-
velopment in Sardinia (CRS4), Via Nazario Sauro 10, 09123 Cagliari,
Italy.

Prof. George F. Pinder
College of Engineering and Mathematics, University of Vermont,
Burlington, Vermont 05405, USA.

Dr. Mario Putti
Dipartimento di Metodi e Modelli Matematici per le Scienze Applicate (DMMMSA), Universitá degli Studi di Padova, Via Belzoni 7,
35131 Padova, Italy.

Dr. Tom E. Reilly
Water Resources Division, U.S. Geological Survey, 431 National
Center, Reston, Virginia 20192, USA.

Dr. Eliyahu Rosenthal
Research Division, Hydrological Service of Israel, P.O. Box 6381,
Jerusalem 91063, Israel.

Dr. Mohsen M. Sherif
Hydrology Department, Water Resources Division, Kuwait Institute
for Scientific Research, P.O. Box. 24885, Safat 13109, Kuwait. On
leave from Irrigation and Hydraulics Department, Faculty of Engineering, Cairo University, Giza, Egypt.

Prof. Shaul Sorek
Water Resources Research Center, J. Blaustein Institute for Desert Research, Ben-Gurion University of the Negev, Sde Boker Campus 84990, Israel.

Ir. Bert Stakelbeek
n.v. PWN Water Supply Company of North-Holland, Postbus 5, 2060 Bloemendaal, the Netherlands.

Prof. Mark Stewart
Department of Geology, University of South Florida, Tampa, Florida 33620-5200, USA.

Prof.Dr.Ir. Jan C. van Dam
van der Horstlaan 9, 2641 RT Pijnacker, The Netherlands.

Dr. Avner Vengosh
Research Division, Hydrological Service of Israel, P.O. Box 6381, Jerusalem 91063, Israel.

Dr. Clifford I. Voss
National Research Program, U.S. Geological Survey, 431 National Center, Reston, VA 20192, USA.

Dr. Alex Yakirevich
Water Resources Research Center, J. Blaustein Institute for Desert Research, Ben-Gurion University of the Negev, Sde Boker Campus 84990, Israel.

Dr. Yoseph Yechieli
Geological Survey of Israel, 30 Malkei Israel St., Jerusalem 95501, Israel.

Dr. David G. Zeitoun
Consultant Hydrology and Water Resources, 9/4 Kosovsky St., Jerusalem 96304, Israel.

Chapter 1

Introduction

J. Bear & A. H.-D. Cheng

1.1 Freshwater Resources

Of all the water on earth, it is estimated that 99.4% (1.4×10^9 km^3) is surface water. Groundwater occurs only as 0.6% (9×10^6 km^3) of the total. However, of the vast amount of surface water, most of it is in the form of saltwater in oceans and inland seas (97%). Fresh surface water accounts for only 2% of the total volume of water.

By excluding saltwater from the above accounting, the proportion between surface and subsurface freshwater is changed to 78% and 22%, respectively. A further breakdown of freshwater shows that most of the surface freshwater is in the form of ice locked in ice-caps and glaciers in polar regions (77% of the total volume of freshwater). The fresh surface water resources that are accessible for human consumption are water in lakes (0.3%) and streams (0.003%). These are dwarfed by the amount of groundwater (22%). As a water crisis is forecasted in the near future [Gleick, 1993], the welfare of the world's population is closely tied to a sustainable exploitation of groundwater.

Historically, surface water has accounted for most of the human consumption, because it is easily accessible (with the exception of arid regions, where groundwater may be the only reliable source of water). Modern development and population growth, however, has greatly increased water demands. Surface water resources are being depleted, and furthermore, contaminated. Alternative water resources have to be sought. In the last half century, the demand for

1

J. Bear et al. (eds.), Seawater Intrusion in Coastal Aquifers, 1–8.
© 1999 *Kluwer Academic Publishers.*

groundwater has been rising steadily. Nowadays, groundwater use amounts to about one-third of total freshwater consumption in the world.

Groundwater resources can be divided into two more or less equal groups: shallow (less than 800 m deep) and deep (more than 800 m) groundwater. Generally speaking, shallow groundwater is easily accessible using conventional water well technologies. Deep groundwater is largely unexploited. However, as the need for new water resources increases, advanced technologies developed in the petroleum industry have been used to reach the deeper groundwater.

Despite its abundance, unregulated extraction of groundwater can easily cause localized problems. In coastal zones, the intensive extraction of groundwater has upset the long established balance between freshwater and seawater potentials, causing encroachment of seawater into freshwater aquifers. As a large proportion of the world's population (about 70%) dwells in coastal zones, the optimal exploitation of fresh groundwater and the control of seawater intrusion are the challenges for the present-day and future water supply engineers and managers.

1.2   Historical Development

Since the famous works of Badon-Ghyben [1888] and Herzberg [1901], extensive research has been carried out, and much progress has been made in many parts of the world in understanding the various mechanisms that govern seawater intrusion. The dominant factors are the flow regime in the aquifer above the intruding seawater wedge, the variable density, and hydrodynamic dispersion.

Over the years, and especially during the 50's and the 60's, a large number of field investigations have also been conducted in many coastal aquifers, providing a basis for understanding the complicated mechanisms that cause seawater intrusion and affect the shape of the zone of transition from fresh aquifer water to seawater. Unfortunately, at that time, computational tools were not available to predict the extent of seawater intrusion under various natural and man-made conditions. In most cases, simplified conceptual models had to be assumed in order to enable analytical solutions for the shape of the seawater wedge. For example, analytical solutions were derived for steady flow in an aquifer that consists of a single homogeneous porous medium overlying a horizontal impervious bottom, assuming that a sharp interface separates the seawater and freshwater subdomains. Another often used assumption was that the flow in the aquifer was essentially horizontal, a statement that is equivalent to the Dupuit assumption. In fact, this assumption lead to the famous *Ghyben-Herzberg relationship*. These analytical solutions provided estimates for the shape of a stationary interface in a coastal phreatic aquifer, with flow everywhere perpendicular to the coast. Some analytical solutions gave the parabolic shape of an assumed sharp interface in an infinitely thick confined aquifer. Strack [1995] used the Dupuit-Forchheimer assumption to derive a solution for sea water intrusion in a horizontal two-dimensional domain.

Obviously, the above paragraph is by no means a literature survey. It was brought up just to emphasize the lack of solutions for more realistic models.

In spite of the rough approximation associated with such analytical solutions, they were very important at that time, as they provided the fundamental relationship between the rate of freshwater discharge to the sea, and the length of the intruding seawater wedge. According to this relationship, the rate of freshwater flow to the sea above the interface determines the length of the seawater wedge intruding into the aquifer. As the this discharge is reduced, e.g., by pumping a larger proportion of the natural replenishment, the length of the seawater wedge will increase, causing wells to start pumping saline water. They also provided information on the relationship between water levels in the vicinity of the coast and the length of the seawater wedge.

Some researchers reached the same conclusions by employing the Hele-Shaw analog to investigate the entire water balance of specific locations in a coastal aquifer (again, in a cross-section perpendicular to the coast), taking into account two-dimensional flow in the vertical cross-section, detailed geology (layers, sub-aquifers, heterogeneity, and anisotropy), natural replenishment, and schedules of pumping. Actually, these two conclusions are essentially coupled to a single relationship between the length of the wedge, the water levels above it, and the rate of freshwater flow to the sea.

This relationship served as a basis for the management of groundwater in coastal aquifers. By controlling the rate of freshwater pumping, it was possible to control the length of the seawater wedge. By controlling water table elevations, e.g., by artificial recharge, it was possible to stop sea water intrusion, or even cause the interface to advance seaward.

Actually, seawater and freshwater are nothing but water with different concentrations of salt. The passage from the portion of the aquifer occupied by the former, to that occupied by the latter, takes the form of a *transition zone*, rather than a sharp interface. Under certain circumstances, depending on the extent of seawater intrusion, and on certain aquifer properties, this transition zone, which is, primarily, a result of hydrodynamic dispersion of the dissolved matter, may be rather wide. Under other conditions, it may be narrow, relative to the aquifer thickness, and may be approximated as a sharp interface.

Obviously, the nature of the mathematical models of seawater intrusion, whether as a sharp interface approximation, or as a transition zone model, called for numerical solutions. Starting in the early 70's such models have been developed and applied. Of special interest were those models that enabled investigations of more realistic coastal aquifer domains, taking into account the three-dimensional flow regime, heterogeneity and anisotropy, etc. As computer hardware and numerical algorithms advance in the last few decades, so do the numerical solution efforts investigating seawater intrusion.

Basically, the management of groundwater in coastal aquifers means making decisions as to the rates, temporal and spacial distributions of pumping and of artificial recharge in such aquifers. Objective functions may be associated with the economic aspects of groundwater utilization. Based on the general trend established by the discussion above, it is obvious that this is an optimization problem: increasing pumping, with its associated economic benefits, will produce more seawater intrusion which causes economic damages (abandoning wells, etc.).

Once seawater has invaded to a distance beyond that is tolerable, restoration of water quality in the invaded zone is generally an expensive or ineffective proposition. It requires a large amount of freshwater flushing for a long period of time. A more effective way is by prevention. Techniques have been developed, e.g., by using artificial recharge to raise water levels closed to the coast. The use of relatively small amount of water to create a freshwater mound close to the coast can arrest the advance of seawater wedge.

Management of coastal aquifers involves more than the management of seawater intrusion into such aquifers. Being, at least in developed regions, heavily populated, coastal regions that overlie the aquifers constitute a permanent threat to the quality of the fresh water in the aquifers. This is especially true when these aquifers are phreatic. Special attention should be given to the danger of groundwater contamination.

1.3 Organization of the Book

The time is ripe for a stock-taking of our present knowledge on this vital subject. A lot of experience has been accumulated in many parts of the world, and new technologies have been developed to combat seawater intrusion. To share this body of knowledge, a panel of internationally renowned experts has been invited to address the various aspects of seawater intrusion.

Chapter 2 reviews the methodologies for geophysical investigations that are used to detect the presence of saltwater in coastal aquifers. In addition to the conventional tools such as electrical resistivity and seismic sounding, of particular interest are the modern technologies based on electromagnetic waves.

Chapter 3 presents the geochemical aspects. The presence of naturally occurring isotopes, or 'environmental tracers', has long provided geologists with an insight into the origin of water. These chemical signatures can be applied to the identification of salinity sources, as not all salt contamination is derived from intruding seawater. It is further demonstrated that, on a worldwide scale, the mixed water in the transition zone carries a geochemical character that can foretell the trend of seawater intrusion.

Chapter 4 discusses the optimal exploitation of fresh groundwater as an element of integrated water management in coastal zones. Management aspects such as data collection, aquifer monitoring as a warning system, modeling, building management scenarios, and legislation, are reviewed. Once contaminated, various remedial measures are suggested to restore fresh groundwater in the aquifer.

Chapter 5 is a rigorous presentation of the physical conceptualization that leads to the various mathematical models. The 2-D and 3-D sharp interface, and the 3-D miscible transport models are laid out in complete detail. Boundary conditions at the various geological and hydraulic situations are clearly defined. This chapter may serve as the foundation for the numerical modeling efforts.

Chapter 6 is an elementary coverage of existing analytical solutions. With the increasing power of numerical methods, the selection is not intended to be comprehensive. Rather, it focuses on solutions that can enhance the physical understanding of the phenomenon, and those that can be of practical use in the initial stages of an investigation and engineering assessment.

Chapter 7 tackles an important, but all too often neglected, issue— the randomness of aquifer parameters (see also Chapter 6). The traditional deterministic approach assumes that the parameter input needed in an analytical or a numerical solution is known with certainty. In reality, however, such physical parameters are never surveyed to an accuracy and a fine enough spatial resolution that can capture the variability of the medium. This input uncertainty creates an output uncertainty that cannot be ignored. This chapter investigate the stochastic location of the seawater-freshwater interface given a random hydraulic conductivity distribution in a stratified aquifer.

Chapter 8 leads off a series numerical contributions. It presents the USGS (U.S. Geological Survey) SHARP computer program, one of the most widely used involving a sharp interface and vertically integrated flow assumption. The program is quasi-three-dimensional by allowing the coupling of layers through leakage. Due to the inherent efficiency of a 2-D sharp interface model, simulating multiple layered aquifer systems in regional scales with sufficient resolution can be achieved.

Chapter 9 presents a two-dimensional, variable density, saturated-unsaturated flow, and dispersive transport code, the USGS SUTRA. The program is claimed to be the most widely used for simulating variable-density, heat and mass transport problems. It is also one of the most popular in seawater intrusion studies. The author discusses in great detail the physical and numerical modeling issues in seawater intrusion using SUTRA. A case study is used to demonstrates the level of attention that is needed to capture the salt-fresh water interacting dynamics in the transition zone.

Chapters 10 offers a three-dimensional computer program that solves the density-dependent, variably saturated flow, with miscible solute transport, using the Eulerian approach. Special attention is paid to the coupling terms, nonlinearities, discretization, and iteration schemes, to ensure the good behavior of the code.

Chapter 11 provides an alternative algorithm, known as the Eulerian-Lagrangian approach, to overcome the numerical difficulties encountered in advection-dominated transport problems, and in aquifer zone where steep hydraulic conductivity gradients are present. Numerical implementation is provided in a vertically integrated 2-D geometry.

Chapter 12 is a comprehensive survey of more than a dozen computer codes that have been used for saltwater intrusion modeling. Despite the exponential increase in CPU speed of modern-day computing equipment, numerical modeling of coupled density dependent flow and dispersive solute transport in real-world geometry still poses a challenge. There does not exist a single code that can outperform all others in accuracy, efficiency, and ease of use, in a wide range of parameter values and problem settings. Trade-offs always exist. This chapter compares the computer codes side-by-side by asking the code developers to answer a sequence of survey questions concerning the physical assumptions and numerical modeling issues. The users are hence presented with the relevant information that can help them to choose the most suitable code for their applications.

Chapter 13 and the remainder of the book are devoted to case studies. The U.S. experience leads this group. To provide a perspective, the history of combating seawater intrusion in California and Florida, the two states that have the largest population served by groundwater, is examined. The readers are then carefully guided through a series of rationales that help to determine the alternative numerical modeling approaches, sharp interface or miscible transport, 2-D or 3-D geometry, etc., using case study examples. Also, engineering approaches to manage or control seawater intrusion are discussed.

Chapter 14 is a case study in the Netherlands. A large part of the country is below sea level and has to be protected by dikes. Seawater intrusion into aquifers is a historical problem that has received considerable attention. This chapter investigates the further impact on aquifer water quality due to the projected sea level rise in the future as a consequence of the global warming.

Chapter 15 is another case study in the Netherlands. It discusses the current industry practice of utilizing deep well injection to preserve fresh groundwater resources.

Chapter 16 concerns the Israeli experience. Under the condition of a deteriorating water quality monitoring well network, the authors demonstrate the semi-empirical approach of using electromagnetic survey data and least square fitting to fill in the gaps left by the inadequate monitoring well network.

Chapter 17 is an investigation of another densely populated region, the Nile Delta of Egypt, where groundwater is a vital part of water resources. A comprehensive study, ranging from hydrogeology, numerical modeling, management, and sea level rise impact, has been conducted.

Finally, the references cited in the chapters are collected in a single bibliography at the end of the book. This collection of nearly 600 references is one of the most comprehensive sets on this subject that exist today.

Chapter 2

Geophysical Investigations

M. T. Stewart

2.1 Introduction

2.1.1 Statement of the Problem

Saltwater intrusion is the mass transport of saline waters into zones
previously occupied by fresher waters. In hydrogeologic systems mass
transport rates and pathways are determined by hydraulic gradients and the distribution of hydraulic conductivity. The configuration of the saltwater interface and its movement are determined by
the configuration of the water table and the spatial variation in hydraulic conductivity. One objective of geophysical surveys in island
and coastal environments is to help define the physical framework
within which saltwater intrusion occurs. Geophysical methods which
can distinguish geologic units on the basis of their physical properties, such as seismic methods, can be very useful even though they
cannot detect saline waters directly.

Saltwater intrusion is distinguished by the movement of waters
with high total dissolved solids (TDS) content into fresher waters.
As the principal ions in seawater, sodium and chloride, are very conductive, water quality in coastal and island environments can be determined by measuring fluid conductivity. In saturated, porous earth
materials fluid conductivity is a dominant factor in determining the
overall bulk conductivity of the fluid/matrix system. This means that
geophysical methods that can be used to estimate bulk conductivity,
or its inverse, bulk resistivity, can be very useful in delineating water
quality variations in coastal and island environments.

9

J. Bear et al. (eds.), Seawater Intrusion in Coastal Aquifers, 9–50.
© 1999 *Kluwer Academic Publishers.*

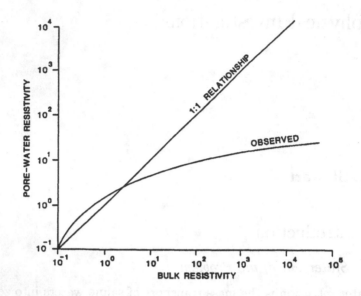

Figure 2.1. Relationship between pore-fluid resistivity and bulk resistivity in ohm-m (modified from Keller and Frischnecht [1966]).

2.1.2 Physical Principles

Geophysical methods measure the spatial distribution of physical properties of the earth, such as bulk conductivity or seismic velocity. These physical properties can be related to hydrologic or geologic features, such as the distribution of water quality or the geometry, position, and properties of geologic units. It is important to note that geophysical methods do not directly determine characteristics such as salinity or lithology. These characteristics must be interpreted from the distribution and magnitude of the physical properties determined by geophysical surveys.

For coastal and island environments, the physical properties of principal interest are bulk conductivity and seismic velocity. Bulk conductivity is a macroscopic property of the fluid/matrix system. Bulk conductivity, and its inverse, bulk resistivity, are determined principally by fluid conductivity and matrix porosity. As porosity and fluid conductivity increase, bulk conductivity increases. As illustrated in Figure 2.1, fluid conductivity is the principal determinant of bulk conductivity for higher TDS pore fluids, while porosity becomes the principal determinant of bulk conductivity when pore

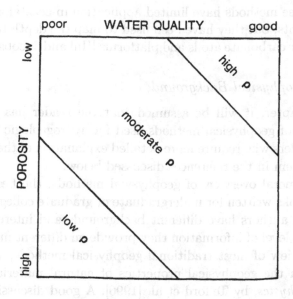

Figure 2.2. Relationship of bulk resistivity, ρ, to variations in porosity and water quality. (From Stewart et al. [1983]).

fluids become very resistive [Keller and Frischneckt, 1966]. This dependence of bulk conductivity or resistivity on porosity and fluid conductivity leads to an ambiguity, as a given bulk conductivity can be derived from a wide range of porosities and fluid conductivities (Figure 2.2). This ambiguity is resolved only when pore waters are very fresh or very saline.

Seismic velocity is dependent principally on the mechanical properties of a material, specifically the values of the elastic constants such as Poisson's ratio, Young's modulus, bulk modulus, and the shear modulus [Telford et al., 1990]. As a porous material becomes more compact, cemented, or lithified, grain-to-grain contacts increase, and mechanical strength and rigidity, as measured by the elastic constants, also increase, leading to an increase in seismic velocity. Variations in seismic velocity can be used to delineate the location and geometry of the boundaries between geologic units, and to make some inferences about their lithology.

Potential field methods, gravity and magnetics, measure variations in the local gravity and magnetic fields, which can be interpreted in terms of variations in density and magnetic susceptibility, respectively. These methods have limited application in coastal and island environments, but they have been used to map the depth to volcanic rocks under carbonate atolls and platforms [Hill and Jacobsen, 1989].

2.1.3   Geophysical Background

In this chapter, it will be assumed that the reader has a general knowledge of geophysical methods used for hydrogeologic investigations. Readers who require more detailed explanation of the methods can find them in the references discussed below.

For a general overview of geophysical methods, the best sources are textbooks written for undergraduate or graduate college courses. As various authors have different backgrounds and interests, texts vary in the level of information they provide on different methods. A good overview of most traditional geophysical methods, with good sections on the geophysical properties of natural materials is *Applied Geophysics*, by Telford et al. [1990]. A good discussion of the physical principles that govern the response of geophysical methods is contained in Beck [1990], *Physical Principles of Exploration Geophysics*. Burger [1992], provides a good introduction to shallow seismic methods, with the emphasis on environmental and engineering applications. Another general text is Kearey and Brooks [1991].

A useful reference for environmental geophysics is Ward [1990a]. This is a collection of 71 papers on applied geophysics. Volume I consists of reviews and tutorials on common geophysical methods. Articles in Vol. I of particular interest for island and coastal studies include those on seismic reflection [Steeples and Miller, 1990], seismic refraction [Lankston, 1990], resistivity [Ward, 1990b], electromagnetic methods [McNeill, 1990], and borehole methods [Daniels and Keys, 1990].

Volume II of Ward [1990a] contains examples of field studies using geophysical methods. Articles in this volume of interest to saltwater intrusion studies include Hoekstra and Blohm [1990], Goldstein et al. [1990], Stewart [1990], Hagemeyer and Stewart [1990], Humphreys et al. [1990], Street and Engel [1990], Barker [1990], Slaine et al. [1990], and Dodds and Dragan [1990].

As EM terrain-conductivity meters, time-domain sounding systems, and ground-penetrating radar (GPR) methods are relatively new, information on these systems is not yet found in most textbooks. Terrain conductivity meters and their applications are discussed in McNeill [1990, 1980], Stewart [1982, 1990], and Stewart and Bretnall [1986]. The principles of TDEM sounding systems are explained in Stewart and Gay [1986], Fitterman and Stewart [1986], Hoekstra and Blohm [1990], and McNeill [1990]. A good introduction to GPR is contained in Davis and Annan [1989].

2.2   Surface Geophysical Methods

As their name states, surface geophysical methods attempt to gain information about the subsurface from geophysical surveys taken at the earth's surface. In general, the vertical and horizontal resolution of these methods both decrease with increasing depths of investigation. The methods and applications discussed in this section are appropriate for depths of investigation of a few meters to a few hundred meters, depending on the method and the survey parameters.

2.2.1   Electrical Methods

Electrical methods are the most widely used geophysical methods in island and coastal environments. This is the result of the increase in bulk conductivity with increasing pore water conductivity. Electrical methods are usually used to locate brackish or saline waters, although they can yield some geologic information as well. The discussion of electrical methods will be divided into two sections, one on DC resistivity methods, the other on electromagnetic methods.

2.2.1.1   DC Resistivity

DC resistivity was the first surface geophysical method used for water quality determination in island and coastal environments [Swartz, 1937, 1939]. It is still one of the more commonly used methods [Chidley and Lloyd, 1972; Fretwell and Stewart, 1981; Gorhan, 1976; Van Dam and Meulancamp, 1967]. Its principal advantages are relatively low instrument cost and simplicity of operation.

The DC method introduces electrical currents into the ground through current electrodes driven into the soil. The resulting electrical potential (voltage) is measured between two potential electrodes. Resistivity is calculated by multiplying the ratio of voltage to current by a geometric factor that varies with the electrode configuration and

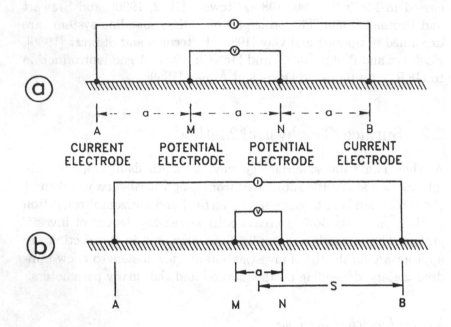

Figure 2.3. DC resistivity electrode arrays; (a) Wenner (b) Schlumberger (From Dobrin and Savit [1988]).

spacing. The calculated resistivity is the true resistivity over homogeneous ground, but is an integrated or apparent resistivity over an electrically heterogeneous earth.

Most DC surveys use a linear, four-electrode array. The most common arrays are the Schlumberger and Wenner (Telford et al. [1990], and Figure 2.3). For vertical electrical soundings (VES), the current electrodes are moved to increasing electrode spacings around a central point. The variation in apparent resistivity with increasing current electrode spacing can be related to vertical variations in bulk resistivity in the subsurface. Operationally, the Schlumberger array is preferred, but the Wenner array allows larger current electrode separations with the portable, battery-powered instruments generally used.

DC resistivity can also be used for profiling to detect lateral changes in bulk resistivity. Traditionally, this involved moving an electrode array with fixed electrode spacings laterally, or completing closely-spaced soundings along a profile. Recent advances in instruments and equipment have made electrical profiling less labor intensive. Multi-conductor cables with appropriately spaced takeouts connect large numbers of electrodes along a profile. A "smart" receiver then cycles through the electrodes sequentially, producing a series of closely-spaced soundings along the profile. This is sometimes called resistivity "imaging". Software which performs 2-D and 3-D resistivity inversions on the field data from these instruments is available [Loke and Barker, 1996a, 1996b].

Vertical electrical sounding data are typically interpreted by calculating the theoretical response for a proposed geoelectric model with specified layer thicknesses and resistivities, and comparing the calculated response to the field data. The geoelectric model is adjusted until there is a reasonable match between the calculated and observed data. Some programs, such as ATO [Zohdy and Bisdorf, 1989; Zohdy, 1989] automate the process, and return a geoelectric model with as many layers as data points. Others, such as RESIX [Interpex, 1995], require the interpreter to select the number of layers, and the program then statistically iterates to a "best-fit" solution. Forward calculations can also be made with appropriate equations entered into a spreadsheet [Sheriff, 1992]. Using a spreadsheet frees the interpreter from the necessity of purchasing or renting a commercial interpretation program, but the iterative solution can be tedious when performed manually.

Resistivity data are prone to equivalence, which means that several different geoelectric models can produce very similar resistivity responses, leading to non-unique or equivalent solutions. Unfortunately, coastal and island environments tend to produce DC data prone to equivalence problems. The large resistivity contrast between the high resistivity, unsaturated or freshwater zones and the low resistivity, saline saturated zones suppresses the expression on the field curves of intervening layers of intermediate resistivity. Equivalence can be reduced by fixing one or more of the geoelectric parameters by using other data. For example, seismic refraction data can be used to fix the thickness of one or more layers, giving more confidence in the interpreted values for layer resistivities. Equivalence analysis determines the range of geoelectric model parameters that will fit the observed data within a specified fitting tolerance. Equivalence analysis is invaluable for assigning confidence levels to derived geoelectric solutions, such as the depth to the saltwater interface.

DC resistivity methods are usually used in island and coastal environments to detect the saltwater/freshwater interface. The sharp increase in TDS at the interface usually produces a distinct drop in bulk resistivity. Interpretation of DC data usually provides a solution with only a few layers, so that the geophysically delineated interface is a sharp boundary. In general, DC methods tend to put the sharp interface near the middle of the transition zone [Stewart, 1990]. The transition zone is seldom well resolved, and attempts to represent it with layers of transitional resistivity usually are not successful [Kauahikaua, 1987]. Kauahikaua concludes that a transition zone must be twice as thick as its depth to be detected.

In Tertiary and Quaternary carbonate terranes, the bulk porosity of the units tends to decrease with increasing geologic age, while the hydraulic conductivity increases [Cant and Weech, 1986]. Both trends are the result of dissolution and reprecipitation of carbonate minerals, creating lower bulk porosity, but larger average pore sizes. The lower porosity of the older, more permeable units creates high bulk resistivities when the units are saturated with freshwater. In a case in southwest Florida, lower porosity, higher permeability carbonate units saturated with freshwater are distinguished from surrounding lower permeability units saturated with brackish water by the distinctive, high resistivity of the permeable, fresh units (Stewart, Layton, and Lizanec [1983], and Figure 2.4).

The depth of investigation of vertical electrical soundings is about 10-20% of the current electrode spacing [Stewart and Bretnall, 1986]. Deep penetration requires long current electrode spacings, and arrays must be expanded along a straight line [Telford et al., 1990]. This can be restrictive on smaller islands or where roads are few. In carbonate terranes with thin soil it can be very difficult to insert the electrodes. Sandy soils found in many coastal regions can have very high contact resistance, making it difficult to introduce electrical currents large enough to produce measurable potentials. In this case, using several electrodes in a small group instead of one current electrode helps, as does watering the current electrodes or driving them deeper in the soil to get below the desiccated surface sands.

It is difficult to confidently resolve more than 3-4 layers with DC soundings, and water quality changes in fine-grained sediments can be difficult to resolve. Complex stratigraphy or decreasing grain size with depth, as found in deltaic and alluvial environments or trans-

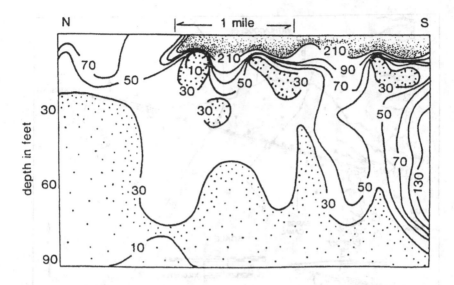

Figure 2.4. Resistivity cross section from southwest Florida, U.S.A. Contours are in ohm-m. Vertical electrical soundings were taken every 0.25 mi along the profile. Chloride ion values are > 250–500 mg/l where bulk resistivity values are < 30 ohm-m. (From Stewart et al. [1983]).

gressive barrier island sequences, respectively, can produce ambiguous results as a result of the difficulty of distinguishing between lithologic and water quality boundaries. Actual estimation of water quality requires calibration with water quality data from wells.

As a field example, a case history from Florida will be used [Hagemeyer and Stewart, 1990]. The Cross Florida Barge Canal was approved by Congress in 1938. When completed, it was to provide a sea level canal across the Florida Peninsula. Only segments of the project were completed before it was deauthorized by Congress. One segment is a sea level canal extending approximately 10 km inland from the Gulf of Mexico to a dam on the Withlacoochee River (Figure 2.5). It was excavated in the 1960's. It is about 25 m wide and about 4 m deep and is dug into the carbonate Floridan Aquifer, which is unconfined in this area.

An earlier, regional geophysical study of the saltwater interface position [Stewart and Gay, 1986] suggested that a major landward reentrant of the interface was present at the canal, but the study was

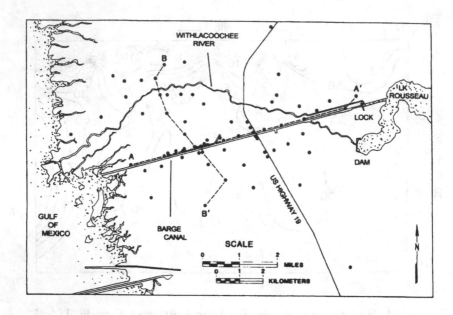

Figure 2.5. Location of vertical electrical soundings (solid circles) in the vicinity of the Cross Florida Barge Canal, a sea-level canal in west-central Florida, U.S.A. (From Hagemeyer and Stewart [1990]).

not detailed enough to outline the position of the interface near the canal. To investigate the effects of the canal on the interface position, 70 vertical electrical soundings were completed. The soundings were inverted using the program developed by Zohdy [Zohdy, 1973, 1975].

Two cross-sections were developed from the interpreted layer thicknesses and resistivities (Figure 2.5). The north-south cross-section, B-B', indicates that there are low resistivity zones associated with the canal and the nearby Withlacoochee River (Figure 2.6). Water quality data indicate that the low resistivity zone under the Withlacoochee River is a natural upconing of mineralized waters high in sulfate derived from intergranular anhydrite at the base of the Floridan Aquifer. The low resistivity zone under the Canal, however, is not connected to the deeper, more mineralized waters, and has higher chloride ion concentrations, indicative of seawater intrusion. The longitudinal, east-west cross-section A-A' (Figure 2.7) shows a low resistivity wedge (< 25 ohm-m) extending inland from the Gulf. Water quality data indicate that groundwater in this zone is elevated in chloride ion content, but not sulfate, suggesting seawater intrusion.

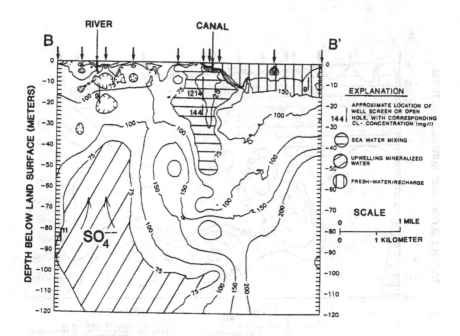

Figure 2.6. North-south geoelectric cross section B-B' near the Cross Florida Barge Canal. See Figure 1.5 for location of section. (From Hagemeyer and Stewart [1990]).

The shape of the intruding seawater wedge suggests that the source of the saline water is saltwater moving landward along the bottom of the canal, and moving downward into the aquifer. This has placed a wedge of saltwater over freshwater, a hydrodynamically unstable situation. The resistivity data, coupled with the water quality data, suggest that the apparent saltwater intrusion is in its early stages, and has not reached hydrodynamic equilibrium with the new hydrologic conditions created by the canal.

2.2.1.2   Frequency Domain Electromagnetic Methods

Electromagnetic methods measure the strength of secondary magnetic fields created by currents in the ground induced by a primary magnetic field. The more conductive the earth, the stronger are the induced currents in the ground, and also the resulting output from the receiver coil. The currents in the ground, termed eddy currents, are induced by time-varying magnetic fields produced by frequency-

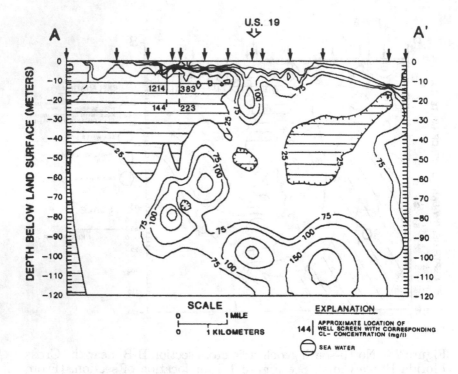

Figure 2.7. East-west geoelectric cross section A-A' along the north side of the Cross Florida Barge Canal. See Figure 1.5 for location of section. (From Hagemeyer and Stewart [1990]).

controlled AC currents in a transmitter coil. For most electromagnetic methods, the transmitter is close to the receiver coil, but one EM method, VLF (very low frequency), uses electromagnetic fields generated by transmitters hundreds of kilometers distant.

The transmitted electromagnetic field is called the primary field. The induced eddy currents in the earth produce the secondary field, the strength of which is related to the strength of the eddy currents, and hence, the terrain conductivity. The receiver coil senses the sum of the primary and secondary fields. The secondary field is usually measured 90° out of phase with the primary field, as the induced electrical field in poorly to moderately conductive units lags the primary magnetic field by about $\pi/2$. The ratio of the out-of-phase or quadrature component of the secondary field to the in-phase component of the primary field is an indication of terrain conductivity.

The depth of investigation of frequency EM methods is principally a function of the frequency of the primary field, with lower frequencies having greater penetration.

2.2.1.3   Airborne EM

Frequency EM methods create eddy currents through electromagnetic induction, so that no contact with the ground is required. This means that frequency EM systems can be flown by fixed wing aircraft or helicopters. The typical airborne system uses several receiver-transmitter coil pairs at varying frequencies. These coil pairs are placed in a "bird", which is towed behind the aircraft at elevations of 25-50 m above the ground. The depth of investigation is determined by transmitter frequencies. Common frequencies range from 56,000 Hz to 200-300Hz, yielding penetration depths of less than a meter to several 10's of meters, respectively. The interpreted output is an apparent resistivity map for each frequency. Interpretation is normally qualitative, aided by comparison of apparent resistivity maps produced by different frequencies [Frazer, 1978].

Successful application of airborne EM methods requires that several criteria be considered. First, the interface should be shallow, no more than a few 10's of meters below land surface. The study area should be relatively free of power lines and buried pipes and cables, as these produce linear anomalies on the resistivity maps. In order to keep the bird height above ground to a minimum, clear flight lines and low relief are helpful. For economic reasons, it is best to use airborne methods to survey large areas in order to spread the mobilization costs of the aircraft over a larger survey, lowering the cost per km of flight line. While airborne surveys work best in undeveloped areas, aircraft mobilization costs in remote areas may be prohibitively expensive.

Sengpiel [1982] used airborne EM to map the extent of a freshwater lens on a sandy barrier island on the North Sea coast. The survey results clearly show the extent of the freshwater lens (Figure 2.8). Fitterman [1996] used a multi-frequency airborne EM survey to locate the freshwater/saltwater boundary in the shallow subsurface in south Florida's Everglades (Figure 2.9). Fitterman's airborne data were calibrated with limited water quality data and surface EM data.

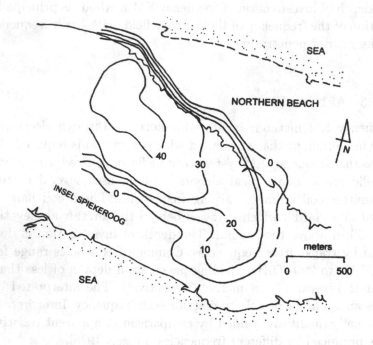

Figure 2.8. Apparent resistivity contours derived from an airborne frequency-domain survey of a barrier island along the North Sea coast. Contours in ohm-m. (After Sengpiel [1982])

2.2.1.4   Loop-Loop EM

Most frequency-domain electromagnetic surveys in island and coastal areas use loop-loop EM methods. These instruments use a transmitter coil to create the primary field, and a receiver coil to measure the resultant secondary field. One set of instruments, horizontal loop EM (HLEM) or "Slingram", uses multiple frequencies. These instruments are capable of conducting soundings by varying frequency, with the lowest frequencies having the deepest exploration depth [Kauahikaua, 1987]. Field data from these instruments are usually receiver coil output in mV, or the ratio of the secondary field to the primary field, in percent. Interpretation of the field data requires conversion to apparent conductivity values. This requires calibrating the instrument's zero levels over highly resistive ground for each frequency, which can be problematic if no calibration site is available. Practical exploration depths are up to 40-60 m.

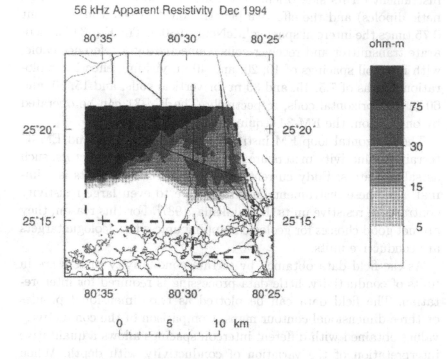

Figure 2.9. Apparent resistivity contours derived from an airborne frequency-domain survey of the Florida Everglades. Contours are in ohm-m. (After Fitterman [1996]).

The frequency-domain EM instruments most commonly used are terrain conductivity meters, often referred to by their commercial names, EM-31 and EM-34. These instruments use a combination of transmitter frequency and intercoil spacing that satisfies the low induction number approximation [McNeill, 1980a; Stewart, 1983, 1990]. This approximation allows these instruments to read directly in conductivity units over their practical operating range of conductivities, about 1-200 mS/m (1000-5 ohm-m). The depth of investigation is controlled by the transmitter frequency and is independent of the terrain conductivity [McNeill, 1980a].

The EM-31 has fixed coils mounted in a boom about 4 m long. In its normal operating position the transmitter and receiver coils are horizontal (vertical magnetic dipoles) and the effective penetration depth is 6 m, or 1.5 times the intercoil spacing. Placing the instrument on its side orients the coils vertically (horizontal magnetic dipoles) and the effective penetration depth is 3 m, or about 0.75 times the intercoil spacing [McNeill, 1980a]. The EM-34 has separate transmitter and receiver coils connected by a reference cable, with intercoil spacings of 10, 20, and 40 m, yielding effective exploration depths of 7.5, 15, and 30 m for vertical coils, and 15, 30, and 60 m for horizontal coils, respectively. The EM-31 can be operated by one person, the EM-34 requires two.

The horizontal loop EM instruments and the EM-31 and EM-34 terrain conductivity meters are best used to locate conductors, such as saline waters. Eddy current induction in resistive units is minimal, and these instruments are insensitive to even large resistivity contrasts in resistive units [Kauahikaua, 1987]. For this reason, they are not good choices for geologic mapping unless the geologic targets are conductive units.

As the field data obtained by terrain conductivity meters are in units of conductivity, little data processing is required for interpretation. The field data can be plotted as two-dimensional profiles or three-dimensional contour maps. Comparison of the conductivity values obtained with different intercoil spacings allows a qualitative interpretation of the variation of conductivity with depth. When mapping the extent of freshwater lenses, topography can strongly affect the field data, as topographic highs, such as dunes, place the instrument farther away from the saline water at depth. The effect of topography is inverted, that is, higher elevations produce lower conductivity readings. A simple correction for topography can be made using the equations for a layered earth given in McNeill [1980a]. At each station a two-layer model is created, with the thickness of the upper layer equal to the station elevation and the conductivity of the upper layer is given a low, estimated conductivity (5 mS/m is a good start). The remaining unknown is the conductivity of the second layer. Solving for this value yields the conductivity of the section below sea level. Different values of the first layer conductivity can be used until the value is found which produces the smallest correspondence between conductivity and land surface elevation. A spreadsheet is a convenient way to make these corrections.

Stewart [1990] describes a similar procedure for estimating the depth to the interface using an EM-34 terrain conductivity meter. This method requires estimates of the conductivity values of the unsaturated zone, the freshwater saturated zone, and the saltwater zone. These values can be obtained with DC soundings. Using a three layer model, where the thickness of layer one is the land surface elevation, the solution solves for the thickness of the second layer, which is the depth of freshwater below sea level. A spreadsheet program can be used to obtain the solution at each station. This method works well when the interface is within 30 m of land surface, the unsaturated and freshwater zones have low conductivities, and the water table is close to sea level. The estimated depth to the interface is very sensitive to the value chosen for the third-layer conductivity, so calibration to water quality data or DC soundings is necessary for good estimates of interface depth.

Alternatively, a forward calculation program, such as EMIX [Interpex, 1988] or PCLOOP [Geonics, 1985] that does not rely on the low induction number approximation can be used to solve for a two or three layer model at each station [Anthony, 1992; Stewart, 1990]. This allows use of the full operating range of the EM-34 instrument, 1-1000 mS/m, and both vertical and horizontal coil configurations. This yields six conductivity readings, one for each coil spacing/orientation combination. While this procedure requires access to a forward modeling program and is more time-consuming than the spreadsheet and low induction number procedure of Stewart [1990], Anthony's [1992] procedure should provide a better estimate of interface depth as it utilizes more data at each station and a full solution for instrument response over a layered earth.

Anthony [1992] notes that the EM-estimated interface position corresponds to chloride ion values of several hundred to several thousand mg/l (seawater is 19,500 mg/l Cl⁻), or the top of the transition zone. Stewart [1990] makes the same observation for his results. In both studies, the chloride ion values at the EM-estimated interface depth vary by an order of magnitude. Clearly, the terrain conductivity method described by Kauahikaua, Stewart, and Anthony can give a good estimate of the depth of the upper part of the transition zone, but cannot give an estimate of water quality at a specific depth. Specifically, the 250 mg/l chloride limit for potable water is a difficult, if not undetectable, geoelectric target.

Because of their simplicity and ease of use, terrain conductivity meters have seen wide use in environmental and hydrogeologic investigations. In island and coastal areas they can be used to outline the coastal position of the interface or to estimate its depth where the interface is within 30 m of land surface (40-60 m for Slingram or MaxMin instruments), relief and topographic elevations are low, and the geologic units which contain the interface have low bulk conductivities when saturated with saltwater. These conditions are met in many small oceanic islands or coastal plain areas. Complex stratigraphy and fine-grained, conductive geologic units such as clays add geologic noise to the conductivity signal produced by water quality variations, and can make locating the interface in such environments difficult.

Kauahikaua describes studies of the freshwater lenses on Laura Island, Majuro Atoll in the Marshall Islands [Kauahikaua, 1987] and Truk [Kauahikaua, 1986]. Laura is a small, triangular island, 1.5 × 3 km, on the west side of Majuro Atoll. Like most atoll islands it has low relief. Kauahikaua used two electrical methods, horizontal loop EM (HLEM; Slingram) and VES using a Schlumberger array. For the HLEM surveys Kauahikaua measured both the in-phase and quadrature response for frequencies of 222, 444, 888, 1777, and 3555 Hz, and a coil separation of 61 m. This yielded exploration depths of between 4 and 30 m. Kauahikaua used an adaptation of a forward-modeling program, MARQLOOPS [Anderson, 1979], to best fit two and three-layer geoelectric models to the data. Kauahikaua concludes that the transition zone cannot be resolved, and bases his interface depths on the two-layer solutions. The Schlumberger data were able to resolve the water table and the saltwater interface. Comparison of the HLEM data with water quality data shows that the EM-interpreted sharp interface at Laura is within the upper part of the transition zone. A cross section across the island (Figure 2.10) shows a good correlation between HLEM and water-quality data.

2.2.1.5   Time-Domain Electromagnetic Soundings

Time-domain (transient) electromagnetic (TDEM) soundings have applications similar to DC soundings, in that they can be used to detect saltwater at depths of 5 m to several hundred meters below land surface. The TDEM method has several significant advantages over DC soundings, notably depths of investigation up to twice the transmitter coil dimension, and the ability to sound through a conductive, near-surface unit, such as a clay confining layer. TDEM equipment, however, is more expensive and complicated to use than

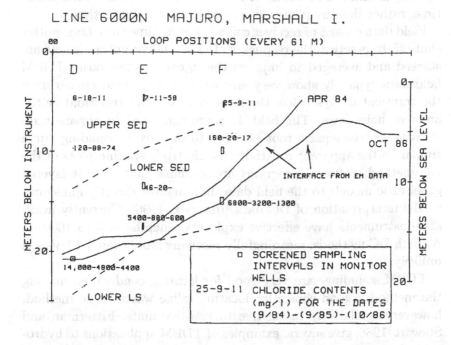

Figure 2.10. North-south cross section of Laura Island, Majuro Atoll, Marshall Islands. EM interface solutions were obtained from MaxMin frequency-domain EM soundings. (From Kauahikaua [1987]).

DC equipment, and the interpretation of TDEM data requires sophisticated interpretation software.

To conduct a TDEM sounding, a square or rectangular transmitter loop, usually a single turn of insulated wire, is charged with an alternating, square-wave current in the range of 3-300 Hz. The primary current in the transmitter coil is abruptly terminated, creating an electromotive force that induces eddy currents in the ground. These currents decay with time, with the location of maximum current density moving downward and outward with increasing time since the transmitter current was shut off. As the transient decay occurs while the primary field is off, the weak magnetic fields created by the eddy currents can be measured with a sensitive receiver coil. As the ring of maximum current density moves downward with time, measurements taken soon after transmitter shut-off are dominated by shallow units, while measurements taken in late time are dominated by deeper units. The depth of investigation becomes a function of time, rather than frequency as in frequency-domain methods.

Field data consist of receiver coil voltage vs. time since transmitter shut-off. In practice, the responses of many individual transients are stacked and averaged to improve the signal to noise ratio. TDEM field data typically show very smooth, low noise field curves until the transient decays below the receiver sensitivity threshold or the ambient noise level. The field data are converted to apparent resistivity vs. the square root of time, to produce a sounding curve similar to the apparent resistivity vs. electrode spacing plots of the DC method. Software programs are available that best fit layered geoelectric models to the field data, in a manner directly analogous to the interpretation of DC data [Interpex, 1996]. Currently available instruments have effective exploration depths of 5 to 1000 m. As with DC methods, meaningfully resolving more than 3-4 layers is unlikely.

TDEM soundings are very good for locating conductors, making the method a good choice for locating saline waters. The method, however, is relatively insensitive to resistive units. Fitterman and Stewart [1986] give several examples of TDEM applications to hydrogeologic problems that illustrate the sensitivity to conductors, and insensitivity to resistors. Resistive units can be mapped indirectly if their boundaries are defined by conductive units. An example is resistive, crystalline bedrock underlying conductive sedimentary rocks. TDEM, however, would be a poor choice for delineating a resistive sand and gravel layer in more conductive alluvial materials.

The depth of investigation of a TDEM sounding has both upper and lower limits. The upper limit is determined by the length of time it takes to terminate the current in the transmitter coil. More rapid turn-off times allow detection of the transient at earlier times. Presently, the EM-47 TDEM instrument has an upper limit of about 5 m using a 5 × 5m-coil. It is difficult to rapidly terminate large currents, so the EM-47 uses small transmitter loops and battery power, limiting the effective penetration depth to about 50-60 m, still a very useful depth for most hydrologic surveys. Deeper penetration depths can be achieved with larger transmitter loops and motor-driven generators. The EM-37 can use 500 m×500 m loops and has an exploration depth of about 1000 m, and a minimum sounding depth of about 30-50 m. Lateral resolution of TDEM soundings is very good, being about the scale of the transmitter loop radius.

Acquiring good TDEM data requires that the operator have a reasonable understanding of the geophysical principles involved. Careful attention to data quality during data acquisition is necessary to get interpretable data. When taken by an experienced operator, TDEM data typically show far less variation as a result of ambient noise and lateral heterogeneities than DC sounding data [Hoekstra and Blohm, 1990].

When used in conjunction with DC soundings to locate the saltwater interface, TDEM soundings tend to place the interpreted sharp interface at shallower depths than DC soundings [Goddard, 1997; Countryman, 1996]. A similar relationship was noted by Stewart [1990] for comparisons of interface depths calculated from terrain conductivity and DC data. Water quality data suggest that electromagnetic methods tend to locate the sharp interface near the top of the transition zone, while DC soundings tend to place the sharp interface near the middle of the transition zone. This may be the result of the higher sensitivity of EM methods to changes in conductivity.

A TDEM survey completed on Long Island, New York, is a good example of the use of TDEM [Maimone et al., 1989]. The objective of the survey was to map the position of the saltwater interface in the upper part of the Magothy Aquifer, the principal source of water for most of Long Island's 2.6 million residents. The survey was complicated by the occurrence of saltwater in the glacial aquifer which overlies freshwater zones of the Magothy Aquifer near the coast. DC resistivity soundings have difficulty resolving layers beneath a highly conductive surface layer, but TDEM soundings can detect deeper units under a shallow conductive layer. TDEM soundings were completed on marshy islands in a saltwater bay on the south side of Long Island. The TDEM soundings were able to distinguish the position of the saltwater wedge in the Magothy Aquifer to depths of about 150 m, using 100 m×100 m transmitter loops.

Hoekstra and Blohm [1990] describe a similar TDEM study of saltwater intrusion in the Salinas Valley, California. At the coast, three clay layers separate four aquifer zones, the surficial aquifer, and the 180, 400 and 900 foot aquifers. Saltwater intrusion has moved farthest landward in the 180 foot aquifer, so mapping the interface position in the 400 foot aquifer required sounding through saltwater-saturated zones of the 180 foot aquifer. Good well control allowed

the depth and thickness of the geoelectric layers to be controlled, so
that good formation resistivity values could be obtained. Transmitter
loops 200 m×200 m allowed depths of investigation of approximately
150 m, well into the 400 foot aquifer. An 8 ohm-m formation resistivity was correlated with a chloride ion concentration of 500 ppm
(Figure 12, Hoekstra and Blohm [1990]), allowing the landward extent of the saltwater to be mapped in the 180 foot aquifer and the
upper part of the 400 foot aquifer.

In Florida, Stewart and Gay [1986] completed a regional reconnaissance of the position of the saltwater interface in the Floridan
Aquifer along a 100 km section of Florida's west coast using TDEM
soundings. Transmitter loops of 80 × 80 m allowed detection of the
saltwater interface to depths of 150 m, and detection of a low porosity zone that marks the bottom of the Upper Floridan Aquifer. This
low porosity zone was detected to depths of 180-200 m. The TDEM
survey revealed major landward re-entrants of the saltwater interface that extend more than 10 km inland from the Gulf of Mexico.
These re-entrants are associated with first magnitude springs, which
discharge freshwater near the coast.

2.2.1.6   Very Low Frequency EM

Very low frequency (VLF) EM uses powerful military VLF radio
transmissions in the range of 15-25 kHz as the primary field. While
these are very low frequencies for radio transmissions, they are rather
high for exploration geophysics. As a result of the high frequency
of the primary field, the effective penetration depth is a function
of terrain resistivity and is approximately $350 (\rho/f)^{1/2}$, where ρ is
terrain resistivity in ohm-m, and f is the transmitter frequency in
Hz [McNeill, 1990]. This is about 10-15 m for 25 ohm-m material
and a frequency of 20,000 Hz.

Conventional VLF is usually used to locate conductive ore bodies
or bedrock fracture zones, but has limited application to saltwater
intrusion studies [McNeill, 1990]. VLF resistivity, however, is a useful
reconnaissance tool for investigations where the interface is within
10-20 m of the land surface. The primary VLF field consists of a
nearly vertical electrical field and a horizontal magnetic field. Over
a perfectly conducting earth, the E-field would be vertical, but as
resistivity increases the E-field tilts away from the transmitter, creating a horizontal, radial electrical component. Using the horizontal

magnetic field as a reference for the primary field strength, the apparent resistivity of the earth varies with the square of the ratio of the horizontal electrical field to the horizontal magnetic field. The magnetic field is measured with a small coil, and the electric field is measured with an electric dipole 10 m long formed by two electrodes pressed into the ground along a line radial to the transmitter. Because resistivity is proportional to the square of the horizontal electrical field, the VLF resistivity method is very sensitive to near-surface resistivity variations, even in high resistivity environments, unlike the other EM methods discussed.

The VLF resistivity method is well suited for reconnaissance level surveys designed to locate and outline freshwater lenses on small oceanic islands or coastal areas where the interface is within 10-20 m of the surface. Primary field strength is adequate in much of the northern hemisphere, particularly in North America and Europe. In parts of the world, such as the central-south Pacific, Africa and South America, signal strength from military transmitters may be too weak. Portable VLF transmitters can be used, but they are impractical for the simple, reconnaissance surveys VLF resistivity is best suited for.

Mackenzie [1990] conducted a VLF resistivity survey of Key West, Florida. Key West is a low-relief, carbonate island at the west end of the Florida Keys. Most of the Key is developed, and conventional EM or DC surveys would be impractical. Mackenzie set up the VLF dipole in yards and parks. Despite the presence of power lines, buried pipes and cables and other sources of interference, the apparent resistivity map plotted directly from the field data (Figure 2.11) outlines the freshwater lens quite well. The highest resistivities are in the area where water quality data indicate the lowest TDS waters to be.

2.2.2 Seismic Methods

Seismic methods utilize differences in mechanical properties of geologic units. They cannot be used to directly detect saltwater, but they can delineate the boundaries of units which may influence the position and movement of the saltwater interface. Advances in instrumentation and processing power available to microcomputers have made shallow seismic reflection surveys practical for engineering and environmental applications, and decreased the acquisition time and interpretive effort for seismic refraction. Both seismic reflection and

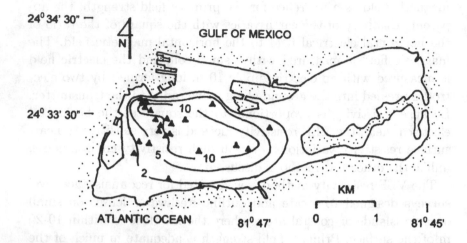

Figure 2.11. VLF resistivity data from Key West, Florida, U.S.A. VLF resistivity is in ohm-m. (After McKenzie [1990]).

refraction remain relatively labor and data processing intensive, and are usually used for studies where detailed information is required. They are, however, a very economic alternative to drilling and often represent the only practical way to obtain detailed subsurface information in remote areas. Because seismic methods detect changes in seismic velocity and are insensitive to fluid conductivity, conjunctive use with electrical methods can be very useful, with the seismic data helping to constrain the ambiguity between porosity and pore water quality in interpreting the electrical methods data.

2.2.2.1 Seismic Refraction

Seismic refraction methods can provide information on the depth, geometry and seismic velocity of geologic units to depths of several tens of meters, using simple seismic sources, if conditions are appropriate. It is difficult to resolve and map more than 2-3 subsurface layers with seismic refraction, so the method is normally used to profile significant geologic boundaries, such as the contact between unconsolidated overburden and the underlying bedrock.

The refraction method relies on refracted wave paths from subsurface boundaries where there is a significant increase in velocity with depth (Figure 4.31, Telford et al. [1990]). Upward refraction of the downward traveling, incident wavefront from the source creates a headwave along the refracting boundary, sending wavefronts back to the surface to the detector array. The field data consist of a travel-time plot of the distance between the source and detector plotted versus the travel time of the first seismic arrival at each detector. Simple interpretation methods can provide information on the depth and dip of the refracting horizon, and the seismic velocity of the layers. Seismic velocity can give some information about the geologic character of a unit, principally the degree of lithification. More sophisticated methods of data acquisition and interpretation, such as the ABC method [Parasnis, 1986] or the general reciprocal method [Lankston, 1990; Palmer, 1980] can provide an estimate of the depth to the refracting horizon at each detector where a refracted arrival was received. By careful arrangement of source and detector positions, a continuous seismic profile of the refracting horizon can be produced.

The most common complications for a refraction survey are the method's inability to resolve thin layers underlain by a high velocity layer, and the fact that no headwave is generated at a boundary where velocity decreases with depth. The thin layer problem can sometimes be recognized on the field records, but the low-velocity layer problem can only be recognized by comparison with well logs. Both field problems result in a miscalculation of the depth of the refractor.

The depth of exploration of the refraction method is roughly 20-30% of the source/detector separation. In the field, however, the depth of exploration is often determined by the strength of the seismic source, and the level of ambient noise. Stronger seismic sources and quieter sites allow longer source/detector separations and deeper exploration depths.

Seismographs used for seismic refraction surveys typically have 12 to 24 channels, allowing one detector or detector array for each channel. Multiconductor cables with appropriately spaced takeouts link the seismograph to the detectors. The detectors, or geophones, have short spikes that are pushed into the soil to couple the geophone to the earth. Shallow refraction surveys can make use of lower frequency seismic energy than seismic reflection surveys, so refraction surveys typically use 40 Hz geophones, as compared with the 100 Hz geophones used for shallow reflection. Modern seismographs "stack" incoming signals. Many records of the same source and detector separation are read into the same memory location. Coherent arrivals from direct and refracted waves stack trough to trough, peak to peak and sum, while incoherent noise arrives at different times on each record, so noise tends to cancel out, greatly increasing the signal to noise ratio.

Seismic sources used for shallow refraction fall into two general classes, weight drop and explosives. Of the two, weight drop sources are the most economical and convenient. The simplest weight-drop source is a 4-6 kg sledgehammer struck against a metal plate laid on the ground. An inertial trigger is taped to the hammer and it triggers the seismograph when the hammer hits the plate. The energy in each hammer blow is low, but many hammer blows can be stacked at each source position. Source to detector spacings of 50-80 m can be obtained with a hammer source, longer at quiet sites or if many blows are stacked. Heavier weight drops provide more energy for greater source/detector separations. A commercially available source is mounted on a small trailer, and uses large industrial rubber belts to drive a heavy weight downward into the ground. Weight drop methods have the advantages of not requiring drill holes or special permitting, as can be the case with explosives.

Explosive sources can be divided into two groups, firearm ammunition sources and traditional explosives. Firearm sources use industrial shotgun blanks or slugs [Pullan and MacAulay, 1987]. A commercial version is called the Betsy Seisgun. Each shot must be placed several feet below the ground, and below the water table if possible. Each shot costs a few dollars, depending on the size of the blank or slug. Because this system is based on firearm ammunition, there are fewer permitting problems for use, except in some urban areas. Transport on commercial aircraft can be a problem, and there are safety concerns, as with all explosives. Traditional explosives use special seismic blasting caps and a firing box that sets off the cap and triggers the seismograph. Explosive charges of varying size, depending on the energy desired, are placed in shot holes several meters deep. Commercial explosives are available, but ANFO, or ammonium nitrate and fuel oil, is an inexpensive, relatively safe, and effective high explosive that can be made from materials available in many parts of the world. Explosive sources are excellent sources of seismic energy, but shot holes must be drilled for each shot, the use of high explosives entails obvious risks, and transportation and storage usually require special permits and facilities.

In order to map the relief of a refracting surface, the surface elevation of each source and detector position must be known to within 10-20 cm. This almost always requires surveying the elevations along each seismic line. Given the need to survey elevations, to stack several records at each source position, and to place source positions at each end of a detector layout, a reasonable rate of advance for a detailed seismic refraction study is about 0.5-1.0 km/day.

The best application for shallow refraction surveying is locating the position and relief of a bedrock surface under unconsolidated materials less than 50 m thick. Seismic refraction can usually resolve no more than two or three boundaries reliably. Sometimes, in clean sands, the water table forms a good refracting boundary and can be easily identified by the distinctive velocity of sound in water (about 1500 m/s). Often, however, a thick capillary zone makes the water table an indistinct seismic boundary. Ambient noise can impede seismic surveys. Common sources of low frequency noise are highways and, in coastal areas, surf. Wind can cause noise problems, particularly if there are trees near the detectors.

An example of an application of seismic refraction is the study of Deke Island, Pingelap Atoll, Pohnpei, reported by Ayers and Vacher [1986]. Deke is considered by Ayers and Vacher to be representative of many small atoll islands. Seismic refraction profiles reveal an essentially two-layer system, with unconsolidated sediments overlying more indurated units. The seismic refraction survey was able to map the boundary between the unindurated and indurated sediments (Figure 2.12). Vertical electrical soundings and water quality data indicate that the freshwater lens is essentially contained within unit V_1, with the top of the transition zone at or near the contact between units V_1 and V_2. In this case, a geologic boundary, which can be mapped on the basis of a contrast in seismic velocity, controls the position of the interface.

2.2.2.2   Seismic Reflection

Seismic reflection uses the reflection of sound waves from subsurface boundaries with significant changes in seismic velocity. It provides information on the geometry of subsurface units to depths of 5-100 m, using simple seismic sources. High resolution reflection surveys can

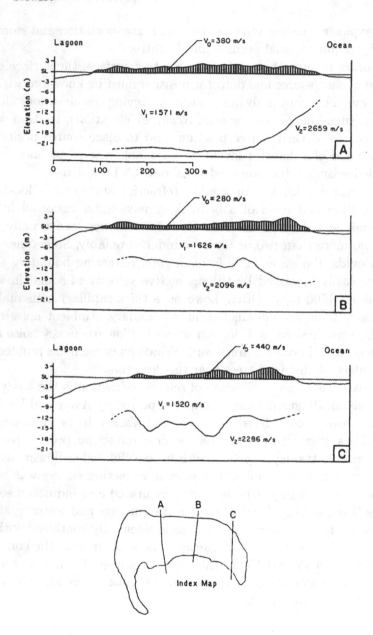

Figure 2.12. Seismic refraction sections across Deke Island, Pingelap Atoll, Pohnpei. Velocities are in m/sec. V_o represents unsaturated, unindurated sediments, V_1 represents saturated, unindurated sediments, and V_2 represents saturated, poorly indurated sediments. (From Ayers and Vacher [1986]).

provide detailed subsurface geologic information, but they require careful field work and extensive and sophisticated data processing.

The reflection method uses seismic energy, which travels from source to detector following a reflected ray path (Figure 4.22, Telford et al., 1990]. Reflecting horizons are boundaries where a significant change in seismic velocity occurs. The ray path geometries, and the resulting equations for converting the two-way travel time to depth, are deceptively simple for planar reflecting horizons [Telford et al., 1990].

Two field methods for obtaining shallow reflection data are commonly used. One is the common offset technique [Hunter et al., 1984]. This method can be used when prominent reflections are clearly visible on the field records. The advantage of this technique is that it requires very little data processing. The disadvantage is that each point on the reflector is sampled only once, so that strong reflections are required. The second method is called the common midpoint (CMP) method [Steeples and Miller, 1990]. By carefully controlling the source/detector geometry, each point on the reflector is sampled several times (Figure 2.13). Summing these multiple reflections for the same mid-point can greatly enhance the record quality and make otherwise faint reflections visible. The number of source/detector pair records summed for a midpoint is called the fold. Theoretically, the signal to noise ratio increases with the square root of the CMP fold [Steeples and Miller, 1990]. The disadvantage of CMP is significantly increased field and processing effort.

Efficient acquisition of good quality, shallow seismic reflection data requires a 12 to 24 (or more) channel seismograph designed for reflection surveys. While refraction surveys can be completed with relatively simple seismographs, reflection surveys require more sophisticated, multichannel units to obtain good records with a reasonable amount of field effort.

A 12-channel seismograph can produce a 6-fold CMP record by placing a source at each detector location. Using two seismic cables, 24 geophones, and a "roll-along" switch allows the survey to proceed along a line, picking up the trailing seismic cable every 12 records. A 24-channel seismograph and two cables can collect 6-fold data and pick up the trailing cable every 24 records. The rate at which the survey advances varies with field conditions, equipment, geophone spacing, CMP fold, and crew size, but high resolution surveys with a

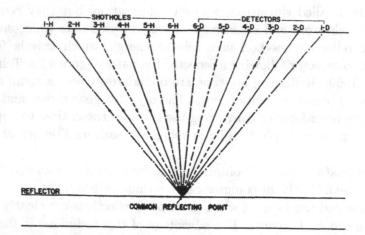

Figure 2.13. Example of source and detector locations for a 6-fold CMP gather from a common point on the reflector. (From Dobrin and Savit [1988]).

well-organized crew can advance at a rate of 150-300 records per day for shallow, high-resolution surveys. Source and detector intervals of 1-5 m are generally used, with 3 m being a good compromise between resolution and rate of data acquisition. The distance between the source and the first detector, termed the offset, is optimized for each survey [Steeples and Miller, 1990], but is typically 5-20 m for shallow surveys.

Seismic sources used in reflection surveying are similar to those used for refraction surveys. Cost and efficiency become more critical, as there are many more source stations ("shots") per km in shallow reflection work than for seismic refraction. Simple hammer or weight-drop sources are inexpensive, and can typically provide data to depths of 30-50 m or greater. Seisguns provide more high-frequency energy [Steeples and Miller, 1990], but the cost of the ammunition can be prohibitive when there are 300-500 shots/km of survey line.

Interpretation of CMP data requires sophisticated software programs that can handle the complex sorting of individual geophone records required to stack the CMP records. These programs also can apply various filters to the data to enhance reflections and apply static and dynamic corrections. It is the development of these processing programs for microcomputers that has made shallow, high resolution, seismic reflection surveys practical for groundwater investigations.

Seismic reflection is best applied in cases where the geometry of shallow geologic units with moderately complex geology influences the configuration of the interface. An example is the contact between younger, less permeable sediments, and older, more permeable units in carbonate islands. This contact typically truncates the bottom of the freshwater lens [Ayers and Vacher, 1986], so that the position of the boundary controls the depth to the interface [Vacher et al., 1992]. A similar boundary occurs where coarse, transgressive beach deposits overlie fine-grained lagoonal sediments in barrier islands. The location and depth of the fine-grained units often control the thickness of the freshwater lens [Wait and Callahan, 1965; Harris, 1967]. Electrical methods are less useful in these transgressive sequences because the fine-grained units are conductive, creating a geoelectric ambiguity between lithology and water quality. In this case, resolving the geologic boundary may be a more effective way of defining the freshwater lens than an ambiguous vertical profile of bulk conductivity.

Shallow, high resolution, seismic reflection methods are not well-suited to reconnaissance surveys. Data acquisition rates are slow, elevations of all source and detector locations must be determined, and considerable effort is expended in data processing. Reflection methods are best applied where detailed information is required, and geologic conditions make electrical methods impractical. For example, numerical modeling of interface movement might require knowledge of the depth to a hydrogeologically significant layer which cannot be resolved electrically. The ability of seismic methods to map significant changes in seismic velocity above or below the interface would make them an excellent choice for such an objective.

Miller et al. [1996], provide an example of a high-resolution reflection survey in a coastal environment. They used reflection surveys to map the principal hydrostratigraphic units in the Coastal Plain aquifers under barrier islands on the Atlantic coast of New Jersey, U.S.A. The reflection data provide information on the depth and dip of the principal units to depths of over 180 m. The seismic source was a 4.5 kg sledge hammer. The seismic sections reveal coast-perpendicular mound and trough structures within the hydrostratigraphic units that may focus groundwater flow (Figure 2.14). This study is a good example of the quality of data and depth of investigation that can be obtained with simple seismic sources.

2.2.3 Ground Penetrating Radar

Ground penetrating radar (GPR) is both a reflection method and an electrical method in that electromagnetic waves are reflected from subsurface boundaries where there is a contrast in electrical properties. The wavelengths of GPR are measured in decimeters to meters, while seismic wavelengths are measured in meters to decameters. Vertical resolution of reflection methods is about 1/4–1/2 the wavelength of the reflected energy, so GPR methods are capable of resolving considerable detail in the subsurface. At 100 MHz, the vertical resolution of GPR is about 0.25 to 0.50 m.

The GPR reflection amplitude of a subsurface boundary is determined principally by the contrast in the dielectric constant across the boundary. The dielectric constant for most minerals is 2-5 units (dimensionless), while the dielectric of water is 80. Clearly, below the water table changes in the dielectric constant are influenced most strongly by changes in porosity. Above the water table both changes in porosity and water saturation affect the dielectric constant.

A significant disadvantage of the GPR method is the rapid attenuation of the transmitted signal. For a transmitted frequency of 100 MHz, the effective penetration depth is roughly 1-2 cm per ohm-m of formation resistivity. Exploration depths can exceed 10 m in resistive materials, but may only be a few centimeters in highly conductive materials.

GPR instruments consist of a receiver and transmitter antenna set. For 100 MHz antennae, each is about a meter long. The antennae are often placed in a skid box, which is dragged across the ground. A controller triggers the outgoing pulse and records the resulting reflections. Many instruments operate on a stacking principle directly analogous to a stacking seismograph to increase the signal to noise ratio.

GPR can be used to map subsurface sedimentary structures in coastal and island areas [Jol et al., 1996]. An example from a beach in Oregon is shown in Figure 2.15 [Jol et al., 1996]. The location of the interface can be estimated by noting where attenuation by the highly conductive saline water masks subsurface reflectors. This

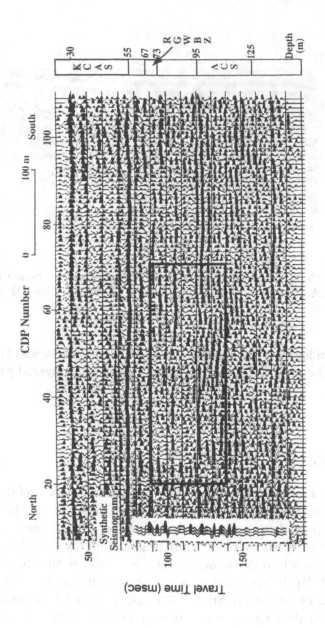

Figure 2.14. Seismic reflection section from a barrier island on the New Jersey, U.S.A., coast. Area in box shows trough and mound structures. KCAS = Kirkwood-Cohansey Aquifer System, RGWBZ = Rio Grande Water Bearing Zone, and ACS = Atlantic City 800-foot Sand. (From Miller et al. [1996]).

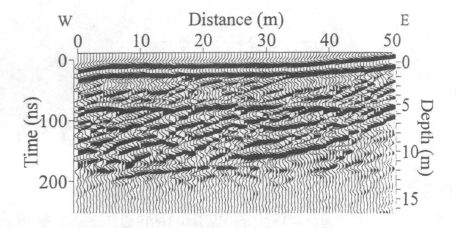

Figure 2.15. GPR profile from a raised boulder-gravel beach in Oregon, U.S.A. The water table is at 5.5 m. (From Jol et al. [1996]).

method of locating the interface, however, uses the absence of a signal to infer the presence of the target, not a good geophysical practice.

2.3   Borehole Methods

While surface geophysical methods can provide very useful subsurface information, the physics of taking measurements at the surface necessarily leads to a decrease in resolution with increasing depth. Borehole methods overcome this disadvantage by introducing geophysical tools into the subsurface in a borehole, so that resolution is no longer depth dependent. Several borehole methods can be used in boreholes cased with plastic (PVC) casing. This allows information on bulk conductivity, porosity, density and lithology to be obtained in boreholes which can be cased through the transition zone, eliminating intrawell circulation and mixing, as in uncased boreholes. Traditional electrical logs, run in uncased boreholes, can also provide detailed information on the vertical variation in water quality if mixing within the borehole does not occur.

2.3.1 Electrical Logs

Traditional electrical borehole tools introduce electrical currents into the borehole fluid and surrounding formation through direct electrical contact between the borehole fluid and the surrounding formation. This requires an open, uncased, fluid-filled borehole. The logging tools have electrodes spaced vertically along the tool or cable. The lateral penetration depends on the separation between the electrodes, with larger separations producing greater lateral penetration, but decreased vertical resolution. Tools with very small electrode separations measure borehole fluid resistivity, while the response of tools with large separations is dominated by the bulk resistivity of the formation outside the borehole. Examples are short-normal or 16″, long-normal or 64″, and fluid-conductivity logs [Telford et al., 1990]. As with surface resistivity methods, bulk formation resistivity measured by borehole tools is influenced by both porosity and fluid conductivity. The response of electrical logs is influenced by materials in and immediately adjacent to the borehole, such as borehole fluids and drilling mud. The effective lateral (radial) penetration of most electrical logging tools used in groundwater surveys is on the order of a few cm to a few decimeters.

Borehole induction logging tools are very similar in operation to loop-loop, frequency EM surface methods (section 2.2.1), except that the transmitter and receiver coils are contained in the logging tool. They are sensitive to conductors, such as saline waters and clay units, but are insensitive to resistivity variations above 1000 ohm-m. A commercial unit, the EM-39, will fit in a 5 cm (2 in), cased borehole. As a result of the response function of the instrument versus radial distance, the EM-39 is quite insensitive to the conductivity of the fluid in the borehole, and is most sensitive to materials about 30 cm out from the tool [McNeill, 1990]. This allows the EM-39 to give an accurate assessment of bulk formation conductivity, even in boreholes filled with conductive fluids or contaminated by drilling muds.

Borehole induction logs are very useful in island and coastal studies because the tool can be used in wells cased with non-conductive (plastic) pipe. Coastal zones are usually groundwater discharge areas, and open boreholes drilled through the transition zone into saltwater can be a source of contamination of the freshwater zone through upward flow of saline or brackish waters in the borehole, particularly when low hydraulic conductivity units are present in the section or freshwater heads have been reduced by pumping. To avoid this problem, and to provide vertically discrete data, water-quality monitoring wells usually have only a short open-hole section. This greatly limits the utility of the well for monitoring the movement of the interface. Induction logs can monitor the entire transition zone with a single well cased through the zone into saltwater. If depth-controlled water quality data are acquired during installation of the well, later changes in logged bulk conductivity can be correlated with changes in water quality [McNeill, 1990], as geologic conditions can be assumed to be constant [Stewart and Hermeston, 1990]. The EM-39 can produce very stable and repeatable measurements, as the instrument zero levels can be accurately set before each survey by holding the instrument in the air. Depths to 200 m can be logged.

2.3.2 Radiometric Logs

Radiometric logs are either passive receptors of naturally-produced radiation (natural gamma log), or contain active radiation sources (neutron and active gamma logs). Like induction logs, they can be run in cased boreholes. An important source of gamma rays detected by a natural or passive gamma logging tool is ^{40}K, a common constituent of clay minerals [Telford et al., 1990]. An active neutron source in a neutron logging tool produces high-energy neutrons, which lose energy and are thermalized in collisions with protons. The back-scattered, thermalized neutrons are measured by a detector. Production of thermalized neutrons is proportional to the proton cross-section of the formation, so the neutron log measures water content or porosity, and is sometimes called a "porosity log" [Telford et al., 1990]. The active gamma log uses an active source of gamma rays, and detects gamma rays backscattered by Compton scattering, which is proportional to the electron density of the formation [Telford et al., 1990]. For this reason, the active gamma log is often called a "density log". Most of the response for the neutron log comes from within 20-50 cm of the tool, and 10-20 cm for the active gamma log.

A significant advantage of the natural gamma logging tool is that it does not contain an active radioactive source, so it does not require special permits or procedures for transportation, use, or storage. Small, "suitcase" gamma tools are available, and when used in conjunction with an induction log, the passive gamma log can help resolve the conductivity ambiguity between fluid resistivity and lithology. If a high conductivity zone has a high gamma count, it is likely to be a clay layer, while a high conductivity zone with a low gamma count probably represents a clean unit saturated with high TDS water. Vertical resolution is quite good. Typical logging tools can resolve the thickness of layers down to about 1 meter, with thinner layers being detected as a single peak.

Neutron and active gamma logs provide useful geologic information, especially for mass transport modeling, which requires porosity data to calculate groundwater velocities and the solute mass per unit volume. Both logs can be run in cased boreholes. Because it provides data on porosity, the neutron log is probably the most useful for hydrogeologic studies, in conjunction with induction logs and passive gamma logs. A significant disadvantage of active radiometric logging tools is that they contain strong radioactive sources. They require careful handling and storage, and often require special permits. Losing an active radiometric logging tool in a well is a serious matter.

2.3.3   Integrated Use of Borehole Logs

Stewart and Hermeston [1990] used the EM-39 borehole induction logging tool to investigate the character of the saltwater/freshwater transition zone in the Tertiary, carbonate Upper Floridan Aquifer along the west coast of central Florida, U.S.A. The logging tool was run in PVC-cased monitoring wells 15 to 30 cm in diameter. The EM-39 tool used had a depth capability of 200 m, using a motor-driven cable reel. At the time of installation of most of the wells, water quality samples were obtained as the drillhole was advanced. These water quality data allow the assessment of the effectiveness of the borehole induction logs for detecting changes in water quality.

Several wells were logged repeatedly, up and down, to test the ability of the induction tool to replicate logging runs. The repeatability of the formation conductivities was within 1-2 mS/m, with much of the error being due to the uncertainty of vertical placement of the tool as a result of about a 0.5% change in length of the cable over the total 200 m cable length. The zero level of the logging tool can be set before each survey by simply holding the tool several meters above the ground. This allows direct comparison of logging runs made at different times or in different wells. Logging runs a month apart in the same well agree with each other within a few mS/m or $< 1\%$ of the observed range in formation conductivities.

An original objective of the project was to determine the formation factor in Archie's Law, which allows the conductivity of the formation fluid to be determined from the bulk formation conductivity [McNeill, 1990; Jackson et al., 1978]. The hope was that general formation factors could be obtained, which would allow an estimation of formation water quality to be made from bulk conductivity values obtained in wells that do not have water quality data. It was found, however, that in the carbonate, karstic Floridan Aquifer, formation factors can vary by a factor of 2 to 3 over vertical distances of less than a meter, and the variation in formation factors could not be correlated between wells only a few km apart. Comparison of the derived formation factor values with neutron, active gamma, and caliper logs from the same boreholes indicates that significant changes in formation porosity, bulk density, and pore size can occur over very short vertical distances in the Floridan Aquifer, and that these changes cannot be consistently related to large, identifiable intervals, such as specific stratigraphic units.

In general, the bulk conductivity values agree well with available water quality data. In every case where the well penetrates into or through the transition zone, the change in bulk conductivity is not the smooth, error function, "S"-shaped curve of a theoretical interface, but is distinctly stepped, with narrow vertical zones where water quality and bulk conductivity change rapidly with depth (Figure 2.16). Examination of neutron, active gamma, caliper, temperature and lithologic logs indicates that the abrupt, step-like changes in water quality and bulk conductivity occur at distinct geologic discontinuities. The geologic character of these discontinuities varies from well to well, and vertically within the same well. Abrupt changes in bulk conductivity occur at transitions from less dense, more porous limestone to more dense, less porous, but more permeable dolostone, where caliper logs indicate an abrupt change in the size or number of solution cavities, or where temperature logs and the thermal gradient suggest an abrupt change in the direction or volume of groundwater flow in the formation. Several surface geophysical surveys have been conducted in the same area, using both DC and EM soundings [Fretwell and Stewart, 1981; Stewart and Gay, 1986; Hagemeyer and Stewart, 1990], but the surface surveys do not resolve the stepped

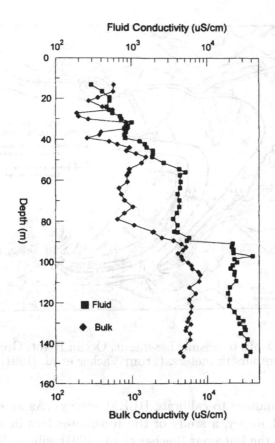

Figure 2.16. Comparison of bulk formation conductivity with fluid conductivity in a monitoring well in west-central Florida, U.S.A. Bulk conductivity values were obtained with an EM-39 borehole induction logging tool. (From Stewart and Hermeston [1990]).

character of the interface.

2.4   Integrated Geophysical Surveys

As discussed earlier, different geophysical methods can be complementary when used together. Some examples are using seismic methods to constrain geologic boundaries when interpreting electrical data, combining borehole induction and passive gamma logs to resolve the ambiguity between porosity and water quality changes, and

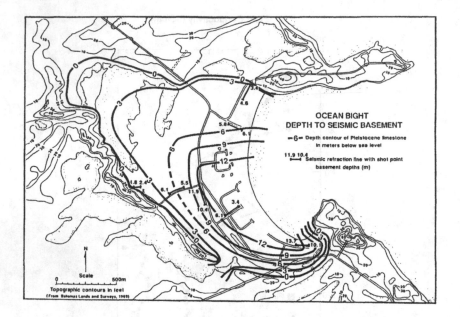

Figure 2.17. Depth to seismic basement, Ocean Bight, Great Exuma, Bahamas. Contours in meters. (From Vacher et al. [1991]).

using DC soundings to calibrate HLEM surveys. As an example of an integrated survey, a study of the freshwater lens in a Holocene sand body in the Bahamas [Vacher et al., 1991] will be discussed.

Ocean Bight is a small embayment on Great Exuma, Bahamas, contained between two headlands formed from well-cemented, carbonate, Pleistocene eolianites. A large, Holocene strandplain, approximately 2 km long and 1 km wide, has formed between the headlands over the last 2,000 years. The strandplain is composed of unindurated to poorly indurated carbonate sands. Because the sands have a relatively low hydraulic conductivity (10 m/d), the Ocean Bight strandplain contains a significant freshwater lens [Little et al., 1976] that supplies several small communities nearby.

As the hydraulic conductivity of the Pleistocene eolianites is an order of magnitude higher than the carbonate sands of the strandplain, it was expected that the bottom of the freshwater lens would be truncated at the Holocene/Pleistocene contact. This is a common situation in the Bahamas [Cant and Weech, 1986], the Florida Keys [Halley et al., 1997; Vacher, 1997], and many Pacific atolls [Ayers and

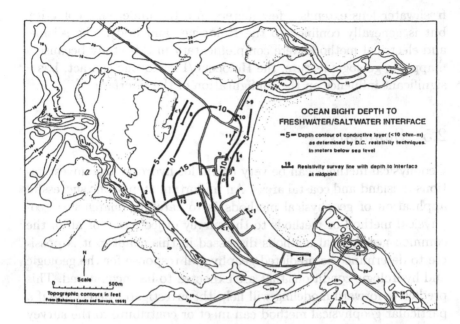

Figure 2.18. Depth to the saltwater/freshwater interface, Ocean Bight, Great Exuma, Bahamas, as determined from resistivity soundings. Contours in meters. (From Vacher et al. [1991]).

Vacher, 1986; Vacher, 1997]. To determine the depth to the geologic contact, a seismic refraction survey was conducted. The seismic data indicate that the thickness of the Holocene sands ranges from 0 m at the edge of the sand body to greater than 12 m near the southeast corner of the Bight (Figure 2.17).

To determine the thickness of the freshwater lens, a combination of EM terrain conductivity and DC resistivity surveys was used. Where the EM and DC data coincide there is good agreement in the depth to the interface determined from the two methods. Unfortunately, the EM instrument malfunctioned early in the survey, and the DC data were principally used to map the thickness of the lens, another argument for using multiple geophysical methods. The electrically-determined lens thickness indicates the interface is deepest in the center of the sand body (Figure 2.18), where the water table reaches a maximum elevation of 0.24 m. Comparison of the seismically-determined depth to bedrock and the electrically-determined lens thickness (Figures 2.17 and 2.18) indicates that the freshwater lens extends a few meters into the underlying eolianite, but is generally confined to the Holocene sand body. The seismic and electrical methods were complementary in this study because a mappable geologic feature, the Holocene/Pleistocene contact, has a significant influence on the configuration of the interface.

2.5 Summary

Geophysical methods can be very useful for hydrogeologic investigations in island and coastal areas. An important step in the successful application of geophysical methods is to carefully match the geophysical method or methods to the survey objectives. For all of the common geophysical methods discussed in this chapter, it is possible to determine the predicted geophysical response for the geologic and hydrologic conditions that are expected to be encountered. This predictive forward modeling will help the investigator determine if a particular geophysical method can meet or contribute to the survey objectives. It will also allow the design of the field procedures and proper selection of instruments, equipment, and survey parameters. This simple, preliminary step can save considerable time and effort later.

The solution of hydrologic problems in island and coastal regions often requires information on both geology and water quality. Providing information on both the physical framework and the variation in water quality usually requires conjunctive use of two or more complementary geophysical, geologic, or geochemical methods. Also, using several methods can provide a redundancy to insure that the survey objectives can be met. An example is the Ocean Bight survey discussed earlier. Although the failure of the EM instrument limited the survey's ability to outline the freshwater lens in detail, the VES data could be used to obtain the general lens thickness and configuration. Given the remote location and the difficulty in getting the instruments to the site, having both instruments available saved the study.

Chapter 3

Geochemical Investigations

B. F. Jones, A. Vengosh, E. Rosenthal & Y. Yechieli

3.1 Introduction

Saltwater intrusion is one of the most wide-spread and important processes that degrades water-quality by raising salinity to levels exceeding acceptable drinking and irrigation water standards, and endangers future exploitation of coastal aquifers. This problem is intensified due to population growth, and the fact that about 70% of the world population occupies coastal plains. Human activities (e.g., water exploitation, including industry and agriculture, reuse of waste water) result in accelerating water development and salinization. The elucidation of the dynamic nature of the fresh-saline water transition zone is of both scientific and practical interest because it reflects or controls the extent of development or exploitation.

The source of salinity in coastal aquifers has been a subject of many studies, but in many cases is still equivocal. Seawater encroachment inland is the most commonly observed reason for the increase in salinity, but other sources or processes can cause an increase. Custodio [1997] listed several saline sources that can affect water quality in coastal aquifers, but which are not directly related to seawater encroachment. These include entrapped fossil seawater in unflushed parts of the aquifer following invasion of seawater during relatively high sea levels, sea-spray accumulation, evaporite rock dissolution, displacement of old saline groundwater from underlying or adjacent aquifers or aquitards through natural advection or thermal convection, leaking aquitards through fault systems, and pollution

J. Bear et al. (eds.), Seawater Intrusion in Coastal Aquifers, 51–71.

from various sources including sewage effluents, industrial effluents, mine water, road de-icing salts, and effluents from water softening or de-ionization plants. In addition, agriculture return flows and leakage of urban sewer systems can contribute salts to phreatic coastal aquifers. For example, Izbicki [1991] found that high levels of chloride in groundwater from the Oxnard Plain near Los Angeles in California were derived beneath irrigation return flow characterized by high B/Cl and I/Cl ratios, and not from modern seawater intrusion. Monitoring and early detection of the origin of the salinity are crucial for water management and successful remediation. Yet the variety of the possible salinization sources, particularly in coastal aquifers that are sensitive to anthropogenic contamination, makes this task difficult.

This chapter summarizes the principal geochemical features of brackish water associated with saltwater intrusion in coastal aquifers and the main processes that control the chemistry of the water in the transition zone of the saltwater encroachment. It shall be shown that the water in this transition zone, on a worldwide scale, has a typical geochemical character that enables one to identify its impact during early states of groundwater salinization.

3.2   World-Wide Phenomena

In many coastal aquifers around the world, modern seawater intrusion commonly occurs because of natural flow controls or because of flows induced by extensive freshwater withdrawals. Ocean water itself is characterized by a salinity of 35 g/l (TDS) whereas internal seas may have higher (e.g., Mediterranean seawater, Red Sea; TDS = 40 g/l) or lower salinities. Seawater in general has a uniform chemistry due to the long residence time of the major constituents, with the following features: predominance of Cl^- and Na^+ with a molar ratio of 0.86, an excess of Cl^- over the alkali ions (Na and K), and Mg greatly in excess of Ca^{2+} (Mg/Ca = 4.5–5.2; Table 3.1). In contrast, continental fresh groundwaters are characterized by highly variable chemical compositions, although the predominant anions are HCO_3^-, SO_4^{2-} and Cl^-. If not anthropogenically polluted, the fundamental cations are Ca^{2+} and Mg^{2+} and, to a lesser extent the alkali ions, Na^+ and K^+. In most cases Ca^{2+} predominates over Mg^{2+}. Further solute differentiation of seawater and groundwaters of other origins is obtained by diagnostic major cation-anion association, which can be

Type	Mediterranean seawater	Salinas Valley California	Coastal Aquifer Israel		
Well		14S/2E-20B1	52B-2	603-5	Hilton North
Ca	459	410	980	172	463
Mg	1,211	126	245	196	1240
Na	12,500	450	2830	890	10600
K	435	12	22	7	392
Cl	21,940	1670	6304	2240	19812
SO$_4$	2,700	212	470	220	2600
HCO$_3$	169	62	206	102	232
Br	74.1	5.4	21.3	7.6	66.9
TDS	39,420	2982	11060	3830	35338
Na/Cl	0.86	0.42	0.69	0.61	0.83
Q*	0.4	3.8	3.7	1.4	0.4
Mg/Cl	0.08	0.11	0.06	0.13	0.09
Ca/Mg	0.2	2.0	2.4	0.5	0.2
K/Cl	0.02	0.006	0.003	0.003	0.02
SO$_4$/Cl	0.05	0.05	0.03	0.04	0.05
Br/Cl	0.0015	0.0014	0.0015	0.0015	0.0015

$Q^* = Ca/(HCO_3 + SO_4)$

Table 3.1. Chemical constituent concentrations (in mg/l units) and ionic ratios (equivalent ratios) in the Meditteranean Sea and in selected groundwater samples from saltwater intrusion zones in the 180-foot aquifer of Salinas Valley (California) and Mediterranean coastal aquifer of Israel.

qualitatively assessed or computed through the program SNORM [Bodine and Jones, 1986]. Seawater solutes are specifically characterized by $Mg > SO_4 + HCO_3$, whereas meteoric waters (dilute or saline), even if dominated by re-solution of marine salts, reflect Na > Cl. In contrast, sedimentary basin fluids can carry significant Ca and perhaps K excess over $SO_4 + HCO_3$ due to diagenetic carbonate or silicate reactions.

The most striking phenomenon that characterizes seawater intrusion is the difference between the chemical composition of the resulting brackish water and the simple mixture of seawater and groundwater. One would expect that the solute composition of seawater would dominate the chemistry of the mixture. In many cases, however, the brackish groundwaters have a Ca-rich composition (i.e., the ratio of $Ca/(SO_4 + HCO_3) > 1$) with low ratios of Na^+, SO_4^{2-}, K^+ and B to chloride plus high Ca/Mg ratios relative to modern ocean water (Table 3.1). Representative case studies in which the chemistry of saltwater intrusion were investigated are presented below.

3.2.1 The USA

The broad transition zone between fresh groundwater and underlying saltwater in the northern Atlantic Coastal Plain of the USA, described according to hydrochemical facies distribution by Back [1966], has been attributed by Meisler et al. [1984] to the effects of eustatic sea-level fluctuations during the Quaternary and late Tertiary. Geochemical study of the waters indicates that the freshwater is of a sodium bicarbonate type. The saltwater in North Carolina is predominantly seawater, but from Virginia northward, it is suggested to be a sodium calcium chloride brine significantly more concentrated than seawater.

Wicks et al. [1995], Sacks et al. [1992] and Wicks and Herman [1996] have combined petrographic, geochemical, and flow modeling to examine the complex coastal zone of western Florida and identify the mixing and consequences of upwelling gypsum-dissolving, carbonate groundwaters into the areas of seawater intrusion.

Magaritz and Luzier [1985] studied the saline/freshwater interface zone in Oregon. They found that most ions, except for Cl, show non-conservative behavior and suggested that the following processes control the chemistry of the investigated groundwater: (1) Ca-Na and Ca-Mg exchange; (2) oxidation of organic matter in sediments; (3) sulfate reduction (probably by bacterial processes) and (4) reaction of HCO_3 and sulfur with iron to produce mineral coatings.

In southern California, U.S. Geological Survey studies [Izbicki, 1991; 1996] have integrated geochemical, isotopic and geophysical techniques to enable an accurate determination of the area affected by seawater intrusion and its rate of encroachment in the Oxnard Plain coastal area northwest of Los Angeles. It was also possible to identify different sources of salinity and its development rather than only direct influx of seawater.

In the coastal basins of central California, in particular the "180-foot" and "400-foot" aquifer systems in the Salinas Valley, salinization of fresh groundwater is a conspicuous aspect of deterioration in groundwater quality. Seawater intrusion has been occurring in the area for several decades and has affected irrigation, domestic and municipal wells. The intrusion has been attributed to extensive withdrawal of water, which has lowered the regional water table below sea level and thereby induced inland migration of seawater into the aquifers from submarine outcrops in Monterey Bay. Intrusion in the shallow 180-foot aquifer has progressed inland about 8 km, whereas intrusion into the deeper 400-foot aquifer extends more than 2 km [Todd, 1989]. Todd [1989] used water quality data to distinguish regional intrusion from well leakage between the aquifers. The encroachment is characterized by increased chloride concentration associated with relatively high calcium and low sodium concentrations. The saline water associated with seawater intrusion in Salinas Valley is characterized by relatively low Na/Cl and high (> 1) Ca/(HCO$_3$ + SO$_4$) and Ca/SO$_4$ ratios relative to seawater [Vengosh et al., 1997].

3.2.2   The Mediterranean

Studies of the carbonate formations in coastal areas of Catalonia, NE Spain, by Custodio et al. [1993], point to the effects of ion exchange and organic matter oxidation in the generation of carbonate dissolution capacity, and suggest that these processes can be more important to karstic development than the mixing of fresh and salt water. In the coastal sediments of the Llobregat delta at Barcelona, Manzano et al. [1990] indicated that the mixing of freshwater from the underlying confined deep aquifer with connate marine pore waters determines groundwater solute composition within the aquitard.

Price and Herman [1991] found in Mallorca that aquifer properties in a coastal Pleistocene limestone were not altered by a modern transitory mixing zone. An excess in Sr and Ca concentration of groundwater over what was expected from conservative mixing with seawater was attributed to limestone dissolution in the vadose zone. Morell et al. [1986] also working in Spain argued that Br is the best indicator for tracing seawater.

Fidelibus and Tulipano [1986] showed that in certain parts of Italy Sr concentrations in saline groundwater were higher than those of seawater. The high Sr concentrations were explained by dissolution-precipitation processes in which Sr remains in solution. Recent seawater was found to intrude shallow aquifers, whereas old seawater, modified by the dolomitization process, occupies the deeper aquifer systems.

Ploethner et al. [1986] showed that in Cyprus saline groundwater had a different chemical composition than seawater and explained it by ion exchange processes. Fossil brines with a salinity 2.5 times that of seawater also were found in the region.

In Israel, saltwater intrusion has led to deterioration in water quality, and many wells completed in the Mediterranean coastal aquifer have been shut down. Mercado [1985] studied the saline groundwater in the interface zone of the Israeli coastal aquifer and found a large discrepancy in solute composition from what would be expected from a simple mixing of freshwater and seawater. The deviation was attributed primarily to processes of cation exchange and carbonate equilibria. The relatively high Ca and low Na contents were explained by Na retention by exchange for Ca and Mg, resulting in low Na/Cl ratios relative to marine values. Subsequently, Vengosh et al. [1991b] showed that the salinization process in the interface zone involved two evolutionary stages: (1) early salinization (Cl < 5000 mg/l) in which the saline water develops a Ca-chloride signature and, (2) a later stage of higher salinities wherein the solute composition is representative of mixing between seawater and a Ca-chloride fluid. In some wells groundwater with a Ca-chloride composition and salinity higher than that of seawater was found. Yechieli et al. [1996] showed that the relationship between conservative constituents (Cl, Br and $\delta^{18}O$) indicate that the main process is mixing between seawater and freshwater, whereas other non-conservative dissolved ion concentrations are modified by water-rock interactions.

3.2.3 Other Areas

Desai et al. [1979] worked on the coastal area of Gujarat, India and suggested exchange of Na^+ for Ca^{2+} resulting in solute Na/Cl ratios as low as 0.5. They also noted low K/Na ratios and low boron contents, which were explained by preferential adsorption of potassium and boron onto clay minerals.

The chemistry of groundwaters in Bermuda, which are a mixture of calcium bicarbonate water with a seawater component, have been investigated in detail by Plummer et al. [1976], and summarized along with the lithologic environment as a special case study by Morse and Mackenzie [1990] because of the wealth of related information available. Using conservative constituents such as chloride, Plummer et al. [1976] were able to calculate a seawater component ranging from 0.6 to 79 percent, and the amount of Ca, Mg, and Sr contributed by the dissolution of carbonate minerals.

Circulation of mixed saline waters and the consequences for carbonate island platforms have been considered at length in studies of the Bahamas by Whitaker and Smart [1993].

Back et al. [1986] studied the mixing zone in the coastal aquifer of Yucatan, Mexico. Using mass balance calculations (PHREEQE code; Parkhurst et al., [1980]) they were able to suggest the following chemical processes: mixing of fresh- and Caribbean seawater, dissolution of carbonate minerals and further precipitation of aragonite and calcite, as well as carbon dioxide dissolution and its further degassing. The geochemical reactions of enhanced carbonate mineral dissolution in the brackish water mixing zone contributed to geomorphic features such as caves and sinkholes [Back et al., 1984].

3.3   Chemical Modifications

As already noted, the chemical composition of saline groundwaters in many locations in coastal aquifers deviate from simple conservative seawater-freshwater mixing (e.g., Appelo and Geirnart [1991]; Sukhija et al. [1996]). Such deviations are attributed either to water-rock interactions [Mercado, 1985; Appelo and Postma, 1993] or to contamination by subsurface brines [Vengosh and Rosenthal, 1994; Vengosh et al., 1991b]. Principal water-rock interactions include ion exchange with clay material [Appelo and Willemsen, 1987] and carbonate dissolution-precipitation processes [Fidelibus and Tulipano, 1986]. Some of the chemical and isotopic parameters which behave conservatively (e.g., Cl, Br, deuterium) can be used to estimate the contribution of the different sources (e.g. Morell et al. [1986]), whereas others give information on the extent of the interactions with the solid matrix (e.g., B and Sr isotopes).

The chemical composition of waters resulting from a simple mixing would appear to be a matter of averaging the compositions of the waters that mix in proportion to their volume contributions to the mixture. While this situation is true for elemental concentrations of conservative solutes, such as sodium or chloride, it is not true where chemical speciation is concerned. This results from the fact that important parameters controlling speciation, such as ion activity coefficients or complexing in solution and sorption or exchange with the sedimentary mineral matrix, do not vary in linear proportion to composition. This situation is of special significance to carbonate or clay mineral diagenesis in coastal aquifer sediments.

Four basic reaction categories are associated with the hydrologic environment characteristic of seawater intrusion: mixing of groundwaters (including fluids associated with evaporites) and seawater, carbonate precipitation and/or diagenesis (e.g., dolomitization), ion exchange and silicate (largely clay) diagenesis, and redox reactions. The contribution of each of these processes and their impact on the chemical composition of saltwater intrusion will be further addressed.

Mixing of normal dilute groundwaters and seawater should be straightforward and tractable using halogens for reference (e.g., Figure 3.1). However, the presence of evaporitic conditions near the edge of the sea, such as the sabkhas of the Trucial coast, Bardawill Lagoon (northern Sinai), Salina Ometepec (Baja California), or the outflow of basinal brines from depth, present special problems with hypersaline fluids and the possible non-conservancy of chloride.

Probably the most widely considered reactions of the mixing zone involve carbonates; most simply, the recrystallization of calcite, and more controversial, the formation of dolomite. The recrystallization of metastable aragonite or magnesian calcite affects porosity and releases minor elements to solution (most notably, Sr^{2+}). For dolomitization, the relative role of activity coefficient depression versus kinetic inhibition, nucleation, and Mg^{2+} or HCO_3^- enhancement, is still controversial.

Ion exchange seems relatively simple in concept, but involvement of all major cations, lack of detail about the control on selectivity in sediments, and simultaneous carbonate interactions, renders the process more complicated than usually considered [Appelo and Postma, 1993]. Furthermore, some of the variation in solute composition attributed to exchange reactions may actually be due to subtle silicate diagenesis (e.g., interstratification) in clay mineral assemblages [Jones, 1986; Kauffman et al., 1998]. Considerable work has been done on normal marine and detrital sediments, but little attention has been paid to volcanic materials highly variable in composition and texture.

Redox reactions can contribute to geochemical changes accompa-

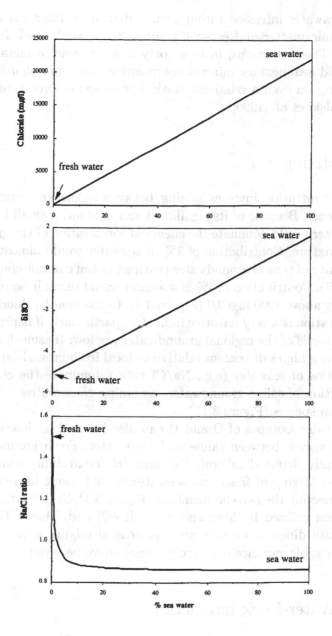

Figure 3.1. Cl, $\delta^{18}O$, and Na/Cl ratios vs. % seawater in mixing between seawater and fresh groundwater with a Cl content of 100 mg/l, $\delta^{18}O = -5‰$, and Na/Cl ratio of 1.5 (typical for regional groundwater in the coastal aquifer of Israel).

nying seawater intrusion through early diagenetic reactions involving organic matter and especially sulfur (e.g., Wicks and Troester [1997]). These reactions, in turn, play a major role in metal solubility and sedimentary mineral composition, and vary significantly depending on even a relatively static hydrologic environment (e.g., Domagalski et al. [1990]).

3.4   Mixing

Seawater intrusion involves mixing between saline and freshwater components. Because of its significant salt content, a small fraction of seawater would dominate the chemical composition of the groundwater mixture. Contribution of 1% of seawater would almost triple the salinity of typical groundwater (with an initial chloride content of 100 mg/l). Contribution of 5% of seawater would result in water with a salinity above 1000 mgCl/l (Figure 3.1). Consequently, chloride ion concentration is a very sensitive indicator, particularly if background salinity levels of the regional groundwater are low. Inasmuch as seawater has a high salt content relative to local fresh groundwater, the ionic ratios of seawater (e.g., Na/Cl ratio) dominate the chemical composition of saline groundwater, assuming conservative behavior of the ion species (Figure 3.1).

The stable isotopes of O and H can also be used to describe the mixing process between saline and freshwater. Fresh groundwater is generally depleted in both ^{18}O and ^2H (deuterium) relative to seawater. Mixing of fresh and seawater should result in a straight line connecting the two end members (Figure 3.1). Such relationships have been utilized by Manzano et al. [1990] and Izbicki [1996] to distinguish different water sources in coastal mixing zones, and to signal possible variance from truly conservative behavior.

3.5   Water-Rock Interaction

3.5.1   Exchange Reactions

The importance of cation exchange in seawater intrusion has been emphasized and treated in considerable detail by Appelo and Postma [1993]. Freshwater generally, but particularly in coastal areas, is dominated by Ca^{2+} and HCO_3^- ions derived primarily from the dissolution of calcite, or secondarily from plagioclase feldspar. Therefore, cation exchangers in aquifers such as clay minerals, organic matter, oxyhydroxides or fine-grained rock materials, have mostly Ca^{2+} adsorbed on the surfaces. In contrast, sediments in contact with seawater have Na^+ as the most prevalent sorbed cation [Sayles and Mangelsdorf, 1977]. When seawater intrudes on a coastal freshwater aquifer, Na^+ replaces part of the Ca^{2+} on the solid surface, as demonstrated in the following equation:

$$Na^+ + \frac{1}{2}Ca\text{-}X_2 \longrightarrow Na\text{-}X + \frac{1}{2}Ca^{2+} \tag{3.1}$$

where X represents the natural exchanger. In such reactions, Na^+ is taken up by the solid phase, Ca^{2+} is released, and the solute composition changes from NaCl to $CaCl_2$ type water [Custodio, 1987; Appelo and Postma, 1993]. Inasmuch as the chloride ion concentration remains unaffected by this reaction, it can be regarded as a reference parameter. Thus as seawater intrudes coastal aquifers containing freshwater, the Na/Cl ratio decreases and the (Ca + Mg)/Cl ratio increases. Under such conditions, the enrichments in calcium and magnesium should be balanced by the depletion of sodium (i.e., Ca + Mg = $-$Na). Accordingly, the decrease in Na/Cl ratio balances the increase in (Ca + Mg)/Cl ratio [Custodio, 1987].

Upon inflow of freshwater a reverse process takes place:

$$\frac{1}{2}Ca^{2+} + NaX \longrightarrow \frac{1}{2}CaX_2 + Na^+ \tag{3.2}$$

Flushing of the mixing zone by freshwater will thus result in uptake of Ca^{2+} and Mg^{2+} by the exchangers with concomitant release of Na^+. This is reflected in the increase of the Na/Cl ratio and a decrease of the (Ca + Mg)/Cl ratio value, and formation of $NaHCO_3$-type fluids. Water quality can thus indicate fluctuations of seawater-freshwater mixing, and the dynamics of ion exchange is reflected in changing ionic ratios. Coastal marine clay-bearing sediments can be the agents for major softening of recharging freshwaters [Hanor, 1980].

When saltwater displaces freshwater, the exchange process is focused in a more concentrated environment because of the general preference of natural exchangers for the divalent cations (in the case of major cations in seawater versus freshwater, specifically Ca^{2+} over

Na^+). Thus, the distribution of water types brought about by mixing of fresh and seawater is related not only to the differences in relative cation domination by Ca^{2+} and Na^+, but also by the nature of monovalent-divalent exchange.

As indicated qualitatively earlier, an exchange constant favors the divalent ion on the exchange sites relative to the solution. This is further complicated, however, by the lack of a straightforward model for the activities of the exchange ions on the solid equivalent to the Debye-Huckel equation which relates activities and concentrations for solutes. Generally, the higher charged ion is preferred more strongly with total solute concentration decrease, as a consequence of the exponent used in the mass action equation to describe mono-divalent exchange, and the calculation of exchangeable ion activities as equivalent fractions with respect to a fixed ion exchange capacity. The value for the equilibrium 'coefficient' for Na/Ca exchange is about 0.4, but this is dependent on both the solid and the water composition, because of the non-ideal behavior of the exchanger [Appelo and Postma, 1993].

The development of a pattern of water types in the mixing zone of seawater and freshwater depends on the amounts of exchangeable cations and their concentrations in solution [Appelo and Postma, 1993]. If this ratio is small, the succession of compositional changes will be relatively restricted when the aquifer material has a low cation exchange capacity or when salinities are high. The relatively high concentrations characteristic of seawater intrusion tend to restrict the thickness of the transition zone and its lateral extent in an aquifer. The sequence of chemical compositions has been demonstrated in column experiments by Beekman and Appelo [1990] and in a mixing cell computer model by Appelo and Willemsen [1987]. In addition to the expected development of a $CaCl_2$ water type with the arrival of the seawater front, a significant increase in Mg^{2+} also takes place. These increased levels are congruent at first, but when the Ca from exchange runs out, the trends of the two ion concentrations diverge—the Ca^{2+} decreases and Mg^{2+} increases towards its higher concentration in seawater. In fact, because of the relatively small amount of exchanged Ca compared to even the relatively low Ca level maintained by $CaCO_3$ saturation in the saline water, the decrease of Ca has already begun before the chloride reaches seawater concentrations [Appelo and Postma, 1993].

Appelo and Postma [1993] pointed out that when seawater intrusion into groundwater is primarily by diffusion rather than advection and the cation exchange capacity of the sediment is low, the effects of ion exchange tend to be considerably reduced and the resulting water compositions will resemble a simple mixture of fresh and saline water.

It should be noted that exchange reactions should result in proportional modifications of Ca and Na. In many cases, as in the saline groundwater from the Yarkon-Taninim aquifer in Israel, the Ca enrichment is not accompanied by Na depletion (i.e., $Ca/(SO_4 + HCO_3) > 1$, Na/Cl = seawater; Starinsky et al. [1995]). Consequently, exchange reactions do not account for such cases, but dolomitization in which Ca^{2+} is enriched and Mg^{2+} is depleted appears a more appropriate explanation.

3.5.2   Carbonate Diagenesis and Dolomitization

Most coastal aquifers are composed of some carbonate materials, either of calcareous clastics in which the cement is made of carbonate phases or of limestone in which calcium carbonate is the predominant mineral. Plummer [1975] and Wigley and Plummer [1976] have provided the clearest demonstrations of speciation effects for the mixing of dilute carbonate groundwaters with seawater. Probably the most often-cited result of such considerations is that mixing of dilute groundwater in equilibrium with calcite with normal seawater that is supersaturated with respect to calcite can produce solutions of intermediate composition that are undersaturated with respect to calcite. Moreover, the mixture products may be undersaturated with respect to calcite but supersaturated with respect to dolomite [Hanshaw et al., 1971; Land, 1973; Badiozamani, 1973; Plummer, 1975; Back et al., 1979; Morse and Mackenzie, 1990]. The extent of calcite undersaturation (and dolomite supersaturation), which results from mixing, is strongly dependent on temperature, the nonlinear nature of the equations governing the chemical equilibria, ionic strength effects, and perhaps most importantly, the partial pressure of CO_2 (P_{CO_2}) of fresh groundwater during its early evolution in the vadose zone. Evaluation of relative carbonate mineral saturation is also heavily dependent on the solubility of the dolomite, which is significantly greater for a disordered, very fine grained, non-stoichiometric form [Hardie, 1987].

The undersaturation of mixed-zone waters may affect the hydraulic conductivity of coastal aquifers. Thus, after mixing with seawater, calcite-saturated groundwaters (even without an external source of CO_2) become aggressive, facilitating further carbonate dissolution and thus increasing the content of dissolved Ca^{2+} and of HCO_3^-. This dissolution process was recognized by Mandel [1964; 1965] and by Schmorak and Mercado [1969]. In areas in which the coastal zone is built mainly of carbonates, this process may lead to karstification and high hydraulic conductivity [Back et al., 1984].

Dissolution features in coastal limestones of the Bahamas, the Yucatan, Greece, and Pacific atolls have been well documented (see references in Back [1986], Morse and Mackenzie [1990], and Ingebritsen and Sanford [1998]). With the discharge from coastal springs, dissolution can be enhanced by increased P_{CO_2} associated with organic oxidation. Sanford and Konikow [1989] have used a fully coupled reaction-transport model to analyze the process of carbonate diagenesis and porosity-permeability development in the mixing zone. They found that over long time intervals, extensive transgressions or regressions of the coastline will cause the transition zone to migrate over large distances, resulting in minor porosity enhancements over large areas rather than a major porosity development at any one location [Ingebritsen and Sanford, 1998]. In contrast to dissolution associated with increased P_{CO_2}, precipitation of calcite resulting from CO_2 degassing (rather than mixing of waters), has been shown by Hanor [1978] through both field experiment and theoretical calculations to be the origin of beachrock in the intertidal to supratidal zone.

The dolomitization process, in which calcite and dolomite are in equilibrium can be described as a transformation reaction such that

$$2CaCO_3 + Mg^{2+} \longrightarrow CaMg(CO_3)_2 + Ca^{2+} \qquad (3.3)$$

resulting in a progressive enrichment of Ca over Mg in solution (i.e., Mg/Ca ratio decreases). Thus saline water affected by dolomitization processes would be characterized by high Ca content and by a Ca-chloride signature (i.e., $Ca/(HCO_3 + SO_4) > 1$; Starinsky [1974; 1983]; Carpenter [1978]).

The fact that activity coefficients for solute Ca and Mg are similar has led to the idea that the Mg/Ca concentration ratio should be fixed for a given T and P in a solution at equilibrium with both calcite and dolomite, and that it can be calculated from the solubility products for the two minerals. However, the data from natural mixtures of seawater and groundwater vary considerably [Carpenter, 1980; Morse and Mackenzie, 1990], indicating non-equilibrium conditions, or control by non-stoichiometric dolomite.

Morse and Mackenzie [1990] have noted that the true influence of reaction rates is largely speculative because the kinetic factors are generally deduced primarily from the presence or absence of dolomite in different environments. Although saline water associated with the seawater-freshwater mixing zone is apparently supersaturated with respect to dolomite, modern marine and coastal sediments contain only relatively rare and minor occurrences of this mineral (i.e., a part of "the dolomite problem"). The formation of dolomite in modern marine sediments and sediment burial to shallow depths is strongly controlled by the reaction kinetics that are slow even at high supersaturation [Morse and Mackenzie, 1990]. The reaction rate of dolomite formation is temperature-sensitive and increases at high ionic strengths, elevated P_{CO_2}, and at high Mg/Ca ratio.

Hanshaw et al. [1971], Badiozamani [1973], and Land [1973] proposed that dolomite formation could result from the mixing of seawater and calcite-saturated fresh groundwater, the proportions varying depending on the degree of disorder and the resultant solubility products used [Plummer, 1975; Hardie, 1987], as well as the P_{CO_2}, pH, temperature, and original meteoric water composition. Ingebritsen and Sanford [1998] noted that the coastal zone of seawater intrusion provides an environment in which seawater can be continuously supplied to reaction sites and that small amounts of dolomite have been attributed to the conditions in some modern mixing zones, but the general lack of dolomite in such environments raises severe doubts about the efficacy of modern mixing mechanisms [Hardie, 1987].

Hardie [1987] has suggested that the occurrence (and type) of dolomite depends on the mechanism of formation, such that direct precipitation of a disordered phase takes place on evaporative concentration, or mixing of high Mg/Ca seawater brines with continental saline waters containing elevated HCO_3 levels. Replacement of precursor calcium carbonate by stoichiometric dolomite requires long reaction times at low temperatures and occurs only where hydrologic systems maintain through-flow of dolomite-supersaturated waters for very extended periods. Ingebritsen and Sanford [1998] indicated that only under certain circumstances will the fluxes be large enough and the driving mechanism in place long enough to deliver the quantity of magnesium necessary for extensive dolomitization. The requirement for Mg^{2+} makes normal or near-normal seawater the most likely dolomitizing fluid [Land, 1985].

Ingebritsen and Sanford [1998, and references cited therein] suggest that the best evidence for modern dolomitization is in the solute data for springs discharging from modern carbonate platforms (e.g., Whitaker and Smart [1994]), but that the process may be kinetically assisted by somewhat elevated temperatures, evaporative concentration, and/or reducing conditions. Certainly the seawater-derived saline water compositions of the Bardawill lagoons in the Sinai [Levy, 1974] suggest extensive exchange of solute Mg for Ca in the intervening calcareous dunes (implying ongoing dolomitization), and, indeed, the well-documented Pleistocene mixing zone dolomite of the Yucatan [Ward and Halley, 1985] is believed to have formed where the waters were 75 to 100 percent seawater.

3.5.3 Adsorption

One of the processes that modifies the chemistry of seawater intrusion is adsorption onto clay minerals in the host aquifer. The elements that are sensitive to adsorption process are potassium, boron and lithium. Vengosh et al. [1991b; 1997] demonstrated that in the coastal aquifers of Israel and central California (Salinas Valley) these elements are relatively depleted in saline water associated with saltwater intrusion. The adsorption process resulted in low K/Cl, B/Cl and Li/Cl ratios in the residual saline groundwater relative to the marine ratios. During adsorption processes, the light isotope ^{10}B, enters preferentially onto adsorbed sites on clay minerals leaving the residual saline water enriched in the heavy isotope ^{11}B. Consequently, saline water associated with saltwater intrusion is characterized by low B/Cl ratios and high $\delta^{11}B$ values ($^{11}B/^{10}B$ ratios normalized to NBS SRM 951 standard) relative to seawater. In the coastal aquifer of Israel, groundwater from the interface zone has $\delta^{11}B$ values as high as 60‰ relative to seawater with a $\delta^{11}B$ value of 39‰ [Vengosh et al., 1994] indicating extensive adsorption processes.

3.5.4   Reduction of Organic Matter

The intrusion of seawater and formation of a relatively static interface zone between overlying fresh and underlying saline water, may produce local low redox conditions, due to decomposition of dissolved organic matter, fine suspended organic particulate, or organic-rich sediments [Schoeller, 1956; Custodio and Llamas, 1976; Hem, 1985]. According to Custodio et al. [1987], this process will cause increased P_{CO_2}, changes in pH, and the reduction of dissolved sulfate to H_2S, resulting in low SO_4/Cl ratios. Such changes shift the calcium carbonate equilibria and most commonly cause dissolution. The resulting increase in the Ca-content is frequently masked by exchange of Ca^{2+} for Mg^{2+} or Na^+ on clays previously equilibrated with more seawater-like cation matrices.

As noted by Whitaker and Smart [1994] for the Bahamas, intense and episodic nature of rainfall, lack of soil cover, well developed karstic fissures and shallow depth of the vadose zone, all contribute to significant inputs of organic matter to the freshwater lens. This generates potential for dissolution considerably greater than that predicted solely by simulations of inorganic mixing between basal freshwater lens waters and underlying saline groundwaters. In the Bahamas, waters become undersaturated with respect to aragonite throughout the mixing zone and even with respect to calcite in the lower part of the zone. In contrast, on Isla de Mona, Wicks and Troester [1997] found that the solute composition of cave-passage waters with no dissolved organic matter (DOC) were determined entirely by mixing and aragonite precipitation, with no sulfate reduction. However, contribution of DOC from a thin unsaturated zone in the coastal plain was associated with CO_2 outgassing, sulfate reduction, and distinct carbonate dissolution in the mixing zone. Whitaker and Smart [1994] have documented that surface-derived organic matter penetrates the aquifer in the Bahamas to considerable depth, supporting both aerobic and sulfate-reducing heterotrophic bacteria. They noted that processes, rates, and distribution of organically mediated carbonate dissolution are controlled by the balance between rates of input and consumption of oxygen and organic matter. Coupled with physical properties of the aquifer, these factors influence the position of the redox interface between oxic and anoxic groundwaters. Reoxidation of reduced sulfur species in this zone then becomes a further means of promoting carbonate dissolution. Such considerations also apply to carbonate cements in clastic aquifers.

3.6   Intrusion of Fossil Seawater

In coastal areas, fossil seawater, entrapped in unflushed parts of the aquifer can also affect the quality of adjacent fresh groundwater. Fossil seawater could have originated from past invasions of coastal aquifers accompanying rises in sea levels. However, only a few studies have attempted to date saline groundwater and to evaluate the timing and rate of seawater intrusion. Among these studies are those of Hahn [1991] in Germany and De Breuck et al. [1991] in Belgium, who found old saline water in coastal aquifers and related it to former sea levels. A more comprehensive analysis evaluating mass transfer of carbon, and sulfur on solute and solid phases and their effects on C-14 dating is given by Izbicki [1996].

Recently, Yechieli et al. [1996] reported the occurrence of old saline groundwater in the coastal aquifer of Israel. The evidences for ancient seawater was primarily from tritium and radiocarbon dating. The tritium content of modern seawater is lower than that of present precipitation. High tritium signals were produced during the 1957-1963 period of high thermonuclear input. The tritium content of saline groundwater associated with saltwater intrusion may reflect linear mixing proportions of the saline and fresh components because tritium is a conservative species. Yechieli et al. [1996] showed that in many cases, both saline and fresh groundwater from the saltwater intrusion zone in the coastal aquifer of Israel have zero tritium content which indicates a relatively long residence time (> 40 years) and low flow rates of the saltwater encroachment inland. In these cases, tritium data rules out modern intrusion of seawater, which has important bearing on modeling of seawater intrusion [Yechieli et al., 1996].

In general, fresh and saline groundwater have a ^{14}C content that is a result of complex contribution from different sources, including the original seawater, soil CO_2, carbon derived from dissolution of carbonate matrix, and carbon derived from decomposition of organic matter. Moreover, the mixing process of saline and fresh groundwater can cause over-saturation and precipitation or dissolution of authigenic carbonate minerals, which might affect the ^{14}C budget [Custodio, 1997]. Yechieli et al. [1996] suggested that interpretation of ^{14}C data must also take into account the processes occurring in seawater while penetrating inland. These include interaction of interstitial (pore) water with bottom sea sediments and with the aquifer matrix. Consequently, ^{14}C interpretation requires additional chemical and ^{13}C supporting data.

In addition to direct dating methods, the age of saltwater intrusion can be indirectly evaluated by a comparison of the chemical composition of the investigated saline water with that of the host aquifer fluid. For example, high Ca concentrations with marine Na/Cl ratios cannot be interpreted as the result of exchange reactions, but rather as the product of dolomitization or silicate authigenesis. In such circumstances, inconsistency between the mineral assemblage of the host aquifer (i.e., absence of dolomite or of authigenic silicates) and the chemical composition of the saline water (i.e., high Ca content) suggests that the saline water probably originated and equilibrated under different hydrological conditions [Fidelibus and Tulipano, 1986]. Vengosh et al. [1991b] argued that the high salinity (greater than seawater) of groundwater associated with saltwater intrusion in several areas of the Israeli coastal aquifer suggests preservation of connate entrapped brines in the Mediterranean coastal aquifer.

Another indirect evaluation of the age of saline groundwater is the δ^{18}O value of saline water. Groundwater with depleted ^{18}O lower than that of modern atmospheric recharge, reflects past replenishment when the δ^{18}O of precipitation was lower. Usually, low δ^{18}O signals are related to glacial-age water which was entrapped in the aquifer [Siegel and Mandle, 1984]. In coastal areas this process can go the other way (i.e., heavier ^{18}O), depending on seasonal and source shifts under different climatic regimes [Plummer, 1993].

3.7 Criteria to Distinguish Saltwater Intrusions

The distinction of different salinization mechanisms is crucial to the evaluation of the origin, pathways, rates and future salinization of coastal aquifers. Discrimination between modern seawater intrusion and relics of entrapped brines within or underlying aquifers also has practical applications for modeling and water-resource management programs.

The interpretation of salinization process should be based upon geological and hydrochemical criteria. Several geochemical criteria can be suggested to identify the origin of salinity, especially detection of seawater intrusion as opposed to other salinity sources in coastal aquifers.

Salinity: Because of the contrast in marine and typical continental anion matrices, the clearest indication of possible seawater intrusion is an increase in Cl-concentration as a proxy for salinity, although other processes may lead to a similar phenomenon. In coastal aquifers, where continuous over-exploitation causes a reduction of the piezometric levels, intrusion of seawater results in a salinity breakthrough. Thus a time-series of chloride concentrations can record the early evolution of relatively rapid salinization processes.

Cl/Br ratios: The Cl/Br ratio can be used as a reliable tracer as both Cl and Br usually behave conservatively (i.e., do not react with the aquifer matrix) except in the presence of very high amounts of organic matter. Seawater (Cl/Br weight ratio = 297) is distinguished from relics of evaporated seawater (hypersaline brines Cl/Br < 297, Dead Sea = 40; Starinsky et al. [1983]), evaporite-dissolution products (over 1000) and anthropogenic sources like sewage effluents (Cl/Br ratios up to 800; Vengosh and Pankratov [1998]) or agriculture-return flows (low Cl/Br ratios). It should be noted that the Cl/Br signal can be modified by degradation of organic matter [Davis et al., 1998].

Na/Cl ratios: As shown above, Na/Cl ratios of saltwater intrusion are usually lower than the marine values (i.e., < 0.86, molar ratio). Thus low Na/Cl ratios, combined with other geochemical parameters, can be an indicator of the arrival of saltwater intrusion, even at relatively low chloride concentrations during early stages of salinization. The low Na/Cl ratio of seawater intrusion is distinguishable from the high (> 1) Na/Cl ratios typical of anthropogenic sources like domestic waste waters.

Ca/Mg, Ca/(HCO$_3$ + SO$_4$) ratios: One of the most conspicuous features of saltwater intrusion is commonly the enrichment of Ca over its concentration in seawater. High Ca/Mg and Ca/(HCO$_3$ + SO$_4$) ratios (> 1) are further indicators of the arrival of seawater intrusion. It should be noted however, that saline water with high Ca can originate by a different mechanism, not necessarily related to base-exchange reaction and modification of modern seawater.

O and H isotopes: Linear correlations are expected from mixing of seawater with ^{18}O depleted groundwater in the correlation of δD versus $\delta^{18}O$ or Cl versus $\delta^{18}O$. Different sources with high salinity (e.g., agriculture return-flows, sewage effluents) would result in different slopes due to evaporation processes that would change the isotopic composition of the saline end-member.

Boron isotopes: The boron isotopic composition of groundwater can be a powerful tool for discrimination of salinization sources, in particular distinguishing seawater from anthropogenic fluid such as domestic waste water. The $\delta^{11}B$ values of saltwater intrusion range over $30\%_{oo}$ to the seawater value ($\delta^{11}B = 39\%_{oo}$), reflecting mixing of freshwater and seawater in coastal areas. Saline groundwater from the coastal aquifer of Israel has high $\delta^{11}B$ values, up to $60\%_{oo}$. The high $\delta^{11}B$ content of saltwater intrusion differs from the boron isotopic composition of sewage effluents ($\delta^{11}B = 0-10\%_{oo}$) and sewage-contaminated groundwater ($5-25\%_{oo}$), and thus can be used to trace the origin of the salinity [Vengosh et al., 1994, 1998].

Acknowledgments: We wish to thank Janet Herman of the University of Virginia, as well as Ward Sanford and Chet Zenone of the U.S. Geological Survey, Water Resources Division, for very helpful comments on the manuscript, plus Marge Shapira for aid in its preparation.

Chapter 4

Exploitation, Restoration and Management

J. C. van Dam

4.1 Introduction

4.1.1 Scope

When dealing with exploitation, restoration and management of fresh groundwater in coastal aquifers the key issue is saltwater intrusion. Saltwater intrusion in groundwater is defined as the inflow of saline water in an aquifer system. This inflow can be in a steady state but mostly it is a transient process. In the latter case the inflowing saline water replaces fresh groundwater which was originally present in the system. The freshwater disappears by outflow at a rate roughly equal to the rate of inflow of saline water. This simultaneous outflow of fresh groundwater can take place either in a natural way, by seepage, or by abstraction. The result is an increase of the volume of saline groundwater and a decrease of the volume of fresh groundwater.

Saltwater intrusion occurs in coastal and deltaic areas all over the world, where the population density is great and many human activities take place. Many interesting cases have been reported in the proceedings of the Salt Water Intrusion Meetings which are held since 1968 at roughly two year intervals [SWIM, 1996 and before]. See also De Breuck [1991] and Jinno et al. [1990].

Saltwater intrusion is, in most cases, recognized by the responsible hydrogeologists. However, the actual extent of the problem is often not enough quantified due to lack of data. As in the process of saltwater intrusion tremendous volumes of fresh groundwater are

J. Bear et al. (eds.), Seawater Intrusion in Coastal Aquifers, 73–125.
© 1999 *Kluwer Academic Publishers.*

replaced by saline groundwater it takes considerable time to reach a new state of equilibrium. Moreover, before any new state of equilibrium is reached the rates of groundwater abstraction have often been gradually increased such that the corresponding states of equilibrium, which still have to be reached, have also changed gradually, often to unacceptable states. Proper measures are not always taken timely because in many cases there is no, or not yet, adequate legislation. So the present interests of e.g. some industrial parties can spoil the future situation.

The flows of fresh and saline groundwater are determined by the levels and gradients of the groundwater table and piezometric levels, which in turn are determined by boundary conditions, such as surface water levels, rates of recharge and abstraction. The boundary conditions change almost everywhere and always over time. This is either due to natural causes as e.g. relative sea level rise or due to human activities such as increasing groundwater abstraction or changes in artificially controlled surface water levels.

Another consequence of the drawdown of groundwater tables and piezometric levels is land subsidence [Carbognin, 1985]. In many coastal regions, especially in and nearby urbanized areas, both the saltwater intrusion and the land subsidence have developed dramatically. Land subsidence and its consequences can easily be assessed by leveling. Saltwater intrusion is invisible, in the subsoil, and the information on the actual salinity distribution in the subsoil is always scarce. There is often not enough information on the extent of saltwater intrusion and consequently no or insufficient insight into measures to be taken and their effects. Once the picture is clear indeed, it becomes evident that intensive and costly measures are required in many cases which are already in an alarming phase. The lessons learned from such cases should also serve for the prevention of new failures elsewhere. The most obvious, but not the only, measures to be taken will often be to reduce or even to stop abstraction of groundwater and, in some cases, to change the locations of abstraction works. For economic, social, legal, and even political reasons, such measures can often not be taken immediately, which can result in further deterioration and even abrupt breakdown of water supplies.

4.1.2   Groundwater as a System

Groundwater can be considered as a dynamic system with recharge, discharge, and changes in storage.

The recharge can be:

- natural, by infiltration and subsequent percolation of precipitation and runoff,
- artificial—for various objectives—by various methods of spreading water at the surface or by means of wells,
- induced from surface water.

Groundwater discharge occurs, under natural conditions, as outflow into surface waters and, artificially, by pumping or by land drainage both by horizontal and by vertical means of drainage.

With respect to changes in storage of groundwater, whether due to natural causes or due to human activities, one should make distinction between elastic storage and phreatic storage. The former is related to a change in pressure which is measured as a change in piezometric levels. It is small in comparison with the latter, which is manifested as a change in the phreatic groundwater table. Where fresh and saline groundwater occur, separated by either a sharp interface or a transition zone, any displacement of this separation, by abstraction or injection, inflow or outflow, involves an opposite change of storage of the fresh and the saline groundwater. According to the principle formulated independently by Badon Ghyben [1889] and by Herzberg [1901] and even long before by Du Commun [1828], the boundaries between groundwater of different salinity change over much greater heights than the groundwater tables. Consequently much greater volumes of water are involved by the vertical change of the interface, or transition zone, than by the change of the groundwater table only.

4.1.3   Origin and Occurrence of Saline Groundwater

Saline water makes up 97.25% of all water on earth. It is present in oceans, seas and estuaries and, as groundwater, in their subsoils and in land areas where seawater has occurred in the geologic history and not been replaced by freshwater so far. This holds in particular for the present coastal and deltaic areas.

The salinity distribution of the groundwater in coastal and deltaic areas is capricious as a result of past and ongoing natural processes as climate change, geologic processes and land subsidence, resulting in changes of the sea level relative to the land surface [van Dam, 1993].

Changes of climate have caused changes in sea levels throughout the geologic history. In the present time the sea level rises; rising temperatures make the seawater expanding and the polar ice caps and glaciers melting. Climate changes can also bring about changes in the rate of natural recharge of the fresh groundwater.

Eustatic and tectonic movements have caused changes of the land surface and sea bottoms with respect to the sea level, either upheaval or downwarping. When the sea level rises with respect to the land level the seawater invades unprotected land areas; this is called a transgression. The opposite, a retreat of the sea, is a regression. Transgressions and regressions have occurred throughout the geologic history, all of them at a very slow rate (in the order of centimeters per century and lasting for many thousands of years).

Due to eustatic and tectonic movements of the earth crust former seas may first have been turned into isolated inland seas (like the present Dead Sea in Israel and Jordan or the Salt Lake in USA) and subsequently dried up by evaporation, so that a body of solid salt was left behind. Such bodies, mostly covered by an impervious cap of anhydrite, were later deformed and covered by younger deposits and are now found at certain depths. They are called salt-domes. The solid salt inside the domes does not come into circulation unless, again by stresses and movements, cracks occur in the cap and there is flow of groundwater along and over the salt-dome which dissolves the solid salt.

Land subsidence, for instance due to compaction of peat layers in the subsoil and subsequent lowering of the controlled water tables, can have local or regional effects on the salinity distribution. By deposition of sediments, carried by rivers, at coasts and in deltas new land is built up in saline water.

As a consequence of the above described processes the salinity of the groundwater generally increases with depth. In some cases there is a sharp transition (i.e. within 5 or 10 m increase in depth) from fresh to saline groundwater; in others the transition is very gradual from fresh, via brackish, to saline groundwater. The present distribution of the salinity in the transition zone depends on the relative magnitudes of the underlying physical processes of flow, dispersion, diffusion and chemical processes throughout the geologic history.

In some cases fresh groundwater is found below brackish or saline groundwater. This is called an inversion, because it is not logic that the heavier brackish or saline groundwater occurs above the lighter fresh groundwater. In stable situations the explanation can only be found in the presence of a highly impermeable layer protecting the fresh groundwater below it from floating up and being mixed with the brackish or saline water above it. The freshwater may be either stagnant fossil groundwater or being recharged presently in the uplands and discharge far away as seepage in the lowlands or even invisibly on the sea bottom. The upward flow takes place through semi-permeable layers if there is an effective piezometric head, i.e. the piezometric level below the semi-permeable layer is higher than the piezometric level above it, both corrected for their respective densities and taking into account the density distribution in between.

In practice the distribution of fresh, brackish and saline groundwater in the subsoil appears to be very complex, with great variations both in vertical and lateral directions. This is partly due to the natural causes, climate changes and eustatic and tectonic movements, as described before, but, in the present time, also due to human activities. The effects lag far behind the causes, because the replacement of large volumes of fresh groundwater by saline groundwater takes considerable time. The effects of salinization and land-subsidence, can become dramatic, especially in and around big cities in coastal and deltaic areas.

4.1.4 Effects of Human Activities on Saltwater Intrusion

The effects of human activities on the salinity distribution in groundwater and seepage water have been recognized since long. In recent literature attention is paid to the description and the quantification of these effects, [Custodio et al., 1987; CHO-TNO, 1980; van Dam, 1976a, 1986, 1992, 1993, 1994 and 1997].

Saltwater problems occur both on the regional or large scale and on the local or small scale. The regional or large scale effects occur in large areas where the interface between fresh and saline groundwater moves slowly and smoothly in upward and/or in inland direction. The large scale displacement is caused by groundwater abstraction and/or by large scale lowering of the controlled groundwater table as in reclamation projects, new polders or for land improvement by drainage, by large excavations, such as borrow-pits for sand and gravel, and by excavations at the inner sides of sand dunes. Land reclamation at the seaside of sand dunes has the opposite effect.

The local or small scale refers, in this context, to the very small area around and below an abstraction well. The phenomenon at this scale is called upconing as the interface deforms locally in the shape of a cone with its top in the lower end of the well screen, resulting in a gradual increase of the salt content of the pumped groundwater which makes it, after some time, unfit for use.

4.1.5 Overview of Possible Problems

Saline and brackish water can not be used directly for public and industrial water supply and for agriculture and horticulture. This problem could technically be solved by desalting, as is done along the coasts of some countries in arid climates. However this is expensive and therefore desalting should only be considered as an ultimate solution in cases where the cheaper conventional (ground)water resources are insufficient. Moreover a by-product of desalting is brine. It is a problem to dispose of the brine; disposal in fresh surface waters will certainly not be tolerated by the responsible water authorities. Transport—e.g. by pipeline—and discharge into saline surface waters as estuaries and seas adds to the costs, and injection of the brine into deep aquifers—if feasible and acceptable from the hydrogeologic point of view—adds also to the costs.

Even for cooling purposes saline and brackish groundwater is not attractive. First of all the salinity might cause damage to the installations, by corrosion, and secondly the same problems are encountered for the disposal of saline and brackish water after its use for cooling purposes as were described for the brine after desalting. So far the conclusion is that saline groundwater should not be pumped for any use, unless it is desalted.

When overpumping fresh groundwater the danger of upconing exists, as already mentioned in section 4.1.4. In the foregoing the problems were related to groundwater abstraction, which is generally done by means of wells in distinct locations. There is however another human activity which can cause equal or even more serious problems. That is control (mostly lowering) of the groundwater table by means of artificial control of the surface water in an area (a polder) or land drainage. This applies in particular in land reclamation projects. In case of land reclamation in lagoons, tidal inlets or lakes the lowering of the water table can even be several meters. It is obvious that such great lowering over sometimes very large areas must have considerable effect on the regional groundwater flow system and can cause an unfavorable redistribution of fresh, brackish and saline groundwater. Artificial lowering of the (ground)water table in an area induces seepage or increase of seepage in that area. The seepage water may be or become brackish or saline. Interesting cases are found in the Netherlands [van Dam, 1976a; CHO-TNO, 1980].

The salinity of the seepage water in reclaimed areas can change with time. In case the seepage water is brackish or saline in the beginning it may gradually become fresh. This process takes centuries. In case the seepage water is initially fresh it may either remain so or, after tens of years at least or even centuries, become brackish or saline. If so, depending on the conditions, this situation can either be permanent or temporary. In the latter case the seepage will ultimately turn back into seepage of freshwater only [van Dam, 1992, 1994; Oude Essink, 1996].

The time to reach a new state of dynamic equilibrium in the distribution of fresh, brackish and saline groundwater after reclamation or artificial lowering of the groundwater table will also be considerable, in the order of tens of years at least or even centuries. This depends also on the areal extent of the land reclamation or groundwater control and on the depth of lowering of the original water level, whether surface water or groundwater.

In this context it should also be mentioned that the natural changes in the salinity distribution due the transgressions and regressions as described in section 4.1.3 may not yet have worked out completely. In other words even in areas without human intervention in the groundwater system, as described above, the present salinity distribution needs not yet to be in the state of dynamic equilibrium.

Sand and gravel borrow-pits bring about great changes in the original hydrogeologic profiles, especially when semi-permeable layers are removed to reach the underlying sand or gravel. Together with newly established water levels, whether constant or variable over the year, natural or artificially controlled, this can have considerable effects on the groundwater flow systems and thus on the amount of seepage and on the salinity distribution in the aquifers.

Excavation of sand at the inner side of sand dunes entails a draw-down of the groundwater table. Land reclamation by artificial raising the land surface, by hydraulic fill, at the seaside of sand dunes has the opposite effect; the groundwater table in the sand dunes will rise. These changes have their effects, in the long run, on the inter-face or the transition zone between fresh and saline groundwater. In case of excavation the corresponding lowering of the groundwater table entails a shrinkage of the freshwater lens. Land reclamation at the seaside, entailing a higher groundwater table throughout the sand dunes, leads to a growth of the freshwater lens, which is not a problem but an advantage.

4.1.6   Principles for Possible Solutions

For the various problems described in section 4.1.5 there is a great variety of technical solutions. Not all of them are economically fea-sible because the effects of some solutions are only at the very long term. Prior to a further elaboration of some solutions in sections 4.2 and 4.3 distinction can be made between solutions for the causes and solutions for the effects.

The saltwater intrusion problems which are caused by overpump-ing of aquifers can also be described as problems where the volume of brackish or saline groundwater increases to the detriment of an equal decrease of the volume of fresh groundwater. This formulation points to possible solutions for elimination or compensation of prob-lems caused by overpumping. Such solutions must either lead directly to a growth of the volume of fresh groundwater at the expense of the brackish or saline groundwater or—which comes to the same—lead to a decrease of the volume of brackish or saline groundwater such that freshwater can occupy the volume which becomes available.

Growth of the volume of fresh groundwater can be achieved di-rectly by an increase of the natural or artificial recharge of the fresh groundwater. This can be realized by increased infiltration at the land surface or in surface waters or by means of recharge wells with well screens in aquifers at any suitable depth. The water for recharge may either come from surface waters or be the pumped groundwa-ter after use and subsequent treatment. Attention should be paid to the question where the displaced brackish and saline groundwater goes and what effects it has. So, for instance, increased seepage of brackish or saline groundwater in adjacent areas will generally not be appreciated. Therefore due attention should also be paid to the layout of the total system of abstraction works and recharge works. Moreover by proper mutual location of such works the dimensions of the volume of freshwater can be favorably influenced.

Pumping the brackish or saline groundwater merely with the aim to reject it meets with the same problems as have already been described in section 4.1.5. This means that such solutions can only be realistic when pumping the brackish or saline groundwater near the coastline or estuaries.

The problems of seepage of brackish or saline groundwater caused by land reclamation and by land drainage are inherent to those activities and occur in the reclaimed and drained areas themselves. Generally the aquifers are thick or deep that it is not realistic to cut off the groundwater flow towards such areas by sheetpiling or grouting. This implies that in reclaimed and drained areas one must resort to fighting the effects because the causes can not be eliminated or compensated.

The harmful effects of brackish or saline seepage on agriculture can be combated or reduced. In those cases where the salinity of the water in the root zone is too high sprinkling of freshwater can be a solution. The salinity of the surface waters in reclaimed or drained areas can be reduced by flushing the surface waters in those areas with freshwater from other sources thus diluting and removing the salt carried by seepage to those areas.

4.2 Optimal Exploitation of Fresh Groundwater in Coastal Aquifer Systems

4.2.1 Exploitation of Fresh Groundwater in Coastal Aquifer Systems as an Element in Integrated Water Management

The exploitation of fresh groundwater in coastal aquifer systems can not be considered separately. It should always be done in connection with other available or potential water resources and with the water requirements, both in terms of quantity and quality, for the various present and future water requirements. Moreover the environmental aspects of all elements of the complete water management plan for an area should be considered and weighed. In other words the exploitation of fresh groundwater in coastal aquifer systems should form part of an integrated water management policy, such that an optimal allocation of the water resources is attained both in terms of economic and of environmental aspects, within the prevailing physical constraints. In the present times there is great concern for, and increasing understanding of, the environmental aspects.

Freshwater resources in coastal areas can be groundwater or surface water, either locally available or from greater distance, or desalted saline groundwater or seawater. It is obvious that a proper exploitation of freshwater resources is only possible if sufficient and reliable data are available. The collection, storage, processing and retrieval of all relevant data is indispensable. This issue will be worked out in the sections 4.2.2 and 4.4.3.

The optimal allocation of the available freshwater resources should be based on the geographical distribution of the water requirements, both in terms of water quantity and water quality. This picture is never constant over time. So any integrated water management plan should also account for future developments and even the control of such developments. In this context the term adaptive water management has also been introduced. The foregoing lines of thought point also to the necessity of adequate legislation. This issue will be worked out in section 4.4.4.

4.2.2 Required hydrogeologic Information

When studying groundwater flow it is necessary to know:

1. the system in which the groundwater flow takes place, i.e. the presence and extent of aquifers and semi-permeable layers, characterized by their thicknesses, permeabilities (or hydraulic resistances), porosities, specific yields and elastic storage coefficients.
2. the contents of the system, viz.:
   - the elevation of the groundwater table,
   - the density distribution of the groundwater, which is a function of its temperature and its chemical and isotopic composition, in particular its salinity distribution.
3. the boundary conditions of the system, such as natural and artificial recharge and connections with surface waters.

More about this subject is found in section 4.4.3.

4.2.3 Objectives

When abstracting fresh groundwater, as an element in integrated water management, the rate of abstraction should not exceed the permissive sustained yield (or sustainable yield or safe yield), defined as "the maximum rate at which water can economically and legally be withdrawn from a particular source for beneficial purposes without bringing about some undesired result" [ASCE, 1961]. In present-day's terminology, the term sustainability is used to indicate the same situation, however with much more concern and care for the environmental aspects (in particular flora and fauna), which are much better understood now and are still subject to intensive studies.

The philosophy of sustainability poses limits to the drawdown of groundwater levels and piezometric levels and to the reduction of natural outflow of groundwater. As a result the sustainable yield is always only a fraction of the recharge (natural or artificially increased). Still this leaves space for discussion because, apart from the area of nature to be maintained, a choice should be made of what type of nature is to be aimed at, either to maintain the present situation, or to restore a historic situation or even to create new nature. Whatever the choice is, it should be made and maintained for a long time ahead. This is because the transition from the present state to the state of nature aimed at takes a long period of time, in the order of several decades. This long lasting natural process should not be interfered by changes in policy .

When abstracting fresh groundwater in coastal areas the same principles hold. However, in addition to groundwater tables and piezometric levels, attention should also be given and choices should be made for the desired distribution of fresh and saline water in the subsoil. This subject will be worked out in more detail in sections 4.2.4 and 4.2.5.

4.2.4 Analysis

From the foregoing it is obvious that any increase of groundwater abstraction, under otherwise unchanged conditions (no artificial recharge, no relative sea level rise), leads to a decrease of the volume of freshwater in the aquifer, simultaneously with an equal increase of the volume of saline water by inflow from the seaside. The rates of

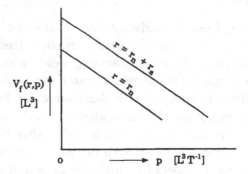

Figure 4.1. Volume of fresh groundwater $V_f(r,p)$ as a function of the rate of abstraction p for different values of the recharge r_n or r_n+r_a.

decrease and increase are also equal. Figure 4.1 presents, schematically, the relationship between the rate of abstraction p $[L^3T^{-1}]$ (p stands for production) and the volume of fresh groundwater in the subsoil $V_f(r,p)$ $[L^3]$, for different rates of recharge r $[L^3T^{-1}]$, whether natural (r_n), or natural and artificial (r_a) together (r_n+r_a). The volume of phreatic or confined groundwater is bounded by its contact with the sea, either at the coastline (for phreatic groundwater) or at some distance off the coastline (for confined groundwater), the impermeable base and a chosen boundary somewhere inland, beyond the presence of the saline groundwater. Part of this volume is fresh, the other part is saline. The recharge r is either of local or regional origin or consists of inflow from the uplands or a combination of the two. The abstraction p is concentrated in wells or in well fields. In case of a great many of small wells, distributed over the area the abstraction may, in this explanatory context, be considered as being uniformly distributed over the surface. The relationships presented in Figure 4.1 hold for a fixed configuration of abstraction works. The relationships drawn in this schematic figure, though drawn as straight lines, are not necessarily linear.

Figure 4.2 shows $V_f(r,p)$ as a function of time in case of a sudden increase, at $t = 0$, of the rate of abstraction p, from p_1 to p_2 and with constant recharge r. From the instant of this increase the curve goes down, from the volume $V_f(r,p_1)$, asymptotically to the volume $V_f(r,p_2)$, according to the relationship of Figure 4.1. The loss in volume of freshwater is equal to the increase in volume of saline

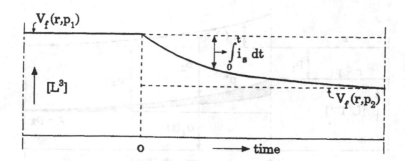

Figure 4.2. Volume of fresh groundwater $V_f(r, p)$ as a function of time in case of a stepwise increase of the groundwater abstraction rate. The accumulated inflow of saline water i_s is indicated.

water, indicated as:

$$\int_0^t i_s(t)\, dt \qquad (4.1)$$

where i_s is the rate of inflow $[L^3 T^{-1}]$ of the saline water. When neglecting the relatively small storage terms, for both the phreatic storage and the elastic storage, the groundwater balance for the total volume of groundwater, both fresh and saline, in the bounded area, reads at any time t:

$$r(t) + i_s(t) = p(t) + o_f(t) \qquad (4.2)$$

where o_f is the rate of outflow of fresh groundwater.

This relationship is graphically presented, for the above-described case, in Figure 4.3. The rate of outflow of the fresh groundwater, o_f, is a loss. This loss could, at least partly, be recovered by temporary pumping an additional rate p^* of fresh groundwater, as tentatively indicated by the hatched area in Figure 4.3. The distribution over time of p^* could be different from what has been drawn, for example stepwise. This volume of water can be pumped only once. This is called mining of fresh groundwater. The sooner this volume of fresh groundwater is taken—before it is (partly) lost by outflow—the better.

This is what has been done in practice, mostly unconsciously. By pumping more fresh groundwater than a permissible fraction of the recharge the volume of fresh groundwater decreased. By the

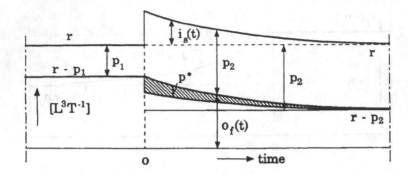

Figure 4.3. Effect of a sudden increase of the rate of abstraction p. The rate p^* is partially lost, the other part can be recovered.

time that some deep wells had to be abandoned because of the rising interface between fresh and saline groundwater new wells were drilled more inland and exploited thus mining another volume of fresh groundwater. This process can be repeated until no further retreat is realistic. The ultimate situation is one where a greater fraction of the recharge may be sustainably abstracted, because less freshwater needs to be sacrificed, as outflow, to maintain a more inland position of the saltwater wedge than in the original situation. The decrease of the volume of fresh groundwater has then been consumed as a one time reserve. This appears to be attractive. However there is a practical limit to this withdrawal. The choice of the permissive ultimate rate of abstraction—which remains always a fraction less than one of the recharge—and the related total volume of fresh groundwater in the subsoil should be based on the following considerations:

1. If abstraction wells have to be abandoned their users should be offered another source of supply, which may be water by pipeline from new wells more inland, which entails costs for construction, operation and maintenance.

2. The ultimate volume of fresh groundwater in the system should not be too small in order to dispose of some reserve for temporary overdraft in periods (seasons or even a succession of extreme years) of less recharge or greater demand or a combination of these two. This is a matter of risk assessment and policy in decision making.

Obviously decisions on the optimal exploitation of fresh groundwater in coastal aquifer systems should be based on economic criteria and on environmental criteria as described in section 4.2.3, taking into account hydrological and environmental risk assessments and the agreed up-to-date policy for integrated water management as described in section 4.2.1.

4.2.5 Location and Capacity of Abstraction Works, Fundamental Aspects

4.2.5.1 Abstraction of Confined Fresh Groundwater, an Illustration

An illustration of the effects of abstraction of fresh groundwater from a confined aquifer is shown in the example of Figure 4.4. The natural outflow is q_o per unit of length perpendicular to the paper (q_o in $[L^2T^{-1}]$). The transmissivity of the aquifer is kD (the hydraulic conductivity k in $[LT^{-1}]$ and the aquifer thickness D in $[L]$), its porosity is ϕ [dimensionless]. When assuming a sharp interface between the fresh and saline groundwater, the length L_1 of the saltwater wedge follows from the expression:

$$L_1 = \frac{\alpha k D^2}{2q_o} \qquad (4.3)$$

where

$$\alpha = \frac{\rho_s - \rho_f}{\rho_f} \qquad (4.4)$$

ρ_a and ρ_f being the densities $[ML^{-3}]$ of the saline and the fresh groundwater.

If, at the instant $t = 0$, the outflow q_o is diminished by an abstraction q_a, somewhere inland, the length of the wedge will ultimately become

$$L_2 = \frac{\alpha k D^2}{2(q_o - q_a)} \qquad (4.5)$$

The difference in volume of saline groundwater ΔV $[L^2]$ between the ultimate and the initial position of the interface is, as both interfaces are parabolas:

$$\Delta V = \frac{\phi D (L_2 - L_1)}{3} \qquad (4.6)$$

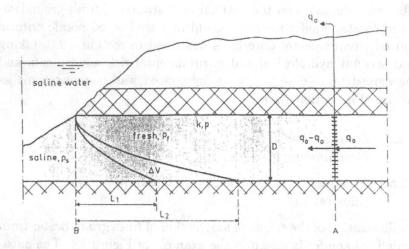

Figure 4.4. Outflow of confined groundwater with position of the salt water wedge with and without abstraction q_a.

or, after substitution of Eqs. 4.3 and 4.5 for L_1 and L_2:

$$\Delta V = \frac{\phi \alpha k D^3}{6} \frac{q_a/q_o}{q_o(1 - q_a/q_o)} \tag{4.7}$$

At $t = 0$, the flow of fresh groundwater just left of section A in Figure 4.4 changes suddenly. From that instant on, this flow changes gradually as tentatively indicated by line a in Figure 4.5 which runs asymptotically to the level $q_o - q_a$. The proper shape of this line depends on the boundary conditions at the right-hand side in Figure 4.4. The area R in Figure 4.5 accounts for the release of storage of phreatic groundwater in the uplands. The volume R is generally small in comparison to the volume ΔV. The release from elastic storage cannot even be made visible on the chosen scale. The recharge is assumed to remain constant. Line a also represents the resulting outflow of groundwater through section B in Figure 4.4. This outflow is the difference between outflow of fresh groundwater and inflow of saline groundwater. The outflow of fresh groundwater in section B changes according to, for instance, line b, also tentatively drawn in Figure 4.5. For continuity, the rate of inflow of saline groundwater in section B is the difference between the ordinates of the lines b and a. The hatched area ΔV in Figure 4.5 represents the volume ΔV of Eq. 4.7.

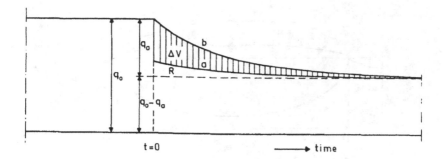

Figure 4.5. Outflow of fresh groundwater (b) and inflow of saline groundwater $(b - a)$ as a function of time t since the onset of abstraction q_a.

The characteristic time T, defined as:

$$T = \frac{\Delta V}{q_a} \tag{4.8}$$

is a measure, and no more than a measure, of the time scale

$$T = \frac{\phi \alpha k D^3}{6 q_o^2 (1 - q_a / q_o)} \tag{4.9}$$

For example, if $\phi = 0.3$; $\alpha = 0.025$; $k = 10$ m/day; $D = 100$ m; $q_o = 1$ m^2/day; $q_a = 0.5$ m^2/day, then $L_1 = 1250$ m, $L_2 = 2500$ m and $T = 25,000$ days ≈ 68.5 years. The time until 90% or 95% of the volume ΔV of fresh groundwater has been replaced by saline groundwater is a multiple of T, say 2 or 3 times T.

From Eq. 4.9, it is clear that the characteristic time is much greater for greater thicknesses of the aquifer than for smaller ones. Moreover, the ratio of the abstraction q_a to the natural outflow q_o plays an important role. Especially for the greater values of this ratio, the characteristic time T can become great.

Remarkably, relative sea level rise has no effect on the length L of the saltwater wedge in confined groundwater. This statement follows from Eq. 4.3 as the depth of the aquifer below mean sea level does not appear in that expression.

4.2.5.2 Abstraction of Phreatic Fresh Groundwater, an Illustration

As in confined groundwater, there is also a limit to the rate of groundwater abstraction in phreatic groundwater. The upper limit is dic-

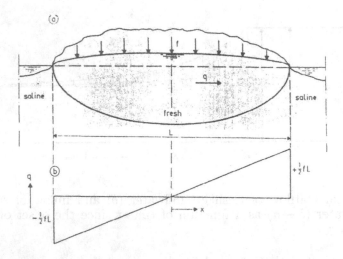

Figure 4.6. (a) Fresh water lens with natural recharge f. (b) Groundwater flow q as a function of location x.

tated by the recharge. The sustainable yield depends on the ultimate position of the interface or transition zone between fresh and saline groundwater. Upconing of saline groundwater must be avoided. The groundwater tables as well as the position of the interface or transition zone are strongly dependent on the location of the groundwater abstraction works. This is illustrated by the following example.

An island of width L and infinite length perpendicular to the profile drawn in Figure 4.6a is subject to natural recharge at a rate f $[LT^{-1}]$. The groundwater table and an assumed sharp interface between fresh and saline groundwater have been drawn. The maximum thickness of the freshwater lens, which is the sum of the elevation h $[L]$ of the groundwater table above mean sea level and the depth H $[L]$ of the interface below mean sea level, occurs in the middle of the island and is:

$$H + h = L\sqrt{\frac{(1+\alpha)f}{4k\alpha}} \qquad (4.10)$$

In other words, the maximum thickness of the freshwater lens is proportional to the width L of the island and to the square root of the recharge f. The latter applies to the effect of a change of the natural recharge e.g. as a consequence of climate change, or by uniformly distributed abstraction of groundwater or by uniformly distributed artificial recharge of groundwater. So, with a natural recharge f and a uniformly distributed abstraction at the rate $\frac{1}{2}f$ the thickness of the freshwater lens will ultimately reduce to roughly 0.7 times the original thickness, i.e. the thickness in the absence of any abstraction. On the other hand uniformly distributed artificial recharge at the rate of $3f$ in addition to the natural recharge f will ultimately yield a lens of only twice the original thickness, which is not very rewarding.

A suitable expression for the characteristic time T for the transition from one state of dynamic equilibrium with recharge f_1 to another state of dynamic equilibrium with recharge f_2 is, similar to Eq. 4.8:

$$T = \frac{\Delta V}{(f_2 - f_1)L} \qquad (4.11)$$

where the difference in volume ΔV $[L^2]$ is:

$$\Delta V = \frac{\pi}{4}\phi L^2 \sqrt{\frac{1+\alpha}{4k\alpha}} \left(\sqrt{f_2} - \sqrt{f_1}\right) \qquad (4.12)$$

so that:

$$T = \frac{\pi}{4}\phi L^2 \sqrt{\frac{1+\alpha}{4k\alpha}} \frac{1}{\sqrt{f_2} + \sqrt{f_1}} \qquad (4.13)$$

For example, if $\phi = 0.3$; $L = 5000$ m; $\alpha = 0.025$; $k = 10$ m/day; $f_1 = 2$ mm/day; $f_2 = 1$ mm/day, then $H + h$ is respectively 226.38 m and 160.08 m for the two recharge rates, and $T = 15,623$ days ≈ 42.8 years.

The characteristic time T can also be expressed in the mean residence times T_1 and T_2 of the groundwater in the states of dynamic equilibrium with natural recharges f_1 and f_2 respectively,

$$T = \frac{\phi(V_2 - V_1)}{L(f_2 - f_1)} = \frac{L f_2 T_2 - L f_1 T_1}{L(f_2 - f_1)} = \frac{f_2 T_2 - f_1 T_1}{f_2 - f_1} \qquad (4.14)$$

so that

$$\frac{T}{T_1} = \frac{f_2/f_1 \cdot T_2/T_1 - 1}{f_2/f_1 - 1} \qquad (4.15)$$

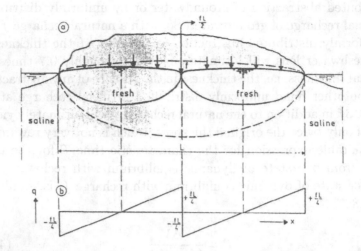

Figure 4.7. (a) Fresh water lens with natural recharge f and abstraction $\frac{1}{2}fL$ in the middle. (b) Groundwater flow q as a function of location x.

As, in this case,

$$\frac{T_2}{T_1} = \sqrt{\frac{f_1}{f_2}} \qquad (4.16)$$

$$\frac{T}{T_1} = \frac{\sqrt{f_2/f_1} - 1}{f_2/f_1 - 1} = \frac{1}{\sqrt{f_2/f_1} + 1} \qquad (4.17)$$

For ratios f_2/f_1 ranging from 0.5 to 2.0, T/T_1 ranges from 0.586 to 0.414. Or, in other words, the characteristic time T ranges from 0.586 to 0.414 times the mean residence time T_1 in the original situation.

The groundwater flow q as a function of the location x in Figure 4.6a is drawn in Figure 4.6b. It should be possible to abstract $\frac{1}{2}fL$, which is half the natural recharge. However, when abstracting by means of one line of wells parallel to the coastlines, in the middle of the island, the groundwater table and the interface will change to ultimate positions as in Figure 4.7a. This figure and the corresponding Figure 4.7b clearly indicate that this is a unrealistic solution as the saline water will enter the well screens. Yet it should be possible to abstract $\frac{1}{2}fL$ or even somewhat more, but of course less than

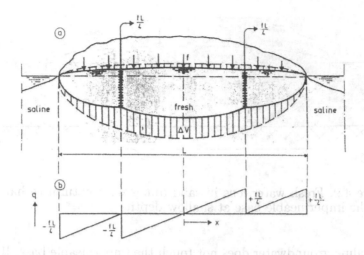

Figure 4.8. (a) Fresh water lens with natural recharge f and abstraction $2 \times \frac{1}{4} fL$ in two lines of wells. (b) Groundwater flow q as a function of location x.

fL. In order to achieve this, two lines of wells must be placed as for instance in Figure 4.8a, showing the groundwater table and the interface. The groundwater flow q as a function of the location has been drawn in the corresponding Figure 4.8b. The groundwater table and the interface of Figure 4.6a have been reproduced as dashed lines in Figures 4.7a and 4.8a for comparison. It appears that volume ΔV as indicated in Figure 4.8a turns from fresh to saline during the transient period from the initial to the ultimate state of dynamic equilibrium. Again, the characteristic time $T = \Delta V / q_a$ is a measure for the transient time.

In order to preserve a large volume of fresh groundwater, ΔV should be small. This can be achieved by placing the lines of wells as close to the coastlines as possible without the danger of upconing of saline groundwater in the wells.

The phreatic cases described so far for an island with parallel coastlines hold equally for the case of sand dunes along the sea coast with the inland side saline groundwater level controlled at mean sea level.

In the phreatic cases dealt with so far the impermeable base was assumed to lie at such a great depth that the interface between fresh

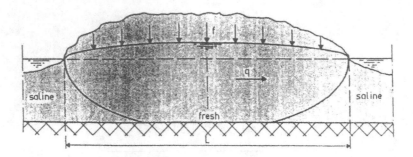

Figure 4.9. Fresh water lens in sand dunes with natural recharge f and the impermeable base at shallow depth.

and saline groundwater does not touch the impermeable base. If the impermeable base lies higher, such that the interface of Figure 4.6 would be intersected, there would be two saltwater wedges, one on either side of the island or the sand dunes, separated by a middle reach where the fresh groundwater would reach to the impermeable base as depicted in Figure 4.9. As a consequence, the transmissivity in the middle reach in Figure 4.9 is somewhat smaller than $k(H+h)$ as in Figure 4.6. The groundwater table will be slightly higher than that in Figure 4.6. The controlled groundwater table at the inland side of the sand dunes is often below mean sea level e.g. due to land reclamation in former lakes or lagoons or simply by lowering the controlled water levels in response to land-subsidence. If, in such a case, the impermeable base is below the interface between fresh and saline groundwater the saline groundwater can no longer be stagnant; it will flow inland underneath the fresh groundwater and occurs as seepage in the low-lying areas behind the sand dunes as indicated in Figure 4.10. The lens is no longer symmetric as in Figure 4.6. It can easily be derived that in this case (Figure 4.10) the groundwater divide has shifted in seaward direction and the deepest point of the interface between fresh and saline groundwater is more inland than in Figure 4.6. This situation occurs in the Netherlands where several polders have been reclaimed at the bottom of former lakes at short distances behind the sand dunes. The controlled water levels in these polders are several meters below mean sea level.

In case the controlled water level at the inland side is below mean sea level and the impermeable base is at shallow depth, as in Figure

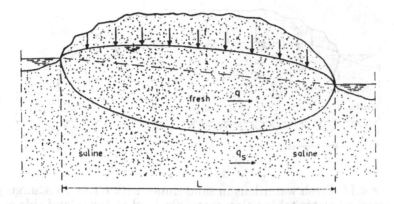

Figure 4.10. Fresh water lens in sand dunes with natural recharge f and a controlled water table below mean sea level at the inland side. Inflow of saline groundwater underneath the fresh groundwater.

4.11 the freshwater lens will reach to the impermeable base thus cutting off the inflow of saline groundwater. This situation occurs, for example, in the delta of the river Vistula in Poland and along the Belgian sea coast.

The foregoing paragraphs clearly illustrate the importance of the depths of the interface or transition zone between the fresh and saline groundwater and the impermeable base. The latter is a given natural fact, which can not be influenced. The interface or transition zone should be as deep as possible, not only to have a big volume of fresh groundwater in storage, but also as a barrier against the inflow of saline water. This is a guideline for the exploitation and restoration of fresh groundwater lenses in sand dunes along a coast.

With respect to the effect of relative sea level rise on a phreatic freshwater lens in an island the preceding text leads to the following conclusion. As long as the relative sea level rise does not lead to a reduction of the width L of the island the shape of the freshwater lens remains the same. The lens as a whole will float upward over the height of relative sea level rise. In case the width L of the island decreases due to relative sea level rise, the width of the lens decreases accordingly and consequently its thickness decreases proportionally as follows from Eq. 4.10. In situations as in Figure 4.10 relative sea level rise brings about saltwater intrusion due to the increase of the difference between the sea level and the controlled water level at the

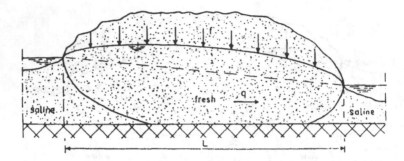

Figure 4.11. Fresh water lens in sand dunes with natural recharge f, a controlled water table below mean sea level at the inland side and the impermeable base at shallow depth. Inflow of saline groundwater cannot occur.

land side of the sand dunes.

4.2.6 Location and Capacity of Abstraction Works, Practical Aspects

In order to avoid upconing of saltwater a greater number of small capacity wells is to be preferred over a few high capacity wells. This holds in particular in case the thickness of the freshwater lens is small. The well screens should be in the fresh groundwater zone, well above the underlying brackish and saline groundwater. So the fresh groundwater is skimmed; this explains the term skimming wells.

Where danger of upconing of brackish and saline groundwater exists the well screens of the abstraction wells should preferably be situated above less permeable layers where these are present and known. Such layers prevent or retard upconing of the underlying brackish and saline groundwater. The consequence of this choice is that the drawdowns of the groundwater table or the piezometric level(s) in the aquifer(s) above the less permeable layers around the wells are greater than when abstracting below such layers. In the latter case the drawdowns are spread out over a larger area. This is generally preferred, for other reasons, in those cases where the aquifers are filled with fresh groundwater only, which is not the case here. In the present context this preference is overruled by the preference for situation of the well screens above less permeable layers as explained above.

Horizontal abstraction works, as galleries and drains can be applied in phreatic groundwater at shallow depths. With such abstraction works there is less risk that brackish or saline groundwater can reach the abstraction works than with vertical wells. This is because with horizontal abstraction works the upconing of brackish or saline groundwater is spread out horizontally instead of concentrated as under vertical wells. This spreading holds, to a lesser extent, also for deep wells equipped with horizontal well screens of several tens of meters length, in different horizontal directions. Such well screens are drilled, with modern techniques, radially from vertical shafts of great diameter.

A proper choice for the number of wells, galleries or drains and their individual capacities needed for a certain total rate of sustainable abstraction must be based on:

- the places where the water is to be consumed, distributed as in agriculture or concentrated as in industry,
- the risk of upconing of saline or brackish groundwater under extreme conditions as long dry spells with little or no recharge—the thinner the freshwater lens the greater the risk,
- the economic aspects of various alternatives, determined by the capitalized costs of construction, operation, energy requirements and maintenance of the wells, pipelines and pumping equipment.

In case of salinization of abstraction wells by upconing—a local or small scale problem—a remedy is to give up the well for some time or even permanently. In the former case abstraction may be resumed after some time, but at a lower pumping rate. Another solution can be the application of scavenger wells, as will be explained in section 4.3.6. The success of any of these measures depends on what is going on in the surroundings. Ongoing large scale saltwater intrusion, due to overpumping, may spoil a well forever.

4.2.7 Mining of Fresh Groundwater

Mining of fossil fresh groundwater, which is presently not recharged, can be considered as a temporary source of freshwater. For economic reasons (the cost of construction, operation and maintenance of abstraction works) it should last for at least a few decades. This one-time reserve could either be used first—pending the availability of other resources later or to postpone the higher costs of more expensive sources such as desalting—or be kept as a reserve for emergencies. The environmental consequences should be studied and can be an impediment. The drawdown of groundwater tables or piezometric levels, for instance, might have geotechnical consequences or effects on the vegetation. Moreover adjacent groundwater flow systems might be affected.

Mining of a fraction of a fresh groundwater body, which is currently subject to natural recharge, is a possibility in the transitional face from the present state to a defined state which is strived after. Of course this possibility exists only in case the present volume of fresh groundwater is greater than that in the defined state which is strived after. See also the relevant texts of the preceding sections.

4.2.8   Excavations and Sand and Gravel Borrow-Pits

Excavations can be made for mining of sedimentary deposits as gravel, sand or clay for various applications, such as raising the land level of a nearby area for housing or industry, for road construction, for manufacturing of concrete or bricks, etc., but also to enable the construction of underground structures as sluices, ship locks, tunnels, garages, basements. In the latter case the excavation is only temporary; the material will, at least partly, be refilled after completing the structure.

If the groundwater table is at small depth below land surface, the excavation can also be executed by dredging. If the excavation is made for the realization of some underground structure it is generally also necessary to draw down the groundwater table temporary by pumping. If the excavation is made at the foot of hills or at the landside of sand dunes along the coast whereby the groundwater table is intersected, the groundwater table will automatically follow the land surface after excavation.

From the foregoing it will be clear that excavations change not only the topography, but also the hydrogeologic profile and, in most cases, also the groundwater table. Changes of the hydrogeologic profile are, in the context of saltwater intrusion, very important, especially in case semi-permeable layers are removed, which offer resistance against upward flow. Changes of the groundwater table bring about changes in the groundwater flow system; a local lowering causes flow towards the depression. This may lead to saltwater intrusion and anyhow to the necessity of pumping the inflowing water.

When the excavation is finished undesirable negative effects, in terms of saltwater intrusion due to the excavation, can be nullified or compensated partly or completely. This can be done in two different ways. Depending on the situation, either of the two or both could be applied. One of these two is to backfill the dredged finer material of the semi-pervious layers which had to be dredged before or during the period of excavation and has been set in depot temporarily. It can be dumped as uniformly as possible to act as a semi-pervious blanket on the bottom and sides of the remaining lake. The other method is to establish such a water level in the remaining lake that the corresponding groundwater flow pattern is more favorable in terms of halting, reducing or even decreasing the saltwater intrusion.

If the excavation is made at the foot of hills or at the landside of sand dunes along the coast the groundwater table is drawn down. As, according to the Badon-Ghyben principle, the interface or transition zone between fresh and saline groundwater underneath the sand dunes is related to the groundwater table, this interface will move upward. This is illustrated in Figure 4.12, when comparing its parts a and b. This is clearly a negative effect as has been explained in the end of section 4.2.5 and the corresponding Figures 4.10 and 4.11.

4.2.9   Land Reclamation and Drainage

Distinction should be made between land reclamation by artificial lowering of the water table on the landside of sand dunes and land reclamation by hydraulic fill on the seaside of sand dunes. Lowering of the water table is applied in lakes, lagoons or marshes behind sand dunes along the coastline in order to create new land for various purposes, such as agriculture, industry and housing. In already existing land areas ongoing land subsidence can necessitate to further lowering of the groundwater table. In land reclamation by hydraulic fill on the seaside of sand dunes new land is created on a level well above sea level.

Land reclamation and drainage often involve the drawdown and control of the phreatic groundwater table over large areas. These activities cause big changes in the groundwater flow pattern. The groundwater will flow towards the created depressions and occur as seepage in the areas with artificially controlled low groundwater tables. Depending on the level of the controlled groundwater table,

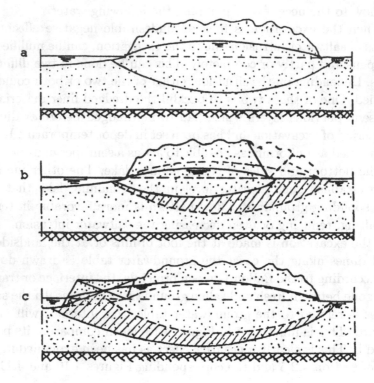

Figure 4.12. The effects of sand excavation at the inner side of sand dunes and of land reclamation at the seaside of sand dunes on the phreatic groundwater table and on the interface or transition zone between fresh and saline groundwater. (a) Original profile. (b) After excavation of sand at the inner side of the sand dunes. (c) After deposition of sand at the seaside of the sand dunes.

the hydrogeologic constants, the horizontal dimensions of the area with the controlled groundwater table and its distance inland from the coastline, this seepage can, sooner or later become saline. The mechanism of saltwater intrusion underneath a phreatic freshwater lens in sand dunes was already illustrated by Figure 4.10, described in section 4.2.5. In case of semi-confined groundwater in a deep aquifer covered by semi-permeable layers with hydraulic resistances against vertical groundwater flow c_s and c_1 $[T]$ as shown in Figure 4.13 the inflow q_s of saline groundwater below the freshwater lens is roughly

$$q_s = kH \cdot \frac{h}{\lambda_s + L + \lambda_1} \qquad (4.18)$$

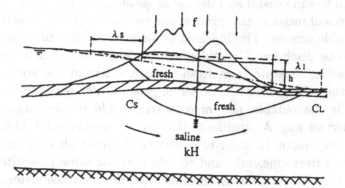

Figure 4.13. Fresh water lens in sand dunes with inflow of saline water between the fresh water lens and the impermeable base in case the lower aquifer is covered by semi-permeable layers.

where kH is an average value of the transmissivity of the lower part of the deeper aquifer, where the saline water flows inland, h is the difference between mean sea level and the controlled groundwater table in the reclaimed area at the inner side of the sand dunes, λ_s and λ_1 are the so-called characteristic lengths or spreading lengths (which are better terms than the often used term leakage factor):

$$\lambda_s = \sqrt{(kHc)_s} \quad [L] \qquad (4.19)$$

and

$$\lambda_1 = \sqrt{(kHc)_1} \quad [L] \qquad (4.20)$$

Depending on the values of c_s and c_1, which can be as much as several or even many thousands of days, and in accordance therewith, λ_s and λ_1, which can be up to several thousands of meters, the presence and the hydraulic resistance of semi-permeable top layers can have a considerable and favorable influence on the reduction of the inflow q_s of saline water.

The effects of the changes in the groundwater flow pattern, in terms of the distribution of fresh and saline groundwater, can be more or less equivalent to those of abstractions by large well fields. As large volumes of groundwater of different densities are involved, which flow slowly, it may take centuries before the ultimate situation is reached. Such situations will often result in permanent seepage of brackish or saline groundwater into the reclaimed areas. As the aquifers in coastal and deltaic areas are generally thick—tens to hundreds of meters—this process can not be stopped by creating impermeable screens. The brackish or saline water has to be accepted and to be discharged by pumping. It should be handled, as much as possible, separate from fresh surface waters in the area. As far as mixing of the brackish or saline groundwater with fresh surface water is unavoidable, the mixed water should be discharged along the shortest way. A pipeline to the sea or a nearby tidal inlet might be too expensive. In order to discharge the brackish water from the surface waters efficiently and to dilute at the same time the salinity in these water courses, flushing with fresh surface water, where available, is a curative method.

Land reclamation by hydraulic fill on the seaside of sand dunes along the coast, which is in fact a broadening of the sand dunes will result in larger and thicker freshwater lenses. This is illustrated in Figure 4.12, when comparing its parts a and c. The thicker freshwater lens leads to a reduction or, depending on the hydrogeologic profile, even a blockade of the saltwater intrusion, as was explained in the end of section 4.2.5 and the corresponding Figures 4.10 and 4.11.

4.2.10 Exploitation of Karstic Coastal Aquifers

Karst is an intriguing and complex phenomenon. Therefore it is not surprising that this subject has been addressed in a great many of publications. The International Association of Hydrogeologists (IAH) has published two volumes of annotated bibliographies on carbonate rocks [LaMoreaux, 1986; LaMoreaux, 1989]. Remarkably only few publications, referred to in these bibliographies, deal with karstic coastal aquifers and saltwater intrusion in such aquifers. The same holds for four other volumes, published by IAH, on karstic aquifers [Back et al., 1992; Burger and Dubertret, 1975; Burger and Dubertret, 1984; Mijatovic, 1984]. The SWIM-proceedings [1996; 1994; 1992; 1990; 1988; 1986 and before] referred to in section 4.1.1, contain several papers on saltwater intrusion in (karstified) carbonate rocks, mainly dealing with cases in the coastal zones of Italy and Spain. The same holds for the IAH-selection of SWIM-papers [De Breuck, 1991].

Carbonate rocks have been formed by deposition of organic matter (shells and corals) i.e. as sedimentary rocks, which have successively been consolidated, and by precipitation of dissolved carbonates. Their primary porosity and permeability is small to moderate. In many cases fracturing has occurred due to tectonic processes. The water flowing through fractures dissolves, gradually and irregularly, the surrounding carbonate material thus forming wider and deeper channels and even caverns. The fractures, fissures, channels and caverns make up the so-called secondary porosity and permeability which is much larger than the primary porosity and permeability. In fact a dual-porosity system is developed. After and beside passing through the primary rock system the discharge of the natural recharge water takes place in the karst channels and openings, which are so wide and large that the flow is often non-Darcian. The water transport and storage in the primary rock system lags behind. In karstic areas surface streams are generally rare. As a consequence of this behavior of carbonate rocks the groundwater flow system is complicated and difficult to know and to describe. For instance, the groundwater divide mostly does not coincide with the topographic divide. The discharge of groundwater is often concentrated in a few springs. Springs occur sometimes also off-shore, i.e. as submarine springs on the sea bottom.

So far this section dealt with karst in general. The special features of saltwater intrusion in karstic aquifers need special attention. Due to the large openings in karstic aquifers the response to impulses is generally very rapid. This does not only hold for the response to natural recharge or human activities, such as groundwater abstraction, but also for the effects of tides, storm surges and variations in atmospheric pressure in terms of saltwater intrusion. Depending on the system, with or without submarine springs, and more in particular the levels of outflow, as determined by the presence and the levels of natural thresholds, saltwater intrusion can take place at the outlets or, more inland, at greater depths. Due to relative sea level rise some springs which occurred long ago at the land surface are submarine at present.

In general, the greater the outflow of karst springs, the lower the salinity of the outflowing water. Mixing of fresh and saline water takes place rapidly, rendering the outflowing freshwater unfit for use. Mixing of fresh and saline water in karst regions takes place under a completely different regime compared to homogeneous and isotropic porous media [Milanovic, 1981]. Not hydrodynamic dispersion but convective flow of fluids dominates the physical mixing. Thus, the fresh-saline relation in karstic regions only partly follows the Badon Ghyben-Herzberg principle. All together, description of this feature with mathematical formulas is complex or even impossible, unless very rough assumptions are permitted.

In order to be able to take the proper measures to prevent or reduce saltwater intrusion and to abstract sustainably a maximum fraction of the recharge the aquifer system must first be understood. From the preceding text it will be clear that this system is very capricious. Not only that; it is difficult, if not impossible, to know. Therefore it will be necessary to resort to empirical relationships, rather than to rely on models of the physical reality, which can not be sufficiently known. Such empirical relationships may be those between the rate of outflow and the salinity of springs on the one hand and depths of preceding rainfall, groundwater levels or piezometric levels and sea-water levels on the other. Even with such empirical relationships the effects of technical measures such as artificial changes of threshold levels, plugging and grouting can not always be predicted correctly, due to unforeseen discontinuities.

Submarine springs occur in many karstic areas all over the world [Schwertfeger, 1981]. The outflows, as a function of the natural recharge, vary strongly with time. Moreover they are difficult to measure. This holds the more for submarine springs at greater depths and at greater distances off the shore. The same holds for the salinity of the outflowing water which is often mixed water. Some submarine springs are designated by the French term estavelles, which means that they act as springs during low tide, when the difference between the level of the sea and that of the groundwater level is maximum and as sinkholes at high tide when the hydraulic head of the sea is greater than that of the groundwater. It is obvious that in estavelles a good deal of mixing takes place. Submarine springs are often groups of several springs close together.

Exploitation of karst water resources can be realized in various ways, depending on the hydrogeologic profile, e.g. by means of wells or by capturing springs. Production wells should be installed in the upper, most permeable, parts of the aquifer, preferably at where the density of fractures, fissures and dissolution channels is greatest. The drawdown of the groundwater table will then be a minimum. Horizontal galleries perpendicular to a system of underground channels developed from fractures and fissures are also a very suitable means for groundwater recovery. The coastal karstic aquifers saltwater intrusion occurs particularly in the dissolution channels. Depending on the mechanism of saltwater intrusion, especially in the case of shallow penetration of saltwater and subsequent mixing with the freshwater in the downstream part of the aquifer system, it may be effective to raise the level of outflow by grouting or by plugging by means of artificial barriers such as concrete dams. Such measures can be without result in case the saltwater, still fed by seawater, comes from greater depths, further inland. Another risk of raising the outflow levels is that other outflows may appear or increase in capacity, thus changing the system and the established empirical relationships.

A very special case is the capturing of springs on the sea bottom. In case such springs occur in shallow water at small distance off the shore an embankment could be built around the spring [Schwertfeger, 1981]. Then the outflowing freshwater will replenish the seawater enclosed in the pond formed by the embankment. Due attention should be given to the controlled water level in the pond and to the presence and behavior or even formation of other nearby springs on the sea bottom.

Any human intervention in aquifers (pumping, capturing, grouting, changing of the drainage base level) changes the flow pattern. This holds for karstic aquifers as well. For karstic aquifers there are moreover long term consequences, caused by changing dissolution patterns, which in turn lead to further changing of flow patterns.

4.3 Measures to Restore Disturbed Fresh Groundwater Systems in Coastal Aquifers

4.3.1 General

When dealing with the restoration of disturbed groundwater systems in coastal areas it is necessary to assess:

1. the present state, in terms of groundwater tables, piezometric levels and salinity distribution and in terms of exploitation, i.e. location and rates of abstraction,
2. the desired state after restoration, in terms of sustainable rates of abstraction and the means and locations thereof, groundwater tables and piezometric levels and the volume of fresh groundwater that should be permanently present in the aquifers

as a strategic reserve for emergencies and to cope with fluctuations in the rates of recharge and abstraction. For considerations and criteria to define the desired state, reference is made to the sections 4.1.6, 4.2.3 and 4.2.4.

Once the present state is sufficiently known and the desired state has been defined, within the natural, technical and economic limits posed by the aquifer system and its exploitation, the necessary actions can be taken for restoration, where needed. The same actions can and must be taken in those cases which need not to be restored, yet where continuity of a controlled exploitation must be assured.

The following measures can or must be taken to achieve these goals:

- reduction of the rates of abstraction, in order not to exceed the sustainable yield,
- relocation of abstraction works—this measure aims at reduction of the losses of fresh groundwater by outflow.
- increase of natural recharge—this measure is to increase the recharge and therewith the sustainable yield.
- artificial recharge—this measure is also to increase the recharge and therewith the sustainable yield.
- abstraction of saline groundwater—this measure aims at increase of the volume of fresh groundwater and at reduction of the losses of fresh groundwater by outflow. The abstracted saline groundwater can under certain conditions be used as a source for desalting.

These measures are discussed in the sections 4.3.2 to 4.3.6 respectively.

4.3.2   Reduction of the Rates of Abstraction

The rates of abstraction can be reduced when the water demand is reduced. This can be reached by a number of measures, such as:

- information to the general public and the industrial community of the necessity of saving water and, if necessary, to forbid certain uses as for instance car washing and garden watering in periods of scarcity, and use for cooling in industry only,
- reduction of losses from the water transport and distribution systems—these losses can be considerable,

- recycling of water in industrial processes, after appropriate treatment before the successive uses,

- reuse of waste water, after some treatment, for other applications such as cooling, irrigation and injection in the subsoil to maintain a barrier against saltwater intrusion,

- reduction of the water requirements for irrigation by choosing crops which require less water and application of water saving techniques such as drip irrigation and canal lining.

4.3.3   Relocation of Abstraction Works

The effect of the location of abstraction works on the sustainable rate of abstraction, as a fraction of the recharge, has been demonstrated in section 4.2.5 for confined and phreatic groundwater respectively.

From the principles outlined there it will be clear that in confined groundwater a more inland position of the abstraction works is more favorable than a more seaward position. This is because in inland direction the thickness of the freshwater lens increases, and the danger of upconing of saltwater decreases accordingly. A more inland position of the abstraction works allows also for a further inland penetration of the saltwater wedge in its permanent ultimate position. With a further inland position of the saltwater wedge, less fresh groundwater will be lost by outflow due to a smaller gradient of the piezometric level of the outflowing fresh groundwater. As explained in sections 4.2.4 and 4.2.7 part of the original volume of fresh groundwater can be mined. So, in defining the desired ultimate position of the saltwater wedge, a choice has to be made between a greater rate of continuous and constant abstraction of fresh groundwater with a higher risk of pumping saline water under extreme conditions versus the opposite combination of the two.

In phreatic groundwater in islands and in sand dunes along a coast the abstraction works should be located at such a short distance from the coastline that the largest possible fraction (but still less than 1) of the recharge can be recovered before it is lost by outflow under the coastline. On the other hand this distance should not be too small as the thickness of the freshwater lens decreases towards the coastline and this thickness must still be great enough under the abstraction works so that, even under extreme situations of temporary overdraft, no upconing of saltwater can occur. A more inland posi-

tion of the abstraction works leads however to a smaller volume of
fresh groundwater for strategic reserve. All this follows from the fundamental aspects for phreatic groundwater dealt with in the second
half of section 4.2.5.

4.3.4   Increase of Natural Recharge

The natural recharge can be increased somewhat by proper land
use (natural vegetation and choice of crops), land tillage practices
(e.g. contour ploughing in sloping areas, terraces), the installation
of check dams and weirs in surface waters, so as to raise the water
levels therein and to divert water to adjacent spreading grounds. The
guiding principle of all these measures is to hold up the water as long
and as much as possible in order to give it more time for infiltration,
rather than to let it run off directly. Most of these measures are also
favorable for erosion control.

With respect to the increase of the natural recharge it is worth
mentioning here that in the present time flood control measures
are also taken in many urban areas, whereby the runoff is retarded
and given more opportunity to infiltrate in open areas and through
more pervious pavements. This is favorable from the point of view
of recharge in quantitative terms, but the quality of the water infiltrating in urban areas may be doubtful.

4.3.5   Artificial Recharge

Though this section on artificial recharge makes part of section 4.3 on
restoration, artificial recharge is not only applied for restoration but
also as an element in the continuous optimal exploitation of aquifers.
Artificial recharge of groundwater is applied for many reasons, such
as to increase the sustainable yield, to control the groundwater table
(for instance for environmental reasons) or the piezometric level (in
order to restrict or to slow down land subsidence), to increase the
volume of fresh groundwater available for emergencies, and/or as a
barrier against inflow of saline groundwater.

Artificial recharge can be realized by (increased) infiltration at
the land surface or in surface waters or by means of recharge wells
with well screens in aquifers at any desired depth. For the recharge
of phreatic groundwater both techniques can be applied. Confined
and semi-confined groundwater in aquifers at some depth can not be

recharged from the land surface or surface waters due to the high hydraulic resistance between the land and water surface and the aquifers at some depth below the land and water surface.

At present there is a growing opposition against artificial recharge by infiltration at the land surface or in surface waters. The objections are the occupancy of large surface areas and undesirable ecological effects due to changes in the phreatic groundwater regime, both in terms of groundwater tables and water quality. The first objection holds particularly in intensively used areas. The second objection holds in particular in case of scarcity of nature and its uniqueness.

A special form of artificial recharge is induced recharge, where groundwater is abstracted along and at short distance from rivers. If the rate of abstraction exceeds the rate of natural flow of groundwater towards the river, inflow of river water is induced and the abstracted groundwater consists partly of river water. This affects the quality of the abstracted groundwater. In the most downstream reaches of rivers and in estuaries the inflowing groundwater may be brackish or even saline [Savenije, 1992; van Dam, 1993], depending on the tidal regime and on the magnitude of the river flow. This should be checked before undertaking any project for induced recharge.

The potential sources of water for artificial recharge are surface water or pumped groundwater after its first use and proper treatment after that. Surface water can be taken from rivers or estuaries, and be transported either by canals or by pipelines. It is obvious that the water should be fresh. Therefore the intakes should be beyond the reach of saltwater intrusion in the rivers and estuaries [Savenije, 1992; van Dam, 1993], or the estuaries should be provided with dams and sluices. The quality of the water at the source should satisfy certain standards. Such standards depend mainly on the use of the water after recharge and subsequent recovery but also on the requirements for transport and subsequent recharge. These two latter aspects in itself may already require some pretreatment. This holds in particular for recharge by means of recharge wells, in order to avoid clogging of these wells.

Apart from pretreatment of the water for recharge, the rate of recharge per well should be limited. Despite pretreatment of the water the capacity of the recharge wells decreases during operation. Therefore regeneration of the recharge wells is needed from time to time. Due to intensive research much progress has been made

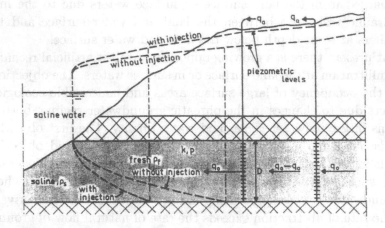

Figure 4.14. Effects of injection of fresh water in confined groundwater in a coastal aquifer on the piezometric surface and the position of the salt water wedge.

both with respect to the pretreatment required for water of different quality and with respect to the techniques of regeneration.

Figure 4.14 illustrates the effects of artificial recharge in confined groundwater, both on the piezometric level and on the interface between fresh and saline groundwater. For clear illustration of the principle, the interface is assumed to be sharp. There is a line of abstraction wells and a line of injection wells of equal capacity q_a, both perpendicular to the drawn hydrogeologic profile. The position of the interface is drawn both for the case without and the case with injection of the abstracted and used water. The dashed lines indicate the piezometric levels without and with injection. The groundwater flow in the different reaches of the aquifer is also indicated for the case of equal abstraction and injection q_a and a natural outflow q_o ($> q_a$). The injected freshwater just serves as a barrier to hold the interface in its original position. Therefore the quality of this water needs not to satisfy strong criteria. Still the used water must undergo some treatment before injection, in order to avoid clogging of the injection wells as much as possible. The principle of a freshwater barrier has been applied, for instance, in California and in Israel.

From the description of the fundamental and practical aspects of abstraction of phreatic groundwater, in the sections 4.2.5 and

4.2.6 respectively, it is clear that, also in the phreatic case more groundwater can be abstracted sustainably and the interface can be pushed down by application of artificial recharge. This latter fact implies that the reserve for temporal overdraft becomes greater. This holds for islands in the sea as well as for sand dunes along the sea coast. If in the latter case the controlled groundwater tables at the land side of the sand dunes are below mean sea level, the inflow of saline groundwater underneath the freshwater lens in the sand dunes is reduced or, depending on the depth to the impermeable base, even halted.

With respect to the mutual location of recharge ponds, canals or wells and the abstraction works the following considerations hold. In order to assure a minimum residence time for improvement of the water quality these works should be located at a minimum distance which can be calculated. The longer the residence time of the infiltrated water the greater also the smoothing out of the fluctuations in the quality of the infiltrated water. A more constant quality of the abstracted water is generally appreciated by the consumers. Fluctuations can even more be flattened by applying variable distances between the lines of recharge works and abstraction works respectively. Therefore, in the design of the combined recharge and abstraction system, one should make use of the topographic features of the terrain also with respect to the aspect of water quality fluctuations.

Another consideration relates to the question whether to locate a line of abstraction works in between two lines of recharge works (or in the center of a more or less elliptic configuration of recharge works) or the other way round, the line of recharge works surrounded by two lines or a more or less elliptic configuration of abstraction works. The latter layout has the advantage that the phreatic groundwater tables will be higher and, accordingly, the depths to the interface greater. This implies the availability of a larger volume of freshwater for emergencies. Moreover, due to the greater thickness the residence times will be somewhat longer and somewhat more spread out. The regime of desired or accepted groundwater tables should also be judged in relation to the topography and the ecological features of the terrain.

Depending on the variation of the water requirements over time, as determined by the climate and weather, and on the availability of water of good quality for artificial recharge, the actual rates of abstraction and artificial recharge of groundwater may vary with time, periodically and incidentally. So the volume of fresh groundwater will also vary with time, and the boundary between fresh and saline groundwater will not only move up and down, but also the transition between fresh and saline will become less sharp due to the effects of dispersion and retardation.

In this context it is interesting to refer to the results that have been obtained in the Netherlands (a.o. Amsterdam waterworks) in pilot projects for injection of freshwater into deeper aquifers by means of wells. One of the aquifers contained brackish water. The spreading of the injected freshwater was carefully monitored in nearby observation wells. The results of the field experiments were compared with computer simulations. The objective of these experiments was to study the growth of the volume of fresh groundwater in storage as a reserve for emergencies and also how to cope with variations in the rates of abstraction and injection depending on the demand and availability of water of good quality. The fact that the transition zone between the injected water and the formation water moves in and out and up and down means that some of the injected fresh-water is lost due to mixing, by dispersion and retardation, with the brackish groundwater. The recoverability is defined as the ratio of the abstracted volume to the injected volume. This ratio, over a longer period (after building up a certain average volume of fresh groundwater), is less than 1.0, a fact which has to be accepted. Investigations must give insight in the optimal layout and operation of the system in order to arrive at the highest recoverability in any particular hydrogeologic profile and situation.

Attention should also be paid to the question where the displaced brackish and saline groundwater goes and what effects it has in the surroundings. So, for instance, increased seepage of brackish or saline groundwater in adjacent areas will generally not be appreciated. Therefore the effects of the combined system of abstraction and recharge in the wider surroundings must be foreseen, predicted and judged before undertaking any artificial recharge project.

Interesting examples of artificial recharge projects are the public water supplies of the cities of The Hague and Amsterdam in the Netherlands. Both cities abstract water from the sand dunes along the Dutch coast. In the 1950's the increased rates of abstraction could no longer be met from the natural recharge. The interface between fresh and saline groundwater moved upward rapidly. Therefore artificial recharge was applied. Water from the river Rhine, and later on also from the river Meuse was transported by pipelines, of 60 and 80 km length, to the sand dunes where it was infiltrated in ponds and canals. So the increased demand could be met and moreover the quality of the infiltrated water improved during its passage through the subsoil by breakdown of certain constituents and by smoothing out of the quality fluctuations of the infiltrated water during its passage through the subsoil.

The increased volumes of fresh groundwater, with thicknesses of the freshwater lenses of over 100 m, are good buffers in case of emergencies such as rupture of a pipeline or calamities making the river water for some time unfit for intake.

A secondary result of the artificial recharge in the sand dunes was the fact that the interface between fresh and saline groundwater was pushed down with the effect that the inflow of saline groundwater underneath was reduced.

In the present time studies are made, both by models and in the field to come to a greater portion of recharge by means of wells directly into the deep aquifers for environmental and for strategic reasons.

4.3.6   Abstraction of Saline Groundwater

The saline or brackish groundwater which is present below fresh groundwater in coastal and deltaic areas can also be abstracted. It can be used for cooling purposes or as a source for desalting. Such abstractions cause the volume of fresh groundwater to grow and the volume of brackish and saline groundwater to decrease. Complete control of the interface is possible by simultaneous abstraction of fresh and saline groundwater, in mutually adjusted proportions. For such watchmakers work a good monitoring system is indispensable.

As was already mentioned in section 4.1.5, disposal of the abstracted saline groundwater can be a problem. The same holds for the brine in case the abstracted saline groundwater is desalted as an additional source of freshwater.

The possibility of abstracting saline groundwater with the objective to reduce the outflow of fresh groundwater, which is otherwise required to maintain the body of fresh groundwater as was outlined in the sections 4.2.5 and 4.2.6, deserves serious attention. In case

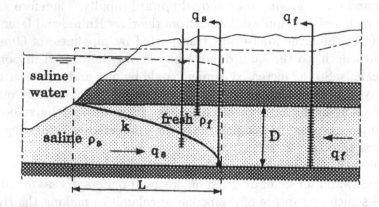

Figure 4.15. Abstraction of saline groundwater in the tip of the salt water wedge below confined fresh groundwater at such a rate that no fresh groundwater is lost by outflow at the seaside.

of scarcity of freshwater it is worthwhile to reduce the loss by outflow of fresh groundwater, theoretically even to zero. The effect of pumping saline groundwater is best illustrated by the extreme, theoretical, situation, in confined groundwater, as presented in Figure 4.15. For clear illustration the interface is assumed to be sharp. The saline groundwater is pumped, theoretically, in the tip of the saltwater wedge at such a rate, in this extreme theoretical situation, that the piezometric level of the fresh groundwater above the saltwater wedge is horizontal. In that case the fresh groundwater above the saltwater wedge is stagnant. There is no loss of fresh groundwater by outflow into the sea and the flow of fresh groundwater can be abstracted totally. The length L of the saltwater wedge, of parabolic shape, and the inflow of seawater q_s are related as follows:

$$L = \frac{\alpha^* k D^2}{2q_s} \tag{4.21}$$

where $\alpha^* = (\rho_s - \rho_f)/\rho_s$ [dimensionless].

As was said before, the situation of Figure 4.15 is an extreme, theoretical, situation, chosen for a clear illustration of the principle only. In reality a drain in the inland tip of the saline water body will attract freshwater as well, due to the effect of dispersion. Therefore the practical solution is a line of wells at some distance seaward of the tip of the body of saline water of Figure 4.15. Such a configuration

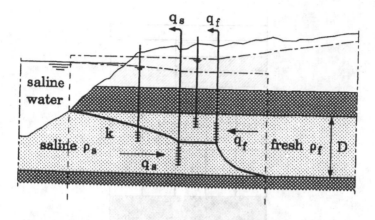

Figure 4.16. Practical configuration for implementation of the principle illustrated in Figure 4.14.

is shown in Figure 4.16. The saline water flows towards that line of wells and is pumped completely by that line of wells. Behind that line of wells the saline water is stagnant. Between the lines of wells for abstraction of saline groundwater q_s and for fresh groundwater q_f the interface is horizontal because there neither the fresh nor the saline groundwater flows. The flow of fresh groundwater (q_f), from recharge, is totally abstracted by the line of wells with well screens in the fresh groundwater. At the inland side of these wells the interface has another parabolic shape as depicted in Figure 4.16. The location of the abstraction wells and the determination of their capacity is and interesting optimization problem.

In the sections 4.1.4 and 4.2.6 distinction was made between regional or large scale effects and local or small scale effects. So far this section dealt with the regional effects only. For the local problem of upconing, the installation and operation of so-called scavenger wells is a good curative solution. Such wells, next to or nearby pumped wells have their well screen at some depth below that of the pumped well in the fresh groundwater. The well screen of the scavenger well is installed in the saline groundwater at some depth below the original depth of the interface or the transition zone between the fresh and saline groundwater. By pumping the saline groundwater, the piezometric level of the saline groundwater is lowered with the effect that there will be no more upconing. The upward cone of saline

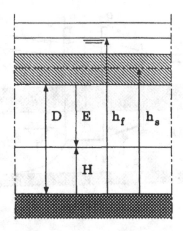

Figure 4.17. Equilibrium depth E of the interface between fresh and saline groundwater in semi-confined groundwater.

groundwater can sink down either by stopping the abstraction of fresh groundwater for some time before pumping both well screens simultaneously or by continuing the abstraction of fresh groundwater and pumping the saline groundwater at a higher rate in the beginning than later on. In both cases the rate of pumping of the scavenger well should be adjusted to the position and movement of the interface or transition zone, which should be monitored frequently or continuously. The disposal of the pumped saline groundwater may be a problem. See section 4.1.5.

Scavenger wells prevent upconing of saline water on the small scale of individual wells, but have large scale effect as well. The large scale effect is twofold; a more favorable distribution of fresh and saline water in the subsoil and less loss of freshwater by outflow to the sea.

In semi-confined groundwater the position of the interface or transition zone depends very much on the piezometric level of the saline groundwater. The effect of a drawdown of the piezometric level of the saline groundwater in a semi-confined situation is best illustrated by Figure 4.17 for the particular situation where both the fresh and the saline groundwater are stagnant. In such a situation there is no vertical flow through the semi-permeable layer and the piezometric levels of the fresh groundwater above and below the semi-permeable layer are equal. Then the following expression holds for the so-called equilibrium depth E $[L]$ the interface below the top of the aquifer:

$$E = D + \frac{h_f}{\alpha} - (1 + \frac{1}{\alpha})h_s \qquad (4.22)$$

where D $[L]$, h_f $[L]$ and h_s $[L]$ are with respect to the reference level, which is chosen at the top of the impermeable base.

From this expression it is clear that any drawdown of the piezometric level of the saline groundwater h_s, for instance by abstraction of saline groundwater, entails an increase of the equilibrium depth which is some 40 to 50 times that drawdown. Of course it takes a very long time to substitute the saline groundwater by fresh groundwater, which should come from recharge, and simultaneous outflow or abstraction of an equal volume of saline groundwater.

Finally it is worth mentioning a study with an experiment carried out for a polder in the deltaic area in the southwestern part of the Netherlands. The polder, which is bounded by a tidal inlet, is subject to seepage of saline groundwater as shown in Figure 4.18. The study deals with a so-called seepage barrier, which consists of a line of artesian wells placed in the polder at short distance from the dike. The effect is that the seepage is concentrated in these wells with free outflow in one ditch instead of occurring all over the polder area which is harmful for agriculture because of the salinity of the water. The saline water in the ditch which is separated from the drainage system of the polder can be pumped out separately. The piezometric level in the aquifer is drawn down from line a, without the seepage barrier, to line b, with the seepage barrier. The seepage barrier reduces the resistance to seepage and thus the total amount of seepage increases. This disadvantage is accepted in exchange for control over the place where the seepage occurs.

4.4 Groundwater Management

4.4.1 General

As was explained in section 4.2.1, the exploitation and restoration of fresh groundwater in coastal aquifer systems should form part of integrated water management, comprising surface water and groundwater, both in terms of water quantity and water quality, and taking into account the demands. This requires cooperation, information, study, planning and legislation.

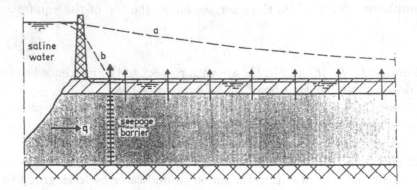

Figure 4.18. Seepage barrier of artesian wells and piezometric levels (a) without and (b) with artesian wells.

The organizational structure for water affairs differs from country to country. Integrated water management can be realized only by co-operation between all authorities which have some responsibility for water. Some of the most important responsibilities are public works, domestic and industrial water supply, agriculture and environment. The responsible authorities can be organized either sectorial or re-gional. Moreover the regional responsibilities are often organized hi-erarchically (state, province, district). The geographical boundaries of the regions where the various sectorial authorities have their re-sponsibilities can be different and overlapping. These boundaries can be either administrative or natural, i.e. based on water divides, the catchment approach. For water affairs this latter approach is very much to be preferred. Regrettably most of the boundaries are ad-ministrative boundaries. This complexity urges for mutual cooper-ation, based on consultation and information. Missing information must be collected in the field, by research and by modeling. Then a water policy must be formulated, based on all available information, including that on the requirements. It is clear that in this process choices will have to be made, taking into account economic, environ-mental and social aspects. Obviously the policies must be reviewed periodically, to take into account new developments. The chosen and agreed policy must be implemented by planning and realization of the necessary works and measures. The measures require proper leg-islation and its observation.

4.4.2   Data Collection and Monitoring

From the sections 4.2.1, 4.2.2 and 4.4.1 it will be clear that a great variety of data is required for integrated water management studies. These are generally collected, processed and stored by many different institutions and organizations. Therefore it is necessary to know what data is available, where and in which format. It is also necessary to have access to the data of other institutions and organizations and to dispose of facilities for calling in those data. This section will be restricted to the data required for groundwater studies in coastal and deltaic areas where saltwater intrusion plays a role. The types of data required relate to:

1. the subsoil

   - the geologic structure of the subsoil
   - hydrogeological constants

2. the natural input

   - climate data, in particular precipitation and evapotranspiration
   - natural recharge

3. water levels

   - groundwater levels and piezometric levels
   - surface water levels

4. groundwater quality

   - chemical and isotopic composition, in particular salinity, but also any other contamination
   - sources of groundwater pollution

5. surface water

   - natural outflow
   - availability and quality of surface water for artificial recharge

6. present and past exploitation

   - present and past abstractions of fresh and saline groundwater
   - present and past artificial recharge, if any

7. water demands, at present and estimates for the future
8. ecology

- flora and fauna, and their relation to the groundwater table regime

9. relative sea level rise.

In the following paragraphs more detailed information is given about the importance and the collection of each of these types of data.

1. Information on the geologic structure of the subsoil comes from the geologists, who obtain their information from boreholes and geophysical prospecting, mainly the geo-electrical methods [van Dam, 1976b] but also from seismic prospecting. Moreover geophysical well-logging in the boreholes gives additional geophysical information. The hydrogeological constants are determined by pumping tests in the same boreholes, but also from water balance and model studies and from the response of groundwater tables and piezometric levels to other impulses such as fluctuations of surface water levels. Remote sensing of the earth surface features is useful in itself, but it can give important indications about the underlying geological structures as well, e.g. the presence of faults.

2. The natural input into the groundwater system comes, apart from lateral inflow, from natural recharge, determined by the climate variables precipitation and evapotranspiration, the soil moisture, the vegetation and the depth of the groundwater table, all of which vary with time. So the natural recharge is a function of time, with a seasonal distribution and stochastic variations over it, thus with extremes. Apart from the short term variations in all these input data, climate change can affect the natural recharge in the long run. This is the domain of the climatologists.

3. It is important to study and follow the groundwater levels and piezometric levels as a function of location and of time as they are indicators of the groundwater flow and of the depth of the interface or the transition zone between fresh and saline groundwater and thus for the volume of fresh groundwater in storage. The observations of groundwater levels and piezometric levels should be carried out with intervals of a few weeks, for instance twice a month. The surface water levels are equally important as boundary conditions for the groundwater system

and as indicators for the streamflow. Depending on the variability of these levels, their observation intervals could vary from one to several days.

4. Groundwater quality is determined by a great number of chemical and isotopic parameters, which determine the suitability of the water for the various uses, directly or after treatment. Moreover the chemical and isotopic compositions are indications of the origin of the water. The most important quality parameter of the water is, in the context of this chapter, its salinity. The groundwater quality can also be influenced by percolation of certain contaminants from human activities at the land surface. As these contaminants can make the water unfit for use it is important to find out the origin of these contaminants as well, so that, possibly, proper measures could be taken. Groundwater quality can be determined from water samples taken from filters at different depths in observation wells. If these filters are in different aquifers, the perforation of the aquitards in between the successive aquifers should be filled up with clay around the tubes of the individual filters in the different aquifers. This is to avoid undesirable connections between the different aquifers, with generally different piezometric levels and different groundwater quality. Such connections give an incorrect picture of the undisturbed reality. The salinity distribution can also be derived roughly from the specific electrical resistivities, determined by geo-electrical prospecting, by electrical well-logging and by so-called salt-watchers, which are systems of electrodes installed permanently in boreholes without permanent steel casing. As groundwater flows slowly, such measurements can be made at relatively large intervals of time, say two to four times a year. There is, however, one exception. This is in case of upconing of saline groundwater just under an abstraction well or in combination with a scavenger well. Then a suitable interval is, for instance, two weeks. When studying karstic aquifers, the application of artificial tracers and the measurement of water temperatures can provide important additional information for proper understanding of the groundwater system.

5. Surface water leaving an area is fed by surface runoff and by outflowing groundwater. It is important to quantify this out-

flow as a term in the groundwater balance of the area under consideration. In case artificial recharge is or will be applied one should know the characteristics and the reliability of the source of water, for instance a river or lake inside or outside the area under consideration. At the same time other present or future claims to that source should be recognized.

6. In order to have a complete and correct picture of the water and salt balance of the groundwater system it is necessary to know the present and past abstractions of fresh and saline groundwater and the present and past artificial recharge, if any. It is necessary to know their locations, the filter depths and their capacities and their salinities, both as functions of time.

7. It is obvious that the present and estimated future water requirements must be known in order to be able to anticipate timely on the future by expanding the capacity of the waterworks, including artificial recharge and abstraction of saline groundwater. If it appears to be impossible to satisfy the future water requirements by expansion of the present waterworks it will be necessary to prepare for other solutions as, for instance, import of water from more remote sources.

8. There is, at present, great concern for the ecological aspects of any change in the groundwater regime. Though in recent times much progress has been made in the understanding of the relationships between the groundwater regime and nature, there is still insufficient knowledge. Therefore all relevant information about flora and fauna, now and in the past, should be documented. Only then the relations can be fully understood and an optimal solution for groundwater exploitation and nature control can be achieved by adequate water control measures. Geographical information systems (GIS) are pre-eminently suitable for recording this type of, geographically distributed, information and to couple this information with the input or output of groundwater model studies as will be described in section 4.4.3.

9. The causes and mechanisms of relative sea level rise have been described in section 4.1.3. In the first half of section 4.2.5 the conclusion has been drawn that relative sea level rise has no effect on saltwater intrusion in a confined groundwater system as long as the natural outflow of groundwater remains

unchanged. In the last paragraph of section 4.2.5 the same conclusion has been drawn for phreatic groundwater in islands or in sand dunes where the controlled groundwater table at the landside remains equal to the sea level, except when the width of the island or the sand dunes decreases due to relative sea level rise. In that case the total recharge decreases proportionally and therewith the thickness of the fresh groundwater lens, also proportional. In case the sea level rises at the seaside of sand dunes relative to the groundwater table at the land side of the sand dunes, inflow of saline groundwater will occur or increase, depending on the depth of the impermeable base, as was also explained in section 4.2.5. The same holds for the semi-confined situation as described in section 4.2.9. So in these last situations it is important to take into account the estimates for the future rates of relative sea level rise [van Dam, 1993; Oude Essink, 1996].

4.4.3 Modeling

This section on modeling is restricted to modeling of groundwater with variable densities as in the case of saltwater intrusion. The effects of any foreseen action, either the realization and exploitation of technical works, such as groundwater abstraction and artificial recharge, or drafting relevant legislation, must be studied beforehand. For groundwater studies a wide variety of models is available as appears from several of the other chapters of the present volume. A complete model to describe saltwater intrusion should be three-dimensional, transient, and account for varied densities and for dispersion. Such a model, if any, is not only complicated, but requires also a lot of input data which are mostly not available. Even in areas rich in data it is never enough and the ever heard complaint of lack of data will remain forever. The modeler will have to learn the art of drawing acceptable conclusions and to indicate the range of their reliability from the available data.

The most recent development in modeling is the coupling of the model with a geographic information system (GIS) for the input data and presentation of the model output. Where no comprehensive sophisticated model and/or computer capacity is available, or in the absence of sufficient knowledge or experience with such models, one can resort to simpler models. Simpler models are, for instance, those

ignoring dispersion and assuming sharp interfaces or, even more simplified, those dealing with only two dimensions, either in a vertical or
in a horizontal plane, and in steady state. It is surprising how much
can be learned from the results of these simpler models. The results obtained by the more sophisticated models are obviously more
detailed because they are based on more detailed input data, but,
after all, they appear to be roughly in line with those of the simpler
models. Irrespective the degree of sophistication of the models, they
enable us to study the sensitivity of the results for variation of the
values of the hydrogeological constants and other input data and to
compare the effects of different ways of groundwater exploitation.
The results are quantifications of what, after all, appears mostly to
be obvious.

4.4.4 Legislation

In addition to the technical measures, legislative measures are required to control and allocate the available fresh groundwater resources properly. By proper rules to control, to restrict or even to
stop abstractions, the sustainability of the exploitation of a groundwater system can be safeguarded. Such rules aim at halting or, where
needed, even pushing back the saltwater intrusion. Before imposing
any restriction on the abstraction of groundwater, the existing abstractions should first be registered. Then licenses could be issued for
the present rates of abstraction until some fixed date in future. After
expiration of this period the given license could be continued or not,
either for the same or for another rate of abstraction. Such a decision depends on the overall picture of the inventory of the existing
abstractions and a master plan for the future in which all available
water resources are allocated to or reserved for the various users,
taking also into account the water quality requirements for different
user categories.

For obtaining a license a levy might have to be paid. This should
not be considered as buying the water but rather as a compulsory
contribution to the cost of the necessary investigations, measures
and works and the administration by the responsible authority.

The specifications in the license may also include the obligation to
measure the rates of abstraction, some groundwater quality parameters, and groundwater tables and piezometric levels at a number of
locations and depths at specified intervals of time and to submit this

information to the responsible authority.

Observation of the rules should be enforced by sanctions. This and the allocation of the fresh groundwater can lead to socioeconomic problems, for instance for small farmers who need water for irrigation. Therefore the water users must be informed and have the opportunity to participate in the decision process.

There should also be proper legislation to prevent pollution of groundwater and surface waters and, where necessary, also to impose measures for restoration of the present water quality.

Chapter 5

Conceptual and Mathematical Modeling

J. Bear

5.1   Introduction

In many coastal aquifers, intrusion of seawater has become one of the major constraints affecting groundwater management. As seawater intrusion progresses, existing pumping wells, especially close to the coast, become saline and have to be abandoned, thus, reducing the value of the aquifer as a source of freshwater. Also, the area above the intruding seawater wedge is lost as a source of water (by natural replenishment). A detailed review of exploitation, restoration, and management of fresh groundwater in coastal aquifers is presented in Chapter 4, as well as in other chapters throughout this volume.

Since the famous works of Badon-Ghyben [1888] and Herzberg [1901], extensive research has been carried out, and much progress has been made in understanding the various mechanisms that govern seawater intrusion. The dominant factors are the flow regime in the aquifer above the intruding seawater wedge, the variable density, and hydrodynamic dispersion. Reviews of the phenomenon of seawater intrusion, and of the research that has been carried out—both theoretical work, and field and laboratory investigations—may be found in many books and publications, and need not be repeated here (e.g., Bear [1972, 1979], Bear and Verruijt [1987], and Reilly and Goodman [1985]). Most aspects of these developments are also presented in a number of chapters throughout this volume.

Over the years, and especially during the 50's and the 60's, a large number of field investigations have been conducted in many parts

J. Bear et al. (eds.), Seawater Intrusion in Coastal Aquifers, 127–161.

of the world, providing a basis for understanding the complicated mechanisms that cause seawater intrusion and affect the shape of the zone of transition from fresh aquifer water to seawater. Unfortunately, at that time, computational tools were not available to predict the extent of seawater intrusion under various natural and man-made conditions. In most cases, simplified conceptual models had to be introduced in order to enable analytical solutions for the shape of the seawater wedge. For example, the conceptual model:

- The aquifer consists of a single homogeneous porous medium formation. The impervious bottom of the aquifer is horizontal.
- A sharp interface exists between seawater and freshwater.
- The interface is stationary, with freshwater being mobile, while seawater remains immobile.
- Flow is two-dimensional in a vertical plane normal to the coast.
- The flow of freshwater is essentially horizontal. In a phreatic aquifer, where this statement is equivalent to the Dupuit assumption, this assumption leads to the famous Ghyben-Herzberg relationship.
- The aquifer (if phreatic) is replenished at a uniform rate.

enabled an analytical solution for the shape of a stationary interface in a coastal phreatic aquifer, in which the flow is everywhere perpendicular to the coast. (e.g., Bear [1972, 1979]). Glover's [1959] and Henry's [1959] solutions for a sharp parabolic interface in an infinitely thick confined aquifer, and Strack's [1995] use of the Dupuit-Forchheimer modeling of seawater intrusion, with the effect of variable density, may also be mentioned here as additional examples of analytical solutions (Chapter 6).

In spite of the rough approximation associated with such solutions, they were very important at that time, as they provided the fundamental relationship between the rate of freshwater discharge to the sea, and the length of the intruding seawater wedge. As this discharge is reduced, e.g., by pumping a larger fraction of the natural replenishment, the length of the seawater wedge will increase, causing wells to start pumping saline water. They also provided information on the relationship between water levels in the vicinity of the coast and the length of the intruding seawater wedge, and on how to create a water level barrier to arrest, or even push back the

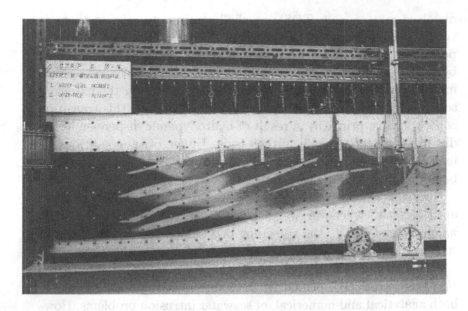

Figure 5.1. Using the Hele-Shaw analog for investigations of seawater intrusion into the coastal aquifer in Israel (Courtesy of Water Planning for Israel, Ltd.)

encroaching wedge. This information is essential for controlling and managing seawater intrusion.

In addition to analytical solutions of simplified models, the Hele-Shaw analog [Bear, 1972] was extensively used, especially in The Netherlands and in Israel (by Water Planning for Israel, Ltd.), to investigate seawater intrusion into coastal aquifers. In the Hele-Shaw analog, or model, the entire vertical cross-section of a coastal aquifer, perpendicular to the coast, was represented, including the detailed geology (layers, sub-aquifers, heterogeneity, and anisotropy), the natural replenishment, wells (actually, drains), etc. (see Figure 5.1). By imposing natural replenishment, and schedules of pumping, the analog provided information on rate and extent of seawater intrusion. This information was used as input information for coastal aquifer management. An optimal management scheme balances the damage caused by seawater intrusion with the advantage of increased pumping.

We often refer to seawater and freshwater as "miscible liquids". Although, actually, both are water, with different concentrations of dissolved matter (salt, TDS), we shall continue to refer to them as two liquids—freshwater, and seawater. Hence, the passage from the portion of the aquifer that is occupied by the former to that occupied by the latter, takes the form of a *transition zone*, rather than a sharp interface. Under certain circumstances, depending on the extent of seawater intrusion, and on certain aquifer properties, this transition zone, which is, primarily, a result of hydrodynamic dispersion of the dissolved matter, may be rather wide. Under other conditions, it may be rather narrow, relative to the aquifer's thickness, and may be approximated as a sharp interface.

Note that in this chapter, since we are considering a problem at field scale, whenever we discuss dispersion, we actually mean *macrodispersion*, i.e., the spreading of a solute due to variations in specific discharge produced by variations in permeability,

Although justified in some cases, the *sharp interface approximation* was introduced, primarily, to enable relatively easy solutions, both analytical and numerical, of seawater intrusion problems. However, numerical solutions of the transition zone problem (obviously, with the effect of density) have been presented in the literature from time to time [Pinder and Cooper, 1970; Lee and Cheng, 1974; Ségol et al., 1975; Ségol and Pinder, 1976; Huyakorn and Taylor, 1976; Frind, 1982a,b; Huyakorn et al., 1987; Voss and Souza, 1987; Diersch, 1988; Galeati et al., 1992; Xue et al., 1995]. Certainly nowadays, with the availability of new improved numerical techniques, including methods for coping with nonlinearities that are inherent in the transition zone model, and with fast and large memory computers (even PCs), numerical solutions of models that take the transition zone into account should not pose special difficulties. There is also no reason to limit the models to (vertical) two-dimensional flow domains. Indeed, a number of models and computer codes that consider seawater intrusion as a solute transport problem have already been developed (e.g., Konikow et al. [1996]).

Plenty of references on analytical solutions and numerical ones, including discussions on specific numerical models, are scattered throughout this book, such as Chapters 6, 8, 9, 10, 11, 12, and need not be repeated here.

The objective of this chapter is to present the conceptual and complete mathematical models that correspond to the three options mentioned above:

- The two- and three-dimensional models of seawater intrusion with a sharp interface (henceforth, to be referred to as 2DSIM and 3DSIM models, respectively).

- Three-dimensional seawater intrusion model with a transition zone (henceforth, to be referred to as 3DTZM model).

In all cases, we shall assume that close to the coast, the aquifer is divided into two sub-aquifers, a lower confined aquifer and an overlying phreatic one, by an impervious or semi-pervious layer that extends some distance land-ward of the coast.

5.2   Three-Dimensional Sharp Interface Model (3DSIM)

Figure 5.2 shows a typical cross-section of a coastal aquifer. Although we show here a two-layered aquifer, the same model can be easily modified to represent more layers. In this cross-section, we note the impervious bottom, the impervious or semi-pervious (= aquitard) layer that extends both seaward and land-ward of the coast, the two sharp interface segments, one for each sub-aquifer, the bottom of the sea, the phreatic surface with accretion from precipitation, the accompanying seepage-face, and pumping and artificial recharge wells.

5.2.1   Conceptual Model

The following conceptual model describes the considered domain and the behavior of fluids in it:

(A-1) The typical cross-section (perpendicular to the coast) of the considered three-dimensional aquifer domain is represented in Figure 5.2.

(A-2) The cross-section shows that close to the coast, the aquifer is divided into two sub-aquifers by an impervious or semi-pervious layer.

(A-3) Freshwater and seawater are assumed to be 'immiscible' liquids, separated by a sharp interface. Under the conditions described in (A-1), a separate interface (GF and CD) will occur in each sub-aquifer.

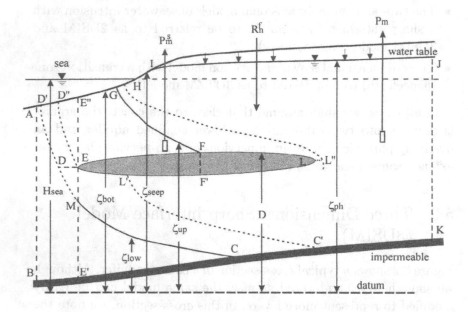

Figure 5.2. A typical cross-section of a two-layered coastal aquifer, with a sharp interface.

(A-4) The sharp, possibly moving, interface divides the aquifer into two sub-domains: one (ABCDEFGA) occupied only by seawater, the other (CDEL'LFGHIJKC) occupied only by freshwater. Obviously, the shape and position of this interface, as it is being displaced, in response to changes in the flow regime in the aquifer, are unknown. In fact, determining them is the objective of the modeling effort.

We could let the interface in the lower sub-aquifer take the form C'L'. This situation may develop as the interface, initially in the form CDD', advances land-ward. When the interface takes the form C'L', no freshwater from the lower sub-aquifer flows to the sea.

Actually, we have delineated the freshwater zone by an arbitrary boundary, DE, along which the sharp interface approximation fails. We could have let the interface take the form CDD', with freshwater flow to the sea above it. However, we then have a problem along D″E, where saltwater will be overlying freshwater—an unstable situation which will not allow for a sharp interface to be maintained. In reality, the domain DED″D′ will be a domain of mixed water; freshwater rising in this zone will mix with seawater.

Another difficulty arises when "toe" of the advancing sharp interface in the upper sub-aquifer, say point F, reaches and surpasses the edge, L, of the impervious or semi-pervious layer. Again, an unstable situation will arise, with seawater being "spilled" into a freshwater domain. Finally, when the intermediate layer is semi-pervious, the interface within it should take the form FL'. However, this, again, means instability, due to the presence of seawater above freshwater. Furthermore, freshwater will rise through the segment EL' (or EF'), such that the domain EFGD"E will contain mixed water. In fact, with the interface in the form CDD', the entire domain FGE"D'DEF will contain mixed water. Bear [1979, p. 402] discusses the effects of a semi-pervious layer on the flow regime in a coastal aquifer.

All these unstable situations only emphasize that, strictly speaking, a model based on a sharp interface as an approximate representation of a narrow transition zone, can be constructed only in limited cases. In order to continue with a model based on the 'sharp interface approximation', let us assume that the domain of interest is bounded by the vertical plane E'MEE", and introduce the constraint that the interface toe, F, will not reach the edge of the layer, L. We shall also assume that the intermediate layer, AFL, is impervious.

In spite of the above considerations, we shall add a few comments concerning leakage through aquitards towards the end of the discussion on the mathematical model.

(A-5) The entire modeled domain is bounded on the seaward side by a vertical plane E'MEE", and on the land-ward side by a vertical plane KJ, at a distance which is beyond the domain of interest in the coastal aquifer.

(A-6) We assume that the saturated zone within the aquifer is bounded from above by a sharp phreatic surface. Accretion from precipitation may take place along the phreatic surface. The detailed distribution of water in the vadose zone above the phreatic surface is neglected, assuming that immediately above the phreatic surface, water is at residual moisture content.

(A-7) The density of each liquid is constant. Both liquids are assumed to be incompressible, except for the contribution of liquid compressibility to the specific storativity (when the latter concept is employed).

(A-8) Isothermal conditions are assumed to prevail.

(A-9) The porous medium within the considered domain may be inhomogeneous and anisotropic. The solid matrix is non-deformable, except for the effect of its compressibility on the specific storativity.

(A-10) Darcy's law describes the flow of both the freshwater and the seawater in the aquifer.

5.2.1.1 Equation of a surface

Let $F(x, y, z, t) = 0$ denote the equation of a surface, and let $\mathbf{u}$ and $\mathbf{n}$ denote the velocity of this surface, and the unit vector normal to it, respectively. Then:

$$\mathbf{n} = \frac{\nabla F}{|\nabla F|}, \qquad \mathbf{u} \cdot \nabla F = -\frac{\partial F}{\partial t} \qquad (5.1)$$

where t denotes time.

5.2.2 *Mathematical Model*

Because of the discontinuity in fluid density across the interface, the entire flow domain is divided into two sub-domains: one occupied only by freshwater, the other only by seawater. In the former, $h^f = h^f(x, y, z, t)$ represents the state variable. In the latter, it is $h^s = h^s(x, y, z, t)$. Within each sub-domain, the state variable has to satisfy the partial differential equation that represents the mass balance for the considered liquid, subject to initial conditions within the sub-domain, and to specified conditions on its boundary.

5.2.2.1 Mass balance for freshwater (f) in the sub-domain occupied by freshwater

$$S_o^f \frac{\partial h^f}{\partial t} = -\nabla \cdot \mathbf{q}^f + \sum_{(n)} R_n^f(\mathbf{x}_n, t)\delta(\mathbf{x} - \mathbf{x}_n)$$
$$- \sum_{(m)} P_m^f(\mathbf{x}_m, t)\delta(\mathbf{x} - \mathbf{x}_m) \qquad (5.2)$$

where S_o^f denotes the specific storativity of the aquifer to freshwater (due to the compressibility of both the water and the porous medium), h^f denotes the piezometric head for freshwater, defined by

$$h^f = z + \frac{p}{\rho^f g} \tag{5.3}$$

where $R_n^f(\mathbf{x}_n, t)$ and $P_m^f(\mathbf{x}_m, t)$ denote, symbolically, the rates of recharge and of pumping at points $\mathbf{x}_n$ and $\mathbf{x}_m$, respectively, $\delta\,(\mathbf{x} - \mathbf{x}_m)$ denotes the *Dirac-delta* function (dims. L^3) at point $\mathbf{x}_m$, p is pressure, ρ^f denotes the density of freshwater, g is gravity acceleration, $\mathbf{q}^f$ denotes the specific discharge vector, with q_i^f denoting the ith component, expressed by Darcy's law in the form

$$\mathbf{q}^f = -\mathbf{K}^f \cdot \nabla h^f, \quad q_i^f = -K_{ij}^f \frac{\partial h^f}{\partial x_j} \tag{5.4}$$

and $K_{ij}^f = K_{ij}^f(x, y, z)$ denotes the ijth component of the tensor of hydraulic conductivity of the porous medium for freshwater $\mathbf{K}^f$:

$$K_{ij}^f = \frac{k_{ij}\rho^f g}{\mu^f} \tag{5.5}$$

in which k_{ij} is the ijth component of the permeability tensor of the porous medium, $\mathbf{k}$, and μ^f denotes the fluid's dynamic viscosity.

In writing Eq. 5.2, we have assumed that $\rho^f = \rho^f(p)$, with the specific storativity (= volume of water released from storage, per unit volume of porous medium, per unit decline in head) defined as $S_o^f = \rho^f g[\phi\beta_p + (1 - \phi)\alpha]$. In this expression, ϕ denotes porosity, β_p denotes the coefficient of fluid compressibility, $\beta_p = (1/\rho^f)(d\rho^f/dp)$, $\alpha = -(1/(1 - \phi))\,d(1 - \phi)/dp$ denotes the coefficient of matrix compressibility, and we have assumed that $|\phi\partial\rho^f/\partial t| \ll |\mathbf{q}^f \cdot \nabla\rho^f|$, such that (e.g., Bear [1979]):

$$\frac{\partial\phi\rho^f}{\partial t} = \phi\frac{\partial\rho^f}{\partial t} + \rho^f\frac{\partial\phi}{\partial t} = \rho^f S_o^f \frac{\partial h^f}{\partial t} \tag{5.6}$$

Similar relationships and definitions apply also to the mass balance for seawater, to be presented below.

5.2.2.2 Mass balance for seawater (s) in the sub-domain occupied by seawater

$$S_o^s \frac{\partial h^s}{\partial t} = -\nabla \cdot \mathbf{q}^s - \sum_{(m)} P_m^s(\mathbf{x}_m, t)\delta\,(\mathbf{x} - \mathbf{x}_m) \tag{5.7}$$

where the symbols, except for the superscripts, have the same meaning as given above. We have included the option of pumping seawater from below the interface. The specific discharge of seawater is expressed by

$$\mathbf{q}^s = -\mathbf{K}^s \cdot \nabla h^s, \quad q_i^s = -K_{ij}^s \frac{\partial h^s}{\partial x_j} \tag{5.8}$$

where

$$h^s = z + \frac{p}{\rho^s g} \tag{5.9}$$

denotes the seawater piezometric head.

5.2.2.3 The flow equations

By inserting the motion equations into the mass balance ones, we obtain two mass balances, often referred to as *flow equations*:

$$S_o^f \frac{\partial h^f}{\partial t} = \nabla \cdot \left(\mathbf{K}^f \cdot \nabla h^f \right) + \sum_{(n)} R_n^f(\mathbf{x}_n, t) \delta \left(\mathbf{x} - \mathbf{x}_n \right)$$

$$- \sum_{(m)} P_m^f(\mathbf{x}_m, t) \delta \left(\mathbf{x} - \mathbf{x}_m \right) \tag{5.10}$$

for the freshwater sub-domain, and

$$S_o^s \frac{\partial h^s}{\partial t} = \nabla \cdot (\mathbf{K}^s \cdot \nabla h^s) - \sum_{(m)} P_m^s(\mathbf{x}_m, t) \delta \left(\mathbf{x} - \mathbf{x}_m \right) \tag{5.11}$$

for the sub-domain occupied by seawater. These two partial differential equations have to be solved for the two variables: $h^f = h^f(x, y, z, t)$ and $h^s = h^s(x, y, z, t)$, each within its respective sub-domain. As we shall see below, the two equations are coupled by boundary conditions on the interface, and have to be solved simultaneously.

5.2.2.4 Initial conditions for the freshwater (D^f) and seawater (D^s) sub-domains

The initial conditions for the two partial differential equations of balance, are:

$$h^f(x, y, z, O) = h_o^f(x, y, z) \quad \text{in} \quad D^f \tag{5.12}$$

$$h^s(x,y,z,O) = h_o^s(x,y,z) \quad \text{in} \quad D^s \qquad (5.13)$$

in which $h_o^f(x,y,z)$ and $h_o^s(x,y,z)$ are the known initial piezometric heads within the respective sub-domains.

Let the initial shape of the interface surfaces in the upper (GF, *up*) and lower (MC, *low*) sub-aquifers be denoted, respectively, by

$$F_{up}(x,y,z,0) = F_{up}^o = 0, \quad F_{low}(x,y,z,0) = F_{low}^o = 0$$
$$(5.14)$$

where F_{up}^o and F_{low}^o are known functions.

5.2.2.5 Boundary conditions for the freshwater sub-domain FGHIJKCMEL'LF

Referring to Figure 5.2, for each boundary segment, we need (i) a statement that expresses the *geometry* of the boundary segment, and (ii) the condition to be satisfied at all points of the considered segment.

a. The interface FG in the upper sub-aquifer

Because this boundary segment is a common boundary to the two sub-domains, D^f and D^s, we need two conditions on it, one condition for each side.

Let $\zeta_{up} = \zeta_{up}(x,y,t)$ denote the elevation of a point on the upper interface, FG, with respect to some datum level. Since the pressure must be the same as this interface is approached from both sides, i.e., at points $z = \zeta_{up}$, we have:

$$\zeta_{up}(x,y,t) = h^s(1+\delta) - h^f\delta \qquad (5.15)$$

where $\delta = \rho^f / \left(\rho^s - \rho^f\right)$. Hence, once we solve the problem for $h^f = h^f(x,y,z,t)$ and $h^s = h^s(x,y,z,t)$, we can obtain the shape of the interface, $\zeta_{up} = \zeta_{up}(x,y,t)$, or $F_{up} = F_{up}(x,y,z,t)$, from the relationship:

$$F_{up}(x,y,z,t) = \zeta_{up}(x,y,t) - h^s(1+\delta) + h^f\delta = 0$$
$$F_{up}(x,y,z,t) = z - \zeta_{up}(x,y,t) = 0. \qquad (5.16)$$

The condition to be satisfied on the interface is that it is a *material surface* with respect to both fluids, i.e., no fluid crosses the interface. We express this condition for the freshwater by:

$$\left(\mathbf{V}^f - \mathbf{u}\right) \cdot \mathbf{n} = 0 \qquad (5.17)$$

where $\phi \mathbf{V}^f = \mathbf{q}^f$.

Expressing $\mathbf{n}$, and $\mathbf{u} \cdot \nabla F$ by Eq. 5.1, and expressing F by Eq. 5.16, we obtain this condition in the form:

$$\phi \frac{\partial \zeta_{up}}{\partial t} = - \left(\mathsf{K}^f \cdot \nabla h^f \right) \cdot \nabla (z - \zeta_{up}) \tag{5.18}$$

or, in terms of the piezometric heads:

$$\phi \delta \frac{\partial h^f}{\partial t} - \phi(1 + \delta) \frac{\partial h^s}{\partial t} = \left(\mathsf{K}^f \cdot \nabla h^f \right) \cdot \{ \nabla z - (1 + \delta) \nabla h^s + \delta \nabla h^f \} \tag{5.19}$$

We note that this boundary condition, for the domain D^f, involves information not only on h^f, but also on h^s. A similar statement applies also to the condition on the same boundary segment, as part of the D^s domain. This couples the two sub-problems and they have to be solved simultaneously.

b. The sea bottom GH

Along this boundary segment, the pressure is prescribed by the elevation of sea level: $p_{bot} = (H_{sea} - \zeta_{bot}) \rho^s g$. Thus, in terms of h^f, the boundary condition is:

$$h^f(x, y, z, t) = H_{sea} \left(1 + \frac{1}{\delta} \right) - \frac{\zeta_{bot}}{\delta} \tag{5.20}$$

where H_{sea} and ζ_{bot} denote the elevations of sea level and of sea bottom, respectively.

c. The seepage face HI

The geometry of the seepage face boundary is known (= ground surface), except for its upper bound, where the phreatic surface is tangent to this boundary segment. Let $\zeta_{seep} = \zeta_{seep}(x, y)$ denote the elevations of points on the seepage face. Then, since the pressure of the emerging water is atmospheric, taken as zero, the condition on this boundary $\zeta_{seep}(x, y) = z$ is

$$h^f(x, y, z, t) = z \tag{5.21}$$

d. The phreatic surface IJ

Like the interface, the shape and position of the phreatic surface are *a-priori* unknown. Since the pressure on the phreatic surface is taken as atmospheric, with $p_{atm} = 0$, the piezometric head at a point on the interface is equal to its elevation, $\zeta_{ph}(x, y, t)$. Thus, the shape of the phreatic surface is expressed by:

$$F_{ph}(x, y, z, t) = \zeta_{ph}(x, y, t) - z \equiv h^f(x, y, \zeta_{ph}, t) - z = 0 \tag{5.22}$$

The condition on this surface is that the flux across it should be continuous. With θ_{wr} denoting the (assumed known) moisture content above the interface (introduced because we wish to avoid the need to deal with flow in the unsaturated zone as part of the modeling effort), this condition takes the form:

$$\phi \left(\mathbf{V}_{sat}^f - \mathbf{u} \right) \cdot \mathbf{n} = \theta_{wr} \left(\mathbf{V}_{unsat}^f - \mathbf{u} \right) \cdot \mathbf{n} \tag{5.23}$$

With $\theta_{wr} \mathbf{V}_{unsat}^f$ denoting the flux, $\mathbf{N} = \mathbf{N}(x, y, t)$, reaching the phreatic surface, say, from precipitation ($\mathbf{N} = -N \nabla z$), or leaving this surface by evaporation, Eq. 5.23 can be rewritten as:

$$\left(\mathbf{K}^f \cdot \nabla h^f + \mathbf{N} \right) \cdot \nabla \left(h^f - z \right) = S_y \frac{\partial h^f}{\partial t} \tag{5.24}$$

where $S_y \equiv \phi - \theta_{wr}$ is the specific yield of the phreatic aquifer.

We recall that we are here in a 'vicious circle': to solve for h^f we need to know F_{ph}, but we cannot know F_{ph} until we have solved the problem for h^f.

e. The external land-side boundary JK

This boundary, described by $F_{ext}(x, y, z) = 0$, is assumed to be sufficiently far from the coast, so that we may assume that the piezometric head along it is a known function, $h_{ext}^f = h_{ext}^f(x, y, z, t)$. Thus, the boundary condition here is

$$h^f = h_{ext}^f \tag{5.25}$$

A horizontal flow and a vertical equipotential surface are often assumed.

Another possibility is that the flux at points on this boundary is known (from available information on the external side of this boundary), say, as a known function $f_2 = f_2(x, y, z, t)$. Then the condition is

$$- \left(\mathbf{K}^f \cdot \nabla h^f \right) \cdot \mathbf{n} = f_2(x, y, z, t) \tag{5.26}$$

where $\mathbf{n}$ denotes the outward unit vector normal to the boundary surface.

f. The impervious bottom KC

This is a no-flux condition:

$$-\left(\boldsymbol{\kappa}^f \cdot \nabla h^f\right) \cdot \mathbf{n} = 0 \qquad (5.27)$$

g. The interface in the lower sub-aquifer CM

Similar to Eq. 5.15, the equation that describes the geometry of this interface segment is:

$$F_{low}(x, y, z, t) = z - \zeta_{low} = z - h^s(1 + \delta) + h^f \delta = 0$$
$$(5.28)$$

The condition on this interface is given by Eq. 5.17, or Eq. 5.18, with some modification of subscripts and superscripts, or by Eq. 5.19.

We note that, in analogy to point I, where the phreatic surface is tangent to the external surface, the location of (the moving) point M on the vertical boundary surface is unknown, until the problem is solved.

h. The sea-side boundary ME

This boundary is introduced artificially, in order to enable us to consider a sharp interface model. In reality, we do not know the piezometric head on this boundary, nor do we know the flux. We have two options for *approximating* the condition along this boundary:

- We assume that the piezometric head at points E and M are dictated by a hydrostatic pressure distribution in the seawater sub-domain. Thus,

$$p_M = (H_{sea} - \zeta|_M)\,\rho^s g, \quad \text{and} \quad p_E = (H_{sea} - \zeta|_E)\,\rho^s g \qquad (5.29)$$

Note that since this is a 3-dimensional model, the elevations of points E and M may vary with x, y (and point M moves with time).

- The sea bottom segment, $D'D''$, is at a constant seawater head. Representing points D' and D'' by a single point at an average elevation $\zeta_{D'D''}$, we may express the freshwater head along this segment by

$$h^f|_{D'D''} = H_{sea}\left(1 + \frac{1}{\delta}\right) - \zeta_{D'D''}\frac{1}{\delta} \qquad (5.30)$$

Then, we express the continuity of freshwater flux towards the sea through the segment ME, assuming horizontal flow, by

$$K^f \frac{\partial h^f}{\partial x} = \alpha_1 \left(h^f - h^f|_{D'D''} \right)$$

$$= \alpha_1 \left[h^f - H_{sea} \left(1 + \frac{1}{\delta} \right) + \zeta|_{D'D''} \frac{1}{\delta} \right] \quad (5.31)$$

This is a boundary condition of the third kind in terms of h^f. We could write this condition without the constraint of horizontal flow through ME. This option (see discussion in Bear [1979], p. 392) is preferable. The transfer coefficient, α_1, has to be determined as part of model calibration. An estimate of the value of α_1, may be obtained from $\alpha_1 = K^s/L_{Msea}$, where L_{Msea} represents the distance from the ME segment to the D'D'' one. We should recall that the point M *is a-priori unknown*, and should be obtained by some iteration process, requiring that ζ_M, as a point on the interface, satisfies Eq. 5.28.

We wish to emphasize again, that we must resort to all kinds of approximations, in order to construct a model based on the sharp interface approximation.

i. The impervious boundary EL'LF

Here, the boundary condition is one of no flux normal to the boundary, as in the case of the impervious bottom, KC.

5.2.2.6 Boundary conditions for the seawater sub-domains FGE''EF and ME'CM

In the way we have eventually set up the sea-side boundary segment ME for the freshwater sub-domain, the seawater sub-domain is made up of two parts: FGE''EF and ME'CM. Once the advancing lower interface bypasses point E, taking the form C'L', the seawater sub-domain will take the form FGABC'L'EF.

a. The interface FG in the upper sub-aquifer

As stated above, the interface is a *material surface* with respect to both fluids. We express this condition for the seawater by:

$$(\mathbf{V}^s - \mathbf{u}) \cdot \mathbf{n} = 0 \quad \text{for } D^s \qquad (5.32)$$

Similar to what was presented for the freshwater domain, this condition can be expressed in the form

$$\phi \frac{\partial \zeta_{up}}{\partial t} = -(\mathbf{K}^s \cdot \nabla h^s) \cdot \nabla(z - \zeta_{up}) \qquad (5.33)$$

or, in terms of the piezometric heads:

$$\phi \delta \frac{\partial h^f}{\partial t} - \phi(1 + \delta)\frac{\partial h^s}{\partial t} = (\mathbf{K}^s \cdot \nabla h^s) \cdot \{\nabla z - (1 + \delta)\nabla h^s + \delta \nabla h^f\} \qquad (5.34)$$

b. The sea bottom GE''

Along this boundary segment, the pressure is prescribed by the sea level elevation: $p_{bot} = (H_{sea} - \zeta_{bot})\rho^s g$, or, in terms of h^s:

$$h^s(x, y, z, t) = H_{sea} \qquad (5.35)$$

c. The sea-side boundary segments EE'' and E'M

Like the segment GE'', this is also a boundary of constant seawater head, described by Eq. 5.35. If, when the interface takes the form C'L', we decide to move the sea-side boundary to AB, the latter will also be a prescribed seawater head boundary specified by Eq. 5.35.

d. The impervious segments EF and E'C

Similar to CK, these are also no-flux boundaries:

$$-(\mathbf{K}^s \cdot \nabla h^s) \cdot \mathbf{n} = 0 \qquad (5.36)$$

where, as for all other boundaries, the unit vector normal to the surface may be obtained from the equation of the surface, using Eq. 5.1.

e. The interface CM in the lower sub-aquifer

The condition on this boundary segment is similar to that on FG. Thus, we may obtain it by replacing ζ_{up} by ζ_{low} in Eq. 5.33, or we may use Eq. 5.34. We should note that the phreatic surface and the interfaces are *free surfaces*, i.e., we do not know *a-priori* where they are. This should be emphasized especially for points F, G, C, M (or C', L').

5.2.2.7 A comment on leakage through semi-pervious layers

We have already discussed the presence of semi-pervious layers (= aquitards) that separate the aquifer close to the coast into two or more sub-aquifers. Because the leakage from a freshwater sub-domain into a seawater one, or vice versa, through such layers violates our assumption of a sharp interface, we have discarded this option, and assumed that such layers are impervious.

However, if we do wish to include leakage through aquitards in our 3-D sharp interface model, let us assume that

(A-11) when water of one kind leaks into a sub-aquifer containing water of another kind, through a semi-pervious layer, it mixes with the latter and loses its identity. The dilution that occurs does not affect the density of water in the invaded sub-domain.

(A-12) The leakage through a semi-pervious layer is always along the vertical, from high to low (referenced) piezometric head.

(A-13) Water is not stored in the semi-pervious layer.

(A-14) The aquitard's hydraulic conductivity is assumed to be constant. However, we assume that the fluid's density varies linearly across the aquitard, and we take this variation into account in determining the (referenced) piezometric head.

Consider point 1, at elevation z_1, in which we have fluid α and point 2, at elevation z_2, in which the fluid is β, with $\alpha, \beta = f, s$. Points 1 and 2 are on both sides of an aquitard, with $B' = z_2 - z_1$ (=thickness of aquitard). With ρ^o denoting the density of a reference fluid, we may write for point i $(i = 1, 2)$ and fluid γ $(\gamma = \alpha, \beta)$:

$$h_i^o = \frac{\rho^\gamma}{\rho^o} h^\gamma - z_i \frac{\rho^\gamma - \rho^o}{\rho^o} \tag{5.37}$$

Under such assumptions, the rate of leakage through an aquitard from point 2 with fluid β, to point 1 with fluid α, is expressed by:

$$q_v|_{2,\beta \to 1,\alpha} = \frac{1}{\sigma} \left(\frac{\rho^\beta}{\rho^o} h_2^\beta - \frac{\rho^\beta - \rho^o}{\rho^o} z_2 - \frac{\rho^\alpha}{\rho^o} h_1^\alpha + \frac{\rho^\alpha - \rho^o}{\rho^o} z_1 \right) \tag{5.38}$$

where σ $(= B'/K')$ denotes the resistance to flow in the semi-pervious layer (=ratio of thickness to hydraulic conductivity), and K' denotes the hydraulic conductivity of the aquitard, related to the reference density.

Choosing $\rho^o = \bar{\rho} = (\rho^\alpha + \rho^\beta)/2$, this equation reduces to

$$q_v|_{2,\beta \to 1,\alpha} = \frac{1}{\sigma \bar{\rho}} \left(\rho^\beta h_2^\beta - \rho^\alpha h_1^\alpha - (\rho^\beta - \rho^\alpha) \frac{z_2 + z_1}{2} \right) \tag{5.39}$$

When the same fluid, say, $\beta (= \bar{\rho})$, is present on both sides of the aquitard, the leakage rate is expressed by

$$q_v|_{2,\beta \to 1,\beta} = \frac{1}{\sigma} \left(h_2^\beta - h_1^\beta \right) \tag{5.40}$$

We recall that in the 3DSIM model considered here, the variables are h^f and h^s. Rigorously, since this is a 3-D model, we should not consider 'leakage'. Instead, we should consider 3-D flow in an inhomogeneous domain. Then, because of the discontinuity in permeability, we should require the (boundary) conditions of equality of pressure and equality of normal flux at such surfaces. This approach, which is undertaken in the discussion on the 3DTZM model, is not appropriate here, as this will force us to introduce a piezometric head for the semi-pervious layer, with uncertainty concerning the density within that layer. Instead, we may assume that the semi-pervious layer acts as a boundary to each of the adjacent (freshwater or seawater) subdomains.

For example, let the leakage take place from the sub-domain occupied by the β-fluid to the one occupied by the α-fluid. Then, in the 3DSIM model, for the latter domain, we add a condition on the aquitard segment of the boundary:

$$- (\rho^\alpha \mathbf{K}^\alpha \cdot \nabla h^\alpha) \cdot \mathbf{n} = \bar{\rho} \; q_v|_{2,\beta \to 1,\alpha} \tag{5.41}$$

where $\bar{\rho} \; q_v|_{2,\beta \to 1,\alpha}$ is obtained from Eq. 5.39.

This completes the discussion on the boundary conditions. The complete 3DSIM model includes the mass balance equations Eqs. 5.10 and 5.11, the initial conditions and the boundary conditions described above. The model is solved simultaneously for h_f and h_s in the respective subdomains. Once a solution has been obtained, Eq. 5.16 is used to obtain the shape and position of the interface. Obviously some kind of iteration procedure is required, as the solution required information on the shape and position of the interface boundary.

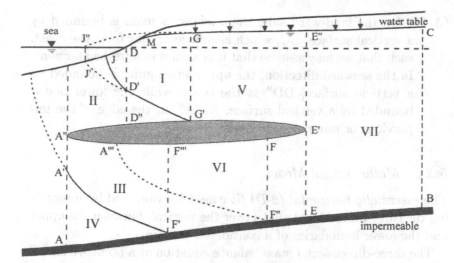

Figure 5.3. A typical cross-section of a two-layered coastal aquifer, with sharp interfaces and essentially horizontal flow.

5.3 Two-Dimensional Sharp Interface Model (2DSIM)

Like the model discussed above, this is also a sharp interface one.

5.3.1 Conceptual Model

Assumptions (A-2), (A-3) and (A-7) through (A-10) are valid also here, and need not be repeated. In addition:

(A-15) Figure 5.3 shows a typical cross-section through a coastal aquifer. In this cross-section, we note seven flow sub-domains: two (I, III) freshwater sub-domains above the interfaces, two ones (II, IV) below the latter, one (VI) of flow in a confined aquifer in the lower sub-aquifer, a phreatic one (V) in the upper sub-aquifer, and a phreatic one (VII) in the undivided portion of the aquifer.

(A-16) We assume that the flow of both the freshwater and the sea-water everywhere within the multi-layered aquifer system, is *essentially horizontal*. This is the Dupuit assumption, valid when the thickness of the considered aquifer domain is much smaller than its horizontal extent.

(A-17) On the land-ward side, the modeled domain is bounded by a vertical surface, BC, which is sufficiently far from the coast, such that we may assume that it is always in a freshwater zone. In the seaward direction, the upper sub-aquifer is bounded by a vertical surface, DD″ at the coast, while the lower one is bounded by a vertical surface, AA′A″, at the edge of the impervious or semipervious layer.

5.3.2   Mathematical Model

The *essentially horizontal (2-D) flow* model is obtained by integrating the three-dimensional model over the vertical, between the upper and the lower boundaries of a considered domain.

The three-dimensional mass balance equation at a point within an α-fluid ($\alpha = s, f$) is Eq. 5.10, without the sources and sinks:

$$S_o^\alpha \frac{\partial h^\alpha}{\partial t} = -\nabla \cdot \mathbf{q}^\alpha \tag{5.42}$$

where $h^\alpha = h^\alpha(x, y, z, t)$.

Let us integrate this equation between two surfaces: $\zeta_t = \zeta_t(x, y, t)$, and $\zeta_b = \zeta_b(x, y, t)$, which bound the domain occupied by a considered α-fluid from above and from below, respectively. We obtain [Bear, 1979, p. 387]:

$$S^\alpha \frac{\partial \overline{h^\alpha}}{\partial t} = -\nabla \cdot (B^\alpha \overline{\mathbf{q}^\alpha}) - \mathbf{q}^\alpha|_{\zeta_t} \cdot \nabla(z - \zeta_t) + \mathbf{q}^\alpha|_{\zeta_b} \cdot \nabla(z - \zeta_b) \tag{5.43}$$

where S^α ($= B^\alpha \overline{S_o^\alpha}$) denotes the storativity within the considered α-fluid domain, $\overline{(..)}$ denotes the average of $(..)$ over the vertical length $B^\alpha \equiv \zeta_t - \zeta_b$, and we have assumed that equipotentials are approximately vertical, i.e., $h^\alpha|_{\zeta_b} \approx h^\alpha|_{\zeta_t} \approx \overline{h^\alpha}$.

The flow through each sub-domain can be expressed by an averaged Darcy law. Again, assuming approximately vertical equipotentials, we obtain:

$$\overline{\mathbf{q}} = -\overline{\mathbf{K}} \cdot \nabla \overline{h^\alpha} \tag{5.44}$$

Henceforth, we shall omit the average symbol

We have to consider three kinds of bounding surfaces:

a. A phreatic surface

On this boundary, we use the boundary condition Eq. 5.24, in which we replace h^f by h^α, and use $F = \zeta_t - z = 0$ as the equation describing the boundary surface. We obtain:

$$(-\mathbf{q}^\alpha + \mathbf{N}) \cdot \nabla (h^\alpha - z) = S_y \frac{\partial h^\alpha}{\partial t} \tag{5.45}$$

Or, with $\mathbf{N} = -N\nabla z$ for precipitation,

$$\mathbf{q}^\alpha \cdot \nabla (z - \zeta_t) + N = S_y \frac{\partial h^\alpha}{\partial t} \tag{5.46}$$

b. An impervious bottom

For a boundary surface described by $F = z - \zeta_b = 0$, the condition is:

$$\mathbf{q}^\alpha \cdot \mathbf{n} \equiv \mathbf{q}^\alpha \cdot \nabla(\zeta_b - z) = 0 \tag{5.47}$$

c. An interface

For an interface at an elevation $\zeta(x, y, t)$ (either ζ_t or ζ_b), the condition is obtained from Eq. 5.19:

$$-\mathbf{q}^\alpha \cdot \nabla(z - \zeta). = \phi\delta \frac{\partial h^f}{\partial t} - \phi(1 + \delta)\frac{\partial h^s}{\partial t} \tag{5.48}$$

These conditions may now be introduced into Eq. 5.43, written *separately* for each of the seven sub-domains within the aquifer, in order to obtain the corresponding mass balance equations.

For sub-domain I:

$$\left(S_y + S_o^f B_{\mathrm{I}}\right) \frac{\partial h_{\mathrm{I}}^f}{\partial t} - \phi(1 + \delta)\frac{\partial h_{\mathrm{II}}^s}{\partial t} + \phi\delta\frac{\partial h_{\mathrm{I}}^f}{\partial t} = \nabla \cdot \left(B_{\mathrm{I}}\mathbf{K}^f \cdot \nabla h_{\mathrm{I}}^f\right) + N \tag{5.49}$$

For sub-domain II:

$$S_o^s B_{\mathrm{II}} \frac{\partial h_{\mathrm{II}}^s}{\partial t} + \phi(1 + \delta)\frac{\partial h_{\mathrm{II}}^s}{\partial t} - \phi\delta\frac{\partial h_{\mathrm{II}}^f}{\partial t} = \nabla \cdot (B_{\mathrm{II}}\mathbf{K}^s \cdot \nabla h_{\mathrm{II}}^s) \tag{5.50}$$

For sub-domain III:

$$S_o^f B_{\text{III}} \frac{\partial h_{\text{III}}^f}{\partial t} - \phi(1+\delta)\frac{\partial h_{\text{IV}}^s}{\partial t} + \phi\delta\frac{\partial h_{\text{III}}^f}{\partial t} = \nabla \cdot \left(B_{\text{III}}\mathbf{K}^f \cdot \nabla h_{\text{III}}^f \right)$$
$$(5.51)$$

For sub-domain IV:

$$S_o^s B_{\text{IV}} \frac{\partial h_{\text{IV}}^s}{\partial t} + \phi(1+\delta)\frac{\partial h_{\text{IV}}^s}{\partial t} - \phi\delta\frac{\partial h_{\text{III}}^f}{\partial t} = \nabla \cdot \left(B_{\text{IV}}\mathbf{K}^s \cdot \nabla h_{\text{IV}}^s \right)$$
$$(5.52)$$

For sub-domain V:

$$\left(S_y + S_o^f B_{\text{V}} \right)\frac{\partial h_{\text{V}}^f}{\partial t} = \nabla \cdot \left(B_{\text{V}}\mathbf{K}^f \cdot \nabla h_{\text{V}}^f \right) + N \qquad (5.53)$$

For sub-domain VI:

$$S_o^f B_{\text{VI}} \frac{\partial h_{\text{VI}}^f}{\partial t} = \nabla \cdot \left(B_{\text{VI}}\mathbf{K}^f \cdot \nabla h_{\text{VI}}^f \right) \qquad (5.54)$$

For sub-domain VII:

$$\left(S_y + S_o^f B_{\text{VII}} \right)\frac{\partial h_{\text{VII}}^f}{\partial t} = \nabla \cdot \left(B_{\text{VII}}\mathbf{K}^f \cdot \nabla h_{\text{VII}}^f \right) + N$$
$$(5.55)$$

Note that in sub-domains I, V, and VII, the value of B is a variable, depending on the elevation of the water table.

Altogether, we have here 7 equations that have to be solved for the 7 variables: h_{I}^f, h_{II}^s, h_{III}^f, h_{IV}^s, h_{V}^f, h_{VI}^f, and h_{VII}^f. Once these are known, one can determine the shape of the upper and lower interfaces, respectively, from

$$F = F(x,y,t) = z - (1+\delta)h_{\text{II}}^s + \delta h_{\text{I}}^f \qquad (5.56)$$

and

$$F = F(x,y,t) = z - (1+\delta)h_{\text{IV}}^s + \delta h_{\text{III}}^f \qquad (5.57)$$

5.3.2.1   A comment on leakage through semi-pervious layers

For the 3DSIM model, we have discussed the possibility that the low permeability layers that may exist close to the coast, or parts of them may be aquitards, rather than impervious, permitting leakage through them. Here, in the case of the 2DSIM model, if we wish to take such leakage into account, we have to add the term(s) that express leakage in the mass balance equations. However, we have to take into account that the freshwater-freshwater and freshwater-saltwater portions of the semi-pervious layer vary continuously, e.g., as points G' and F' are displaced.

Thus, we may add the term $+ q_v|_{f,\text{VI}\to f,\text{V}}$, obtained from Eq. 5.39, to the r.h.s. of Eq. 5.53 to take care of leakage of freshwater through the aquitard into the freshwater sub-domain V.

Obviously, the same term, but with a negative sign must be added to the r.h.s. of Eq. 5.54. Similarly, we may add a term $+ q_v|_{f,\text{III}\to s,\text{II}}$ to the r.h.s. of Eq. 5.50, and the same term, but with a negative sign to the r.h.s. of Eq. 5.51, to account for freshwater from sub-domain III leaking into the seawater sub-domain II. Note that for each sub-region above or below the aquitard, we must add the terms expressing leakage to all sub-domains connected with it.

5.3.2.2   Initial and boundary conditions

Initial conditions are required for each of the 7 piezometric head variables, $h_i = h_i^\alpha(x, y, t)$, $\alpha = f, s$, and $i =\text{I}, \dots, \text{VII}$, each within its respective sub-domain.

We need boundary conditions for each of the 7 sub-domains. These include external boundaries, as well as internal ones, GG' and $F'''F'$. On the latter, we require equality of pressure and equality of the normal flux, as each boundary is approached from both sides. We note that these two boundary surfaces are moving as seawater intrusion progresses. We shall not specify these conditions in detail.

5.3.2.3   Boundary condition along BC

This boundary may be one of prescribed head:

$$h_{\text{VII}}^f = f_1(x, y, t) \qquad (5.58)$$

or one of prescribed flux:

$$\left(-\mathbf{K}^f \cdot \nabla h_{\text{VII}}^f\right) \cdot \mathbf{n} = f_2(x, y, t) \qquad (5.59)$$

where $f_1(x, y, t)$ and $f_2(x, y, t)$ are known functions. We recall that according to the Dupuit assumption employed here, the flux is constant along the vertical.

We have also to maintain the condition of continuity of fluxes along the boundary E''E'E:

$$\left[\left(B_\mathrm{V}\mathbf{K}^f \cdot \nabla h_\mathrm{V}^f\right) + \left(B_\mathrm{VI}\mathbf{K}^f \cdot \nabla h_\mathrm{VI}^f\right)\right] \cdot \mathbf{n} = \left(B_\mathrm{VII}\mathbf{K}^f \cdot \nabla h_\mathrm{VII}^f\right) \cdot \mathbf{n} \tag{5.60}$$

It may be of interest to mention here that the above condition may lead to a contradiction to the Dupuit assumption employed here. According to this assumption, the flux remains constant along the vertical, and hence the distribution of the flux leaving sub-domain VII between sub-domains V and VI should be according to the respective thicknesses, B_V and B_VII. The condition imposed by Eq. 5.60 may lead to a different result. However, it is correct to impose this condition which takes into account the resistance that the flows encounter in the various sub-domains.

5.3.2.4   Boundary condition along DD' and A''A'

A discussion on the condition to be imposed on this kind of boundary was presented earlier, for the 3DSIM model (boundary ME). The same discussion is also applicable here. For DD', we use the flux condition:

$$\begin{aligned} K^f \nabla h_I^f &= \alpha_1 \left(h_I^f - h^f|_{DD'}\right) \\ &= \alpha_1 \left[h_I^f - H_{sea}\left(1 + \frac{1}{\delta}\right) + \frac{\zeta|_{DD'}}{\delta}\right] \end{aligned} \tag{5.61}$$

A similar condition can be written for A''A'.

5.3.2.5   Boundary condition along D'D'' and A'A

We assume a constant seawater head:

$$h_\mathrm{II}^s = H_{sea}; \quad h_\mathrm{IV}^s = H_{sea} \tag{5.62}$$

It is also possible to impose on these boundary segments a flux condition similar to Eq. 5.61, written for seawater fluxes.

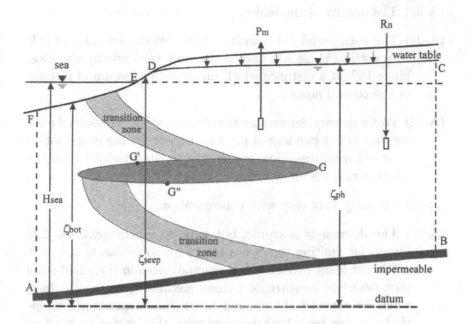

Figure 5.4. A typical cross-section of a two-layered coastal aquifer, with transition zones.

5.4   Transition Zone Model (3DTZM)

5.4.1   Conceptual Model

Here, we have a single flow domain (Figure 5.4), containing a single liquid phase, with a varying concentration of dissolved matter. We'll continue to model only the saturated zone. Assumptions (A-2), (A-6), (A-8) and (A-10) are valid also here, and need not be repeated.

(A-18) The typical cross-section (perpendicular to the coast) of the considered three-dimensional aquifer domain is represented in Figure 5.4.

(A-19) Freshwater and seawater are assumed to be a single liquid phase, referred to as water, with a variable concentration of dissolved matter. No sharp interfaces exist. Instead, we shall observe transition zones across which the concentration of dissolved matter will vary from that of freshwater to that of seawater.

(A-20) The density of the infiltrating water is specified.

(A-21) The entire modeled domain is bounded on the seaward side by a vertical plane AF, and on the land-ward side by a vertical plane BC, at a distance which includes the domain of interest in the coastal aquifer.

(A-22) Water density depends on the concentration of dissolved matter, but is independent of pressure except for the contribution of liquid compressibility to the specific storativity (when the latter concept is employed).

(A-23) Viscosity may vary with concentration.

(A-24) The domain is isotropic, but may be inhomogeneous. The reason for limiting the discussion to isotropic domains is that we do not know enough about macrodispersion (i.e., field-scale dispersion) in anisotropic porous medium domains. In fact, when we consider the domain to be 'inhomogeneous', we include in this term both heterogeneity that is due to random spatial variations of permeability (that manifests itself in the spreading referred to as 'macrodispersion'), and well defined permeability variations, e.g., in the form of layers of different permeability. The former always exist, even when a domain is declared homogeneous, while the latter may exist or not in a considered domain.

(A-25) Dissolved matter does not adsorb on the solid, and does not undergo any chemical reaction or decay.

5.4.2 Mathematical Model

Since we are considering here a single liquid, water, there is no need for a superscript to denote this fact.

a. Mass balance equation for the water phase

The mass balance for the water takes the form

$$
\frac{\partial \phi \rho}{\partial t} = -\nabla \cdot (\rho \mathbf{q}) + \sum_{(n)} \rho_{Rn} R_n(\mathbf{x}_n, t) \delta (\mathbf{x} - \mathbf{x}_n)
$$

$$
- \sum_{(m)} \rho P_m(\mathbf{x}_m, t) \delta (\mathbf{x} - \mathbf{x}_m) \tag{5.63}
$$

where all symbols have already been explained earlier.

The relationship between water density, pressure, and concentration, may take the form

$$\rho = \rho^o \exp\left\{1 + \beta_p(p - p^o) + \beta_c(c - c^o)\right\} \tag{5.64}$$

where c is the concentration of dissolved matter (TDS), ρ^o, p^o, and c^o are reference values of ρ, p, and c, respectively, $\beta_p = (1/\rho)\partial\rho/\partial p$ denotes the coefficient of water compressibility (at constant temperature and concentration), and $\beta_c = (1/\rho)\partial\rho/\partial c$ is a coefficient that introduces the effect of concentration changes on fluid density (at constant pressure and temperature). Another possible constitutive relation is the linearized approximation of Eq. 5.64

$$\rho = \rho^o \left\{1 + \beta'_p(p - p^o) + \beta'_c(c - c^o)\right\} \tag{5.65}$$

with $\beta'_p = (1/\rho^o)\partial\rho/\partial p$ and $\beta'_c = (1/\rho^o)\partial\rho/\partial c$. In what follows, we shall assume that for the range of pressures considered here, $\beta'_c\Delta c \gg \beta'_p\Delta p$, so that we may employ the approximation:

$$\rho = \rho^o \left\{1 + \beta'_c(c - c^o)\right\} \tag{5.66}$$

Actually, this is not a good approximation, as it does not take into account the effect of changes in concentration on the volume of the solution.

Making use of Eq. 5.64, we may rewrite the mass balance equation Eq. 5.63 in the form

$$\rho S_{op}\frac{\partial p}{\partial t} + \rho\phi\beta_c\frac{\partial c}{\partial t} = -\nabla\cdot(\rho\mathbf{q}) + \sum_{(n)}\rho_{Rn}R_n(\mathbf{x}_n, t)\delta(\mathbf{x} - \mathbf{x}_n)$$

$$-\sum_{(m)}\rho P_m(\mathbf{x}_m, t)\delta(\mathbf{x} - \mathbf{x}_m) \tag{5.67}$$

where the specific storativity with respect to pressure changes, S_{op}, is defined as $S_{op} = \phi\beta_p + \alpha(1 - \phi)$, with $\alpha = -\frac{1}{(1-\phi)}\frac{\partial(1-\phi)}{\partial p}$ denoting the coefficient of solid matrix compressibility.

b. Flux equation for the water

For a variable density fluid, Darcy's law takes the form

$$\mathbf{q} = -\frac{\mathbf{k}}{\mu}\cdot(\nabla p + \rho g\nabla z) \tag{5.68}$$

where **k** denotes the permeability tensor, and μ is the fluid's viscosity. The latter varies with the concentration, say, in the form:

$$\mu = \mu^o \{1 + \beta_\mu(c - c^o)\} \tag{5.69}$$

where μ^o denotes a reference dynamic viscosity, and β_μ is a coefficient that introduces the effect of concentration changes on the fluid's dynamic viscosity.

Obviously, it is possible to continue to refer to pressure, p, as the state variable describing liquid motion. However, it is often more convenient, primarily for numerical purposes to introduce a piezometric head, $h' = h'(x, y, z, t)$, as a variable associated with a reference density, ρ^o:

$$h' = z + \frac{p}{\rho^o g}, \qquad \frac{\partial p}{\partial t} = \rho^o g \frac{\partial h'}{\partial t}, \qquad \nabla p = \rho^o g \nabla(h' - z) \tag{5.70}$$

It is convenient to use the density of freshwater as reference density.

Rewritten in terms of h', Darcy's law takes the form:

$$\mathbf{q} = -\frac{\rho^o g \mathbf{k}}{\mu} \cdot [\nabla h' + \epsilon \nabla z] = -\frac{\rho^o g \mathbf{k}}{\mu} \cdot [\nabla h' + \beta_c'(c - c^o)\nabla z] \tag{5.71}$$

where $\epsilon = (\rho - \rho^o)/\rho^o = \beta_c'(c - c^o)$.

Introducing the (constant) hydraulic conductivity, $\mathbf{K}^o = \rho^o g \mathbf{k}/\mu^o$, related to the reference density and the reference dynamic viscosity, and expressing the effect of concentration on viscosity by Eq. 5.69, we may rewrite Darcy's law in the form

$$\mathbf{q} = -\frac{\mathbf{K}^o}{1 + \beta_\mu(c - c^o)} \cdot [\nabla h' + \beta_c'(c - c^o)\nabla z] \tag{5.72}$$

c. Flow equation for the water

By inserting the flux equation into the mass balance equation, we obtain the flow equation in the form

$$\rho S_{op} \frac{\partial p}{\partial t} + \rho \phi \beta_c \frac{\partial c}{\partial t} = \nabla \cdot \left[\rho \frac{\mathbf{k}}{\mu} \cdot (\nabla p + \rho g \nabla z) \right]$$
$$+ \sum_{(n)} \rho_{Rn} R_n(\mathbf{x}_n, t) \delta(\mathbf{x} - \mathbf{x}_n) - \sum_{(m)} \rho P_m(\mathbf{x}_m, t) \delta(\mathbf{x} - \mathbf{x}_m) \tag{5.73}$$

Or, in terms of the (referenced) piezometric head, h', and the (constant) hydraulic conductivity, $\mathbf{K}^o$ $(= \rho^o g \mathbf{k}/\mu^o)$,

$$S_o^o \frac{\partial h'}{\partial t} + \phi \beta_c \frac{\partial c}{\partial t} =$$

$$\frac{1}{\rho} \nabla \cdot \left[\mathbf{K}^o \cdot \left(\frac{\rho}{1 + \beta_\mu (c - c^o)} \right) (\nabla h' + \beta_c' (c - c^o) \nabla z) \right]$$

$$+ \sum_{(n)} \frac{\rho_{Rn}}{\rho} R_n(\mathbf{x}_n, t) \delta (\mathbf{x} - \mathbf{x}_n) - \sum_{(m)} P_m(\mathbf{x}_m, t) \delta (\mathbf{x} - \mathbf{x}_m) \tag{5.74}$$

where $S_o^o = \rho^o g S_{op}$.

In a model of seawater intrusion, where concentrations are restricted to the range c^o–c_s, with c_s denoting the concentration of seawater, it may be convenient to introduce a normalized, or relative concentration $c^* = (c - c^o)/(c_s - c^o)$, which varies in the range 0–1.

Equation 5.74 contains the two variables of the problem: $h'(x, y, z, t)$ and $c(x, y, z, t)$, but we still have a third variable, ρ, which is defined by Eq. 5.66.

Actually, Darcy's law expresses the *specific discharge relative to the (possibly moving at velocity V_s) solid*, $\mathbf{q}_r (= \phi(\mathbf{V} - \mathbf{V}_s))$, rather than $\mathbf{q}(= \phi \mathbf{V})$ as written above. We shall continue to overlook this discrepancy. It may be corrected by defining the specific storativity as $S_{op} = \phi \beta_p + \alpha$ (see Bear and Verruijt [1987], p. 63).

d. Mass balance equation for the dissolved matter (TDS)

The mass balance for the dissolved matter (salt) can be expressed in the form:

$$\frac{\partial \phi c}{\partial t} = -\nabla \cdot (c\mathbf{q} - \phi \mathbf{D}_h \cdot \nabla c)$$

$$+ \sum_{(n)} c_{Rn} R_n(\mathbf{x}_n, t) \delta (\mathbf{x} - \mathbf{x}_n) - \sum_{(m)} c P_m(\mathbf{x}_m, t) \delta (\mathbf{x} - \mathbf{x}_m) \tag{5.75}$$

where $\mathbf{D}_h$ (components D_{hij}) denotes the coefficient of hydrodynamic dispersion, and c_{Rn} is the concentration of the water recharged through a well at point $\mathbf{x}_m$. Recall that here 'dispersion' means 'macrodispersion', or 'field scale dispersion'.

The coefficient of hydrodynamic dispersion, $\mathbf{D}_h$, is the sum of the coefficient of mechanical dispersion, $\mathbf{D}$, and the coefficient of molecular diffusion in a porous medium, $\mathcal{D}^*$. In terms of components, we write

$$D_{ij} = a_{ijkm}\frac{V_k V_m}{V} + \mathcal{D}_{ij}^* \tag{5.76}$$

where the a_{ijkm}s are components of the (fourth rank) dispersivity tensor. For *an isotropic porous medium*:

$$D_{hij} = a_T V \delta_{ij} + (a_L - a_T)\frac{V_i V_j}{V} + \mathcal{D}^* \delta_{ij} \tag{5.77}$$

where a_L and a_T are the longitudinal and the transversal dispersivities (= macrodispersivities), respectively, and δ_{ij} denotes the Kronecker Delta.

However, since we are dealing here with a field-scale problem, where dispersion is controlled primarily by the scale of heterogeneity of the aquifer, due to permeability variations, we may often neglect the effect of molecular diffusion, compared to that of dispersion (except, perhaps, in zones where velocities are very small).

By combining Eq. 5.63 with Eq. 5.75, neglecting molecular diffusion and the effect of pressure variations on ρ, we obtain:

$$(1 - \beta_c c)\left(\phi\frac{\partial c}{\partial t} + \mathbf{q}\cdot\nabla c\right) = \nabla\cdot(\phi\mathbf{D}\cdot\nabla c)$$
$$+ \sum_{(n)} R_n\left(c_{Rn} - c\frac{\rho_{Rn}}{\rho}\right) \tag{5.78}$$

Altogether, we have to solve Eqs. 5.74 and 5.78 for h' and c.

Note that in comparison with the density-independent flow equation, we have here two additional terms that describe the coupling between flow and transport: $\mathbf{K}\beta_c(c - c^o)\cdot\nabla z$ and $\phi\beta_c\frac{\partial c}{\partial t}$ (in addition to the effects of the fluid's variable density on some parameters). The former expresses natural convection caused by the density difference along the vertical, thus resulting in recirculation within the transition zone. The latter expresses the change in the mass of the dissolved matter with time, due to the change in concentration. The transport and the flow equations are also coupled by the velocity, which appears as part of the coefficient of hydrodynamic dispersion. The two equations are, thus, nonlinear and coupled, and must be solved simultaneously.

5.4.2.1   Initial conditions

The initial conditions for the two partial differential equations of balance, are:

$$h'(x,y,z,O) = h'_o(x,y,z), \quad c(x,y,z,O) = c_o(x,y,z)$$

$$(5.79)$$

in which $h'_o(x,y,z)$ and $c_o(x,y,z)$ are the known initial piezometric head and the known initial concentration within the considered domain.

5.4.2.2   Boundary conditions for the flow equation

a. Impervious bottom AB

$$q_n = \mathbf{q} \cdot \mathbf{n} = 0 \qquad (5.80)$$

where the flux $\mathbf{q}$ is expressed by Eq. 5.72.

b. The land-side boundary BC

When this boundary is within the zone occupied by freshwater, and will remain there, the piezometric head there may be defined as $h^f = z + p/\rho^f g$ (recalling that we cannot define a piezometric head in a zone of variable density). Let this head be a constant, equal to the elevation of the phreatic surface at point B, H_{ph}. From the definitions of h^f, and h', it follows that along this boundary

$$h'(x,y,z) = H_{ph}(x,y)\frac{\rho^f}{\rho^o} - z\frac{\rho^f - \rho^o}{\rho^o} \qquad (5.81)$$

where z denotes the elevation of a point on this surface. Instead of a constant, H_{ph}, we could select any known function, $H_{BC}(x,y,z,t)$. For $\rho^f = \rho^o$, the above condition will then reduce to

$$h' = H_{BC}(x,y,z,t) \qquad (5.82)$$

In the case of specified flux, the boundary condition will be

$$q_n = \mathbf{q} \cdot \mathbf{n} = \mathbf{q}_{ext} \cdot \mathbf{n} \qquad (5.83)$$

where the flux $\mathbf{q}$ is expressed by an appropriate form of Darcy's law. For $\rho^f = \rho^o$, this condition will reduce to

$$-\mathbf{K}^o \cdot \nabla(h') \cdot \mathbf{n} = 0 \qquad (5.84)$$

c. The phreatic surface CD

The phreatic surface, as a boundary of a flow domain, has already been discussed above. That discussion, concerning the shape of the phreatic surface, and the condition to be satisfied on it, are valid also here. Thus, Eq. 5.22 is valid, with h^f replaced by h', and the condition of continuity of mass flux across this boundary takes the form:

$$\rho\,(\mathbf{q} - \phi\mathbf{u}) \cdot \mathbf{n} = \rho_N\,(\mathbf{N} - \theta_{wr}\mathbf{u}) \cdot \mathbf{n} \tag{5.85}$$

or:

$$(\rho_N\mathbf{N} - \rho\mathbf{q})\,\nabla(h - z) = (\phi\rho - \theta_{wr}\rho_N)\,\frac{\partial h'}{\partial t} \tag{5.86}$$

In this equation, we have now to insert an appropriate expression for $\mathbf{q}$, and for $\mathbf{N}$. If we assume that *at the phreatic surface*, $\rho \approx \rho_N$, we may rewrite the last equation in the form:

$$(\mathbf{N} - \mathbf{q}) \cdot \nabla(h' - z) = S_y\frac{\partial h'}{\partial t} \tag{5.87}$$

which is identical to Eq. 5.24.

We recall that we are here in a 'vicious circle': to solve for h' we need to know F_{ph}, but we cannot know F_{ph} until we have solved the problem for h'.

d. The seepage face DE

Here water emerges to the atmosphere (assuming $p_{atm} = 0$), and the condition is

$$h'(x, y, z, t) = z, \quad z = \zeta_{seep} \tag{5.88}$$

where $\zeta_{seep}(x, y)$ denotes the elevations of points on the seepage face.

e. The sea bottom EF

Along this boundary segment (of known geometry), the pressure is dictated by the elevation of the sea level. In terms of h', the condition is

$$h'(x, y, z, t) = H_{sea}(t)\frac{\rho^s}{\rho^o} - \zeta_{bot}(x, y)\frac{\rho^s - \rho^o}{\rho^o} \tag{5.89}$$

f. The sea-side boundary FA

Here we assume that seawater is always present on the external side of the boundary, so that the condition is of specified head:

$$h'(x, y, z, t) = H_{sea}(t)\frac{\rho^s}{\rho^o} - z\frac{\rho^s - \rho^o}{\rho^o} \quad (5.90)$$

where z denotes the elevation of points on this surface.

5.4.2.3   Boundary conditions for the solute transport equation.

a. Impervious bottom AB

Along this boundary, $\mathbf{q} \cdot \mathbf{n} = 0$, and the total solute flux is zero. The condition is, therefore,

$$\mathbf{D}_h \cdot \nabla c = 0, \quad D_{hij}\frac{\partial c}{\partial x_j} n_i = 0 \quad (5.91)$$

b. The land-side boundary BC

We assume that this boundary is sufficiently far from the area of seawater intrusion, such that the concentration there remains that of freshwater (which is also the initial condition there):

$$c = c_{fresh} \quad (5.92)$$

c. The phreatic surface CD

This condition is the same as Eq. 5.86, but with c replacing ρ, and adding the dispersive flux:

$$\{c_N\mathbf{N} - c\mathbf{q} + \phi\mathbf{D} \cdot \nabla c\} \nabla(h' - z) = (\phi c - \theta_{wr}c_N)\frac{\partial h'}{\partial t} \quad (5.93)$$

d. The seepage face DE

We assume that as water emerges to the atmosphere through the seepage face, its concentration remains unchanged. By equating the total flux on both sides, we obtain the condition:

$$(\mathbf{D} \cdot \nabla c) \cdot \mathbf{n} = 0 \quad (5.94)$$

where $\mathbf{n}$ is obtained from the equation of the seepage face

e. The sea bottom EF

We assume that the sea is a *well mixed zone*, at the concentration c_s, with a 'buffer zone' between the aquifer and the sea (see Bear and Bachmat [1990], p. 257). The condition of no-jump in the normal component of the total flux takes the form:

$$c_s \mathbf{q} \cdot \mathbf{n} + \alpha^*(c_s - c) = (c\mathbf{q} - \phi \mathbf{D} \cdot \nabla c) \cdot \mathbf{n} \qquad (5.95)$$

where α^* plays the role of a transfer coefficient, which combines the effects of dispersion and diffusion through the buffer zone. We may rewrite this condition in the form

$$(\mathbf{q} \cdot \mathbf{n} + \alpha^*)(c_s - c) = -\phi(\mathbf{D} \cdot \nabla c) \cdot \mathbf{n} \qquad (5.96)$$

This condition remains valid for both seaward and land-ward flow through this boundary. However, the introduction of the buffer zone, introduces also another coefficient to be estimated in the calibration process.

Freshwater drains to the sea through one part, while seawater enters the aquifer through another part. The two parts are separated by a point of zero flux.

If we assume that on the sea-side of the boundary segment through which water leaves the aquifer, the concentration is c rather than c_s, then the condition of flux equality reduces to

$$(\mathbf{D} \cdot \nabla c) \cdot \mathbf{n} = 0 \qquad (5.97)$$

For the inflow segment, we obtain Eq. 5.96 without the buffer zone, i.e.,

$$\{\mathbf{q}(c_s - c) + \phi(\mathbf{D} \cdot \nabla c\} \cdot \mathbf{n} = 0 \qquad (5.98)$$

f. The sea-side boundary FA

Analogous to the land-side boundary, the condition Eq. 5.92, and since this boundary is assumed to be located sufficiently far from the transition zone, we have there

$$c = c_s \qquad (5.99)$$

5.5   Conclusion

We have thus achieved our objective of presenting the complete models that can be used for predicting seawater intrusion into coastal aquifers, in response to natural replenishment pumping and artificial recharge operations:

- The 3-D sharp interface model (3DSIM).
- The 2-D sharp interface model (3DSIM).
- The 3-D transition zone model (3DTZM).

We have tried to state all the assumptions that underlie the construction of each of these models. As we have already emphasized in the introductory remarks, with the computational tools available today, there is no reason to use models based on the sharp interface approximation. A number of such models and their numerical solutions are presented in this volume.

Chapter 6

Analytical Solutions

A. H.-D. Cheng & D. Ouazar

6.1 Introduction

Due to their simplified physical assumptions and geometry, analytical solutions normally do not directly solve "real-world" problems. Nevertheless, they serve a number of important purposes. First, they are useful as instructional tools. For a hydrogeologist investigating saltwater intrusion it is imperative that he attains a clear understanding of the mechanical trend of the flow. Elegant analytical solutions are most useful in presenting such fundamental insights, while numerical solutions are often not. In fact, a person without such physical insights should not be entrusted with a powerful numerical tool to solve complicated problems, as such a person can have blind spots that harbor catastrophic consequences.

Despite their simplifying assumptions, analytical solutions can also be used as a tool for first-cut engineering analysis in a feasibility study. More sophisticated models normally require hydrological and hydrogeological information that is either not available or not known to the required resolution at the time of initial investigation. A sophisticated model without the support of reliable input data does not provide more accurate result. In fact, before a large scale site investigation, or a comprehensive numerical modeling, it is often a good idea to perform some rudimentary calculations based on basic information and simplifying assumptions.

Another contribution of analytical solutions is to serve as benchmark problems for testing numerical algorithms. Numerical meth-

J. Bear et al. (eds.), Seawater Intrusion in Coastal Aquifers, 163–191.

ods, especially newly developed ones, need to be validated against programming and other errors before put into use.

With these uses in mind, we present in this chapter a few analytical solutions that are of historical and practical importance. To stay in focus, this chapter is not a comprehensive compilation of existing analytical solutions. In fact, only analytical solutions that are simple enough to be used as instructional tools to illustrate the physical concepts are included. The more complicated problems can always be dealt with using numerical solution.

6.2   Ghyben-Herzberg Solution

More than a century ago Badon-Ghyben [1888] and later Herzberg [1901] independently found that saltwater occurred underground, not at sea level as expected for static water bodies, but at a depth below sea level of about forty times the freshwater head above sea level. This relation, known as Ghyben-Herzberg relation, has been widely used.

We define the piezometric head in the saltwater zone and freshwater zone respectively as

$$h_s = \frac{p}{\rho_s g} + z \qquad (6.1)$$

$$h_f = \frac{p}{\rho_f g} + z \qquad (6.2)$$

where p is the pore pressure, ρ_s and ρ_f are the saltwater and freshwater density, and g is gravity acceleration. As understood by Muskat [1937] in the form of oil/water interface and later by Hubbert [1940], the dynamic equilibrium requires that the pressure p be continuous across the saltwater/freshwater interface. The pressure can be eliminated between Eqs. 6.1 and 6.2 to give the condition

$$\rho_f h_f = \rho_s h_s + (\rho_s - \rho_f)\xi \qquad (6.3)$$

in which $\xi = \xi(x, y) = -z$ is the depth of interface below the datum (see Figure 6.1). In solving a boundary value problem involving a freshwater zone and a saltwater zone, Eq. 6.3 is a condition that must be satisfied at the interface.

For aquifers that are relatively shallow such that flowlines are nearly horizontal, it is customary to invoke the Dupuit assumption

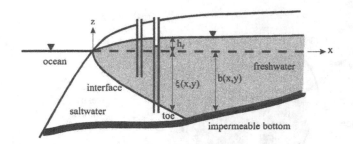

Figure 6.1. Saltwater-freshwater interface.

[Bear, 1979] which assumes that h_s and h_f does not change in the vertical direction, z. Equation 6.3 is no longer restricted to head measured at the interface and can be written as

$$\xi = \frac{\rho_f}{\rho_s - \rho_f} h_f - \frac{\rho_s}{\rho_s - \rho_f} h_s \qquad (6.4)$$

This offers the opportunity of predicting the interface depth ξ from water level in observation wells. Referring to Figure 6.1, if two wells are located near to each other, but one is deep enough to be opens in the saltwater region and the other shallow enough to be in the freshwater region, then h_s and h_f can be observed at the same location. Equation 6.4, as suggested by Lusczynski [1961], provides a better estimate of ξ as compared to the Ghyben-Herzberg relation to be introduced below. However, despite the advantage, this method is less practical as it requires extra wells that penetrate deeper into the saltwater region. This requires the rough knowledge of interface location for the proper placement of well screen, and higher cost.

Ghyben and Herzberg made the further assumption that the saltwater is stagnant. The pressure in the saltwater region becomes hydrostatic, $p = -\rho_s g z$. With the datum shown in Figure 6.1, the saltwater head is zero and Eq. 6.4 reduces to

$$\xi = \frac{\rho_f}{\rho_s - \rho_f} h_f \approx 40 \, h_f \qquad (6.5)$$

'In the above and throughout the chapter we adopt these values: $\rho_f = 1.000$ g/cm^3 and $\rho_s = 1.025$ g/cm^3. Equation 6.5 is known as the Ghyben-Herzberg relation. The prediction of interface location requires only the water level in freshwater wells.

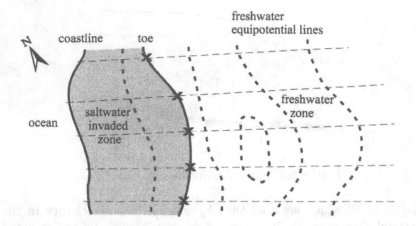

Figure 6.2. Tracking saltwater toe location by Ghyben-Herzberg relation.

With the Ghyben-Herzberg relation, it is possible to map the saltwater-freshwater interface using the following steps in a field investigation:

1. A network of shallow wells are deployed for observation of water table height.

2. Contour lines showing the freshwater head h_f are drawn using an interpolation technique, such as kriging.

3. The interface depth $\xi(x, y)$ is represented by the same set of contour lines, but with amplified contour values according to Eq. 6.5.

4. The location of aquifer bottom $b(x, y)$ (see Figure 6.1) is found from geological maps.

5. The intersection of the two surfaces $\xi(x, y)$ and $b(x, y)$ is sought, which represents the saltwater toe location. Instead of finding the intersection in a three-dimensional space, it might be more efficient to find it in a number of two-dimensional vertical sections cut along lines perpendicular to the coast and connect these points together, see Figure 6.2.

The above procedure is applicable for confined as well as unconfined aquifers.

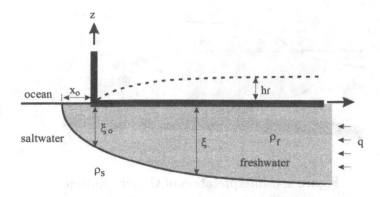

Figure 6.3. The Glover problem.

6.3   Glover Solution

The Glover [1964] problem has a geometry as shown in Figure 6.3. The aquifer is homogeneous, confined at the top, and unbounded at the bottom. The saltwater is stagnant. The steady-state freshwater head satisfies the two-dimensional Laplace equation in the vertical plane

$$\nabla^2 h_f = 0 \qquad (6.6)$$

The freshwater head is found as

$$h_f = \sqrt{\frac{\Delta s\, q}{K}} \sqrt{x + \sqrt{x^2 + z^2}} \qquad (6.7)$$

where q is the freshwater outflow rate per unit length of coastline, K is hydraulic conductivity, and

$$\Delta s = \frac{\rho_s - \rho_f}{\rho_f} = \frac{\Delta \rho}{\rho_f} \qquad (6.8)$$

is the difference between seawater and freshwater specific gravity. The interface depth is given by

$$\xi(x) = \sqrt{\frac{2\,q\,x}{\Delta s\, K} + \frac{q^2}{\Delta s^2 K^2}} \qquad (6.9)$$

Substituting Eqs. 6.9 and 6.7 into the interface dynamic condition Eq. 6.3, it can be shown that $h_s = 0$ at $z = -\xi$, confirming the statement of stagnant saltwater.

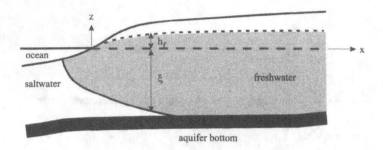

Figure 6.4. Interpretation of Glover solution.

The size of the outflow face as shown in Figure 6.3 are found from Eq. 6.9 and

$$x_o = \frac{q}{2\Delta sK} \qquad (6.10)$$

$$\xi_o = \frac{q}{\Delta sK} \qquad (6.11)$$

The freshwater head at $z = 0$ is

$$h_f = \sqrt{\frac{2\Delta s\,q\,x}{K}} \qquad (6.12)$$

We observe that the solution geometry in Figure 6.3 is unrealistic. To make the solution more useful, the geometry is usually interpreted to a more practical configuration as shown in Figure 6.4. The aquifer is changed form confined at the top to unconfined with a freshwater head given by Eq. 6.12. The outflow face terminates at a certain depth under the sea. An impermeable bottom exist that intersects the interface to form a saltwater wedge that terminates at a certain distance inland. These conditions are in violation with the exact mathematical statement that defines the solution. However, these discrepancies are expected to be small.

The Glover solution can be further simplified by using the Dupuit assumption. In such case, the freshwater flow equation becomes one-dimensional. Such a solution was given by Todd [1953] with the interface depth given as

$$\xi(x) = \sqrt{\frac{2\,q\,x}{\Delta s\,K}} \qquad (6.13)$$

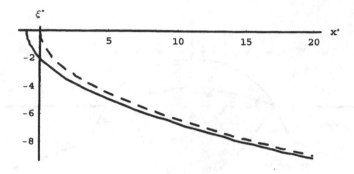

Figure 6.5. Comparison of Glover solution (solid line) and Todd (dashed line) solution.

Comparison with Eq. 6.9 shows that the Dupuit assumption does not allow an outflow face. In Figure 6.5 we plot the two solutions using normalized spatial coordinates of $x^* = 2\Delta s K x/q$ and $\xi^* = 2\Delta s K \xi/q$. We observe that the difference between the two solutions are small after about 5 to 10 distances of the outflow face. This provides the justification of using the Dupuit assumption.

6.4   Fetter Oceanic Island Solution

Consider an circular oceanic island shown in Figure 6.6. Supported by a constant surface recharge w, a freshwater lens floats on top of saltwater. Such solution is provided by Fetter [1972] by taking the Dupuit assumption for freshwater flow, and Ghyben-Herzberg relation for the interface location. The governing equation for freshwater flow is one-dimensional in the radial direction and the solution of freshwater head is

$$h_f^2 = \frac{w \Delta s \left(R^2 - r^2\right)}{2K \left(1 + \Delta s\right)} \tag{6.14}$$

where R is the radius of the island. The saltwater/freshwater interface depth is then given by Eq. 6.5.

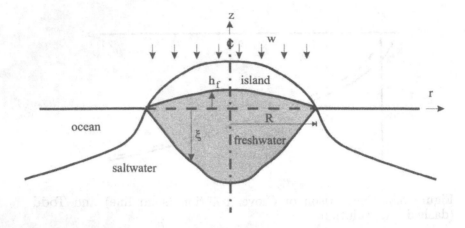

Figure 6.6. Circular island.

6.5 Strack Pumping Well Solution

The previous solutions deal with the situation where freshwater exists on top of saltwater. Referring to Figures 6.7 and 6.8, this zone will be designated as zone 2. Inland of the saltwater toe, there exist a freshwater only zone, designated as zone 1. The governing equations for these two zones are different. For a flow domain that covers both zones, this can cause a slight inconvenience in deriving the mathematical solution. To achieve a more efficient solution algorithm, Strack [1976] proposed a single potential for the two zones. The potential and it derivative are continuous across the zones such that only a single governing equation is needed.

In the formulation, the Dupuit assumption is applied to the freshwater flow, and the Ghyben-Herzberg relation is utilized to define the interface depth. The freshwater head is a function of horizontal coordinates only, $h_f = h_f(x, y)$. Assuming that the aquifer is homogeneous, a general governing equation which is valid for both zone 1 and zone 2, and for confined as well as unconfined aquifers, can be written:

$$\nabla \cdot (b\nabla h_f) = 0 \qquad (6.15)$$

where $b = b(x, y)$ is the thickness of freshwater layer shown in Figures 6.7 and 6.8. We note that the thickness corresponds to different

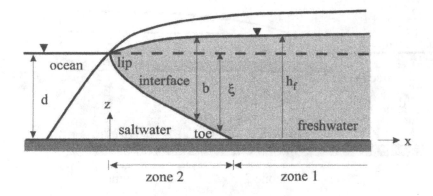

Figure 6.7. An unconfined aquifer.

quantities under different conditions:

$$
\begin{aligned}
b &= h_f & &\text{unconfined aquifer, zone 1} \\
&= h_f - d + \xi & &\text{unconfined aquifer, zone 2} \\
&= B & &\text{confined aquifer, zone 1} \\
&= \xi - d + B & &\text{confined aquifer, zone 2} \quad (6.16)
\end{aligned}
$$

where B is the confined aquifer thickness which is a constant, and d is the ocean depth above the aquifer bottom. In zone 2, the Ghyben-Herzberg relation in the current notations is written as:

$$ h_f - d = \Delta s\, \xi \qquad (6.17) $$

Following Strack [1976] approach, the following potential functions are defined (see also Taigbenu et al. [1984]), for unconfined aquifer:

$$
\begin{aligned}
\phi &= \frac{1}{2}\left[h_f^2 - (1 + \Delta s)\, d^2\right] & &\text{zone 1} \\
&= \frac{(1 + \Delta s)}{2\Delta s}(h_f - d)^2 & &\text{zone 2} \quad (6.18)
\end{aligned}
$$

and for confined aquifer:

$$
\begin{aligned}
\phi &= B\, h_f + \frac{\Delta s\, B^2}{2} - (1 + \Delta s)\, B\, d & &\text{zone 1} \\
&= \frac{1}{2\Delta s}\left[h_f + \Delta s\, B - (1 + \Delta s)\, d\right]^2 & &\text{zone 2} \quad (6.19)
\end{aligned}
$$

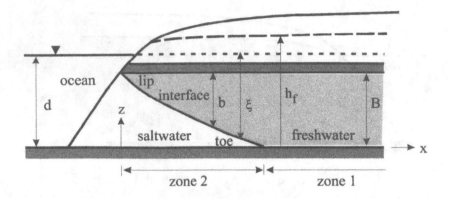

Figure 6.8. A confined aquifer.

We note these functions and their first derivatives are continuous across the zonal interface.

At the interface, i.e. at the "toe" of the saltwater wedge (Figures 6.7 and 6.8), the potentials take these values, for unconfined aquifer:

$$\phi_{\text{toe}} = \frac{\Delta s(1 + \Delta s)}{2} d^2 \qquad (6.20)$$

and for confined aquifer:

$$\phi_{\text{toe}} = \frac{\Delta s}{2} B^2 \qquad (6.21)$$

With these properties, the potentials defined in Eqs. 6.18 and 6.19 satisfy the same Laplace equation in two horizontal spatial dimensions

$$\nabla^2 \phi = 0 \qquad (6.22)$$

At the freshwater outflow lip, $\xi = 0$, we observe the boundary condition

$$\phi = 0 \qquad (6.23)$$

for both the confined and unconfined aquifer. With other appropriate boundary conditions, an elliptic boundary values problem can be formed based on Eq. 6.22. Once the solution is found, the location of the toe can be tracked using Eqs. 6.20 and 6.21.

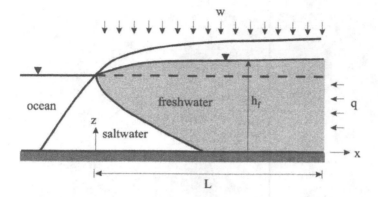

Figure 6.9. Tracking saltwater toe in a one-dimensional problem.

As an example, consider a vertical cross-section of an unconfined aquifer with a constant surface recharge w as shown in Figure 6.9. The governing equation is

$$\frac{d^2\phi}{dx^2} = -\frac{w}{K} \qquad (6.24)$$

The boundary conditions are:

$$\phi = 0 \qquad \text{at } x = 0$$
$$\frac{d\phi}{dx} = \frac{q}{K} \qquad \text{at } x = L \qquad (6.25)$$

where q is the freshwater volume outflow rate per unit length of coastline. Solution of the above is

$$\phi = -\frac{w}{2K}x^2 + \left(\frac{q}{K} + \frac{wL}{K}\right)x \qquad (6.26)$$

Substituting the potential value defined in Eq. 6.20 into the above, the toe location can be solved

$$x_{\text{toe}} = \frac{q}{w} + L - \sqrt{(\frac{q}{w} + L)^2 - 2\frac{\phi_{\text{toe}}K}{w}} \qquad (6.27)$$

where ϕ_{toe} is as defined in Eq. 6.20. For the special case that the recharge rate is zero, $w = 0$, the toe location is given by

$$x_{\text{toe}} = \frac{\phi_{\text{toe}}K}{q} \qquad (6.28)$$

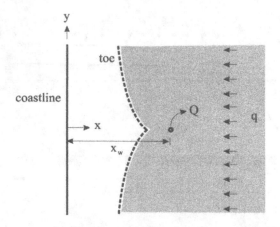

Figure 6.10. Pumping well in a coastal aquifer.

Since no recharge is involved, the solution can be interpreted for both confined and unconfined aquifers.

In the second problem, we consider a two-dimensional geometry of a horizontal coastal plain with a straight coastline (Figure 6.10). A pumping well with discharge Q_w is located at a distance x_w from the coast. There also exists a uniform freshwater outflow of rate q. The aquifer can be either confined or unconfined. Solution of this problem can be found by the method of images and is given by Strack [1976]

$$\phi = \frac{q}{K} x + \frac{Q_w}{4\pi K} \ln \left[\frac{(x - x_w)^2 + y^2}{(x + x_w)^2 + y^2} \right] \tag{6.29}$$

The toe location x_{toe} can be solved from

$$\phi_{\text{toe}} = \frac{q}{K} x_{\text{toe}} + \frac{Q_w}{4\pi K} \ln \left[\frac{(x_{\text{toe}} - x_w)^2 + y^2}{(x_{\text{toe}} + x_w)^2 + y^2} \right] \tag{6.30}$$

where ϕ_{toe} is as defined in Eqs. 6.20 and 6.21. To facilitate the presentation of result, we introduce the following normalization

$$X = \frac{x}{x_w} \tag{6.31}$$

$$Y = \frac{y}{x_w} \tag{6.32}$$

$$X_{toe} = \frac{x_{toe}}{x_w} \tag{6.33}$$

$$\Phi_{toe} = \frac{\phi_{toe}\,K}{q\,x_w} \tag{6.34}$$

$$Q^* = \frac{Q_w}{q\,x_w} \tag{6.35}$$

Equation 6.30 then becomes

$$\Phi_{toe} = X_{toe} + \frac{Q^*}{4\pi}\,\ln\left[\frac{(X_{toe}-1)^2 + Y^2}{(X_{toe}+1)^2 + Y^2}\right] \tag{6.36}$$

The physical meaning of X, Y and X_{toe} is quite clear. To understand the meaning of Φ_{toe} we notice the following. If the pumping well does not exist, $Q_w = 0$, the saltwater would invade a constant distance of (see Eq. 6.28)

$$x_{min} = x_{toe}\,|_{Q_w=0} = \frac{\phi_{toe}K}{q} \tag{6.37}$$

It is clear that Φ_{toe} is the ratio of minimum invasion distance x_{min} (corresponding to $Q_w = 0$) to the well location x_w. The range of interest for our investigation of Φ_{toe} is $0 \le \Phi_{toe} \le 1$, because for $\Phi_{toe} > 1$, the well would be invaded even without the pumping. The quantity Q^* is the ratio of well discharge to the freshwater outflow in one width of x_w. In other words, the pumping well sweeps an equivalent width of Q^*x_w of seaward freshwater outflow.

Equation 6.36 can be rewritten into the following form:

$$Y = \sqrt{-X_{toe}^2 + 2\left(\frac{e^{4\pi X_{toe}/Q^*} + e^{4\pi \Phi_{toe}/Q^*}}{e^{4\pi X_{toe}/Q^*} - e^{4\pi \Phi_{toe}/Q^*}}\right)X_{toe} - 1} \tag{6.38}$$

The trajectory of (X_{toe}, Y) defines the toe location. In Figure 6.11 we plot the toe location in the X–Y (dimensionless) plane for $\Phi_{toe} = 0.3$ and various Q^* values, $Q^* = 0$, 0.5, 1.0, 1.3, 1.4546 and 1.5. Due to symmetry, only the upper plane $Y \ge 0$ is shown. We note that the coastline is located at $X = 0$, and the well is situated at $(1,0)$. For the case $Q^* = 0$, the toe location is plotted in dashed line, which is located at $X = 0.3$ ($= \Phi_{toe}$). As pumping rate increases, the toe moves inland toward the well. It is of interest to observe that

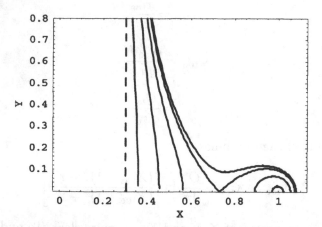

Figure 6.11. Saltwater toe locations, from left to right corresponding to pumping rate $Q^* = 0$ (dashed line), 0.5, 1.0, 1.3, 1.4546 and 1.5 (solid lines), and $\Phi_{\text{toe}} = 0.3$ for all cases.

surrounding the well there exist a region where the potential is low enough, due to the cone of depression, such that saltwater can exist. However, if this region is isolated from the sea, there is no pathway for the saltwater to intrude and the region is occupied by freshwater. As the pumping rate increases to reach a critical level, which for the present case is $Q^* = Q_c^* = 1.4546$, this region is in point contact with the seaward intruded region. An unstable situation develops as a slight increase of pumping causes a jump of the toe position and it reaches behind the well. This is observed in the case of $Q^* = 1.5$.

It is of interest to find the critical well discharge Q_c^* beyond which the well is invaded. This result has been obtained by Strack [1976] as,

$$\Phi_{\text{toe}} = \sqrt{1 - \frac{Q_c^*}{\pi}} + \frac{Q_c^*}{2\pi} \ln \left(\frac{1 - \sqrt{1 - \frac{Q_c^*}{\pi}}}{1 + \sqrt{1 - \frac{Q_c^*}{\pi}}} \right) \qquad (6.39)$$

Given a Φ_{toe} value, as defined in Eqs. 6.20, 6.21 and 6.34, Q_c^* can be solved from the above. Corresponding to a critical well discharge Q_c^*, there is a critical maximum intrusion distance of

$$X_{\max} = \sqrt{1 - \frac{Q_c^*}{\pi}} \qquad (6.40)$$

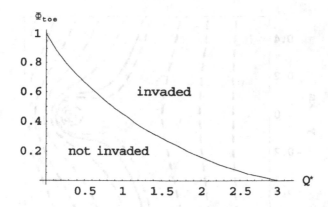

Figure 6.12. Critical intrusion condition as Φ_{toe} versus Q_c^*.

at $Y = 0$.

In Figure 6.12 we plot Eq. 6.39 in the Q^*–Φ_{toe} plane. The non-dimensionalization enables a universal design chart for all parameter values. For each case under investigation, given the input parameters Q_w, q, x_w, K, Δs, d (unconfined aquifer) and B (confined), the dimensionless pair (Q^*, Φ_{toe}) can be evaluated. If the point (Q^*, Φ_{toe}) falls under the (Q_c^*, Φ_{toe}) curve, the well will not be invaded. Otherwise it will be invaded. This chart can be used as a first-order estimate of allowable pumping rate in coastal aquifers.

6.6   Superposition Solution

The governing equation 6.22 and the geometry of a semi-infinite coastal plain bounded by a straight coastline allows the solution by superposition and the method of images. Equation 6.29 is a super-position of a uniform flow and a pumping well with an inverse mirror image to satisfy the boundary condition at the coast. For multiple pumping wells, Eq. 6.29 can be extended to

$$\phi = \frac{q}{K} x + \sum_{i=1}^{n} \frac{Q_i}{4\pi K} \ln \left[\frac{(x - x_i)^2 + (y - y_i)^2}{(x + x_i)^2 + (y - y_i)^2} \right] \quad (6.41)$$

where (x_i, y_i) are well coordinates, and Q_i is the pumping rate of well i. Q_i can be taken as positive or negative to represent pumping and recharging wells, respectively.

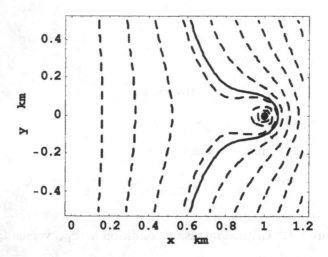

Figure 6.13. Saltwater intrusion with a single pumping well.

One method of preventing saltwater intrusion is to recharge fresh-water to form a local mound that serves as a barrier against intrusion. We can demonstrate the effect by placing a recharge well in front of a pumping well. Consider an unconfined aquifer with the following data: $K = 100$ m/day, $q = 0.6$ m^3/m·day, $d = 14$ m, $\Delta s = 0.025$, and a pumping well located at $(x_1, y_1) = (1000$ m, 0 m$)$ with a dis-charge $Q_1 = 720$ m^3/day. Figure 6.13 plots the freshwater potential ϕ as evaluated from Eq. 6.41 in equipotential lines (in dashed lines). The toe location is defined by the equipotential line of the value $\phi_{toe} = 2.51$ m^2, which is marked by the solid line in Figure 6.13. Using either Eq. 6.39 or Figure 6.12, we find the critical pumping rate to be $Q_c = 641$ m^3/day. The well is expected to be invaded. As observed from Figure 6.13, it is indeed so.

We next add a recharge well at $(x_2, y_2) = (500$ m, 0 m$)$. For the two case of $Q_2 = -120$ m^3/day and -180 m^3/day we plot respectively the toe location (solid lines) in the upper and lower diagrams of Fig-ure 6.14. We observe that in both cases the pumping well is protected from saltwater toe encroachment, although in the first case the salt-water front is about to break through and there is saltwater under the recharge well. The net gain of freshwater is found by summing the discharge of the wells. We obtain 600 m^3/day and 540 m^3/day respectively. In both cases we find that the efficiency of freshwater

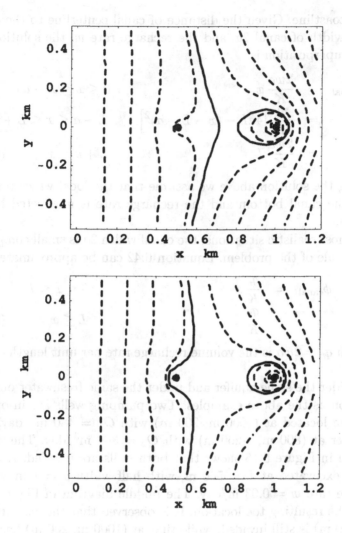

Figure 6.14. Location of saltwater toe caused by a pumping well at (1000 m, 0 m) with $Q_1 = 720$ m^3/day and a recharge well at (500 m, 0 m) with $Q_2 = -120$ m^3/day (upper figure) and -180 m^3/day (lower figure).

extraction is less than that of a single pumping well extracting at just below the critical rate, $Q_c = 641$ m^3/day. However, it is a common practice to use lower quality water for recharge. Hence the net gain of high quality groundwater is positive.

In the next example we consider a recharge canal that runs parallel to the coastline. Given the distance of canal centerline to the coast L, the width of canal $2a$, and the recharge rate w, the solution for use in superposition is

$$
\begin{aligned}
\phi_{\text{canal}} &= \frac{2wa}{K}\, x & 0 \leq x \leq L - a \\
&= \frac{w}{2K}\left[4aL - (x - L - a)^2\right] & L - a \leq x \leq L + a \\
&= \frac{2wa}{K}\, L & L + a \leq x \qquad (6.42)
\end{aligned}
$$

In using the solution above we assume that the local water table is below the canal bottom and the recharge rate is unaffected by its position.

For most realistic situations, the canal width $2a$ is small compared to the scale of the problem. Equation 6.42 can be approximated as

$$
\begin{aligned}
\phi_{\text{canal}} &= \frac{q_r}{K}\, x & 0 \leq x \leq L \\
&= \frac{q_r}{K}\, L & L \leq x \qquad (6.43)
\end{aligned}
$$

in which $q_r = 2wa$ is the volume recharge rate per unit length of the canal.

Consider the same aquifer and under the same freshwater outflow condition as the above examples. Two pumping wells are in operation, one located at (1000 m, 200 m) with $Q_1 = 480$ m^3/day, and the other at (800 m, -200 m) with $Q_2 = 360$ m^3/day. The upper diagram in Figure 6.15 shows that both wells are invaded. A canal is then excavated at $L = 500$ m, with half width $a = 1$ m, and a recharge rate $w = 0.03$ m/day. The middle diagram in Figure 6.15 shows the resulting toe location. It is observed that the well at (800 m, -200 m) is still invaded, while that at (1000 m, 200 m) becomes protected. If the recharge rate is increased to $w = 0.09$ m/day, we find from the bottom diagram of Figure 6.15 that both wells are now protected by the recharge canal.

In Figure 6.16 we demonstrate the effect of positioning the canal. The canal recharge is fixed at $w = 0.075$ m/day. In the upper diagram, the canal is located at 500 m from the coast, and in the lower diagram, the distance is 900 m. We observe that the well at (800 m, -200 m) is invaded in the first case, and is protected in the second. Indeed, the recharge canal is not very effective if placed far too seaward from the pumping well.

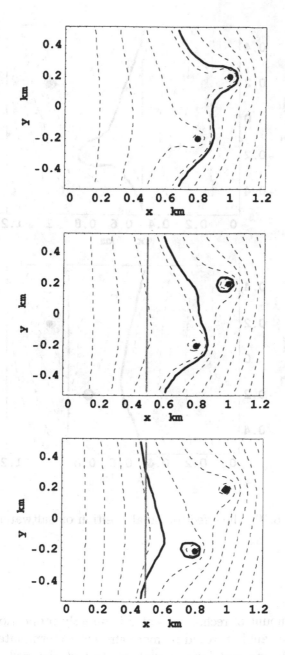

Figure 6.15. Saltwater intrusion due to two pumping wells and a recharge canal. Canal recharge is respectively $w = 0$ (upper diagram), 0.03 m/day (middle diagram) and 0.09 m/day (lower diagram).

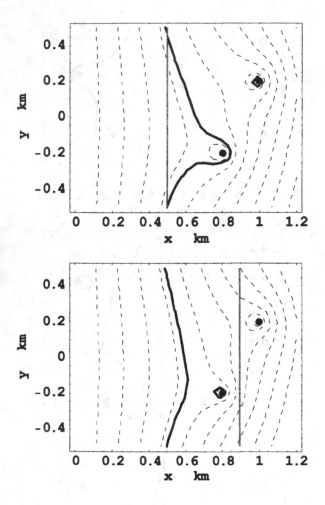

Figure 6.16. The effect of canal position on saltwater toe.

As the amount of recharged water is directly proportional to the length of the canal, it would be more efficient to terminate the canal at a finite length just long enough to protect the well. Assume a ditch that is a straight line and parallel to the coast, with length 2ℓ. Its mid-point is located at (L_x, L_y), and the volume recharge rate per unit length is q_r. Again using the method of images, the solution

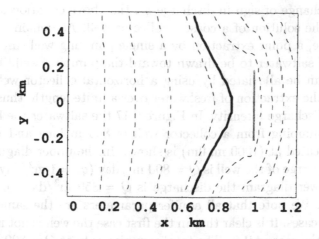

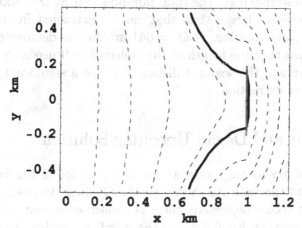

Figure 6.17. Pumping by collector well, $q_r = 1.00$ m^2/day for upper diagram and $q_r = 1.06$ m^2/day for lower diagram.

is obtained as

$$
\phi_{\text{ditch}} = \frac{q_r}{4\pi K}\Bigg[2(x + L_x)\left(\tan^{-1}\frac{y - L_y + \ell}{x + L_x} - \tan^{-1}\frac{y - L_y - \ell}{x + L_x}\right)
$$

$$
-2(x - L_x)\left(\tan^{-1}\frac{y - L_y + \ell}{x - L_x} - \tan^{-1}\frac{y - L_y - \ell}{x - L_x}\right)
$$

$$
+(y - L_y + \ell)\log\frac{(y - L_y + \ell)^2 + (x + L_x)^2}{(y - L_y + \ell)^2 + (x - L_x)^2}
$$

$$
-(y - L_y - \ell)\log\frac{(y - L_y - \ell)^2 + (x + L_x)^2}{(y - L_y - \ell)^2 + (x - L_x)^2}\Bigg] \qquad (6.44)
$$

With a change of sign in discharge q_r, the above equation also represents the solution of a coastal collector well. As demonstrated in the above, a point extraction by a single pumping well can cause a wedge of saltwater to be drawn toward the pumping well. This situation can be alleviated by using a horizontal collector well which spreads the extraction of freshwater over a finite length, thus reducing the discharge intensity. In Figure 6.17 the saltwater toe location due to pumping from a collector well of 800 m long and with its center located at (1000 m, 0m) is shown. In the upper diagram, the total discharge of the well is $Q = 800$ m^3/day ($q_r = 1$ m^2/day), while in the lower diagram the discharge is $Q = 850$ m^3/day ($q_r = 1.06$ m^2/day). We note that all aquifer parameters are the same as the previous cases. It is clear that in the first case the well is not invaded, while in the second it is. The total pumping rate at $Q = 800$ m^3/day is only somewhat larger than that can be extracted from a single well at the same location, at $Q_c = 641$ m^3/day, as demonstrated before. It seems that the length of the collector well needs to be much greater than x_w, the seaward distance, to see a significant increase in freshwater extraction.

6.7   Bear and Dagan Upconing Solution

Pumping of freshwater causes a cone of depression in the freshwater head around the well. As shown by the Ghyben-Herzberg relation, underneath such a depression there is a much enhanced upconing of saltwater, given by Eq. 6.5. Once the interface touches the screened portion of the well, the well is invaded by saltwater.

Although the Ghyben-Herzberg relation can be used to calculate the extent of upconing, its result is generally unreliable due to the inherent Dupuit (horizontal flow) assumption which is incongruent with the upconing geometry. Hence the problem of upconing should be examine in higher spatial dimensions. Although there exist a few more complicated solutions of this kind, we present below only two simple formulae provided by Dagan and Bear [1968] using perturbation technique.

Consider a geometry shown in Figure 6.18, which represents both a two-dimensional x–z plane with a line drain (collector well), and an axisymmetric r–z plane with the well as a point. Both the freshwater on top and the saltwater below has no boundary. The well or

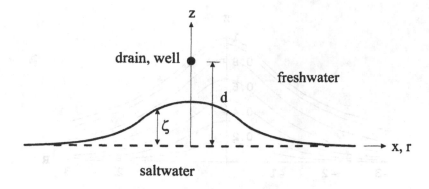

Figure 6.18. Upconing of saltwater interface.

the drain is installed at a height d above the initially undisturbed interface (dashed line).

For a point well pumping at discharge Q_w, the transient interface location ζ is give in dimensionless form as

$$Z = \frac{1}{\sqrt{R^2 + 1}} - \frac{1}{\sqrt{R^2 + (1 + T)^2}} \qquad (6.45)$$

where

$$Z = \frac{2\pi \Delta s\, K\, d\, \zeta}{Q_w}$$

$$R = \frac{r}{d}$$

$$T = \frac{\Delta s\, K\, t}{n d(2 + \Delta s)} \qquad (6.46)$$

in which n is the porosity. Figure 6.19 plots the successive locations of the upconing interface in dimensionless form for various dimensionless time values. We note that under steady state condition $t \to \infty$ the height of upconing directly underneath the well ($r = 0$) is $Z = 1$, or

$$\zeta = \frac{Q_w}{2\pi \Delta s\, K\, d} \qquad (6.47)$$

For the well not to be invaded, obviously the maximum rise ζ_{max} must be less than d, hence the maximum pumping rate is

$$Q_{max} < \frac{2\pi K d^2}{\Delta s} \qquad (6.48)$$

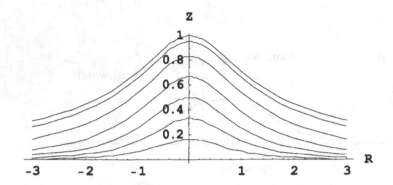

Figure 6.19. Successive upconing position for $T = 0.2$, 0.5, 1, 2, 5, 20, and ∞.

However, the solution is based on perturbation, and linearization of boundary condition by placing it at the initial interface location. It should not be applied to the limit as shown in Eq. 6.48. Experiment conducted by Bear and Dagan [1964a] shows that the interface becomes unstable when maximum rise $\zeta_{\max}$ reaches to about $d/2$. For safety considerations, $\zeta_{\max} = 0.3d$ is often used. Hence Eq. 6.48 should be modified to

$$Q_{\max} < \frac{0.6\pi\,Kd^2}{\Delta s} \tag{6.49}$$

For the case of a line drain, the upconing height is

$$Z = \ln\frac{X^2 + (1+T)^2}{1 + X^2} \tag{6.50}$$

where

$$
\begin{aligned}
Z &= \frac{2\pi\Delta s\,K\,\zeta}{q_w} \\
X &= \frac{x}{d} \\
T &= \frac{\Delta s\,K\,t}{nd(2+\Delta s)} \tag{6.51}
\end{aligned}
$$

We note that an equilibrium interface position can not be attained in this case.

6.8   Stochastic Solutions

The solutions presented so far are of deterministic nature which assumes that the parameters needed in the solutions can be gathered with high degree of accuracy. Any deviation in the outcome prediction due to input information uncertainty is likely to be small and can be neglected. In the real would, however, this is often not the case—aquifer boundaries, hydrogeological parameters, hydrological conditions, pumping rates, etc. are difficult to determine due to lack of measurement and the need to project into the future. In the interest of water resources planning, the input uncertainties should be estimated and expressed in terms of statistical quantities of mean and standard deviation. The stochastic solution can then quantify the outcome uncertainty in terms of the same statistical measures.

Despite their importance, there are few stochastic solutions of saltwater intrusion. Chapter 7 presents a stochastic solution of sharp interface location due to the random layering of hydrogeological parameters. In this section, a different kind of solutions is presented. It is assumed that the aquifer properties are homogeneous, yet uncertain. Also, the recharge rate, groundwater extraction, freshwater outflow, and sea level are random variables. The sharp interface location is then predicted as a mean and a standard deviation.

Given a random variable ζ and its expectation (ensemble mean) $\bar{\zeta}$, its perturbation from the mean is denoted as $\zeta' = \zeta - \bar{\zeta}$. We assume that the perturbation from the mean is a small quantity. This assumption allows the approximate solution to be obtained based on Taylor's series expansion. Although this condition appears to be somewhat restrictive, practical applications have shown that the deviation can be quite large with good results still obtained [Gutjahr and Gelhar, 1981; Dagan, 1985].

Set f as a function dependent on a number of random variables expressed as a vector $\zeta = (\zeta_1, \zeta_2, \ldots, \zeta_n)$. The function can be expanded around the mean of ζ based on Taylor's series [Cheng and Ouazar, 1995; Indelman et al., 1996]

$$
\begin{aligned}
f(\zeta) = f(\bar{\zeta} + \zeta') = f(\bar{\zeta}) + \sum_{i=1}^{n} \frac{\partial f(\bar{\zeta})}{\partial \bar{\zeta}_i} \zeta_i' \\
+ \frac{1}{2} \sum_{i=1}^{n} \sum_{j=1}^{n} \frac{\partial^2 f(\bar{\zeta})}{\partial \bar{\zeta}_i \partial \bar{\zeta}_j} \zeta_i' \zeta_j' + O(\zeta')^3 \quad (6.52)
\end{aligned}
$$

Using this equation we can approximate the mean of $f(\zeta)$ as

$$\overline{f(\zeta)} \approx f(\overline{\zeta}) + \frac{1}{2}\sum_{i=1}^{n}\sum_{j=1}^{n}\frac{\partial^2 f(\overline{\zeta})}{\partial\overline{\zeta}_i\,\partial\overline{\zeta}_j}\,\sigma^2_{\zeta_i\zeta_j} \qquad (6.53)$$

where $\sigma^2_{\zeta_i\zeta_j}$ is the covariance between the variables ζ'_i and ζ'_j. When $i = j$, the covariance becomes the variance $\sigma^2_{\zeta_i}$. From Eqs. 6.52 and 6.53, the variance of f is found as

$$\sigma^2_f \approx \sum_{i=1}^{n}\sum_{j=1}^{n}\frac{\partial f(\overline{\zeta})}{\partial\overline{\zeta}_i}\frac{\partial f(\overline{\zeta})}{\partial\overline{\zeta}_j}\,\sigma^2_{\zeta_i\zeta_j} \qquad (6.54)$$

For a covariance between two functions $f(\zeta)$ and $g(\eta)$ we have

$$\sigma^2_{fg} \approx \sum_{i=1}^{n}\sum_{j=1}^{m}\frac{\partial f(\overline{\zeta})}{\partial\overline{\zeta}_i}\frac{\partial g(\overline{\eta})}{\partial\overline{\eta}_j}\,\sigma^2_{\zeta_i\eta_j} \qquad (6.55)$$

Armed with the above tools, a number of problems can be investigated. For example, consider the problem shown as Figure 6.9. With all input parameters known for certain, the deterministic toe location is given by Eq. 6.27, which is rewritten here with the substitution of Eq. 6.20

$$x_{\text{toe}} = \frac{q}{w} + L - \sqrt{\left(\frac{q}{w} + L\right)^2 - \frac{\Delta s(1 + \Delta s)Kd^2}{w}}$$

$$(6.56)$$

Assume that some of the input parameters are uncertain, such as the freshwater outflow rate q, the recharge w, the sea level d, and the hydraulic conductivity K. With their mean values given, the mean toe location is not simply using $\overline{q}$, $\overline{w}$, $\overline{d}$ and $\overline{K}$ in Eq. 6.56. The solution of the mean and the variance of the toe location is in fact given as follows [Naji et al. 1998]

$$\frac{\overline{x_{\text{toe}}}}{L} = \overline{\alpha} - \sqrt{\overline{\alpha}^2 - \overline{\beta}}$$

$$+ \frac{1}{8(\overline{\alpha}^2 - \overline{\beta})^{3/2}}\left(4\overline{\beta}\,\sigma^2_\alpha + \sigma^2_\beta - 4\overline{\alpha}\,\sigma^2_{\alpha\beta}\right) \qquad (6.57)$$

$$\frac{\sigma^2_x}{L^2} = \frac{1}{4(\overline{\alpha}^2 - \overline{\beta})}\left[4\left(2\overline{\alpha}^2 - \overline{\beta} - 2\overline{\alpha}\sqrt{\overline{\alpha}^2 - \overline{\beta}}\right)\sigma^2_\alpha + \sigma^2_\beta \right.$$

$$\left. + 4\left(\sqrt{\overline{\alpha}^2 - \overline{\beta}} - \overline{\alpha}\right)\sigma^2_{\alpha\beta}\right] \qquad (6.58)$$

where

$$\overline{\alpha} = 1 + \frac{\overline{q}}{\overline{w}\,L}\left(1 + \frac{\sigma_w^2}{\overline{w}^2} - \frac{\sigma_{wq}^2}{\overline{w}\,\overline{q}}\right) \tag{6.59}$$

$$\overline{\beta} = \frac{\Delta s(1 + \Delta s)\overline{K}\,\overline{d}^2}{\overline{w}\,L^2}\left[1 + \frac{\sigma_w^2}{\overline{w}^2} + \frac{\sigma_d^2}{\overline{d}^2}\right] \tag{6.60}$$

$$\sigma_\alpha^2 = \left(\frac{\overline{q}}{\overline{w}\,L}\right)^2\left(\frac{\sigma_w^2}{\overline{w}^2} + \frac{\sigma_q^2}{\overline{q}^2} - 2\frac{\sigma_{wq}^2}{\overline{w}\,\overline{q}}\right) \tag{6.61}$$

$$\sigma_\beta^2 = \left[\frac{\Delta s(1 + \Delta s)\overline{K}\,\overline{d}^2}{\overline{w}\,L^2}\right]^2\left(\frac{\sigma_w^2}{\overline{w}^2} + 4\frac{\sigma_d^2}{\overline{d}^2} + \frac{\sigma_K^2}{\overline{K}^2}\right) \tag{6.62}$$

$$\sigma_{\alpha\beta}^2 = \left(\frac{\overline{q}}{\overline{w}\,L}\right)\left[\frac{\Delta s(1 + \Delta s)\overline{K}\,\overline{d}^2}{\overline{w}\,L^2}\right]\left(\frac{\sigma_w^2}{\overline{w}^2} - \frac{\sigma_{wq}^2}{\overline{w}\,\overline{q}}\right) \tag{6.63}$$

The toe location is then defined given the deterministic parameters L and Δs, the mean values $\overline{q}$, $\overline{w}$, $\overline{d}$ and $\overline{K}$, the variances σ_q^2, σ_w^2, σ_d^2 and σ_K^2, and the covariance σ_{wq}^2. In the above we have invoked physical reasoning that the following correlations do not exist, $\sigma_{wK}^2 = \sigma_{qK}^2 = \sigma_{dK}^2 = \sigma_{qd}^2 = \sigma_{wd}^2 = 0$, to somewhat shorten the above expressions.

To give an example, we assume a mean freshwater outflow rate $\overline{q} = 1$ m^3/m·day, at a distance $L = 1,000$ m, a mean recharge rate $\overline{w} = 0.005$ m/day, a mean hydraulic conductivity $\overline{K} = 70$ m/day, and a mean sea level $\overline{d} = 20$ m. The sea level is deterministic, $\sigma_d^2 = 0$. However, w, q, and K are random variables. We assume that the coefficients of variation, defined as the normalized standard deviation, $c_w = \sigma_w/\overline{w}$, $c_q = \sigma_q/\overline{q}$, and $c_K = \sigma_K/\overline{K}$, take the following values: $c_w = c_q = c_K = 0.01, 0.10, 0.15, 0.20$ and 0.30. We further assume that the recharge and the freshwater outflow rate are not correlated, $\sigma_{wq} = 0$. We hence have sufficient information to evaluate the analytical solution defined in Eqs. 6.57–6.63. The results are shown in Table 6.1.

For comparison, we note that the toe location calculated using the deterministic formula, Eq. 6.56, by the substitution of the mean parameters, $\overline{q}$, $\overline{w}$, $\overline{K}$, etc., yields $x_{\text{toe}} = 61.3$ m. We observe in Table 6.1 that at small variances this value is approached. However, as the variances of the parameters increase, the mean intruded distance deviates from the deterministic estimate. Hence the common practice of using mean parameter values in a deterministic formula does not produce the correct prediction of mean toe location. In this

c_w, c_q, c_K	$\overline{x}_{\text{toe}}$ (m)	σ_x (m)
0.01	61.3	0.82
0.10	61.8	8.23
0.15	62.4	11.6
0.20	63.2	16.4
0.30	65.6	24.3

Table 6.1. The mean toe location and its standard deviation.

particular case, the deterministic prediction is non-conservative, i.e., it underestimates the mean intrusion distance. We also observe in Table 6.1 that the standard deviation of toe location increases with the standard deviation of the input. In the worst case, the standard deviation is 24.3 m for a mean toe location of 65.6 m.

As a second example, we examine the Glover solution Eq. 6.9, which is rewritten here as $x = x(\xi)$:

$$x = \frac{\Delta s\, K}{2q} \xi^2 - \frac{q}{2\Delta s\, K} \tag{6.64}$$

Assume that q and K are random variables, the mean of interface location is given by

$$\overline{x} = \frac{\Delta s\, \overline{K}}{2\overline{q}} \xi^2 - \frac{\overline{q}}{2\Delta s\, \overline{K}} + \frac{1}{2} \frac{\overline{q}}{\Delta s\, \overline{K}} \left[\frac{\Delta s^2 \overline{K}^2 \xi^2}{\overline{q}^2} \frac{\sigma_q^2}{\overline{q}^2} - \frac{\sigma_K^2}{\overline{K}^2} \right] \tag{6.65}$$

and the variance is

$$\sigma_x^2 = \frac{1}{4} \left(\frac{\Delta s\, \overline{K}}{\overline{q}} \xi^2 + \frac{\overline{q}}{\Delta s\, \overline{K}} \right)^2 \left(\frac{\sigma_q^2}{\overline{q}^2} + \frac{\sigma_K^2}{\overline{K}^2} \right) \tag{6.66}$$

From physical consideration, we have assumed $\sigma_{qK}^2 = 0$.

As an illustration, the following values are used: $\overline{K} = 69$ m/day and $\overline{q} = 3.9$ m^3/day, $c_K = c_q = 0.2$, $\sigma_{qK} = 0$. Given a depth ξ, the mean interface location $\overline{x}$ and its variance σ_x^2 can be evaluated following Eqs. 6.65 and 6.66. In Figure 6.20, the interface is shown as the mean location plus/minus one standard deviation. It is of interest to observe that the standard deviation of the interface location is quite small near the outflow face, but increases significantly further inland. With the presence of an aquifer bottom, the determination of the toe location can have a relatively large uncertainty when the toe is at a large distance inland. For more detail and other solutions of this kind, the reader is referred to Naji et al. [1998].

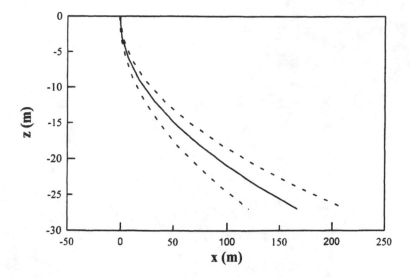

Figure 6.20. Mean interface location (solid line) ± standard deviation (dashed lines).

Acknowledgments: The authors wish to acknowledge the partial support provided by the projects "An intelligent integrated saltwater intrusion modeling system," through the CDR program, and "Monitoring and modeling of saltwater intrusion, implemented to Gaza Strip and Morocco," through the MERC program of the U.S. Agency for International Development during the preparation of the manuscript. The assistance of Ms. Boutaina Bouzouf and Mr. Mohammed Oussaa in conducting an initial literature review is appreciated. Part of the stochastic solutions section is extracted from a paper of which Mr. A. Naji is a co-author.

Chapter 7

Steady Interface in Stratified Aquifers of Random Permeability Distribution

G. Dagan & D. G. Zeitoun

7.1 Introduction

Models of salt and fresh waters flow in coastal aquifers serve as important tools for assessing the extent of saltwater intrusion and for planning the rational exploitation of water resources. In a few circumstances the seawater is separated from the overlaying freshwater body by a relatively narrow zone which can be approximated by a sharp interface. Then, the aim of modeling is to determine the shape of the interface and the flow field in the two water bodies.

A few mathematical approaches have been developed in the past in order to solve the equations of flow and transport. The first, and most general, regards the entire fluid as one of variable density which depends on salt concentration. The system of equations governing the process are Darcy's law, conservation of mass and the equation of transport of salt regarded as a solute. Examples of numerical solutions of such a system are those of Ségol and Pinder [1976] for 2D and Huyakorn et al. [1987] for 3D. This approach is of great numerical complexity and it faces difficulties in modeling the transition zone, of high concentration gradients. Furthermore, it may affected by numerical dispersion which leads to excessive mixing.

In the second category of solutions, a sharp interface approximation is adopted and the basic equations are Darcy's Law, mass conservation in the two fluids and continuity of pressure across the interface. The ensuing problem of a free boundary is still a nonlinear, difficult one. A few analytical solutions were obtained in the past

J. Bear et al. (eds.), Seawater Intrusion in Coastal Aquifers, 193–211.

for steady flow, e.g. Henry [1959] and Bear and Dagan [1964a], or by linearization, Dagan and Bear [1968]. A few examples of more general, numerical, solutions are those of Mercer et al. [1980b], Liu et al. [1981] and Taigbenu et al. [1984].

The third and simplest approach is the one adopting the additional Dupuit assumption, of hydrostatic pressure distribution along the vertical in each fluid, applying to the case of a shallow interface. Its well known consequence is the Ghyben-Herzberg relationship (see, e.g. Bear [1979]) between the freshwater head and the depth of the interface in steady flow. The unsteady flow problem is still nonlinear and besides a simple analytical solution obtained by Keulegan [1954], numerical procedures, e.g. Shamir and Dagan [1971] and Wilson and Da Costa [1982], have to be adopted.

All the above solutions were applied to homogeneous formations or to heterogeneous ones made up from a few distinct units of well defined properties. Thus Rumer and Shiau [1968] and Mualem and Bear [1974] have considered stratified aquifers of a few horizontal layers of given contrasting permeabilities.

Actual formations are known to display spatial variability of their properties. Thus, the permeability is generally found to change by orders of magnitude and in an irregular manner in space over scales much larger than the pore-scale. Heterogeneity has been shown to play an important role in transport of solutes, leading to "macrodispersion". To account for this seemingly erratic variation and the uncertainty affecting its values, the common approach is to regard permeability as random and to characterize it statistically. This randomness propagates to the flow variables and concentrations through the equations of flow and transport, which are of a stochastic nature (see, for example Dagan [1989]). The present contribution, based on Dagan and Zeitoun [1998a, 1998b], investigates the effect of random heterogeneity upon seawater intrusion. It treats two relatively simple problems which have been solved analytically in the past for homogeneous formations.

Our basic assumption is that heterogeneity manifests in perfect layering, with permeability a random function of depth. This picture agrees with most field findings for sedimentary aquifers, which are made from thin lenses of different depositional characteristics. Generally, layering is not perfect in the horizontal direction, but as far as the interface shape is concerned, the effect of the assumption is presumably small if the anisotropy ratio is sufficiently small. We also adopt the Dupuit assumption, which may be valid for shallow flow and interface of mild slope.

Under these conditions we solve first the two-dimensional, planar, problem of seaward freshwater flow above a body of immobile salt-water (in a homogeneous aquifer the interface shape is the Dupuit parabola) for which we get an exact analytical solution. Subsequently we address the problem of axisymmetric upconing beneath a well pumping freshwater. Our twofold aim is to acquire some basic understanding of the effect of heterogeneity on one hand and to obtain a few simple results of immediate application to field problems on the other hand.

7.2   Mathematical Formulation

Starting with the planar problem, we consider an aquifer of thickness D (Figure 7.1), bounded by an impervious horizontal bottom at $z = 0$ and an impervious top layer at $z = D$, while x is an inland horizontal coordinates. For the sake of generality we shall derive the equations of unsteady flow. With $z = \xi(x,t)$ the equation of the interface, Darcy's law for the horizontal components of the specific discharges reads

$$q_f = -K_f(z)\frac{d\Phi_f}{dx}; \qquad (\xi < z \le D) \qquad (7.1)$$

$$q_s = -K_s(z)\frac{d\Phi_s}{dx}; \qquad (0 \le z < \xi) \qquad (7.2)$$

where $\Phi_f = p_f/\rho_f g + z$ and $\Phi_s = p_s/\rho_s g + z$ are the pressure heads, p is the pressure, ρ is density and K is the hydraulic conductivity. Here and in the sequel the indices f and s stand for fresh and saltwater, respectively.

The hydraulic conductivities may be written in terms of the permeability k, the dynamic viscosity μ and the fluid density ρ as follows

$$K_f(z) = \frac{k(z)\,\rho_f\,g}{\mu_f}; \qquad K_s(z) = \frac{k(z)\,\rho_s\,g}{\mu_s} \qquad (7.3)$$

with k modeled as a stationary random function of z. We express it as $k(z) = k_A[1 + \mathcal{K}(z)]$ where $k_A = \langle k \rangle$ represents the constant

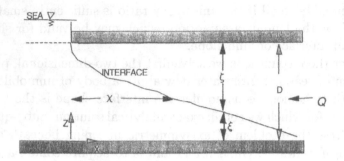

Figure 7.1. Definition sketch of an interface in steady flow.

arithmetic mean and $\mathcal{K}(z) = [k(z) - k_A]/k_A$ is the normalized fluctuation.

Thus, at second order $\mathcal{K}$ is characterized by

$$\langle \mathcal{K}(z) \rangle = 0; \quad \langle \mathcal{K}(z_1)\mathcal{K}(z_2) \rangle = \sigma^2 \, \rho(z_1 - z_2); \quad I = \int_0^\infty \rho(z) \, dz \tag{7.4}$$

where the brackets $\langle \, \rangle$ denotes ensemble averaging, $\sigma^2 = \sigma_k^2/k_A^2$ is the variance and $\rho(z)$ is the autocorrelation of $\mathcal{K}$, while I is its integral scale.

In an actual field application, the information about the statistical structure of k may be obtained, for instance, from measurements of permeability along the vertical in an exploratory well. Assuming that measurements $k_i = k(z^{(i)})$ $(i = 1, 2, \ldots, N)$ are carried out at N points, the univariate distribution of k may be determined empirically by depicting its histogram and fitting to it a distribution, say lognormal. The empirical values are then determined as $k_A \simeq (1/N) \sum_{i=1}^N k_i$ and $\sigma^2 \simeq (1/N) \sum_{i=1}^N (k_i - k_A)^2$ for $N \gg 1$. As for the autocorrelation ρ, it can be derived by building the raw variogram based on pairs of measured values and fitting to it, say, an exponential function (see, for instance, Kitanidis [1997]).

The total discharges in the freshwater zone Q_f and in the saltwater zone Q_s are obtained by integrating Eqs. 7.1 and 7.2 over each domain as follows

$$Q_f = \int_\xi^D q_f(z) \, dz = -K_{Af} \left[D - \xi + \int_\xi^D \mathcal{K}(z) \, dz \right] \frac{d\Phi_f}{dx} \tag{7.5}$$

$$Q_s = \int_0^\xi q_s(z)\, dz = -K_{As} \left[\xi + \int_0^\xi \mathcal{K}(z)\, dz \right] \frac{d\Phi_s}{dx} \qquad (7.6)$$

where $K_{Af} = k_A\, g\, \rho_f/\mu_f$ and $K_{As} = k_A\, g\, \rho_s/\mu_s$. The mass conservation of the incompressible fluid in each zone of the confined aquifer yields

$$\frac{dQ_f}{dx} = -\frac{dQ_s}{dx} = 0 \qquad (7.7)$$

Finally, the pressure continuity at the interface $p_f(x,\xi) = p_s(x,\xi)$ yields by the definition of the pressure heads

$$\xi = \frac{\rho_s}{\rho_s - \rho_f}\Phi_s - \frac{\rho_f}{\rho_s - \rho_f}\Phi_f \qquad (7.8)$$

Dupuit assumptions implies that the heads are independent of z, i.e. $\Phi_f = \Phi_f(x)$ and $\Phi_s = \Phi_s(x)$. Generally, substitution of Eqs. 7.5, 7.6 and 7.8 in Eq. 7.7 yields two differential equations for Φ_f and Φ_s that have to be solved with appropriate initial and boundary conditions while ξ is given by Eq. 7.8. Instead, we consider here simpler cases in which the interface expression can be determined directly from an unique partial differential equation (see, e.g. Bear and Dagan [1964b] for a homogeneous aquifer). Denoting by $Q(t) = Q_f(x,t) + Q_s(x,t)$ the total given discharge of the fluid, which according to Eq. 7.7 is a function of t solely, leads with the aid of Eqs. 7.5 and 7.6 to

$$K_{Af}\, J(\xi, D)\frac{d\Phi_f}{dx} + K_{As}\, J(0, \xi)\frac{d\Phi_s}{dx} = -Q \qquad (7.9)$$

where for brevity

$$J(a,b) = b - a + \int_a^b \mathcal{K}(z)\, dz \qquad (7.10)$$

Differentiation of Eq. 7.8 with respect to x and insertion of $d\Phi_s/dx$ in Eq. 7.9 leads to expressions of $d\Phi_f/dx$ and $d\Phi_s/dx$ in terms of $d\xi/dx$. Subsequently, substitution of Q_f and Q_s in Eqs. 7.5 and 7.6, and of the latter in Eq. 7.7 yields the following equation for $\xi(x,t)$

$$\frac{\partial}{\partial x}\left[\frac{1}{A(\xi) + B(\xi)}\frac{\partial \xi}{\partial x}\right] + Q(t)\frac{\partial}{\partial x}\left[\frac{A(\xi)}{A(\xi) + B(\xi)}\right] = 0 \qquad (7.11)$$

where

$$A(\xi) = \frac{\rho_s}{(\rho_s - \rho_f)K_{As}} \frac{1}{J(0,\xi)}$$

$$B(\xi) = \frac{\rho_f}{(\rho_s - \rho_f)K_{Af}} \frac{1}{J(\xi,D)} \qquad (7.12)$$

For a homogeneous aquifer, i.e. $\mathcal{K} \equiv 0$, $J(0,\xi) = \xi$, $J(\xi, D) = D - \xi$, we recover the previous formulation of Bear and Dagan [1964b].

We shall show in the sequel that in the case of a heterogeneous aquifer it is advantageous to express the interface equation in the explicit form $x = \chi(z)$ rather than $z = \xi(x)$. By using the relationships between derivatives

$$\frac{d\chi}{dz} \frac{d\xi}{dx} = 1 \qquad (7.13)$$

we get in Eq. 7.11

$$\frac{d}{dz}\left[\frac{1}{A(z) + B(z)}\left(\frac{d\chi}{dz}\right)^{-1}\right] - Q\frac{d}{dz}\left[\frac{A(z)}{A(z) + B(z)}\right] = 0;$$

$$(0 \le z \le D) \qquad (7.14)$$

The dynamic viscosity of the freshwater is slightly different from that of the saltwater, i.e. $\mu_s \simeq \mu_f$, and Eq. 7.14 can be further simplified by using the single parameter for the mean hydraulic conductivity $K'' = (\rho_s - \rho_f)K_{Af}/\rho_f \cong (\rho_s - \rho_f)K_{As}/\rho_s$. With the substitution of A, B, Eq. 7.14 yields the final form of the equation satisfied by $\chi(x)$

$$\frac{d}{dz}\left[\frac{J(0,z)\,J(z,D)}{\partial\chi/\partial z}\right] + \frac{Q}{K''}\frac{dJ(0,z)}{dz} = 0; \quad (0 \le z \le D; \ t \ge 0)$$

$$(7.15)$$

where it is reminded that $J(a,b) = b - a + \int_a^b \mathcal{K}(z)\,dz$.

The stochastic differential equation Eq. 7.15, supplemented by appropriate boundary and initial conditions, and for given statistical moments of $\mathcal{K}$, constitutes the starting point for obtaining solutions of the interface motion in the sequel.

7.3 Solution of Planar Problem

7.3.1 General

We consider here the case of saltwater under rest, i.e. $\Phi_s =$ const and in this case $Q = Q_f =$ const, $Q_s = 0$ and from Eq. 7.15 we get

$$\frac{d}{dz}\left[J(0,z)\left(\frac{J(z,D)}{d\chi/dz} + \frac{Q}{K''}\right)\right] = 0; \qquad (0 \le z \le D) \tag{7.16}$$

Since at $z = 0$, $J(0,z) = z + \int_0^z \mathcal{K}(z')\, dz' = 0$, integration in Eq. 7.16 yields the general solution

$$\frac{d\chi}{dz} = -\frac{K''}{Q}\left[D - z + \int_z^D \mathcal{K}(z')\, dz'\right] \tag{7.17}$$

Consistent with Dupuit assumption we neglect the seepage face and assume that the interface originates at $x = 0$, $z = D$, i.e. the sea serves as a sink (Figure 7.1), and the boundary condition is $\chi(D) = 0$. This approximation becomes accurate for a shallow interface, i.e. for $\chi(0) > D$ [Bear and Dagan, 1964a].

With the change of variable $\zeta = D - z$ (see Figure 7.1) and after integration of the last equation with $\chi(0) = 0$, the exact solution for the random $\chi(\zeta)$ may be written as follows

$$\chi(\zeta) = \frac{K''}{Q}\left[\frac{\zeta^2}{2} + \int_0^\zeta d\zeta' \int_0^{\zeta'} \mathcal{K}(\zeta'')\, d\zeta''\right]; \quad \text{where } K'' = \frac{\rho_s - \rho_f}{\rho_f}K_{Af} \tag{7.18}$$

Equation 7.18 permits one to compute the various moments of the nonstationary random function $\chi(\zeta)$. Thus, ensemble averaging of Eq. 7.18 yields for the average position of the interface

$$\langle \chi(\zeta) \rangle = \frac{K''}{Q}\frac{\zeta^2}{2} \tag{7.19}$$

which is precisely the Dupuit parabola obtained previously for a homogeneous medium, provided the constant hydraulic conductivity is taken equal with K_{Af}. This is consistent with the well known result that for flow parallel to the bedding the effective permeability is the arithmetic mean.

The autocovariance $C_\chi(\zeta_1, \zeta_2) = \langle[\chi(\zeta_1)-\langle\chi(\zeta_1)\rangle][\chi(\zeta_2)-\langle\chi(\zeta_2)\rangle]\rangle$ is obtained by subtracting first Eq. 7.19 from Eq. 7.18 and ensemble averaging the product of χ fluctuations, leading to

$$C_\chi(\zeta_1, \zeta_2) = \sigma^2 \left(\frac{K''}{Q}\right)^2 \int_0^{\zeta_1} \int_0^{\zeta_2} \int_0^{\zeta'} \int_0^{\zeta''} \rho(u - v)\, du\, dv\, d\zeta'\, d\zeta'' \tag{7.20}$$

It is reminded by Eq. 7.4 that σ^2 and ρ are the variance and autocorrelation of $\mathcal{K}(\zeta)$, respectively. The variance of χ is given by $\sigma_\chi^2(\zeta) = C_\chi(\zeta, \zeta)$ and the mean position of the toe of the interface and its variance are $\langle\chi(D)\rangle = (K'' D^2/2Q)$ and $\sigma_\chi^2(D) = C_\chi(D, D)$, respectively.

These general results are applied next to a few typical problems encountered in practice.

7.3.2 Applications

The interpretation of the results of the previous subsection is that in a stratified aquifer of random permeability distribution the interface deviates from its mean value and its shape cannot be predicted in a deterministic manner in the given realization, but only in terms of probability. It is, therefore, important to establish an interval of confidence for the interface location (Figure 7.2). Toward this aim we select a particular permeability covariance function

$$\rho(\zeta) = \exp\left(-\frac{|\zeta|}{I}\right) \tag{7.21}$$

To simplify notations we introduce the dimensionless variables $\chi' = \chi/D$, $\zeta' = \zeta/D$, $Q'' = Q/K''D = (Q\rho_f)/[K_{Af}(\rho_s - \rho_f)]$ and $I' = I/D$. It is seen that $\langle\chi'\rangle = \zeta'^2/2Q''$ in Eq. 7.19 and $C_{\chi'}$ in Eq. 7.20 depend on the dimensionless parameters $\sigma = \sigma_k/k_A$, $I' = I/D$ and Q''.

Substitution of Eq. 7.21 in Eq. 7.20 leads to the following close form expression of C_χ

$$C_{\chi'}(\zeta_1', \zeta_2') = \frac{\sigma^2 I'^4}{Q''^2}\left[-\frac{1}{3}\left(\frac{\zeta_1'}{I'}\right)^3 + \left(\frac{\zeta_1'}{I'}\right)^2 \frac{\zeta_2'}{I'} - \frac{\zeta_1'}{I'}\frac{\zeta_2'}{I'} + \frac{\zeta_2'}{I'} - \frac{\zeta_1'}{I'}\right.$$
$$+1 - \frac{\zeta_2'}{I'}e^{-\zeta_1'/I'} - \frac{\zeta_1'}{I'}e^{-\zeta_2'/I'} + e^{(\zeta_1'-\zeta_2')/I'}$$
$$\left. -e^{-\zeta_2'/I'} - e^{-\zeta_1'/I'}\right]; \qquad (\zeta_1' \le \zeta_2') \tag{7.22}$$

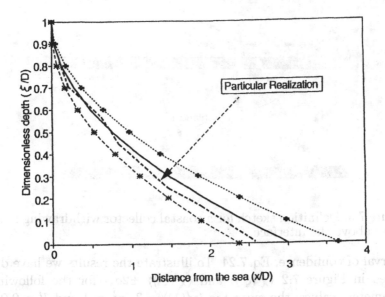

Figure 7.2. The mean interface profile $\langle \chi' \rangle$ (Eq. 7.19) and the intervals of confidence $\pm 2\,\sigma_\chi$ (Eq. 7.23) for $\sigma_Y = 1$, $I' = I/D = 0.01$, $Q/K''D = 1/6$.

In particular the variance of the interface horizontal coordinate is given by

$$\sigma_{\chi'}^2(\zeta') = C_{\chi'}(\zeta,\zeta) = \frac{\sigma^2 I'^4}{Q'^2}\left[\frac{2}{3}\left(\frac{\zeta'}{I'}\right)^3 - \left(\frac{\zeta'}{I'}\right)^2 - 2\left(\frac{\zeta'}{I'}\right)e^{-\zeta'/I'}\right.$$
$$\left. -2\,e^{-\zeta'/I'} + 2\right]; \qquad (0 \le \zeta' \le 1) \qquad (7.23)$$

To illustrate this results we further assume that k is lognormal and with $Y = \ln k$ we have $\sigma^2 = \exp(\sigma_Y^2) - 1$. Then, the coefficient of variation is given by

$$CV_\chi(\zeta') = \frac{\sigma_{\chi'}}{\langle \chi' \rangle} = \frac{2I'^2}{\zeta'^2}\left\{\left[\exp\left(\sigma_Y^2\right) - 1\right]\left[\frac{2}{3}\left(\frac{\zeta'}{I'}\right)^3 - \left(\frac{\zeta'}{I'}\right)^2\right.\right.$$
$$\left.\left. -2\left(\frac{\zeta'}{I'}\right)e^{-\zeta'/I'} - 2\,e^{-\zeta'/I'} + 2\right]\right\}^{1/2} \quad (7.24)$$

As a first application, we consider the problem of predicting the shape of the interface for given $D, Q,\ \rho_s, \rho_f, K_{Af},\ \sigma_Y^2$ and I. The best we can do is to determine the mean profile Eq. 7.19 and the

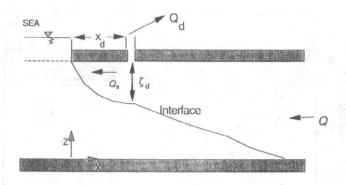

Figure 7.3. Definition sketch for a coastal collector withdrawing fresh water above the interface.

interval of confidence, Eq. 7.24. To illustrate the results we have depicted in Figure 7.2 $\langle \chi'(\zeta') \rangle$ and $\langle \chi'(\zeta') \rangle \pm 2\sigma_{\chi'}$ for the following parameters values: the mean toe $\langle \chi'(1) \rangle = 3$, $\sigma_Y^2 = 1$ and $I' = 0.01$. The small value of $I' = I/D$ is consistent with the assumption that the statistical moments of $Y = \ln k$ can be identified from the available aquifer data, say by analyzing measurements of conductivity values from a well. In particular, the coefficient of variation of the toe location Eq. 7.24 for such a small I' is approximately equal to $CV_\chi(1) \cong [8(e^{\sigma_Y^2} - 1)I/3D]^{1/2} = 0.21$ (Figure 2) and the uncertainty is quite large.

Another application of interest is related to pumping of freshwater above the interface (Figure 7.3) by a "coastal collector" located at $x = x_d$. We assume that out of a total fresh-water discharge Q a portion Q_d is drained, leaving $Q_s = Q - Q_d$ to flow to the sea and as a result the interface raises at a depth ζ_d (Figure 7.3). The relevant question in this case is what is the maximum available Q_d provided that the interface does not raise above $\zeta = \zeta_d$. An appropriate model has to take into account the fact that pumping is carried out by a drain or by wells, which create a local upconing (see the recent discussion by Dagan [1995]). Nevertheless, within the limitations of Dupuit assumption, we wish to answer this question in a stochastic framework, for a random stratified aquifer. Thus, we recast Eq. 7.18 in the following form

$$\frac{Q_s}{K''} = \frac{1}{x_d} \left[\frac{\zeta_d^2}{2} + \int_0^{\zeta_d} d\zeta' \int_0^{\zeta'} \mathcal{K}(\zeta'') \, d\zeta'' \right] \qquad (7.25)$$

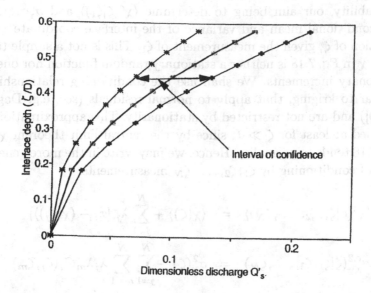

Figure 7.4. The dependence of the fresh water discharge to the sea Q'_s and its interval of confidence $\pm 2\sigma_{Q_s}$ (Eq. 7.26) on the elevation of the upconed interface ζ'_d beneath the coastal collector at x'_d ($\sigma_Y = 1$, $I' = I/D = 0.01$, $Q/K''D = 1/6$).

relating the random discharge Q_s to the given interface depth ζ_d at $\chi = x_d$, for a random permeability field. From Eq. 7.25, for an exponential ρ in Eq. 7.21 and by analogy with Eq. 7.18 we obtain for the dimensionless $Q'_s = Q_s/K''D$

$$\langle Q'_s \rangle = \frac{\zeta'^2_d}{2\,x'_d}; \qquad \frac{\sigma_{Q_s}}{\langle Q_s \rangle} = CV_\chi(\zeta'_d) \qquad (7.26)$$

where CV_χ is given by Eq. 7.24. To illustrate the results we have depicted in Figure 7.4 the dependence of $\langle Q'_s \rangle \pm 2\sigma'_{Q'_s}$ on ζ'_d for the same parameters values as in Figure 7.2. It is seen that the fresh-water discharge to the sea is affected by relatively large uncertainty, depending on the location of the collector.

Finally, we consider the prediction of the interface location for given $Q, D, K_{Af}, \rho_s, \rho_f$ and I, supplemented by a measurement of ζ_1 at a given $x = x_1$. This is the case in which, besides the information on the fresh-water discharge to the sea and on the statistics of k, we also measure the interface depth in a given cross-section by a piezometer. We treat this problem in the context of conditional probability, our aim being to determine $\langle \chi^c(\zeta|\zeta_1) \rangle$ and $\sigma_\chi^{c,2}(\zeta|\zeta_1)$, the conditional mean and variance of the interface coordinate χ as function of ζ, given the measurement of ζ_1. This is not a simple task since χ in Eq. 7.18 is neither a stationary random function nor one of stationary increments. We shall use the conditioning relationships, similar to kriging, that apply to normal residuals (see, e.g. Dagan [1989]) and are not restricted by stationarity. This approximation is justified at least for $\zeta \gg I$, since by the central limit theorem χ in Eq. 7.18 tends to normality. Hence, we may write in the more general case of conditioning by $\zeta_1, \zeta_2, \ldots, \zeta_N$ measurements

$$\langle \chi^c(\zeta|\zeta_1, \zeta_2, \ldots, \zeta_N) \rangle = \langle \chi(\zeta) \rangle + \sum_{j=1}^{N} \lambda_j \left[x_j - \langle \chi(\zeta_j) \rangle \right]$$

$$\sigma_\chi^{c,2}(\zeta|\zeta_1, \zeta_2, \ldots, \zeta_N) = \sigma_\chi^2(\zeta) - \sum_{j=1}^{N} \sum_{m=1}^{N} \lambda_j \lambda_m C_\chi(\zeta_j, \zeta_m)$$

$$\sum_{j=1}^{N} \lambda_j C_\chi(\zeta_j, \zeta_m) = C_\chi(\zeta, \zeta_m) \tag{7.27}$$

where the unconditional $\langle \chi \rangle$ and C_χ are given by Eqs. 7.19 and 7.23, respectively. For the sake of illustration we consider $N = 1$, leading in Eq. 7.27 to

$$\lambda = \frac{C_\chi(\zeta, \zeta_1)}{\sigma_\chi^2(\zeta_1)}$$

$$\langle \chi^c(\zeta|\zeta_1) \rangle = \langle \chi(\zeta) \rangle + \frac{C_\chi(\zeta, \zeta_1)}{\sigma_\chi^2(\zeta_1)} [x_1 - \langle \chi(\zeta_1) \rangle]$$

$$\sigma_\chi^{c,2}(\zeta|\zeta_1) = \sigma_\chi^2(\zeta) - \frac{C_\chi^2(\zeta, \zeta_1)}{\sigma_\chi^2(\zeta_1)} \tag{7.28}$$

To illustrate the results we consider the impact of a measurement upon the mean interface profile in Figure 7.5, the parameters being the same as before, i.e. $\langle \chi'(1) \rangle = 3$, $I' = 0.01$, $\sigma = 1$, with conditioning at $x'_d = 1, \zeta'_1 = 0.4$. As expected from Eq. 7.28 the mean interface passes through the conditioning point while $\sigma_\chi^{c,2}(\zeta_1|\zeta_1) = 0$. However, the more interesting application is the reduction of uncertainty of the toe location, i.e. the ratio $\sigma_\chi^{c,2}(D|\zeta_1)/\sigma_\chi^2(D)$ which is represented in Figure 7.6 as function of ζ'_1 for the same parameters values as in the previous figures. The impact of a measurement upon

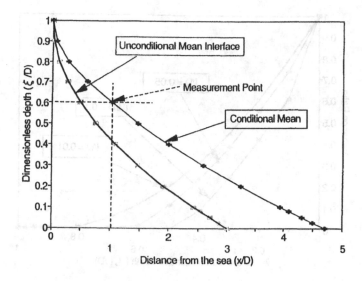

Figure 7.5. The mean interface profile $\langle \chi'^{c}(\zeta'|\zeta_1') \rangle$ and its interval of confidence (Eq. 7.28), conditioned on a measurement $\zeta_1' = 0.4$ at $x_1' = 1$ ($\sigma_Y = 1$, $I' = I/D = 0.01$, $Q/K''D = 1/6$).

variance reduction of toe location becomes more significant as the toe cross-section is approached. For the largest $I' = I/D = 0.1$ considered, the conditional variance is practically zero when $\zeta_1' \geq 0.5$. This effect is related in principle to the fact that for such large correlation scales, measurements affect a large interval of ζ. Furthermore, the unconditional variance Eq. 7.20 "builds up" as one moves upstream from the sea boundary in a nonlinear manner.

7.4   Upconing of Interface Beneath a Pumping Well

A typical problem is one of freshwater being pumped by a well or by a battery of wells above an interface. Then the problem is to determine the admissible discharge such that the salt concentration will not exceed a certain level. This requirement is translated within the sharp interface model by limiting the upconing to say 1/3 of the distance between the wells screen and the undisturbed interface (see Dagan and Bear [1968], and the recent discussion about exploitation of the Geneva aquifer in Florida [Dagan, 1995]).

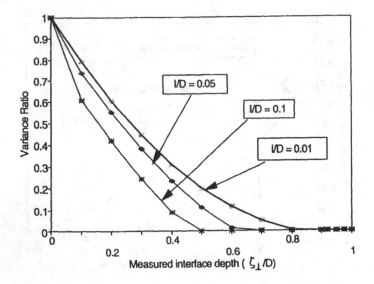

Figure 7.6. The reduction of the variance of the toe coordinate $\sigma_\chi^{c,2}(D|\zeta_1)/\sigma_\chi^2(D)$ (Eq. 7.28) as function of the measured interface depth ζ_1' ($\sigma_Y = 1$, $I' = I/D = 0.01$, $Q/K"D = 1/6$).

We consider upconing of an interface (Figure 7.7), defined by $\zeta(r)$, where ζ is the distance from the impervious upper boundary and r is a radial coordinate. For steady flow, with the salt-water body under rest and under the Dupuit assumption, the equation satisfied by ζ is as follows

$$Q = 2\pi K"r \frac{d\zeta}{dr} \int_0^\zeta [1 + \mathcal{K}(Z)]\, dZ \qquad (7.29)$$

which generalizes Eq. 7.5 for radial flow.

The boundary condition now is $\zeta = \zeta_R$ for $r = R$ (Figure 7.7), i.e. the interface depth is given at some distance from the pumping well. In practice the pumping well is partially penetrating and it withdrawals water above the interface. Then, the flow near the well and near the interface peak is three-dimensional. However, if the interface rise is moderate and the peak elevation is well below its maximal elevation, at which saltwater breakthrough occurs (see e.g. Dagan and Bear, 1968), Dupuit assumption is an acceptable approximation. Then, the r_w parameter of Figure 7.7 has to be selected as the distance from the well axes beyond which Dupuit assumption is

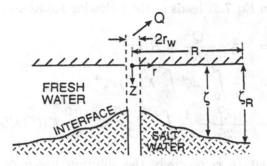

Figure 7.7. Definition sketch for interface flow toward a well pumping fresh-water.

obeyed. Due to the logarithmic dependence of ζ upon r, the precise definition of r_w is not crucial. Furthermore, our main aim is to assess on a comparative basis the effect of heterogeneity upon ζ.

Finally, for the upconing case we shall derive the interface mean $\langle\zeta\rangle$ and variance σ_ζ^2, for given Q and ζ_R. We compute now the upconed interface depth ζ at any $r < R$ by solving Eq. 7.29. We could obtain exact expressions of the moments of $r = \chi(z)$ in Eq. 7.16. However, in applications one is interested mainly in the explicit dependence of the water-table elevation ζ upon r. To obtain the latter directly, we pursue a perturbation expansion in σ as follows

$$\zeta(r) \;=\; \zeta_0(r) + \zeta_1(r) + \zeta_2(r) + \dots ;$$
$$\text{in which } \zeta_0 = \mathbf{O}(1), \; \zeta_1 = \mathbf{O}(\sigma), \; \zeta_2 = \mathbf{O}(\sigma^2),\dots \quad (7.30)$$

The equations satisfied by ζ_0, ζ_1 and ζ_2 are obtained by substituting Eq. 7.30 in Eq. 7.29 and expanding at different orders as follows

$$\zeta_0 \, d\zeta_0 \;=\; \frac{Q}{2\pi K'' r} dr; \qquad (r = R, \; \zeta_0 = \zeta_R)$$

$$d(\zeta_0 \zeta_1) \;=\; -\left[\int_0^{\zeta_0} K(z)\, dz\right] d\zeta_0; \qquad (r = R, \; \zeta_1 = 0)$$

$$d(\zeta_0 \zeta_2) \;=\; -\zeta_1 d\zeta_1 - d\left[\zeta_1 \int_0^{\zeta_0} K(z)\, dz\right]; \; (r = R, \; \zeta_2 = 0) \quad (7.31)$$

Integration in Eq. 7.31 leads to the following solutions

$$\zeta_0^2 = \zeta_R^2 - \frac{Q}{\pi K''} \ln(R/r)$$

$$\zeta_1 = \frac{1}{\zeta_0} \int_{\zeta_0}^{\zeta_R} d\zeta' \int_0^{\zeta'} \mathcal{K}(\zeta'') d\zeta''$$

$$\zeta_2 = -\frac{\zeta_1^2}{2\zeta_0} - \frac{1}{\zeta_0^2} \int_{\zeta_0}^{\zeta_R} \zeta' \int_0^{\zeta'} d\zeta'' \int_0^{\zeta_0} \mathcal{K}(\zeta'') \mathcal{K}(\eta) d\eta \quad (7.32)$$

It is seen that ζ_0 is precisely the solution based on Dupuit assumption, pertaining to a homogeneous aquifer of conductivity K''. Adopting again an exponential autocorrelation Eq. 7.21 we get after integration

$$\sigma_\zeta^2 = \langle \zeta_1^2 \rangle = \frac{\sigma^2 I^2}{\zeta_0^{'2}} f(\zeta_0', \zeta_R') \quad (7.33)$$

where $\zeta' = \zeta/I$ and

$$f(\zeta_0', \zeta_R') = 2\left[\zeta_0'(\zeta_R' - \zeta_0')^2 - \frac{(\zeta_R' - \zeta_0')^3}{3} - \frac{(\zeta_R' - \zeta_0')^2}{2} - \zeta_R' + \zeta_0' \right.$$
$$\left. + 1 + e^{-\zeta_0'}(\zeta_R' - \zeta_0')(1 - e^{-\zeta_R' + \zeta_0'}) - e^{-\zeta_R' + \zeta_0'} \right] \quad (7.34)$$

Similarly, we get for the dimensionless mean interface profile

$$\langle \zeta' \rangle = \zeta_0' + \langle \zeta_2' \rangle; \quad \langle \zeta_2' \rangle = -\frac{\sigma^2}{2(\zeta_0')^3} f(\zeta_0', \zeta_R') - \frac{\sigma^2}{(\zeta_0')^2} g(\zeta_0', \zeta_R') \quad (7.35)$$

in which

$$g(\zeta_0', \zeta_R') = 2\zeta_0'(\zeta_R' - \zeta_0') - \zeta_R' + \zeta_0' + e^{-\zeta_0'}(\zeta_R' - \zeta_0')$$
$$- e^{-\zeta_R'} + e^{-\zeta_0'} + e^{\zeta_0' - \zeta_R'} - 1 \quad (7.36)$$

To illustrate the results we have plotted in Figure 7.8 the ratio between $\langle \zeta_2' \rangle$, the correction to the mean interface profile due to heterogeneity Eq. 7.35, and ζ_0', the solution for a homogeneous aquifer of conductivity $K'' = K_{Af}(\rho_s - \rho_f)/\rho_f$, as function of ζ_0', for $\zeta_R' = 100$. It is seen that the mean profile is slightly higher due to heterogeneity. A larger effect is experienced by the coefficient of variation $\sigma_\zeta/(\sigma \zeta_0)$ in Eqs. 7.33 and 7.34, which is represented in Figure 7.9.

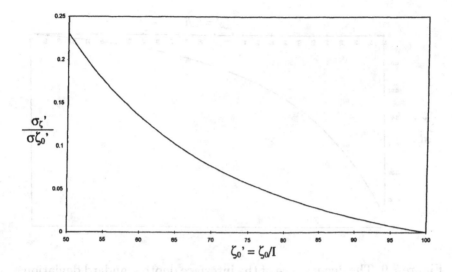

Figure 7.8. The relative interface depth correction $\langle \zeta_2' \rangle / \zeta_0'$ (Eq. 7.35) as function of ζ_0' for $\zeta_R' = 100$.

To further illustrate the use of these results, let's assume that for $\zeta_R' = 100$, the discharge $\tilde{Q}$ has been selected based on the zero-order approximation $\tilde{\zeta}_0$, as if the aquifer were homogeneous, and from the requirement $\tilde{\zeta}_0' = (2/3)\zeta_R' \cong 66.7$ at a given r/R. This leads in Eq. 7.32 to $[\tilde{Q}/(\pi K''I^2)]\ln(R/r) = 100^2 - 66.7^2 = 5.55 \times 10^3$. In reality the aquifer is layered, with $\sigma = 1$, and the question is what should be the pumped fresh-water discharge Q such as to make the probability of $\zeta' \leq 66.7$ equal to 5%?

We assume that ζ' is normal of mean $\zeta_0' + \langle \zeta_2' \rangle$ and variance $\sigma_{\zeta'}$ and seek the value of ζ_0' for which $\zeta' - (\zeta_0' + \langle \zeta_2' \rangle) = -1.65\,\sigma_{\zeta'}$ for $\zeta' = 66.7$. Since both $\langle \zeta_2' \rangle$ in Eq. 7.35 and Figure 7.8, and $\sigma_{\zeta'}$ in Eq. 7.33 and Figure 7.9 are nonlinear functions of ζ_0', we have to iterate until the above conditions are satisfied. The result is $\zeta_0' \cong 72.3$ and from Eq. 7.33 we find $[Q/(\pi K''I^2)]\ln(R/r) = 100^2 - 72.3^2 = 4.77 \times 10^3$. Hence, we get the final result $Q/\tilde{Q} = 0.86$, i.e. a reduction of 14% of the admissible fresh-water discharge. This is an upper bound and the effect is probably smaller for an aquifer of an anisotropic three-dimensional heterogeneous structure. Another question is whether the same criterion of admissible upconing in a homogeneous aquifer should be applied to a heterogeneous one, but this topic is beyond

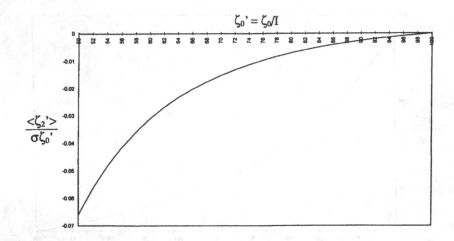

Figure 7.9. The dependence of the interface depth standard deviation $\sigma_{\zeta'}/\zeta_0'$ (Eq. 7.33) upon ζ_0' for $\zeta_R' = 100$.

the scope of the present study.

7.5   Summary and Conclusions

The present study aims at assessing the influence of heterogeneity, related to the random spatial variability of permeability, on the shape of the interface between salt and fresh waters in coastal aquifers. To achieve simple results of an analytical nature, a few assumptions have been adopted: the existence of a sharp interface, shallow flow model (Dupuit assumption) and a layered structure with permeability a random and stationary function of depth only.

We have been able to arrive at an exact closed form solution for the statistical moments of the interface coordinate in two-dimensional steady flow of fresh-water to the sea. It generalizes the well-known Dupuit parabola for a homogeneous aquifer. We have found that the mean profile is the same parabola, provided the constant effective conductivity is taken equal to the arithmetic mean of the spatially variable one. The main finding, however, is the coefficient of variation of the interface coordinate and particularly that of the toe. This can be quite large, depending on the permeability variance and integral scale. Thus, prediction of interface shape and of saltwater intrusion in heterogeneous aquifers is affected by uncertainty and our results permit one to determine the intervals of confidence. We have been able also to assess, by using conditional probability, the impact of measurements of interface position on reducing this uncertainty.

We adopted the same approach to investigate the impact of heterogeneity upon the upconing of the interface beneath a well pumping fresh-water. We were able to determine the mean profile and its variance by a perturbation expansion in the conductivity variance.

From a theoretical standpoint, many extensions of the present study that account for factors we neglected are possible, e.g. incorporating the effect of field scale pore-scale dispersion, of heterogeneity of a three-dimensional structure and solving for two- or three-dimensional flows for which Dupuit assumption does not apply. Such extensions may require, however, using more involved numerical methods.

From a field applications perspective it seems to us that assessing the impact of heterogeneity upon the interface shape by analyzing field data on both permeability variability and interface location is of definite interest. The present results may, nevertheless, serve for rough estimates of the uncertainty of prediction of salt-water intrusion.

Acknowledgments: The support of the CDR Program of the US Agency for International Development (AID) is acknowledged with gratitude.

Chapter 8

USGS SHARP Model

H. I. Essaid

8.1 Introduction

In many coastal settings, aquifer systems consist of sequences of layers with varying hydraulic properties. An idealized cross section through a layered coastal aquifer system extending offshore to a submarine canyon outcrop is shown in Figure 8.1. Under natural, undisturbed conditions an equilibrium seaward hydraulic gradient exists within each aquifer, with excess freshwater discharging to the sea (Figure 8.1a). In the uppermost, unconfined aquifer the freshwater flows out to sea across the ocean floor. In the lower, confined aquifers the freshwater discharges to the sea by leaking upward through the overlying layers and/or by flowing out the canyon outcrop. Within each layer a wedge-shaped body of denser seawater will develop beneath the lighter freshwater.

Any change in the flow regimen within the freshwater region caused by changes in discharge or recharge inland induces movement of the freshwater-saltwater interface. Reduction in freshwater flow toward the sea causes intrusion of saltwater into the aquifers as the interface moves inland. In a layered system, saltwater can enter an aquifer by flowing through the aquifer outcrop and/or leaking through the confining layers (Figure 8.1b). The ease with which saltwater can move into, or out of, an aquifer system affects the rate of interface movement in response to changes in freshwater discharge.

This chapter describes SHARP, a quasi-three-dimensional, numerical finite difference model to simulate freshwater and saltwater flow

213

J. Bear et al. (eds.), Seawater Intrusion in Coastal Aquifers, 213–247.

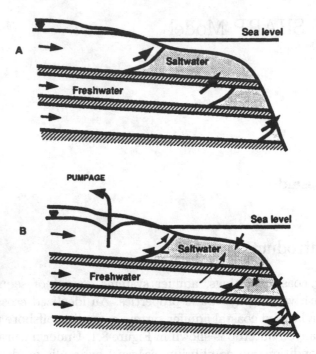

Figure 8.1. Idealized cross-section of a layered coastal aquifer system showing paths of freshwater discharge and potential paths for saltwater intrusion: (a) steady-state system with constant freshwater discharge offshore, (b) transient system with intruding saltwater and inland interface movement.

separated by a sharp interface in layered coastal aquifer systems [Essaid, 1990a, b]. The SHARP model facilitates regional simulation of coastal groundwater conditions and includes the effects of saltwater dynamics on the freshwater system.

8.1.1 The Sharp Interface Modeling Approach

When the width of the freshwater-saltwater transition zone is small relative to the thickness of the aquifer, it can be assumed that the freshwater and saltwater are separated by a sharp interface (see Chapter 5). Sharp interface models couple the freshwater and saltwater flow domains through the interfacial boundary condition of continuity of flux and pressure. In three dimensions this boundary condition is highly nonlinear [Bear, 1979]; however, assuming horizontal aquifer flow and integrating the flow equations over the vertical simplifies the problem.

The sharp interface modeling approach, in conjunction with vertical integration of the aquifer flow equations, facilitates regional scale studies of coastal areas. This approach does not give information concerning the nature of the transition zone but does reproduce the regional flow dynamics of the system and the response of the interface to applied stresses. The disperse interface approach models the mixing of freshwater and saltwater in the transition zone by solving the density-dependent solute transport equations (see Chapters 5 and 9). Volker and Rushton [1982] compared steady state solutions for the disperse and sharp interface approaches and showed that as the coefficient of dispersion decreases the two solutions approach each other.

8.1.2   Vertical Integration of the Coupled Freshwater-Saltwater Flow Equations

Within each aquifer of a layered coastal system the freshwater and saltwater domains are coupled by the common boundary at the interface. Within each flow domain the equation of continuity may be vertically integrated over the domain thickness, reducing determination of freshwater and saltwater heads to a problem in two dimensions (see Chapter 5). Introducing boundary conditions and accounting for source/sink terms, the vertically integrated equations for freshwater and saltwater flow in a coastal aquifer become, respectively:

$$
\underbrace{S_f B_f \frac{\partial \phi_f}{\partial t}}_{(1)} + \underbrace{n\alpha \frac{\partial \phi_f}{\partial t}}_{(2)} + \underbrace{\left[n\delta \frac{\partial \phi_f}{\partial t} - n(1+\delta)\frac{\partial \phi_s}{\partial t} \right]}_{(3)} =
$$

$$
\underbrace{\frac{\partial}{\partial x}\left(B_f K_{fx} \frac{\partial \phi_f}{\partial x} \right)}_{(4)} + \underbrace{\frac{\partial}{\partial y}\left(B_f K_{fy} \frac{\partial \phi_f}{\partial y} \right)}_{(4)} + \underbrace{Q_f}_{(5)} + \underbrace{Q_{lf}}_{(6)} \quad (8.1)
$$

$$\underbrace{S_s B_s \frac{\partial \phi_s}{\partial t}}_{(1)} + \underbrace{\left[n(1+\delta)\frac{\partial \phi_s}{\partial t} - n\delta\frac{\partial \phi_f}{\partial t} \right]}_{(3)} =$$

$$\underbrace{\frac{\partial}{\partial x}\left(B_s K_{sx} \frac{\partial \phi_s}{\partial x} \right)}_{(4)} + \underbrace{\frac{\partial}{\partial y}\left(B_s K_{sy} \frac{\partial \phi_s}{\partial y} \right)}_{(4)} + \underbrace{Q_s}_{(5)} + \underbrace{Q_{ls}}_{(6)} \quad (8.2)$$

where ϕ_f and ϕ_s are the vertically averaged fresh and saltwater heads (L), S_f and S_s are the fresh and saltwater specific storages (L^{-1}), B_f and B_s are the fresh and saltwater thicknesses (L), K_{fx} and K_{sx} are the fresh and saltwater hydraulic conductivities in the x direction (LT^{-1}), K_{fy} and K_{sy} are the fresh and saltwater hydraulic conductivities in the y direction (LT^{-1}), Q_f and Q_s are the fresh and saltwater source/sink terms (LT^{-1}), Q_{lf} and Q_{ls} are the net fresh and saltwater leakage across the top and bottom of the aquifer (LT^{-1}), n is the effective porosity, α is equal to one for an unconfined aquifer and zero for a confined aquifer, $\delta = \gamma_f/(\gamma_s - \gamma_f)$, and γ_f and γ_s are the fresh and saltwater specific weights $(ML^{-1}T^{-2})$. For typical values of γ_f and γ_s, $\delta = 40$.

In Eqs. 8.1 and 8.2 the type 1 terms represent the change in elastic storage within each domain. The type 2 term represents the change in freshwater storage due to drainage at the water table, and the type 3 terms represent the change in storage within each domain due to movement of the interface. The divergence of the fluxes in the x and y directions is represented by the type 4 terms. Sources and sinks to the aquifer are given in the type 5 (recharge, pumpage) and type 6 (leakage) terms. In general, the change in storage due to fluid displacement at the water table and interface (terms 2 and 3) is greater than that due to elastic storage (term 1).

Equations 8.1 and 8.2 are coupled, parabolic partial differential equations that must be solved simultaneously for the freshwater head ϕ_f and the saltwater head ϕ_s. Once the freshwater and saltwater head distributions are known, the interface elevation (ζ_1) at any x–y location in the aquifer can be obtained from:

$$\zeta_1 = (1+\delta)\phi_s - \delta\phi_f \quad (8.3)$$

In regions away from the interface, only one type of fluid is present in the aquifer, and the flow is described by a single equation without the interface (type 3) storage terms.

The type 6 terms in Eqs. 8.1 and 8.2 represent the net leakage of freshwater and saltwater, respectively, across overlying and underlying confining layers. If the confining layers are impermeable, these terms will equal zero. The leakage through a confining layer can be calculated by applying Darcy's Law in one dimension (Figure 8.2) if the effects of storage within the confining layer are negligible and flow through the confining layer is essentially vertical [Bredehoeft and Pinder, 1970]. When the waters on either side of the confining layer have the same density, Darcy's law can be formulated in terms of hydraulic head differences across the layer. However, if waters of different density occur on either side of the confining layer, vertical density gradients become important and Darcy's Law must be formulated in terms of pressure:

$$q_l = -\frac{k'}{\mu}\left(\frac{p_a - p_b}{\Delta z} + \rho g\right) \tag{8.4}$$

where q_l, is the vertical leakage (positive upward) (LT^{-1}), k' is the vertical permeability of the confining layer (L^2), μ is the fluid dynamic viscosity $(ML^{-1}T^{-1})$, p_a and p_b are the fluid pressures above and below the confining layer $(ML^{-1}T^{-2})$, ρ is the fluid density (ML^{-3}), and g is gravitational acceleration (LT^{-2}).

Evaluating the pressures above and below the confining layer and making use of the definitions for hydraulic head $(\phi = z + p/\gamma)$, specific weight $(\gamma = \rho g)$, and freshwater hydraulic conductivity of the confining layer $(K' = k'\rho_f g/\mu)$, Eq. 8.4 can be rewritten as:

$$q_l = -\frac{K'}{B'\gamma_f}\left[\gamma_a\left(\phi_a - z_a\right) - \gamma_b\left(\phi_b - z_b\right) + \gamma B'\right] \tag{8.5}$$

where B' is the thickness of the confining layer (L), ϕ_a and ϕ_b are the hydraulic heads above and below the confining layer, respectively (L), and z_a and z_b are the elevations of the top and bottom of the confining layer (L). When freshwater occurs on one side of the confining layer and saltwater occurs on the other, there is a transition from one type of water to the other through the confining layer. The density above and below the confining layer is determined by the type of water present; however, the density distribution within the confining layer depends on the direction and rate of flow. These are unknown until the equation is solved. For simplicity, the specific weight in the gravity term in Eq. 8.5 is approximated by $\gamma \approx \bar{\gamma} = (\gamma_a + \gamma_b)/2$,

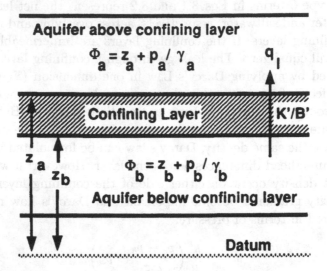

Figure 8.2. Leakage through a confining layer.

where γ_a and γ_b are the fluid specific weights above and below the confining layer.

Rearranging, the general form of the leakage term becomes:

$$q_l = -\frac{K'}{B'}\left[\frac{\gamma_a}{\gamma_f}\phi_a - \frac{\gamma_b}{\gamma_f}\phi_b + \frac{(\gamma_b - \gamma_a)}{\gamma_f}\frac{(z_b + z_a)}{2}\right] \qquad (8.6)$$

where K'/B' is the leakance of the confining layer (T^{-1}). The first two terms in Eq. 8.6 represent the equivalent freshwater heads above and below the confining layer. The third term incorporates the effect of the vertical density gradient. For the case of waters with equal density above and below the confining layer, this equation reduces to $q_l = -K'/B'(\phi_a - \phi_b)$.

8.2   Finite Difference Approximation of Freshwater and Saltwater Flow Equations

SHARP is a quasi three-dimensional, numerical model that solves finite difference approximations of Eqs. 8.1 and 8.2 to simulate coupled freshwater and saltwater flow separated by a sharp interface

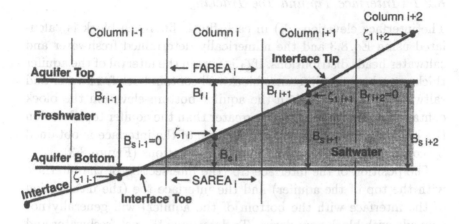

Figure 8.3. Cross-section through a model aquifer layer showing discretized freshwater thickness (B_f), saltwater thickness (B_s), and interface elevation (ζ_1).

in layered coastal aquifer systems. The model is quasi-three dimensional because each aquifer is represented by a layer in which flow is assumed to be horizontal. This chapter presents a summary of the major features of the model and examples of its use. Further details of the model development and its use are given by Essaid [1990a, b].

An implicit finite difference discretization scheme that is central in space and backward in time is used to solve Eqs. 8.1 and 8.2 for each model layer. The aquifer bottom elevation and the freshwater and saltwater thicknesses are discretized in each layer (Figure 8.3). Spatial discretization is achieved using a block-centered finite difference grid that allows for variable grid spacing. In the central difference approximations for the space derivatives the thicknesses at the grid block boundaries are linearly interpolated, and the conductivity terms are estimated using the harmonic mean of nodal values. At blocks containing pumped wells the amount of freshwater and saltwater extracted depends on the position of the interface relative to the elevation of the screened interval of the well. The rate of freshwater and/or saltwater extraction from a block, relative to the total fluid extraction rate, is determined linearly on the basis of the proportion of screen penetrating the freshwater and saltwater zones relative to the total open interval of the well.

8.2.1   Interface Tip and Toe Tracking

The interface elevation (ζ_1) in each finite-difference block is calculated using Eq. 8.3 and the numerically determined freshwater and saltwater heads distributions. If ζ_1 is within the interval of the aquifer thickness, the finite-difference block will contain both freshwater and saltwater. If ζ_1 is less than the aquifer bottom elevation the block contains freshwater, and if ζ_1 is greater than the aquifer top elevation the block contains saltwater. The shape of the interface is obtained by connecting the discretized interface elevations (Figure 8.3).

The position of the interface tip (the intersection of the interface with the top of the aquifer) and the interface toe (the intersection of the interface with the bottom of the aquifer) will generally not coincide with block boundaries. To determine the net freshwater and saltwater leakage into a block, the extent of freshwater and saltwater in contact with the top and bottom of the block must be known. This is achieved by determining the positions of the interface tip and toe within the finite difference grid for each aquifer. The tip and toe are located by linearly projecting a line defined by the interface elevations calculated in adjacent blocks until it intersects the top and bottom of the aquifer (Figure 8.3). Once the interface is located, the fraction of the top of a block in contact with freshwater (FAREA) and the fraction of the bottom of the block in contact with saltwater (SAREA) are determined. These fractions are used in the leakage calculations.

8.2.2   Leakage Calculations

The leakage across the overlying and underlying confining layers at each block may be calculated once tip and toe positions have been determined. Two methods can be used to allocate leakage between model layers. They differ in the manner in which leakage is allocated when freshwater in one aquifer leaks into saltwater in another aquifer, or vice versa. The two methods are identical in the case of single-layer problems. Comparisons of simulations using the two different methods are presented in later sections.

In method 1 leakage (complete mixing) it is assumed that when freshwater leaks into saltwater or saltwater leaks into freshwater the amount of leakage is small relative to the water in place, and the water mixes instantaneously and is incorporated into the flow zone to which it leaks. This means, for example, that when freshwater leaks into saltwater it becomes part of the saltwater domain. This assumption is reasonable for low rates of leakage but, if there is significant vertical flow it may not yield realistic results. For example, if inland pumpage reduces freshwater heads significantly in areas having saltwater in the overlying aquifer, downward saltwater leakage into freshwater can be induced. Because the downward leaking saltwater is incorporated into the freshwater flow zone, it erroneously acts as a source of water for the freshwater zone and prevents the interface from moving inland.

Method 2 (restricted mixing) limits the mixing of freshwater and saltwater to overcome the limitations of method 1. To prevent erroneous sources of water, saltwater is not allowed to leak into freshwater zones and downward leakage of freshwater into saltwater is not allowed. To approximate mixing, upward leakage of freshwater is distributed between the overlying freshwater and saltwater zones based on the amount of freshwater in the overlying block as represented by the value of FAREA. If FAREA = 1.0 in the overlying block, all freshwater leakage goes into the overlying freshwater zone. If FAREA = 0.5, half goes into the freshwater zone and the other half is incorporated into the saltwater zone. When FAREA = 0.0, all freshwater leakage is incorporated into the overlying saltwater zone.

Once the individual freshwater and saltwater leakage components are calculated, the net freshwater and saltwater leakage for a finite-difference block is given by the sum of the leakages across the top and bottom.

8.2.3   Equation Solution

The finite-difference approximations of Eqs. 8.1 and 8.2 are nonlinear because the freshwater and saltwater thicknesses and areas are time-dependent, changing with the position of the interface and water table. The equations are linearized within the iterative solution technique by evaluating the coefficients at the previous iteration level (picard iteration). The system of coupled finite-difference equations representing freshwater and saltwater flow at each node are solved in the residual form using the strongly implicit procedure for three-dimensional, two-phase flow [Stone, 1968; Weinstein et al., 1969, 1970]. This solution procedure results in the calculation of a freshwater and saltwater head for each finite difference block. The interface

elevation is then calculated at each block using Eqs. 8.3 and used
to determine the thicknesses of the freshwater and saltwater flow
domains.

8.3   General Model Features and Use

The SHARP model accommodates multiple aquifers with spatially
variable porous media properties. The uppermost aquifer of the system may be confined, semi-confined, or unconfined with areally distributed recharge. Temporal variations in recharge and pumping are
accounted for by multiple pumping periods. Model input can be in
any consistent set of units with time in seconds. All elevations must
be relative to sea level.

8.3.1   Input Parameters

SHARP requires all of the input parameters typically required by a
finite-difference groundwater flow model. However, because it solves
both freshwater and saltwater flow equations, it has additional input
requirements. The fresh and saltwater specific gravities and dynamic
viscosities must be specified. Freshwater hydraulic conductivities are
specified and saltwater hydraulic conductivities are calculated in the
model. Fresh and saltwater specific storages, effective porosity and
confining layer leakance values must be specified.

For interface tip and toe tracking, SHARP requires elevations of
the base of each layer and the thickness of the layer. Offshore bathy-
metric elevations are required to correctly represent offshore boundary conditions.

8.3.2   Initial Conditions

The amount of input information needed to specify initial conditions
depends on the type of simulation. Model runs may be initialized in
two manners, as a new run or as a continuation run. In both cases,
initial saltwater heads are calculated by the model based on the
initial freshwater head and interface elevation.

There are two options for initializing new runs. In the first option,
initial freshwater heads and interface elevations must be specified
and are used to calculate initial saltwater heads. For the second

option, only the initial freshwater heads are needed and initial interface elevations are calculated as $\zeta_1 = -\delta\phi_f$ (the Ghyben-Herzberg equilibrium interface, see section 6.2). The resulting initial saltwater heads will be zero.

For continuation runs, the model output from a previous simulation defines the initial values of freshwater head, interface elevation, and interface projection factors for determining the interface tip and toe. A continuation run could be a transient simulation initialized using a previously obtained steady-state solution, or a run that continues a previous transient simulation.

Problems may arise with interface projection and movement if initial values are not properly specified. When initializing new runs, it is advisable to use values of freshwater head and interface elevation that result in saltwater heads close to zero (even in blocks where saltwater is not present). These values determine the initial distribution of freshwater and saltwater in the aquifers and if a problem is initialized with isolated pockets of saltwater in the freshwater zone, or vice versa, there will be no outlet for that water.

Another consideration when specifying initial conditions is the initial position of the interface. If the interface is initialized to a very unstable position (for example, a vertical interface), computational instabilities could arise.

8.3.3 Boundary Conditions

The SHARP model can simulate fresh and saltwater constant flux (no-flow or prescribed flux), constant head, and head-dependent leaky boundaries. By use of these boundary conditions it is possible to accommodate the variety of onshore and offshore settings encountered in coastal systems. No-flow boundaries are specified using inactive nodes. At unconfined blocks in the uppermost aquifer, constant flux can be specified by the recharge matrix. For all layers, the pumpage matrix can be used to specify a constant flux. A positive pumpage value represents extraction of water from a block and a negative value represents addition of water to the block.

Constant freshwater head blocks and constant saltwater head blocks can be specified. At constant saltwater head nodes, the initial freshwater head and interface elevations must be specified to give the desired saltwater head using $\phi_s = (\zeta_1 + \delta\phi_f)/(1 + \delta)$. The saltwater head will then remain fixed at its initial value for the duration of the simulation, although freshwater head and interface location at this node may change. Fixing both freshwater and saltwater head at a node fixes the interface position in that node.

The leaky, head-dependent boundary condition can be used to represent streams, springs, or offshore areas. This type of boundary condition may only be used in blocks specified as confined. The flux at head-dependent boundaries, q_{hd}, is calculated as follows: $q_{hd} = K'/B'(\phi_{fh} - \phi_{aq})$. The leakance, K'/B', and fixed head value, ϕ_{fh}, are specified by the user. The head in the aquifer, ϕ_{aq}, is calculated in the model solution. Thus, as the head in the aquifer changes, the flux across the boundary changes. The leakance value at the node can be adjusted to obtain the desired degree of interaction between the aquifer and the fixed head node. If the leakance at a head-dependent boundary block is sufficiently large, the block will act as a constant head block with a head equal to the overlying fixed head value. If this type of boundary is used offshore, the fixed head value must be the freshwater head equivalent to the overlying column of saltwater, $\phi_f = -Z/\delta$, where Z is the bathymetry or elevation of the ocean floor relative to sea level. The water in the overlying inactive block is assumed to be saltwater.

8.3.4   Steady-State Simulations

Steady-state simulations can be obtained only by running a transient simulation until it achieves steady state; the interface position must be allowed to move gradually as steady state is approached. The approach to steady state can be accelerated in several ways. First, the initial conditions should be set close to the expected steady-state solution. The specific storage of the aquifers should be very small or zero to eliminate elastic storage terms from Eqs. 8.1 and 8.2. The movement of the interface can be accelerated by using a very small porosity (as small as 0.0001). In this manner, the interface storage terms in Eqs. 8.1 and 8.2 become small, and less water must be moved as the interface changes position. Though the steady-state solution is not a function of specific storage or porosity, transient solutions are. Therefore, after obtaining the steady-state solution the values of specific storage and porosity must be set to their estimated values for subsequent transient simulations.

8.3.5   Discretization

Caution should be exercised in choosing both the temporal and spatial discretization. The model is quite sensitive to time-step size. If time steps are too large compared with the rate of interface movement, the solution will become unstable. To obtain accurate interface projections, it is advisable to use smaller blocks near the interface. Where the interface is present in only one block, the projection of the tip and toe becomes less accurate. If the interface tip or toe oscillates at a block boundary, smaller block sizes or smaller time steps may be needed to stabilize the solution.

8.3.6   Mass Balance Calculations

A mass balance is calculated for each layer in the system. Values are reported as cumulative volumes for the total duration of the simulation and as rates for the time step under consideration. Influx of water into a layer by leakage is divided into components derived from overlying and underlying freshwater and saltwater. The global mass balance for all layers is obtained by summing the inflows and outflows across the boundaries of each layer.

8.4   Model Evaluation

In the following sections several example problems are described that allow us to illustrate and evaluate the utility of the SHARP model and the sharp interface modeling approach. In the first example, SHARP results are compared to an analytical solution of a steady-state interface in a layered coastal aquifer. In the second example, SHARP results are compared to results from a density-dependent transport model for a layered system. Finally, a comparison is made between the sharp interface approach with coupled freshwater-saltwater flow (SHARP) and the sharp interface approach that incorporates the Ghyben-Herzberg approximation.

8.4.1   Steady-State Interface in a Layered Coastal Aquifer

Mualem and Bear [1974] presented an approximate analytical solution for the steady-state shape of an interface in a coastal aquifer when a thin horizontal semi-confining layer is present (Figure 8.4).

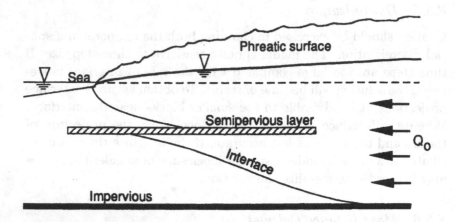

Figure 8.4. The case of an interface in a coastal aquifer with a thin horizontal semi-confining layer.

Parameter	Value
ρ_f	1.0 g/cm^3
ρ_s	1.025 g/cm^3
K_f	0.10 m/s
K'	8.0×10^{-4} m/s
n	0.1

Table 8.1. Aquifer parameters used in the case with a thin horizontal semi-confining layer.

Their solution is based on the Dupuit assumption and a linearization of the flow equations. Mualem and Bear [1974] also made two simplifying assumptions regarding the leakage conditions in the region where saltwater is present above the semi-confining layer: (1) the hydraulic head above the semi-confining layer in Eqs. 8.5 and 8.6 (ϕ_a) was constant, and (2) the freshwater leaking through the semi-confining layer from below was incorporated into the freshwater flow zone above. This is approximately equivalent to the restricted mixing leakage option (method 2) of SHARP.

The geometry of the test problem is shown in Figure 8.5, and the parameters used in the simulation are given in Table 8.1. To compare the numerical solution with the analytical solution, method 2 leakage calculations (restricted mixing) were used to approximate Mualem and Bear's assumptions. Figure 8.6a shows good agreement

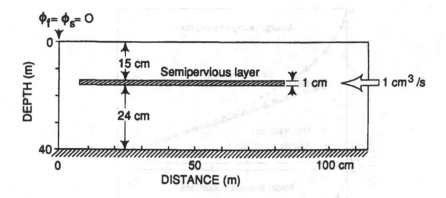

Figure 8.5. The simulated geometry and boundary conditions for the case with a thin horizontal semi-confining layer.

between the two solutions. The same problem was then simulated using the method 1 leakage conditions (complete mixing), and the results are shown in Figure 8.6b. With this method, the freshwater flowing through the semi-confining layer leaks into the overlying salt-water zone. The saltwater zone is no longer static, as was assumed in the analytical solution. The interface below the confining layer is slightly deeper, whereas the interface above the layer extends further inland. This result is caused by: (1) the leakage of freshwater into the overlying saltwater, and (2) flow in the saltwater zone, which is ac-tually a mixing zone. In Hele-Shaw laboratory experiments, Mualem and Bear [1974] observed that there was a clear boundary between the freshwater zone and the mixing zone but that no clear boundary could be distinguished between the mixing zone and the saltwater zone. In the complete mixing simulation, the position of the interface separates the zone containing freshwater from the zone containing any mixed water. For the restricted mixing simulation, the position of the interface approximates the middle of the mixing zone.

8.4.2 Comparison of Sharp Interface and Disperse Interface Solutions

Hill [1988] compared the sharp interface and disperse interface solu-tions for a generalized cross-section through the ground-water system in the coastal area of Cape May County, New Jersey. The comparison was made using complete mixing leakage (method 1) calculations in

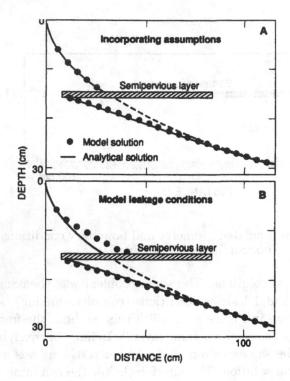

Figure 8.6. Comparison of numerical and analytical solutions for the case with a thin horizontal semi-confining layer (dashed line represents the position of the interface in the absence of a semi-confining layer): (a) restricted mixing leakage (method 2), (b) complete mixing leakage (method 1).

the sharp interface model. The disperse interface simulation was carried out using the density-dependent, convective-dispersive transport model SUTRA [Voss, 1984a] (see also chapter 9).

The geometry of the simulated cross-section is shown in Figure 8.7. The system consists of an unconfined aquifer overlying two confined aquifers. No-flow boundary conditions were imposed across the bottom and the landward vertical boundaries, a constant sea level head was imposed at the seaward vertical boundary. A leaky, head-dependent boundary was imposed on the upper boundary in SHARP simulations and a constant-head boundary condition in SUTRA simulations. Onshore, the overlying fixed head was a parabolic distribution varying from sea level at the shore to 12 feet above sea level onshore. Offshore, the overlying fixed head was the equivalent

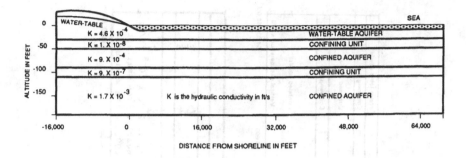

Figure 8.7. Cross-section modeled for the comparison of sharp inter-face and disperse interface solutions.

Parameter	Value
Aquifer horizontal to vertical anisotropy	100
Confining bed horizontal to vertical anisotropy	10
n	0.1
δ	40
Maximum longitudinal dispersivity	25 ft
Minimum longitudinal dispersivity	2.5 ft
Transverse dispersivity	2.5 ft

Table 8.2. Parameters used in the comparison of the sharp and dis-perse interface solutions.

freshwater head for the overlying column of seawater. Other parame-ters used in the simulation are given in Table 8.2, and further details of the convective-dispersive simulation are given by Hill [1988].

Figure 8.8 compares the results of Hill's SHARP simulations with complete mixing, results of an additional SHARP simulation ob-tained using the restricted mixing leakage method, and the SUTRA results. The position of the sharp interface is compared to the zone between the 0.4 and 0.6 concentration contours obtained from the SUTRA simulations. This zone contains approximately half freshwa-ter and half saltwater, and will be referred to as the 0.5 zone.

In general, for all SHARP simulations, the interface is closer to land than the 0.5 zone obtained from the SUTRA simulations. Also, the sharp interface has a slope that is gentler than that of the 0.5 zone. This is because mixing leads to circulation of saltwater near the transition zone (Figure 8.8b), a process that cannot be reproduced

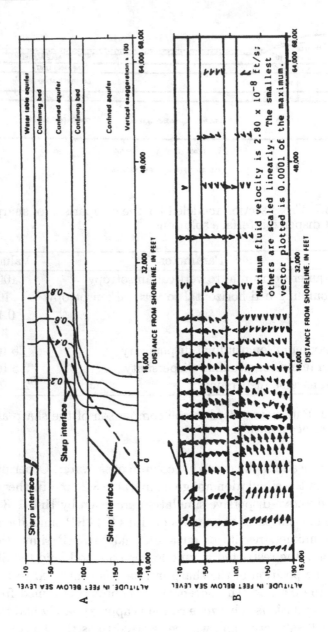

Figure 8.8. Results of steady-state simulations with sharp- and dis-
perse-interface approaches: (a) lines of equal chloride concentration
from the SUTRA simulation and the sharp interface positions from
the SHARP simulations (solid lines are interface with method 1 leakage, dashed lines are interface with method 2 leakage), (b) fluid velocity vectors from SUTRA.

using the sharp interface approach. Volker and Rushton [1982] have shown that the toe position predicted by the sharp interface tends to be farther inland than the actual transition zone because the effects of dispersion are neglected. Thus, the sharp interface solution gives a more conservative estimate of saltwater intrusion. Hill [1988] reported that as the coefficient of hydrodynamic dispersion was reduced the orientation of the SUTRA concentration contours approached that of the sharp interface.

In the upper unconfined aquifer, the sharp interface is considerably landward of the 0.5 zone for both complete and restricted mixing SHARP simulations. In this unconfined aquifer, the interface separates the zone where considerable freshwater is being discharged to the sea from the zone of relatively stagnant water offshore (Figure 8.8b). In the SUTRA steady-state simulation, the offshore zone of relatively stagnant water has been flushed of saltwater by leakage of freshwater from below. However, this zone would not be a useful source of freshwater, because flow directions would quickly be reversed by inland pumping, resulting in saltwater influx. This suggests that the sharp interface may be a reasonable indicator of actual availability of freshwater resources.

In the middle aquifer, the sharp interface very roughly passes through the 0.5 zone for both leakage methods, but has a different slope because dispersion is not accounted for. The interface is closer to the shore in the complete mixing simulation because there is some loss of freshwater by downward leakage into the underlying saltwater domain.

The interface in the lowermost aquifer is considerably landward of the 0.5 zone when the complete mixing leakage method is used. With the restricted mixing method, the tip of the interface is close to the 0.5 zone but, the toe is still considerably landward. In the complete mixing case, some freshwater from the middle confined aquifer leaks down into the saltwater zone of the bottom aquifer and is mixed into the saltwater domain. This shrinks the freshwater zone in the lowermost aquifer. In the restricted mixing calculations, downward freshwater leakage into saltwater is not allowed and loss of freshwater to the underlying saltwater domain does not occur. The toe of the interface in this case, however, is still landward of the 0.5 zone as the sharp interface approach does not reproduce the circulation of saltwater in the transition zone. The sharp interface position approximately corresponds to the boundary between the freshwater flow vectors and the zone of saltwater recirculation.

In general, the sharp interface approach results in a conservative estimate of the position of the interface separating freshwater and saltwater [Hill, 1988]. The assumptions of the complete mixing leakage calculations are appropriate for systems dominated by horizontal flow components but deteriorate when vertical flow approaches the magnitude of horizontal flow. For high leakage cases the restricted mixing leakage approach leads to better results.

8.4.3 Comparison of the SHARP and the Ghyben-Herzberg One-Fluid Sharp Interface Approaches to Modeling Transient Behavior

When a freshwater lens floating on saltwater is pumped, the freshwater removed from the system comes from: (1) drainage from the water table; (2) upward movement of the interface where freshwater is being replaced by saltwater; and (3) water released due to elastic storage effects. At steady state, the position of the interface can be estimated using the Ghyben-Herzberg approximation (section 6.2) that assumes that horizontally flowing freshwater is floating on static saltwater. During transient periods, however, the behavior of the freshwater-saltwater flow system is controlled by both the fresh and saltwater flow dynamics. Significant amounts of saltwater must move into or out of the aquifer to accommodate interface movement. If the net recharge to a system is reduced such that freshwater head drops by one meter, a new steady state will be reached only when the freshwater-saltwater interface has moved up forty meters (for $\rho_f = 1.0$ g/cm^3 and $\rho_s = 1.025$ g/cm^3). Conversely, if freshwater heads increase one meter the interface must ultimately become forty meters deeper.

Wentworth [1941], in studying the hydrology of the island of Oahu, Hawaii, divided groundwater storage in the freshwater body into two components: he referred to the part of the lens above sea level as 'top storage', and the part below sea level as 'bottom storage'. If the thicknesses of top and bottom storage are not in the equilibrium ratio predicted by the Ghyben-Herzberg principle, water will flow into the deficient part. In his explanation of this type of system, Wentworth used the analogy of two reservoirs connected by a pipe offering resistance to flow. The movable boundary of the bottom storage reservoir is not a free moving air-water boundary, but a freshwater-saltwater boundary in the porous rock. Movement of this boundary or interface can only be achieved by inducing flow of saltwater. Wentworth reasoned that the resistance to flow through the rock, and the fact that forty units of saltwater must move for each unit change in freshwater head at the top, would damp the motion of the interface relative to the motion of the water table. Sudden or abrupt changes in recharge or discharge would first cause changes in head in the top storage part of the aquifer. Slower changes in bottom storage and interface position would take place in response to the longer term trends. The transient behavior of such systems is a function of aquifer storage properties and the ease with which freshwater and saltwater can move through the aquifer. This is controlled by several factors: aquifer effective porosity and specific storage; the transmissivity of the fresh and saltwater zones of the aquifer; the seaward and landward boundary conditions; and the vertical anisotropy of the system [Essaid, 1986].

Sharp interface models fall into two categories: those that model coupled freshwater and saltwater flow (two-fluid approach), and those that model freshwater flow only (one-fluid approach). The sharp interface models which simulate flow in the freshwater region only incorporate the Ghyben-Herzberg approximation and assume that at each time step saltwater adjusts instantaneously to changes in the freshwater zone, so that an equilibrium interface position is achieved. The saltwater flow equation is not solved.

In the freshwater flow equation, Eq. 8.1, the freshwater interface storage term $n\delta(\partial\phi_f/\partial t)$ represents the rate at which freshwater would be released from storage at the interface if saltwater heads adjusted to equilibrium instantaneously (i.e. the Ghyben-Herzberg approximation). The coupling term $n(1 + \delta)(\partial\phi_s/\partial t)$ represents the impact of flow in the saltwater zone on the rate of interface movement and is equal to zero for the one-fluid approach. At early times, the time derivatives of freshwater and saltwater head will have the same sign and the magnitude of the freshwater interface storage term is reduced by an amount equal to the magnitude of the coupling term, leading to a larger freshwater head response than would be predicted using the one-fluid sharp interface approach. In many cases the two-fluid sharp interface model is more appropriate than the one-fluid model for investigating short-term responses.

8.4.3.1 The Waialae Aquifer, Southeastern Oahu, Hawaii

The Hawaiian island of Oahu is underlain by an extensive freshwater lens. A thick sequence of thin bedded basaltic lava flows forms
the principal aquifer in which freshwater floats on top of saltwater
[Visher and Mink, 1964; Takasaki and Mink, 1982]. Inland groundwater is unconfined, but towards the coast the groundwater is confined
by a caprock of reef and marine sedimentary deposits. The seaward
flow of the freshwater is impeded by the caprock, allowing a thick
freshwater lens to develop. The Waialae aquifer of southeastern Oahu
is bounded on the north by the major rift zone of the Koolau volcano
and on the west by the Kaau rift. Another zone of dikes forms the
eastern boundary. The Waialae aquifer is confined by the caprock
from the coast to about 4 km inland; the caprock thickness increases
seaward. The caprock and aquifer extend offshore for about 30 km
where they end at a rift zone.

The Waialae area was first modeled by Eyre [1985] using the one-
fluid sharp interface model AQUIFEM-SALT [Voss, 1984b]. The
aquifer parameters used by Eyre were adapted to the freshwater-
saltwater flow model SHARP and the system was extended seaward
to incorporate the appropriate saltwater zone boundary conditions
[Essaid, 1986]. A comparison of the models follows.

8.4.3.2 Results of the One-Fluid Sharp Interface Model

Eyre [1985] simulated the transient response of the Waialae freshwater lens to inland pumpage using the one-fluid sharp interface
approach. A graph of simulated and observed heads from 1937 to
1975 at the Waialae shaft observation well is shown in Figure 8.9.
The simulation results were obtained using a constant mean annual
recharge and a specific yield of 0.1. The fluctuating observed head
values deviated from the relatively smooth simulated trend by about
plus or minus 0.25 m. Eyre [1985] repeated the simulation using annually adjusted recharge values that were obtained by adjusting the
distributed basin recharge according to the deviation of rainfall from
the mean value. The simulated amplitude of head variations with a
specific yield of 0.1 is significantly less than that observed but simulations with specific yields of 0.02 and 0.05 gave rise to larger head
changes (Figure 8.10). The simulation with a specific yield of 0.05
comes closest to matching the magnitude of observed head fluctuations.

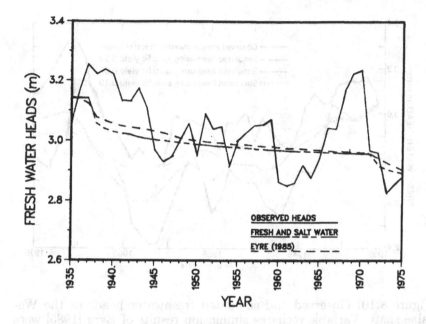

Figure 8.9. Observed and simulated freshwater heads at the Waialae shaft. Constant recharge simulation results are from SHARP and the one-fluid Ghyben-Herzberg sharp interface model [Eyre, 1985].

This Ghyben-Herzberg one-fluid sharp interface modeling approach is very sensitive to the value of specific yield because it assumes that the interface responds instantaneously to head changes. When the true interface does not respond rapidly to head changes, a good fit between observed and simulated heads can only be obtained by using a value of specific yield that is lower than the true aquifer value.

8.4.3.3   SHARP Results

To investigate the influence of saltwater flow on the transient response of the system, the analysis of the Waialae aquifer was repeated using the SHARP model, incorporating both freshwater and saltwater flow. To allow a realistic seaward boundary condition on the zone of saltwater flow, the aquifer and caprock were extended seaward 32 km to an offshore rift zone where a no-flow boundary was imposed. Model parameters are listed in Table 8.3.

The model was calibrated by adjusting leakance values to match the predevelopment steady state head distribution. The transient re-

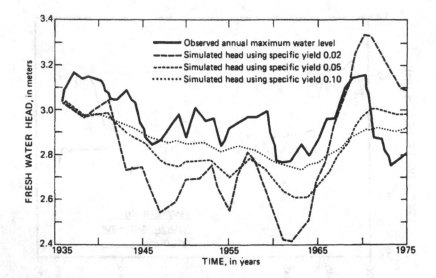

Figure 8.10. Observed and simulated freshwater heads at the Waialae shaft. Variable recharge simulation results of Eyre [1985] were
obtained using specific yields of 0.1, 0.02, and 0.05.

Parameter	Value
δ	40
K_f	5.8×10^{-3} ft/s
K'/B'	2.12×10^{-8} s^{-1}
n	0.1 and 0.05
Mean annual recharge	1.0×10^{-5} ft/s

Table 8.3. Aquifer parameters used in the case with a thin horizontal
semi-confining layer.

sponse for constant recharge is shown in Figure 8.9. It differs from
Eyre's results in that initial head drops are steeper due to the influence of transient saltwater flow. The sharp simulation was repeated
using the annually adjusted recharge values, and the results are
shown in Figure 8.11. This modeling approach reproduces the observed amplitude of the head changes with an effective porosity of
0.1 and 0.05, although the positions of simulated peaks and troughs
do not quite match those observed.

The solution obtained with the SHARP two-fluid approach is not
significantly affected by changing effective porosity (specific yield)

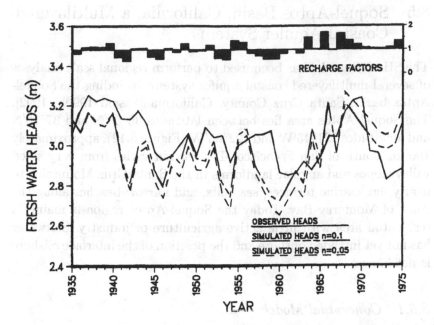

Figure 8.11. Observed and simulated freshwater heads at the Waialae shaft. Variable recharge simulation results were obtained using SHARP and effective porosities of 0.1 and 0.05. The bar graph represents the recharge factors that multiplied the mean basin recharge to obtain the annual recharge values.

values. This is because lowering effective porosity does not significantly affect the saltwater gradients induced by changes in onshore freshwater flow. The short term response of freshwater heads is highly sensitive to the saltwater flow dynamics and is not as sensitive to effective porosity. The lack of sensitivity to effective porosity makes this solution more robust than that obtained with the one-fluid sharp interface model.

The one-fluid Ghyben-Herzberg sharp interface approach can be used to model long-term responses of freshwater lenses, as well as short-term responses in aquifers where saltwater can flow in and out easily. However, in many common hydrogeologic settings it is necessary to incorporate the flow dynamics of the saltwater zone by using a coupled freshwater-saltwater flow approach in order to realistically model short-term stresses and transition periods.

8.5   Soquel-Aptos Basin, California, a Multilayered Coastal Aquifer System

The SHARP model has been used to perform regional scale analyses of several multilayered coastal aquifer systems including the Soquel-Aptos basin, Santa Cruz County, California [Essaid 1990b, 1992]. The Soquel-Aptos area lies between latitudes 36°55'N and 37°10'N and longitudes 121°45'W and 122°05'W (Figure 8.12), approximately 100 km south of San Francisco. The terrain varies from very steep valley slopes and angular landforms in the Santa Cruz Mountains to nearly flat marine terraces, sea cliffs, and narrow beaches along the coast of Monterey Bay. Today the Soquel-Aptos region is mainly a residential area with little active agriculture or industry. Saltwater has not yet intruded onshore and the position of the interface offshore is not known.

8.5.1   Conceptual Model

The hydrogeologic conditions of the Soquel-Aptos basin were translated into a conceptual model representing the onshore and offshore geometry, boundary conditions, and physical parameters. Figure 8.13 shows the areal extent of the model and the spacing of the finite difference grid (43 rows by 36 columns). In the onshore area most grid blocks have dimensions of 610 by 610 m, with increasing spacing toward the southeast and southwest to facilitate incorporation of the offshore boundaries.

8.5.1.1   System Geometry

The principal hydrogeologic unit of interest in the Soquel-Aptos basin is the Purisima Formation. Using exploratory wells, geophysical logs, and other geologic evidence, Luhdorff and Scalmanini Consulting Engineers [1984] were able to delineate distinct Purisima subunits throughout the basin. This geologic framework was adopted for the numerical model and the Purisima was represented by five model layers. Distinct, finite thickness confining layers could not be identified, so the system was modeled with vertical leakance representing the hydraulic connection between the layers.

Basin-wide structural contour and thickness maps were constructed by integrating onshore and offshore geologic information. The sys-

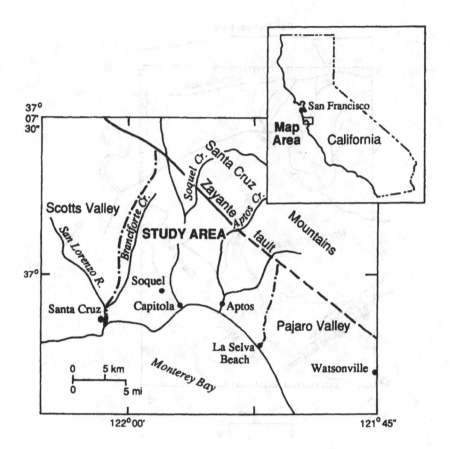

Figure 8.12. Location of the Soquel-Aptos basin.

tem geometry was delineated by using outcrop data [Hickey, 1968], well logs [Luhdorff and Scalmanini, 1984], and offshore seismic stratigraphic sections [Greene, 1977]. The lowermost subunit A has variable thickness (maximum 250 m) because the basement was an uneven erosion surface. Overlying subunits (B, C, and D) have relatively uniform thicknesses (76 m, 37 m, and 30 m, respectively) throughout the basin, except in outcrop areas where the original thicknesses have been reduced due to erosion. Subunit E increases in thickness from west to east, reaching a maximum thickness of 183 m.

The extents of the subunit outcrops were defined by intersections between the structural contours for the Purisima subunits and the land surface topography (Figure 8.14). The older units are exposed

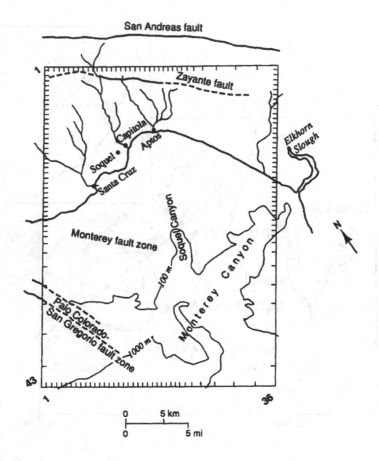

Figure 8.13. The SHARP model grid area for the Soquel-Aptos case study.

in the western area and are overlain by younger units toward the southeast. The thickness, elevation, and areal extent of each layer were discretized over the 43 by 36 grid.

8.5.1.2   Boundary Conditions

The lateral boundaries of the system were delineated on the basis of physical features (Figure 8.15). The Purisima Formation decreases in thickness westward and pinches out along the northwestern border of the area. This pinchout was prescribed as a no-flow boundary for all layers. The Zayante fault was set as a no-flow boundary, as was the

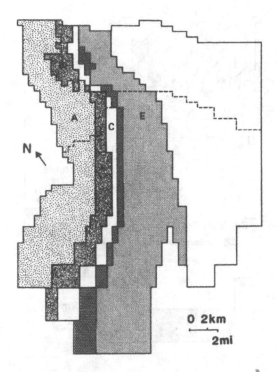

Figure 8.14. Discretized outcrop areas of Soquel-Aptos model layers. Dashed line is the shoreline.

offshore Palo Colorado-San Gregorio fault zone (see Figure 8.13 for fault locations). At the southeastern edge of the Soquel-Aptos area there is no physical boundary so the model boundary was extended 8 km toward the southeast and an estimated flow line was represented by a no-flow boundary. This boundary was placed far enough away so as not to influence the solution in the area of interest. The offshore canyon outcrop was represented by constant head nodes. Saltwater heads were fixed at zero at this boundary and freshwater heads were fixed at the equivalent freshwater head representing the overlying column of saltwater.

The base of the system is relatively impermeable. Onshore, recharge enters each layer through its outcrop area, and groundwater discharges to streams as base flow. Offshore, freshwater discharges to the sea through ocean floor outcrops. It is very difficult to accurately determine the quantity and distribution of these fluxes. Toth

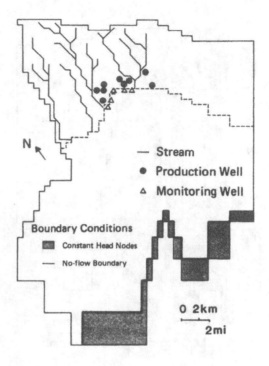

Figure 8.15. Boundary conditions and major features of the So-quel-Aptos model area.

[1963] showed that topography is commonly the driving force for ground-water flow and that the water table is generally a subdued replica of the land surface. To achieve this in the model, the on-shore and offshore outcrops of each subunit were simulated using head-dependent boundary conditions (see section 8.3.3). The specified fixed head was the average topography onshore and it was the freshwater head equivalent to the overlying column of saltwater offshore. The leakance represented the vertical flow resistance. By specifying the upper boundary condition in this manner water could leak into or out of each layer, allowing use of the model to estimate the quantity of recharge to, and discharge from, the system.

In the southeastern part of the model area, subunit E is overlain by the Aromas Sand which has undergone more than 60 m of draw-down in some areas due to pumpage in the Pajaro basin [Bond and Bredehoeft, 1987]. A time-dependent water table elevation was used to represent the fixed head at this head-dependent boundary, with a

Layer	Hydraulic Conductivity (m/s)	Maximum Thickness (m)	Specific Storage (m^{-1})	Leakance (s^{-1})
A	10^{-6} to 10^{-5}	250	10^{-6} to 10^{-7}	10^{-9} to 10^{-13}
B	10^{-6}	76	10^{-7} to 10^{-6}	10^{-9} to 10^{-11}
C	10^{-6}	37	10^{-6}	10^{-9} to 10^{-10}
D	10^{-6}	30	10^{-7} to 10^{-6}	10^{-9} to 10^{-10}
E	10^{-6}	183	10^{-5}	10^{-9} to 10^{-11}

Table 8.4. Summary of aquifer parameters used in the Soquel-Aptos simulation.

linear decrease in heads from predevelopment conditions to present day water levels.

8.5.2 Model Calibration

Model values of aquifer parameters (conductivity, specific storage, leakance, and effective porosity) were initially chosen on the basis of available data and then adjusted to obtain simulated water levels which satisfactorily matched observed water levels. Estimates of recharge to the system and base flow to the streams were also made and used to constrain the calibration process [Essaid, 1990b, 1992]. This trial-and-error process of model calibration does not lead to a unique solution but can be constrained on the basis of hydrologic information. The calibration process leads to an understanding of the factors determining the system's response and behavior. A sensitivity analysis [Essaid, 1990b, 1992] showed that hydraulic conductivity and leakance were the critical properties. Changes in conductivity affected the head gradients in the basin and the drawdowns at the wells. Leakance values controlled the amount of water moving through the system (recharge and discharge) and the amount of vertical leakage to the wells. From these simulations it also became apparent that the freshwater flow system achieved a nearly steady state freshwater head distribution within a one-year time step so that specific storage was not a critical parameter. Simulations were also insensitive to changes in effective porosity which was assigned a constant value of 0.1 in all units. Table 8.4 summarizes the aquifer parameters used in the model.

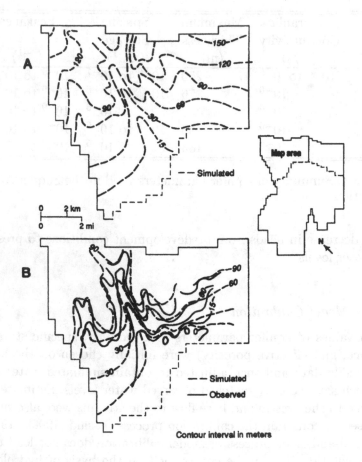

Figure 8.16. Composite onshore water level maps of the Soquel-Aptos model area showing: A. simulated predevelopment water levels, and B. simulated and observed water levels for April 1981.

8.5.3 Results of SHARP Simulations

Initial conditions for the 1930-1985 period were obtained by simulating predevelopment conditions and allowing the system to achieve steady state. The simulated onshore predevelopment water levels are shown in Figure 8.16a, a composite water level map constructed by taking water levels from each unit's outcrop area (e.g., unit A in the west, unit E in the east). In the near-surface groundwater flows mainly to the streams and toward the coast under the influence of topography.

Transient conditions were simulated for the period from 1930 to 1985, using annual time steps through 1980 and monthly time steps for 1981 through 1985. Records of the Soquel Creek Water District production wells tapping the Purisima Formation were used to apportion historical pumpage to individual wells as they went into production. Wells penetrating several layers were simulated by locally increasing the leakance between the layers, allowing the model solution to determine the flow contribution of each layer.

Bloyd [1981] constructed a ground-water level map representing the near-surface flow system in April 1981 using water levels from the shallow Purisima subunits, except near the coast where production wells tap the deeper units. The simulated shallow unit water levels for April 1981 are compared with the observed water levels of Bloyd [1981] in Figure 8.16b. In the northern part of the area, water levels remained relatively close to predevelopment conditions through 1981. At the coast, however, pumpage has modified the predevelopment flow field. The below sea level cones of depression reflect heads in the lower subunits tapped by the deep production wells. These cones of depression now capture some of the water that previously flowed to streams and offshore.

To help understand how development has modified fluxes through the system, the simulated annual recharge, discharge, and pumpage relations for the period from 1930 to 1985 are shown in Figure 8.17. Prior to development, the recharge to the system was 0.50 m^3/s, of which 0.47 m^3/s discharged onshore to creeks and to the overlying Aromas Sand. Only 0.03 m^3/s of the water flowed offshore to the sea, yet this small proportion was sufficient to maintain the freshwater-saltwater interface position offshore. With development and an increase in pumpage in the basin the 1981 onshore and offshore discharges had decreased to 0.43 m^3/s and 0.01 m^3/s, respectively. The decrease in ground-water discharge to the streams is illustrated in Figure 8.17 by the decrease in base flow to Soquel Creek corresponding to increased ground-water pumpage.

The freshwater flow system is most active onshore where there is strong topographic relief driving flow toward the streams. Water that is not discharged to the streams flows offshore. This flow is small but has been sufficient to prevent saltwater intrusion. Most of the water that has been developed from the basin comes from captured base flow and additional induced recharge.

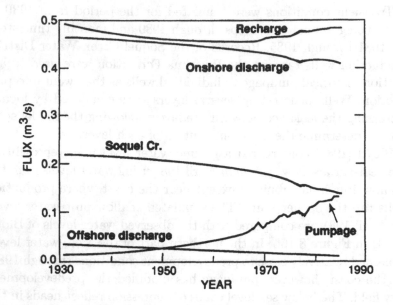

Figure 8.17. Simulated 1930 to 1985 fluxes through the Soquel-Aptos basin.

The 1930 to 1985 simulations showed almost no movement of the offshore interface, despite significant changes in the flow system. This suggests that interface response is quite slow and takes place over long time frames. The slow response of the saltwater zone is a result of the low horizontal and vertical hydraulic conductivities that impede the flow of saltwater into the interface zone.

8.6   Conclusions

Disperse and sharp interface approaches have been used to analyze saltwater intrusion in coastal aquifers. The disperse interface approach explicitly represents the transition zone, where there is mixing of freshwater and saltwater due to the effects of hydrodynamic dispersion. The sharp interface approach simplifies the analysis by assuming that freshwater and saltwater do not mix and are separated by a sharp interface.

Each of these approaches has advantages and limitations and can only be employed successfully under the appropriate conditions. In general, studies using the disperse interface approach have generally been limited to two-dimensional vertical cross-sections due to computational constraints. The sharp interface approach, in conjunction with vertical integration of the flow equations, allows problems to be reduced by one dimension. This approach does not give information concerning the nature of the transition zone; however, it will reproduce the general response of the interface to applied stresses. SHARP, a quasi-three-dimensional model that simulates freshwater and saltwater flow separated by a sharp interface in layered coastal systems is a tool that can be used to conduct regional simulations of coastal ground-water conditions.

Chapter 9

USGS SUTRA Code — History, Practical Use, and Application in Hawaii

C. I. Voss

9.1   Introduction

The U.S. Geological Survey's SUTRA code is the most widely used simulator in the world for seawater intrusion and other variable-density groundwater flow problems based on solute transport or heat transport. Since its initial release in 1984, the SUTRA code has also been widely used for many other types of problems. It is routinely employed for hydrogeologic analyses published in both white and gray scientific literature, and for many practical engineering studies that are not widely distributed. It is used in university courses to teach hydrologic concepts and modeling.

The SUTRA code is a flexible tool for hydrogeologic analysis offering the means to numerically simulate a variety of subsurface physical processes and situations. It robustly and reliably provides simulation results for well-posed problems. While the SUTRA code can be used to simulate saturated-unsaturated systems with subsurface energy transport or single-species reactive solute transport, this chapter focuses only on seawater intrusion. These systems employ the SUTRA code's capability to simulate the processes of saturated variable-density flow with non-reactive solute transport.

The topics discussed in the first half of this chapter are: the history of the development and use of SUTRA, the mathematical and numerical basis for the code, and finally, practical aspects of using the code for variable-density flow and transport simulation. The second half of the chapter is a case study of the major coastal aquifer

J. Bear et al. (eds.), Seawater Intrusion in Coastal Aquifers, 249–313.

system on the island of Oahu, Hawaii, which reviews a number of SUTRA-simulation-based analyses of the system. The case study considers: groundwater flow and aquifer hydraulics, dynamics of the transition-zone between freshwater and seawater, water ages and geochemistry, and management of a potable water supply. Many aspects of both the modeling approach used and of the physical behavior of the Oahu system learned from the analyses are readily applicable to other coastal aquifer systems and other variable-density flow problems.

9.2   History of SUTRA

SUTRA (acronym for **S**aturated-**U**nsaturated **TRA**nsport code) [Voss, 1984a] began in 1981 as an effort to create a finite-element code to solve a very general set of single-phase subsurface fluid flow and single-species transport problems. At the time, the U.S. Geological Survey (USGS) had available only the finite-difference SWIP code (subsurface waste injection program) [Intercomp, 1976; Intera, 1979] for salt-water intrusion problems, and no published finite-element code for transport. Initially SUTRA was written as a three-dimensional (3D) code, also including two- and one-dimensional (2D and 1D) mesh types.

The code was written to solve the basic equations presented by Bear [1979] which cover most types of groundwater flow and transport physics known. The two equations solved are a fluid mass balance for unsaturated and/or saturated groundwater flow, and a unified energy and solute mass balance. The unified balance provides a single equation that describes either solute or energy transport simply by changing the definition of parameters in the equation.

In creating SUTRA, a good deal of effort was spent on numerical methods development, searching for computationally-economical finite-element methods (FEM) competitive with finite-differences' (FD) efficiency. However, no FD variant of FEM was found that provides the robustness and direction-independence of regular FEM. The only numerical economy implemented in SUTRA was use of a diagonalized time-derivative term. Otherwise, a standard Galerkin FEM was initially used, based on developments by Professor George F. Pinder and students at Princeton University (of which the SUTRA author was one) in the 1970s. Later, numerical problems with this method when applied to variable-density flow required development of some modifications, as described in this chapter.

The coding style of SUTRA is modular with each routine serving a special function, and coding simplicity is stressed over code efficiency in order to allow users to make easy changes and additions of processes to the code. A robust direct solver was selected to avoid the convergence difficulties associated with iterative matrix solvers, despite the fact that such a solver would be less efficient for problems with large meshes. The code is written in standard Fortran-77.

The original 3D-SUTRA code was tested in 1982. Nonetheless, after running the code on some 3D-test problems, the 3D capability was removed and SUTRA was reduced to 2D. On computers generally available at that time, a single time step of a small cube-shaped flow and transport test problem on a 3D mesh consisting of ten elements on an edge (1000 elements total) would take 20 minutes of computer time. Thus, users without super-computer access could not have applied the code to any real 3D problem with the density of discretization required for accurate solution of the balance equations. In order to avoid forcing users to carry out 3D-simulation analysis with large errors due to poor discretization, only a 2D code was publicly released by USGS.

The earliest application of the 1982 SUTRA code was in modeling the major coastal aquifer of Oahu, Hawaii, as part of the Regional Aquifer System Analysis (RASA) program of the USGS. This hydraulic modeling study is described later in this chapter. The southern Oahu aquifer contains a very thin transition zone (TZ) between freshwater and seawater in comparison with the aquifer thickness. This was a fortunate first field case for SUTRA as it required significant numerical improvement. It was originally impossible to simulate the thin TZ in cross section (with Galerkin FEM); regardless of how low the dispersion parameters were set, the simulated TZ would become very wide. This resulted in a series of test problems to isolate the reason for the difficulty. One such problem consisted of a closed box in cross-section initially containing static freshwater over static seawater separated by a one-element-wide TZ. When simulated though time the TZ grew in width although no stresses or flows were imposed on the box. This test case is now recommended as a test for any new variable-density code, as it highlights a potential shortcoming of the standard Galerkin FEM. The solution to this numerical problem implemented in SUTRA is the consistent velocity algorithm [Voss and Souza, 1987] which matches the spatial variability of the pressure gradient with that of the density-gravity product found in Darcy's law. If the spatial variability of these terms is not matched, then any numerical method (finite-element or other) may generate false spurious vertical flow in regions of vertical density gradients. This revision to SUTRA allowed the simulation to reproduce the relatively thin TZ of Oahu, Hawaii [Souza and Voss, 1987]. It also provided a series of test cases now generally accepted as basic tests for new variable-density codes [Voss and Souza, 1987].

The USGS released the 2D-version of SUTRA in 1984 [Voss, 1984a]. As with most USGS-produced codes, SUTRA is in the public domain. This implies that it is free to use and redistribute by any user, and that it may be freely modified and redistributed as long as the modification is clearly indicated and the original code is referenced as a USGS product. Since SUTRA's release, many users indeed have made modifications to the code, as originally intended by USGS. They have included a variety of new processes, such as addition of: simultaneous multiple solute and energy transport, evapotranspiration, chemical reactions, leakage through semi-confining layers, areal recharge, concentration-dependent permeability, and non-linear density dependence on concentration. The first revision of SUTRA was released in 1990, correcting the few bugs that had been found in six years of use and adding some new features.

Pre- and post-processing for SUTRA began with printer-paper output of nodewise values that were to be contoured by hand. Pre-processing for generation of meshes and SUTRA input data was limited to simple regular meshes in the utility codes distributed with SUTRA by USGS. In 1987, automatic plotting of contours and velocity vectors given in SUTRA simulation results became possible with the USGS code, SUTRAPLOT [Souza, 1987], although many users continued to use their own procedures, plotting packages, or commercial packages such as SURFER™.

Since SUTRA's initial release, passive 'by-request-only' distribution of SUTRA by USGS was supplemented by more-aggressive marketing and redistribution by commercial firms. More-sophisticated pre- and post-processing interfaces were developed by universities and commercial firms worldwide. These developments led to extensive worldwide use of the code, and SUTRA has arguably become the most-widely-used code for variable-density groundwater problems including seawater intrusion.

More-recent developments include a complete graphical user interface (GUI) for SUTRA that provides a geographical information system (GIS) type of functionality together with automatic mesh generation [Voss et al., 1997]. The SUTRA-GUI consists of public-domain USGS codes that connect SUTRA with the commercial Argus ONE™ package that provides the meshing and GIS-like capabilities. This GUI provides SUTRA users with new possibilities and analytic power, making it easy to set up, run, and view results for even complex SUTRA models. A second minor revision was released in 1997 that provides a few new minor features. SUTRA and the SUTRA-GUI have become available electronically on the USGS World Wide Web site: *http://water.usgs.gov/software/*

SUTRA has been found to be a highly reliable code that can solve any well-posed variable-density problem, though it is not optimized for computational speed. Development of new numerical methods is a low priority in planned near-future further development of SUTRA. Computer speed and size have been the only real limiting factors for practical application since the code was released, and will likely continue to be so. Computer speed and size have increased since SUTRA's 1984 release much more than any extra speed or efficiency given by new numerical or matrix-solver methods developed since then. Now, complex 2D SUTRA simulations are done regularly on the ubiquitous personal computer. The numerical effort for a simulation (i.e. computer time and space required) is determined by the density of spatial discretization and the number of time steps simulated. Lack of sufficiently powerful computers to most users has been the major reason for USGS restraint in releasing a 3D upgrade of SUTRA. A 3D upgrade is under development in the hope, however, that continued evolution of computer power will indeed allow well-discretized 3D transport simulations to be carried out on personal computers in the near future.

9.3   Uses of SUTRA

SUTRA is a very general code, it is in the public domain, and it robustly provides solutions; these characteristics have led to application of the code to evaluate a wide variety of problems. Although an exhaustive listing of SUTRA applications is not possible to provide in this chapter, some of the more recent examples of the types of problems to which SUTRA has been applied are described. This review serves as background for SUTRA application to variable-density flow and seawater intrusion studies. Most of the following examples of SUTRA applications are from the 'white' literature (such as refereed journals), though by far, the bulk of SUTRA applications are reported in the 'gray' literature (such as government, university and consulting reports).

SUTRA has been applied to evaluation of groundwater flow, for example, in a crystalline rock fracture zone [Andersson et al., 1991], and in areal field cases (e.g. Cherkauer and McKereghan [1991]). Unsaturated zone flow studies (e.g. Herbert [1992], McReanor and Reinhart [1996]) using SUTRA are less common than saturated media studies. This is because the SUTRA manual [Voss, 1984a] takes a conservative view of this type of code application, warning users that simulation of such non-linear problems is difficult, although the code capably handles such problems.

SUTRA has also been applied to evaluation of solute transport. This includes areal contaminant transport in field cases (e.g. Cherkauer et al. [1992]), cross-sectional organic liquid contaminant transport [Abriola et al., 1993], bioremediation of jet-fuel contamination [Chapelle et al., 1996], and effects of heterogeneous fields on transport [Duffy and Lee, 1992; Saiers et al., 1994; Bajracharaya and Barry, 1997]. Other solute-transport applications of SUTRA include evaluation of tracer tests [Hyndman et al., 1994; White, 1994; Lessoff and Konikow, 1997; Haggerty et al., 1997], and analysis of well-aquifer interaction [Reilly and Gibs, 1993; Reilly and LeBlanc, 1998]. Energy transport applications of SUTRA with nearly constant fluid density include flow and transport through an embankment dam [Johansson, 1991] and temperature regimes in shallow groundwater [Bundschuh, 1993].

Modifications to SUTRA have been made for a variety of purposes including addition of equilibrium chemical reactions [Lewis et al., 1986, 1987], adding kinetic reactions [Brogan, 1991; Haggerty,1992; Smith et al., 1997; Sahoo et al., 1998], and calculation of path lines [Cordes and Kinzelbach,1992]. Model comparisons have been done using SUTRA as a standard model for transport method development [Syriopoulou and Koussis,1991] and for seawater intrusion

[Ghassemi et al., 1993]. Also, SUTRA has been used in model verifications for other constant-density codes [Liu and Dane, 1996] and for other variable-density codes [Senger and Fogg, 1990; Oldenburg and Pruess, 1995; Schincariol and Schwartz, 1994; Williams and Ranganathan, 1994].

SUTRA has been widely used for inverse modeling and optimization in groundwater systems. An Inverse SUTRA, called SUTRA^{-1}, was developed by Piggott et al. [1994, 1996]. Examples of SUTRA applications to calibration and parameter estimation include estimation of solute retardation [Rogers, 1992], transport model calibration and contaminant source identification [Wagner, 1992], mass transfer rate calibration for remediation [Brogan and Gailey, 1995], and estimating hydraulic conductivity fields [Harvey and Gorelick, 1995].

Applications of SUTRA to hydraulic optimization include, hydraulic management [Makinde-Odusola and Mariño, 1989], temporal hydraulic management [Jones, 1990], stochastic hydraulic management [Chan, 1993, 1994], and field applications of aquifer management (e.g. Hallaji and Yazicigil [1996]). Applications to optimization of aquifer remediation schemes include consideration of parameter uncertainty [Wagner and Gorelick, 1987], consideration of heterogeneity [Wagner and Gorelick, 1989; Gailey and Gorelick,1993], consideration of chemical kinetics [Haggerty and Gorelick, 1994], spatial optimization using neural networks [Rogers and Dowla, 1994; Johnson and Rogers, 1995; Rogers et al., 1995], spatial optimization using a genetic algorithm [Cedeño and Rao Vemuri, 1996], and temporal and spatial optimization [Rizzo and Dougherty, 1996]. Optimization of sampling design was considered by Wagner [1995], and Nordqvist and Voss [1996], and groundwater quality monitoring was considered by Smith and Ritzi [1993], all using SUTRA.

Perhaps the widest use of SUTRA has been for subsurface systems wherein fluid density is variable and is dependent on temperature or solute concentration. Applications for variable-density flow and solute transport include regional brine displacement [Lahm et al., 1998], and nuclear waste site evaluations, such as risk at the WIPP site, New Mexico, USA [Helton, 1993], hydrogeologic evaluation of the Sellafield area, U.K. [Heathcote et al., 1996], potential sites in Germany [Schelkes et al., 1991], and sites in Sweden's Precambrian shield [Voss and Andersson, 1993; Provost et al., 1998]. SUTRA has been applied to evaluate sedimentary basins, such as the arid Great

Basin [Duffy and Al-Hassan, 1988], the Palo Duro basin [Senger, 1993], both in the USA, and for analysis of flow and transport in deforming accretionary sub-sea prisms [Screaton et al., 1990; Bekins et al., 1995]. Variable-density flow near salt domes and evaporites has been evaluated with SUTRA by Ranganathan and Hanor [1988, 1989], Ranganathan [1992], Vogel et al. [1993], Schelkes and Vogel [1992], and Klinge et al. [1992]. Variable-density flow near saline lakes has been analyzed using SUTRA by Rogers and Dreiss [1995a, b], Narayan and Armstrong [1995], Simmons [1997], and Simmons and Narayan [1997, 1998]. Simmons [1997] also applied SUTRA to a new laboratory-experiment-based benchmark problem that generates a complex variable-density flow field from evaporation at a saline lake. Variable-density flow with energy transport, where fluid density depends on temperature rather than solute concentration, is exemplified by SUTRA application to study of carbonate platform dolomitization by Jones et al. [1997] and Sanford et al. [1998], as well as by Screaton and Ge [1997] for heat transport in an accretionary prism.

Finally, many examples of the use of SUTRA for variable-density flow with seawater intrusion studies exist. A few examples of these SUTRA applications are: coastal groundwater systems [Souza and Voss, 1987; Bush, 1988; Price and Woo, 1990; Achmad and Wilson, 1993; Smith, 1994; Gangopadhyay and Das Gupta, 1995; Emekli et al., 1996; Nishikawa, 1997; Oki et al., 1998], island groundwater systems [Ghassemi et al., 1990a, b], atoll island groundwater systems [Oberdorfer et al., 1990; Underwood et al., 1992], saltwater upconing near wells [Reilly and Goodman, 1987], and generic studies of seawater intrusion [Reilly, 1990].

9.4   SUTRA Equations and Algorithms

9.4.1   Physics and Governing Equations

The SUTRA code is merely a numerical solver of two general balance equations for variable-density single-phase saturated-unsaturated flow and single-species (solute or energy) transport based on Bear [1979]. The general fluid mass balance equation that is usually referred to as the 'groundwater flow model' is:

$$\left(S_w\rho S_{op} + \varepsilon\rho\frac{\partial S_w}{\partial p}\right)\frac{\partial p}{\partial t} + \left(\varepsilon S_w\frac{\partial \rho}{\partial U}\right)\frac{\partial U}{\partial t}$$

$$-\nabla\cdot\left[\left(\frac{\mathbf{k}k_r\rho}{\mu}\right)\cdot(\nabla p - \rho\mathbf{g})\right] = Q_p \qquad (9.1)$$

where S_w is the fractional water saturation [1], ε is the fractional porosity [1], p is the fluid pressure [kg/m·s^2], t is time [s], U is either solute mass fraction, C [kg$_{solute}$/kg$_{fluid}$], or temperature, T [°C], $\mathbf{k}$ is the permeability tensor [m^2], k_r is the relative permeability for unsaturated flow [1], μ is the fluid viscosity [kg/m·s], $\mathbf{g}$ is the gravity vector [m/s^2], and Q_p is a fluid mass source [kg/m^3·s]. Additionally, ρ is the fluid density [kg/m^3], expressed as:

$$\rho = \rho_o + \frac{\partial \rho}{\partial U}(U - U_o) \qquad (9.2)$$

where U_o is the reference solute concentration or temperature, ρ_o is fluid density at U_o, $\partial\rho/\partial U$ is the density change with respect to U (assumed constant), and S_{op} is the specific pressure storativity [kg/m·s^2]$^{-1}$

$$S_{op} = (1 - \varepsilon)\alpha + \varepsilon\beta \qquad (9.3)$$

where α is the compressibility of the porous matrix [kg/m·s^2]$^{-1}$, and β is the compressibility of the fluid [kg/m·s^2]$^{-1}$. Fluid velocity, $\mathbf{v}$ [m/s], is given by:

$$\mathbf{v} = -\frac{\mathbf{k}k_r\rho}{\varepsilon S_w\mu}\cdot(\nabla p - \rho\mathbf{g}) \qquad (9.4)$$

The solute mass balance and the energy balance are combined in a unified solute/energy balance equation usually referred to as the 'transport model':

$$[\varepsilon S_w\rho c_w + (1 - \varepsilon)\rho_s c_s]\frac{\partial U}{\partial t} + \varepsilon S_w\rho c_w\mathbf{v}\cdot\nabla U$$

$$-\nabla\cdot\{\rho c_w[\varepsilon S_w(\sigma_w\mathbf{I} + \mathbf{D}) + (1 - \varepsilon)\sigma_s\mathbf{I}]\cdot\nabla U\}$$

$$= Q_p c_w(U^* - U) + \varepsilon S_w\rho\gamma_1^w U + (1 - \varepsilon)\rho_s\gamma_1^s U_s$$

$$+\varepsilon S_w\rho\gamma_0^w + (1 - \varepsilon)\rho_s\gamma_0^s \qquad (9.5)$$

where: c_w is the specific heat capacity of the fluid [J/kg·°C], c_s is the specific heat capacity of the solid grains in porous matrix [J/kg·°C],

ρ_s is the density of the solid grains in porous matrix [kg/m^3], σ_w is the diffusivity of energy or solute mass in the fluid (defined below), σ_s is the diffusivity of energy or solute mass in the solid grains (defined below), **I** is the identity tensor [1], **D** is the dispersion tensor [m^2/s], U^* is the temperature or concentration of a fluid source (defined below), γ_1^w is the first-order solute production rate [s^{-1}], γ_1^s is the first-order sorbate production rate [s^{-1}], γ_0^w is an energy source [J/kg·s], or solute source [kg$_{solute}$/kg$_{fluid}$·s], within the fluid (zero-order production rate), γ_0^s is an energy source [J/kg·s], or solute source [kg$_{solute}$/kg$_{fluid}$·s], within the fluid (zero-order production rate).

In order to cause the equation to represent either energy or solute transport, the following definitions are made:
for energy transport

$$U \equiv T, \qquad U^* \equiv T^*, \qquad \gamma_1^w = \gamma_1^s = 0,$$

$$\sigma_w = \frac{\lambda_w}{\rho c_w}, \qquad \sigma_s = \frac{\lambda_s}{\rho c_w}$$

for solute transport

$$U \equiv C, \qquad U_s \equiv C_s, \qquad U^* \equiv C^*,$$

$$\sigma_w = D_m, \qquad \sigma_s = 0, \qquad c_s = \kappa_1, \qquad c_w = 1$$

where λ_w is the fluid thermal conductivity [J/s·m·°C], λ_s is the solid-matrix thermal conductivity [J/s·m·°C], T is the temperature of the fluid and solid matrix [°C], C is the concentration of solute (mass fraction) [kg$_{solute}$/kg$_{fluid}$], T^* is the temperature of a fluid source [°C], C^* is the concentration of a fluid source [kg$_{solute}$/kg$_{fluid}$], C_s is the specific concentration of sorbate on solid grains [kg$_{solute}$/kg$_{grains}$], D_m is the coefficient of molecular diffusion in porous medium fluid [m^2/s], κ_1 is a sorption coefficient defined in terms of the selected equilibrium sorption isotherm (see Voss [1984a] for details).

The dispersion tensor is defined in the classical manner:

$$D_{ii} = |\mathbf{v}|^{-2} \left(d_L v_i^2 + d_T v_j^2 \right)$$
$$D_{ij} = |\mathbf{v}|^{-2} \left(d_L - d_T \right) v_i v_j \tag{9.6}$$

for $i = x, y$ and $j = x, y$, but $i \neq j$, with

$$d_L = \alpha_L |\mathbf{v}|$$
$$d_T = \alpha_T |\mathbf{v}| \tag{9.7}$$

where d_L is the longitudinal dispersion coefficient $[m^2/s]$, d_T is the transverse dispersion coefficient $[m^2/s]$, α_L is the longitudinal dispersivity $[m]$, and α_T is the transverse dispersivity $[m]$.

A very useful ad-hoc generalization of the dispersion process is included in SUTRA allowing the longitudinal and transverse dispersivities to vary in a time-dependent manner at any point depending on the direction of flow. The practical implications of this are discussed below in section 9.6. The generalization is:

$$\alpha_L = \frac{\alpha_{L\,max}\alpha_{L\,min}}{\alpha_{L\,min}\cos^2\theta_{kv} + \alpha_{L\,max}\sin^2\theta_{kv}}$$
$$\alpha_T = \frac{\alpha_{T\,max}\alpha_{T\,min}}{\alpha_{T\,min}\cos^2\theta_{kv} + \alpha_{T\,max}\sin^2\theta_{kv}} \qquad (9.8)$$

where $\alpha_{L\,max}$ is the longitudinal dispersivity for flow in the maximum permeability direction $[m]$, $\alpha_{L\,min}$ is the longitudinal dispersivity for flow in the minimum permeability direction $[m]$, $\alpha_{T\,max}$ is the transverse dispersivity for flow in the maximum permeability direction $[m]$, $\alpha_{T\,min}$ is the transverse dispersivity for flow in the minimum permeability direction $[m]$, θ_{kv} is the angle from the maximum permeability direction to the local flow direction [degrees].

These relations Eq. 9.8 provide dispersivity values that are equal to the square of the radius of an ellipse that has its major axis aligned with the maximum permeability direction. The radius direction is the direction of flow.

In the case of seawater intrusion (i.e. variable density saturated flow with non-reactive solute transport of total dissolved solids or chloride, and with no internal production of solute), Eqs. 9.1, 9.4, and 9.5 are much simplified. The fluid mass balance is:

$$\rho S_{op}\frac{\partial\rho}{\partial t} + \varepsilon\frac{\partial\rho}{\partial U}\frac{\partial U}{\partial t} - \nabla\cdot\left[\frac{k\rho}{\mu}\cdot(\nabla p - \rho\mathbf{g})\right] = Q_p \qquad (9.9)$$

The fluid velocity is given by:

$$\mathbf{v} = -\frac{k\rho}{\varepsilon\mu}\cdot(\nabla p - \rho\mathbf{g}) \qquad (9.10)$$

and the solute mass balance is:

$$\varepsilon\rho\frac{\partial U}{\partial t} + \varepsilon\rho\mathbf{v}\cdot\nabla C - \nabla\cdot[\varepsilon\rho(D_m\mathbf{I} + \mathbf{D})\cdot\nabla U] = Q_p(C^* - C) \qquad (9.11)$$

9.4.2 Numerical Methods

The numerical technique employed by the SUTRA code to solve the above equations uses a modified two-dimensional Galerkin finite-element method with bilinear quadrilateral elements. Solution of the equations in the time domain is accomplished by the implicit finite-difference method. Modifications to the standard Galerkin method that are implemented in SUTRA are as follows. All non-flux terms of the equations (e.g. time derivatives and sources) are assumed to be constant in the region surrounding each node (cell) in a manner similar to integrated finite-differences. Parameters associated with the non-flux terms are thus specified nodewise, while parameters associated with flux terms are specified elementwise. This achieves some efficiency in numerical calculations while preserving the accuracy, flexibility, and robustness of the Galerkin finite-element technique. Voss [1984a] gives a complete description of these numerical methods as used in the SUTRA code.

An important modification to the standard finite-element method that is required for variable-density flow simulation (as mentioned in section 9.2) is implemented in SUTRA. This modification provides a velocity calculation within each finite element based on consistent spatial variability of pressure gradient, ∇p, and buoyancy term, $\mathbf{g}$, in Darcy's law, Eq. 9.10. Without this 'consistent velocity' calculation, the standard method generates spurious vertical velocities everywhere there is a vertical gradient of concentration within a finite-element mesh, even with a hydrostatic pressure distribution [Voss, 1984a; Voss and Souza, 1987]. The spurious velocities make it impossible to simulate a narrow TZ between freshwater and seawater with the standard method, irrespective of how small a dispersivity is specified for the system.

The two governing equations, fluid mass balance and solute mass (or energy) balance, are solved sequentially on each iteration or time step. Iteration is carried out by the Picard method with linear half-time-step projection of non-linear coefficients on the first iteration of each time step. Iteration to the solution for each time step is optional. Velocities required for solution of the transport equation are the result of the flow equation solution (i.e. pressures) from the previous iteration (or time step for non-iterative solution).

SUTRA is coded in a modular style, making it convenient for sophisticated users to modify the code (e.g. replace the existing direct banded-matrix solver) or to add new processes. Addition of new terms (i.e. new processes) to the governing equations is a straightforward process usually requiring changes to the code in very few lines. Also to provide maximum flexibility in applying the code, all boundary conditions and sources may be time-dependent in any manner specified by the user. To create time-dependent conditions, the user must modify subroutine BCTIME. In addition, any desired unsaturated functions may be specified by user-modification of subroutine UNSAT.

Two numerical features of the code are not often recommended for practical use: upstream weighting and pinch nodes. Upstream weighting in transport simulation decreases the spatial instability of the solution, but only by indirectly adding additional dispersion to the system. For full upstream weighting, the additional longitudinal dispersivity is equivalent to one-half the element length along the flow direction in each element. Pinch nodes are sometimes useful in coarsening the mesh outside of regions where transport is of interest by allowing two elements to adjoin the side of a single element. However, pinch nodes invariably increase the matrix bandwidth possibly increasing computational time despite fewer nodes in the mesh.

9.4.3   Testing Variable-Density Aspects of the Code

The SUTRA code has been tested for its ability to solve a number of basic bench-mark-type problems that partly prove its accuracy in variable-density simulation. In approximate order of increasing complexity, these tests are:

1. A closed rectangular region containing freshwater above highly saline water, separated by a horizontal TZ one element wide and with a hydrostatic pressure distribution [Voss and Souza, 1987]. No simulated flow may occur in the cross section under transient or steady state conditions.

2. The same rectangular cross section with uniform horizontal flow imposed by specifying hydrostatic pressures along one vertical edge somewhat higher than along the opposite edge. Transverse dispersivity and molecular diffusivity are set to zero. The correct transient and steady state solution maintains the TZ in the same row of elements as exists initially.

3. The Henry seawater intrusion problem on a coarse mesh [Voss and Souza, 1987] and on a very fine mesh [Ségol, 1993]. Note that until Ségol [1993] improved Henry's approximate semi-analytical solution [Henry, 1964], this test was merely an inter-comparison of numerical code results. No published results had matched the Henry solution. Ségol's result provides a true analytic check on a variable-density code.

4. The Elder natural convection problem [Voss and Souza, 1987], a transient problem consisting of dense fluid circulating downwards under buoyancy forces from a region of high specified concentration along the top boundary. Results of this test can only be used in comparison among numerical codes.

5. The HYDROCOIN salt-dome problem (e.g. see Konikow et al. [1997]), a steady state and transient problem, initially consisting of a simple forced freshwater flow field sweeping across a salt dome (specified concentration located along the central third of the bottom boundary). This generates a separated region of brine circulating along the bottom. Again, this result is for comparison of codes.

6. The salt-lake problem [Simmons, 1997], a transient system in which a complex evaporation-driven density plume evolves downwards in an area of upward-discharging groundwater in a Hele-Shaw cell experiment. Results are compared with laboratory measurements.

9.5   Application of SUTRA to Seawater Intrusion

Application of SUTRA for simulation of variable-density fluid problems, such as seawater intrusion, requires attention to a variety of aspects of setup peculiar to variable-density simulation, in addition to aspects normally required for setup of constant-density simulations. These aspects include mesh design, spatial and temporal discretization, iteration, boundary conditions, and initial conditions.

9.5.1   Boundary Conditions

Normally, for SUTRA, when no pressure value, source, or sink is specified along a model boundary, the boundary becomes closed to fluid flow. When either one of these is specified, flow may occur across the boundary at that point.

SUTRA requires specification of pressures rather than hydraulic heads when the simulation deals with variable fluid density. The most typical specification of pressure along a vertical (or any non-horizontal) boundary in a cross-sectional model is a hydrostatic pressure distribution. This condition allows there to be horizontal flow across the boundary, but disallows vertically oriented flow at the boundary. This condition applies very well along the downward sloping sea bottom when it is specified as the seaward boundary of an aquifer. Commonly, because aquifers extend well seaward of a coast, or to great distances from the study area of hydrologic interest, the model boundary must be placed at an arbitrary position within the aquifer and not at a natural hydrologic boundary. Arbitrary positions may be 'just offshore' or at the 'last observation well' near the coast. If such an arbitrary vertical boundary is placed sufficiently far from the area of interest, such that any flow occurring across the boundary would be approximately horizontal, and if the vertical distribution of fluid density (or solute concentration) is known or may be reasonably assumed, then a hydrostatic pressure specification is appropriate. Inland boundaries may be at natural hydrogeologic boundary locations, or also may have to be placed arbitrarily within the aquifer (e.g. at an observation well).

The specified hydrostatic pressure distribution must be matched to the vertical distribution of fluid density expected along the boundary, and the fluid density need not be constant. To calculate the pressure distribution, the incremental pressure increases ($\rho_i g \triangle E_i$, where ρ_i is the mean density of fluid in vertical increment $\triangle E_i$ between adjacent nodes) should be simply summed, starting at the top of the boundary with the desired top pressure, and working downwards through the nodes along the boundary.

Note that along such a boundary, the only requirement is that the vertical pressure gradient is hydrostatic. A higher or lower 'hydraulic head' along the boundary relative to that found inside the modeled area (that will cause inflow or outflow) may be applied. This is done by specifying a higher or lower starting pressure as the top pressure of the boundary prior to calculating the hydrostatic distribution.

The vertical distribution of fluid density at such an arbitrary boundary is not always well known. After some simulation time, the fluid density (solute concentration) just inside the hydrostatic pressure boundary differs significantly from the distribution assumed when first specifying the condition. In this case, it would be preferable to specify a self-consistent boundary condition. Such a condition would match the specified hydrostatic pressure distribution at the boundary with whatever the fluid density distribution is (as it changes with time) just inside the model boundary instead of with the arbitrarily chosen density distribution. Provost et al. [1998] used a modification to SUTRA user-programmed subroutine BCTIME for time-dependent boundary conditions that implements such a self-consistent condition.

When specifying pressure at a node in a SUTRA simulation, the solute concentration of any fluid that may enter the model domain at that location must also be specified. This is quite different from specifying the solute concentration of the fluid within the model itself (SUTRA's "specified concentration boundary condition"). The fluid that enters at the specified pressure node first must mix with fluid existing within the model at that point before the model concentration can be determined. Specifying concentration directly in a model is possible, but is usually not realistic for representing field situations; rather, this condition is most useful for generic analyses or for matching analytic solutions of the transport equation. Another means of adding solute to a model is by use of SUTRA's "solute source" which contributes solute mass without water to the system. This is useful when the rate of solute mass flux into a region is known, such as from dissolution of subsurface salts.

A special consideration for cross-sectional simulation of variable-density flow concerns numerical precision. When pressures are specified along a boundary, they must be given with the complete precision of values stored in the computer being used, usually 15 significant digits of computer 'double-precision' [Voss and Souza, 1987]. The reason for this is as follows. When fluid velocities are calculated by SUTRA, very small spatial differences in pressure (i.e. from node to node) may give rise to fluid flow at any depth. Pressures near the top of a cross-sectional model may be near 0 Pa, while pressures at depth are often 10^5 to 10^7 Pa due to hydrostatic load of the water column. For example, a lateral difference of 10^{-2} Pa at any depth must give rise to the same lateral velocity in SUTRA. If, however, due to insufficient numerical precision in the higher pressures, this small difference is lost or not accurately represented at depth, then significant errors would occur in the simulated flow field. Thus, specified pressures (e.g. hydrostatic pressures) along a boundary must be calculated using the exact relation for fluid density with exact parameters used by SUTRA (Eq. 9.2) and must be entered to full 15-digit precision in SUTRA's input data for specified pressure. If approximate, rather than exact, pressures are entered into the input data, then undesired vertical flows will be generated at the boundary in the simulation.

9.5.2   Initial Conditions

Another special consideration for cross-sectional simulation of variable-density flow concerns initial conditions for pressure and concentration. Because the pressure distribution in a variable-density fluid depends on the density distribution, the initial pressures for a simulation may not be specified independently of the initial concentrations specified. The initial pressure field must be consistent with the concentration field, and simultaneously consistent with the boundary conditions (sources, sinks and specified pressures) affecting the pressures.

Creation of a set of consistent pressures requires an additional preliminary run of the SUTRA code. This preliminary SUTRA run is a steady-state pressure solution using the desired flow-field boundary conditions, the desired initial concentration field, and arbitrary (e.g. blank) initial pressures. The resulting steady-state pressures calculated by SUTRA are consistent with the initial concentrations and boundary conditions. The procedure to be followed after this run is that the arbitrary pressures in the initial conditions input file should be replaced with the new consistent pressures (using a text editor to move and replace the data). The SUTRA initial conditions input file then contains initial conditions that may be used to start a series of transient simulations or the additional preliminary runs described below.

A further complication occurs when the initial concentration distribution for the time period to be simulated is not well known. If concentrations (and fluid densities) are known at only a few points within the region to be modeled and are assigned assumed initial values in-between, then false, most likely unrealistic, flows will be calculated by SUTRA. This difficulty is exacerbated by our unavoidable ignorance of how the concentration distribution evolved to the state at which it exists as an initial condition. The measured state of the concentration distribution may indeed be transient, having evolved over time with various environmental and hydrogeologic stresses. In such a transient situation, it is difficult, or even impossible, to generate an initial condition for concentration that is useful.

One practical solution is to run a second preliminary simulation. This simulation may be started at some simple initial condition (e.g. no solute in aquifer, or aquifer contains high-concentration solute throughout, etc.). It should be run in transient mode (with both pressure and concentration solved on each time step) until such time when the simulation result is similar to that observed in the field. This simulation result may then be used as initial conditions for the desired transient analysis. Of course, the hydrogeologic parameters of the system used in the model may not be representative of reality and this could lead to errors in further analysis. The ambiguity may be mitigated to some extent by a sensitivity analysis that considers the impact on the second preliminary run results of changes in the model parameters.

Another practical (and more commonly used) solution is to assume that the known conditions of initial concentration are approximately in steady state. This also requires a second preliminary run to establish the steady-state conditions. This run may be started with any concentration distribution, but starting with a guess that is closer to the correct steady solution will result in faster convergence and shorter real simulation times. The SUTRA simulation should be transient (pressure and concentration solved each time step) and should use the set of boundary conditions applicable at the time at which the concentration field is desired. The simulation is continued until the concentration distribution no longer changes (i.e. it reaches steady state). Most often, this is the longest and most difficult simulation of an entire modeling analysis, as it begins with poor initial conditions and must end with completely self-consistent steady-state conditions for pressure and concentration. Subsequent transient simulations use this result as an initial condition. These are quite often easier to carry out and suffer less from any numerical instability. This approach, though appealing and simpler than the one suggested just above, because it allows the assumption that the system is at steady state, suffers from the same ambiguity as before.

It therefore also requires sensitivity runs to determine the impact of the underlying assumed model structure on the aspect of the system being studied by the simulations.

9.5.3   Spatial Discretization

As in constant-density simulation, modeled domains must be discretized sufficiently to represent the spatial distributions of hydrogeologic parameters. It is important to note that one layer of elements is insufficient to discretize a geologic unit within which flow and transport results are of interest. Many element layers are required to properly represent distributions and physical processes in the layer. Additional discretization may be required in regions of high fluid flow and to accommodate a mesh Peclet number criterion [Voss, 1984a]. In variable-density simulations, however, some of these considerations are modified somewhat.

For most constant-density transport simulations with SUTRA, in order to avoid general oscillations in the concentration distribution, finite element size must be limited. The limit is at most four times the longitudinal dispersivity (where size is measured along the direction of fluid flow). This is the mesh Peclet number criterion presented by Voss [1984a]. In the case of variable-density simulation, this requirement may often be relaxed. Oscillation occurs only in regions where there is a significant concentration gradient along the flow direction. Many seawater intrusion situations exhibit very low concentration gradients along the flow direction, due to the fact that variable-density flow physics causes freshwater to flow parallel to, rather than toward, the inland margin of intruding seawater. Thus, in many cross-sectional models where flow tends to be parallel both to the land surface and the intruding seawater wedge, the lateral element size may safely violate the mesh Peclet number criterion by ten times or more. To carefully reproduce narrow TZs, vertical discretization often needs to be finer than lateral discretization. Narrow TZs generally occur in the above-described flow systems when transverse dispersion is the major mixing process, or in some coastal systems where low-amplitude tidal pumping generates the zone. In other types of variable-density systems where longitudinal mixing is more significant, discretization must be adjusted according to the mesh Peclet number criterion as usual.

In cross-sectional variable-density simulation, freshwater flow may converge to a narrow discharge area (for example, in a freshwater lens discharging at the coast). In such areas, the flow velocity increases significantly near the discharge. This is the most likely area for another type of numerical instability to develop in a SUTRA simulation despite the fact that it was properly discretized for the mesh Peclet number criterion. The instability may typically appear as a stationary region of negative solute concentration around the discharge area. This situation is most often alleviated by finer spatial discretization (along the flow direction) near the discharge and by finer time discretization. Occasionally, the time or space discretization required to remove the instability would be impracticably small. Then two practical alternatives are suggested: 1) Higher values of dispersivity may be applied near the discharge to remove the instability, or 2) the simulation result may be used with the instability remaining (i.e. ignore the instability), if indeed it does not disturb the rest of the simulated field, and if the area of interest is sufficiently far away from the instability.

A mesh designed for use in simulation analysis of a coastal seawater intrusion problem should be tested for its ability to represent the desired TZ as generated by the physics represented by the SUTRA code rather than by numerical discretization errors. For coastal seawater intrusion problems with two water types (freshwater and seawater), when transverse dispersivity and molecular diffusivity are both set to zero, the exact steady-state solution contains a perfectly sharp interface between freshwater and seawater. For the type of finite-elements used in SUTRA, a perfectly sharp interface is not possible to represent in the mesh, and for this problem, a more gradual transition will be the result. Such a simulation will give an indication of the minimum width TZ the mesh can represent. If this minimum width is much smaller than that of the TZ required for the analysis, then the discretization across the zone is fine enough. If the minimum zone has about the same width as the required zone, then the simulation is not reproducing the variable-density flow and transport physics sufficiently well; rather, the TZ that appears is caused by poor discretization. In this case, further mesh refinement is required in the direction across the TZ, usually vertically.

9.5.4   Iteration and Time Discretization

Iteration is often required for SUTRA simulations of systems with non-linear sorption or unsaturated flow. Iteration is sometimes required for simulation of variable-density flow. Note that iteration refers to re-solving each time step a number of times in order to resolve physical or chemical non-linearities. It does not refer to iterations that may be required to solve the matrix equations for the SUTRA unknowns when using an iterative matrix solver. The non-linearity in variable-density flow is primarily systematic (dependent on the coupling of the flow field with the fluid density distribution) rather than parametric (as in unsaturated flow where parameter values themselves are highly dependent on the state of the system). The parametric non-linearity is generally mild as density is nearly a linear function of concentration. (In variable-density systems with energy transport, the parametric non-linearity is greater due to the exponential dependence of fluid viscosity on temperature, but it is still only a minor non-linearity.)

There is no a-priori rule that determines when to use iterative solution for a variable-density flow problem. The practical approach is to try a simulation twice: with and without iteration. If iteration provides a different result, then iteration must be used for similar simulations. The same philosophy applies to selection of time-step size. A simulation should be tried with a reduced time step (e.g. halved). If the result is different, then an even smaller time step should be tested and used. Often in seawater intrusion problems, the density distribution changes slowly enough over each time step such that iteration is not required. Thus, smaller time steps may be used as a substitute for iterations.

Convergence of the iterative process in SUTRA is considered to have occurred when the absolute change in all pressures and concentrations between subsequent iterations drops below user-selected values. For typical variable density problems, the pressure may be considered converged when changes are less than the pressure equivalent of 1 cm of head (100 Pa) and concentration changes are less than 1/100 of the maximum system concentration (for normal seawater, 3.57×10^{-4} kg/kg for total dissolved solids concentration). More stringent convergence criteria may be used if needed. Most often, when iterative solution is allowed for a new SUTRA seawater intrusion simulation, 5 to 20 iterations may be required for the first few time steps. The number of iterations typically decreases to one per time step (i.e. the equivalent of no iteration) after 10 to 20 time steps.

9.6   Setting Up a Coastal Seawater Intrusion Simulation

9.6.1   Model Simplicity and Problem Definition

The guiding principle of successful numerical modeling is to start out as simply as possible and add complexity to the model only as needed and as warranted. The temptation to make the initial model complex, by including many details in order to make it appear visually and technically realistic, must be resisted by the modeler. Access to a numerical code and fast computer that will allow a model to be made complex is neither a license nor an imperative to make it so. Usually, most of the useful result that can be obtained in application of numerical model analysis to a generic or field problem can be gained with highly simplified and approximate representations of the problem by the model. When the model is kept simple, the analysis may be accomplished more swiftly and with considerably less human and computational effort.

The other key to successful model analysis is the clear definition of one or more questions to answer, to be defined before embarking on the analysis. Typically, questions relate to what the future state of the system will be (e.g. How much will the salinity increase at this well-field if pumping continues at today's levels?). In contrast, the most useful questions relate to defining the controlling factors that determine the system behavior of interest (e.g. Which hydrogeologic parameter determines the rate of movement of the potable water limit towards the supply well?). Little is achieved by calibrating a model to represent 'today's' conditions in an aquifer system and making predictions with the calibrated model. Such blind application of a model is a poor (albeit popular) use of numerical modeling and results from the lack of a meaningful goal for the analysis. The main goal should be to gain new understanding of how the system of interest functions. The combination of a well-defined goal and parsimonious (i.e. as simple as possible) modeling is usually a guarantee of a useful result.

9.6.2   Model Setup

Model Domain. The spatial cross-sectional domain of the SUTRA model should be selected to include the areas of interest for seawater intrusion, and is ideally extended to natural hydrogeologic boundaries such as practically impermeable units, groundwater divides, or surface-water bodies. Often, it is not possible to end the model domain at natural boundaries, as discussed above in section 9.5.1. Many coastal models have a seaward vertical boundary located at an arbitrary point offshore.

Thickness. When using SUTRA for seawater intrusion in 2D cross-section, the model "Thickness" refers to the direction perpendicular to the plane of the vertical section. While 'unit-thickness' is often used (e.g. 1 m), this may require arbitrary allocation of sources and sinks such as vertical recharge and pumping wells to the section. The difficult question is, how much of the aquifer pumping comes out of the particular section being modeled. The arbitrary assignment of some or all pumping in an aquifer to an arbitrarily thick cross section is the main ambiguity that exists in 2D modeling of a 3D system. An equivalent, but more appealing, approach may be to make the model thickness equal to the entire width of the aquifer system (this need not be spatially constant in the section). Then it is natural to include all sources in the aquifer, such as vertical recharge and pumping within the section. This will give an aquifer-width-averaged simulation that will likely underestimate the amount of seawater intrusion occurring near any specific well at a given pumping rate. In contrast, taking all pumping from a unit-thickness section would likely overestimate the intrusion. In face of this dilemma, a practical 2D approach must be based on trying a range of source values assigned to the section considered and observing the resulting range of simulation responses.

Permeability. Variable-density flow fields are very sensitive to permeability distribution. Regions of constant permeability differing by as little as about ten times in permeability will likely host separately circulating flow cells. Thus, initial parameterization of permeability in a new model should consist of, at most, two or three regions of uniform permeability.

Dispersivity. Dispersivity should usually be constant throughout the model. Prior to simulation and when lacking field values (as is the usual case), a range of values possible for the system should be decided upon. Initial simulations may be carried out with the higher values to improve numerical stability of the first runs. Sometimes dispersivity has an unimportant effect on results (for example, when the width of the simulated TZ is not of concern). Then it may simply be set to a value that guarantees spatially-stable concentration solutions (e.g. see Voss and Andersson [1993] and Provost et al. [1998]).

SUTRA allows both the longitudinal and transverse dispersivity values to vary depending on direction of flow (Eq. 9.8), and SUTRA is apparently still unique among available transport codes in this respect. Flow-direction-dependent dispersivity is indeed a useful generalization of the classical dispersion paradigm. For example, it may be expected that flow parallel to layering in a layered medium and flow perpendicular to the layering do not necessarily have the same longitudinal dispersivity. Further, effective values of dispersivity are understood to be transport-scale dependent, and typical values of longitudinal dispersivity are approximately one-tenth of the transport reach. For seawater intrusion, as considered in cross-section with SUTRA, solutes may travel kilometers inland from the sea within a given aquifer layer and experience an effective longitudinal dispersivity of hundreds of meters. After reaching inland, the solutes may be induced to travel upwards to an overlying aquifer layer when shallow wells are turned on; this transport occurs over the distance of aquifer thickness of a few tens of meters. The dispersivity of hundreds of meters for lateral flow is not at all appropriate or meaningful for quantifying transport over a vertical distance of tens of meters. Rather, a dispersivity of a few meters is more reasonable. Thus, for simulating this system using SUTRA, longitudinal dispersivity for flow in the horizontal direction would be specified to be about 100m and for flow in the vertical direction would be specified to be about 1m. Equivalent flexibility is available in a SUTRA simulation for transverse dispersivity; however, physical arguments for choosing different values are not as straightforward as for longitudinal dispersivity and its value is usually held constant (i.e. independent of flow direction).

Porosity. An appropriate value of porosity is not always easy to select for transport modeling in heterogeneous aquifers. Grain-size analyses and other laboratory measurements give a value of 'total porosity', whereas the transport equation (Eqs. 9.5 and 9.11) employs the porosity associated with the fluid that moves ('effective porosity'). These two values can be quite different in heterogeneous media. Flow may occur within only a few percent of the total pore volume (i.e. effective porosity of a few percent for fractures or permeable sand stringers), while most of the water volume exists in pores within low-permeability regions located between the flowing channels (with a total porosity of tens of percent). For relatively fast flow through such a medium, the existence of the static water may be neglected and porosity may be set to the value of effective porosity. If not available from a large-scale tracer test, this porosity value must be determined in an ad-hoc manner or by trial and error. In contrast, for very slow-flow through such a system, the flow in permeable channels is not much faster than through the low-permeability regions, and solute concentrations may equilibrate between flowing channels and relatively-immobile zones through solute diffusion. In this case, an appropriate value of effective porosity tends to be higher, approaching the value of total porosity.

In some variable-density systems, there is a wide range of fluid velocity with the lowest values usually found in deep dense fluids. In such a case, and when the medium is heterogeneous, there is no one value of effective porosity that is appropriate everywhere, although the average permeability, total porosity and structure may be constant throughout. In the fast-flowing portion of the system, the nearly immobile water does not participate in the transport. In the slow-flowing portion, the immobile water may act as source or sink of solutes with transport through the immobile water by slow flow or diffusion. Thus to some extent, the appropriate value of effective porosity depends on fluid velocity which itself may be time-dependent. SUTRA is not based on a dual-porosity type of formulation. It allows only time-constant (although possibly spatially dependent) porosity to be specified. Thus, care must be exercised when interpreting transport results for such heterogeneous systems.

Water Table in Cross Section. When a water table is required in a cross-sectional model, there are two ways of implementing it in SUTRA. The first way requires modification of the code to add a specific yield term to nodes along the top edge of the mesh where the water table occurs. This approach is described in Souza and Voss [1987]. The second way involves use of the unsaturated-zone capability of SUTRA to model the water table naturally as an unsaturated

zone. The details of the unsaturated zone are not at all of importance when using this approach to track the water table. Thus, both
functions that must be set for SUTRA unsaturated-zone simulation,
the saturation-pressure relationship and the relative permeability-
saturation relationship, should be made simple linear functions. The
first line may connect a saturation of one at a pressure of zero with
a non-zero residual saturation (e.g. 0.01) at a selected negative pressure. This negative pressure value should be selected to be about
the hydrostatic pressure value occurring in a column of water with
height of one to a few times the vertical element spacing. The second
line may connect a relative permeability of one at a saturation of one
with a relative permeability of zero at the selected residual saturation. Because the resulting unsaturated problem is nearly linear, only
a few iterations per time step are required and the computational
effort needed for the variable-density simulation is not appreciably
impacted.

9.6.3   Running the Simulation

Usually for variable-density simulation, SUTRA should be set to
solve for both pressure and concentration on each time step. When
the first pressure solution (e.g. for creating natural initial conditions
as described above) is obtained, it must be checked for conformity
with the specified pressure boundary conditions. The objective is
to set the SUTRA parameter, GNUP, such that about seven digits
of the calculated pressure at each specified pressure node match the
value specified in SUTRA's input. Only one node need be checked, for
example, the one at the bottom of a hydrostatic seaward boundary.
To adjust the match, increasing/decreasing GNUP by one order of
magnitude will cause one more/less digit of calculated pressure to
match.

The objective of a first simulation of a variable-density problem
is usually to create a natural steady-state pressure field for use as
an initial condition for subsequent runs (described above). This run
requires only one steady-state pressure step. The next simulation is
most often used to find the steady-state pressure and concentration
solution to a static set of boundary conditions, usually prior to development of water resources from the aquifers being studied. This
run is a transient non-iterative solution, which may require significant computational time. By checking observation node values or

plotting concentration spatially, it may be determined whether the concentration field has reached steady state by the end of the run. If not, the run may be restarted using the former result as an initial condition.

If unacceptable numerical stability appears during the simulation, then mesh or time-step refinement may be required as described in sections 9.5.3 and 9.5.4. Another approach to circumvent the instability may be to increase the dispersivities for this run, and continue the simulation to near steady conditions. Then, reduce the dispersivity sequentially in a series of subsequent runs, using each previous run's result as initial conditions for the next until a steady solution with the desired dispersivity has been achieved.

If other types of problems arise, it may be useful to plot and check the velocity field, as these plots often help to isolate an error in the model's input. Note that plots of the pressure are most often uninteresting, a series of near-horizontal lines with values increasing linearly with depth that give no indication of the flow direction. This is due to the basic nature of variable-density flow, wherein no true potential field (i.e. hydraulic head) can be defined. A velocity plot gives much more of the desired information on the flow field.

When a consistent steady-state solution has been obtained, transient simulations may then be run with different system stresses, for example, with the addition of withdrawals from the aquifer. For time-dependent boundary conditions or sources, the user must program SUTRA subroutine BCTIME and re-compile the code.

9.7   Application of SUTRA to Major Coastal Aquifer of Southern Oahu, Hawaii

9.7.1   Introduction to Oahu Groundwater

The main population center of the volcanic Hawaiian island chain in the north-central Pacific Ocean, as well as the city of Honolulu, is on the island of Oahu (population about 900,000) (Figure 9.1). Oahu depends on groundwater for its entire potable water supply, most of this produced from coastal aquifers. The growth and continued existence of this large city on a small oceanic island is possible only because of a number of specific circumstances associated with the groundwater supply. Oahu has high rainfall, highly perme-

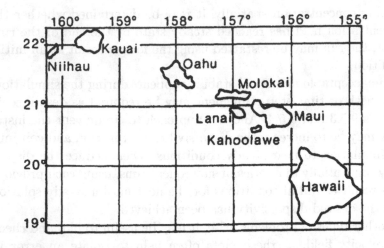

Figure 9.1. Map showing location and relative positions of Hawaiian islands (after Hunt et al. [1988]).

able aquifers, and a coastal semi-confining layer above many of the aquifers. The coastal confining unit keeps heads high at the coast, and creates a very thick freshwater lens. This combination of high recharge, high permeability and impeded discharge provides an ample supply of fresh groundwater for both drinking and irrigation. Aside from anthropogenic contamination, seawater, a natural contaminant, exists at the bottom of each of Oahu's coastal aquifers, and water supplies are potentially threatened by significant seawater intrusion.

Development of water supply from aquifers on Oahu first began in about 1880. Fortunately, the U.S. Geological Survey effectively maintained records of withdrawals and groundwater levels for the entire history of groundwater development on Oahu. This information provides the basis for analysis of the hydrology of the aquifers. The data include a fortuitous aquifer-wide pumping test, when the bulk of pumping (irrigation wells) shut down during a 1958 strike at Oahu's sugar mills. Additionally, the Board of Water Supply of the City and County of Honolulu and the State of Hawaii placed deep observation wells and have monitored these regularly. These are used to track the movement of the transition zone between freshwater and seawater (TZ). This type of information provides a basis for analysis of seawater intrusion in the aquifers.

A major issue in Oahu that remains officially unresolved today is how much groundwater may be safely produced from each aquifer. This is a contentious issue as limited availability of freshwater may potentially limit further economic growth. On the other hand, over-development of groundwater may lead to deleterious impacts on local groundwater supplies from seawater intrusion. Such intrusion may be difficult to reverse over a short period and may cause economic damage.

The premise of studies described here is that a careful quantitative analysis of aquifer hydraulics and the physics of seawater intrusion will give Oahu water managers the most reliable scientific basis possible for deciding on water allocation and for managing Oahu's aquifer systems. The aquifer of interest in these studies is the largest Oahu aquifer, the Pearl Harbor area aquifer of southern Oahu. Although the focus here is on a particular regional aquifer, many results on the dynamics of transition zones are general and apply to other coastal aquifers.

The approach taken in this analysis is to apply the SUTRA code to first evaluate aquifer hydraulics, and then to evaluate seawater intrusion by study of the TZ. The results presented here are extracted from six works: Hunt et al. [1988] (hydrogeology of Hawaii), Souza and Voss [1987] and Voss and Souza [1987] (analysis of aquifer hydraulics), Souza and Voss [1989] (aquifer management), Voss and Souza [1998] (transition-zone dynamics), and Voss and Wood [1994] (geochemistry and groundwater ages).

9.7.2   Hydrogeology

Oahu was built by lavas extruded from two shield volcanoes and their associated rift zones (Figure 9.2), the Waianae on the west (approximately 3 million years old) and the younger Koolau on the east (approximately 2 million years old). The rift zones are characterized by a concentration of sub-vertical dikes through which basalts were extruded at the surface. The bulk of the island mass is composed of dike-free basalts formed by sub-aerial lava flows that make up a gently dipping stratified aquifer fabric extending from land surface to considerable depth. These layers rest upon a base of submarine extruded pillow basalts, which, because of subsidence of the island, are now at a depth of more than 2 km below sea level [Andrews and Bainbridge, 1972]. Although some interfingering of Waianae and Koolau

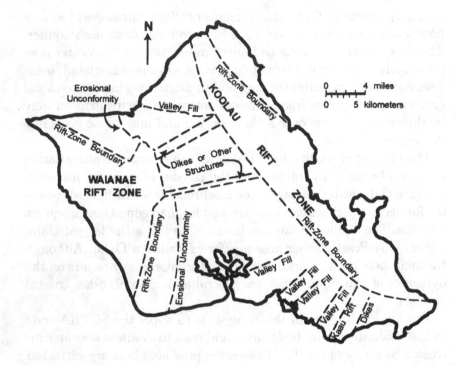

Figure 9.2. Map showing hydrogeologic barriers on Oahu (after Hunt et al. [1988]).

basalts is thought to have occurred, the bulk of Koolau basalts rest upon the weathered flank of the older Waianae volcano. While the eastern side of the Koolau volcano and both sides of the Waianae volcano are severely eroded, the Koolau basalt flows in the central saddle of the island have not been eroded extensively. Much of the coastal portions of the island are covered by a sedimentary wedge (referred to as 'caprock') composed of layers of marine and terrestrial sediments and clays as well as coral reef and organic debris. The largest caprock extends offshore 30 km to 40 km south of Oahu. The caprock forms a barrier that inhibits the free discharge of fresh groundwater to the ocean.

The layered basalts are highly transmissive to groundwater and have hydraulic conductivity typically 100 to 1000 m/d. The basalts are also highly vesicular and porous (up to 50%). Connected porosity, through which significant water flow can occur, is less than 10%. The largest lava flows can be tens of kilometers long, up to 1 km wide and up to 10 m thick. Current examples of such flows may be found on the volcanically active island of Hawaii. The well-connected, highly conductive zones occur in rubbly material between overlapping lava flows, while the core of the lava flows themselves are relatively impermeable. The stack of tabular units may be 100 to 1000 times less conductive vertically than horizontally and the caprock is 1000 to 10000 times less conductive than the layered basalts [Souza and Voss, 1987]. In view of a geologic history of rapid building of the volcanoes, the stack of lava flows likely makes up an anisotropic but homogeneous aquifer fabric on a regional scale.

A number of geologic barriers control the regional flow of groundwater on Oahu (Figure 9.2). The rift zones contain swarms of intersecting low-permeability vertical dikes which cut across layered basalts and impound water in numerous compartments [Takasaki and Mink, 1985; Hunt et al., 1988]. Sediment-filled valleys scoured into the layered basalts also act as barriers to flow through the layered basalts. Another barrier is the contact zone between Waianae and Koolau basalts (erosional unconformity in Figure 9.2), wherein the weathered surface of the Waianae basalts likely has lower conductivity than unweathered layered basalts. The caprock acts to impede groundwater discharge near the coast, raising hydraulic heads in the layered basalt aquifers. Except beneath the caprock, aquifers are unconfined. Two other barriers to flow are inferred from precipitous drops in groundwater levels below the central saddle of Oahu [Dale and Takasaki, 1976]. To date, it has not been determined whether these barriers are dikes, weathered ridges of the Waianae volcano, faults, or other low-permeability structures.

Groundwater recharge is concentrated in the high rainfall areas along the eroded volcanic ridges. In areas where rainfall is more than 1.3 m/yr, roughly half of the rainfall may recharge the aquifers [Dale and Takasaki, 1976]. The high-recharge areas are coincident with the rift zones and central saddle area of Oahu. In some areas, rainfall can exceed 9 m per year. The resulting groundwater levels display sharp elevation changes at geologic barriers. In the Waianae rift zone, the highest measured water level is 490 m, in the Koolau rift zone the highest level is 300 m, the central saddle area has a water level of 85 m, while between the coast and inland geologic barriers, water levels rarely exceed 10 m. Water levels are nearly flat within compartments formed by barriers. Thus, various water bodies and barriers have

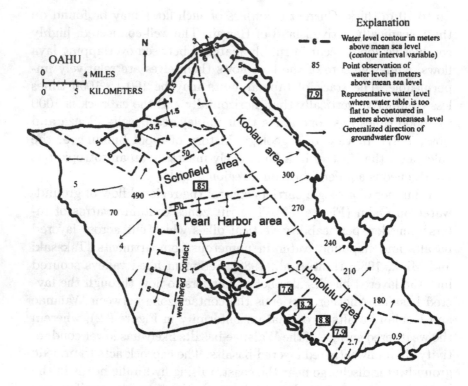

Figure 9.3. Water levels in principal Oahu aquifers and inferred fresh-water flow directions (modified from Hunt et al. [1988]).

been delineated on Oahu largely on the basis of water level (Figure 9.3). These include the Koolau area aquifers consisting of water in dike compartments, the Schofield area aquifer in subaerial Koolau basalts in the Waianae-Koolau saddle, the Honolulu area aquifers partly separated by scoured filled valleys, and the large Pearl Harbor area aquifer in subaerial Koolau basalts.

The coastal water bodies with low-elevation water tables are called 'basal-water bodies'. A basal-water body is a freshwater lens that freely floats on intruded seawater (Figure 9.4), with an intervening zone of transition containing a mixture of freshwater and saltwater. The Pearl Harbor area body, the largest continuous water body on Oahu, is separated from the Honolulu area water bodies to the east by a valley-fill barrier (Figure 9.3). It is bounded by the Koolau dike-impounded water bodies in the rift zone to the northeast, by the Schofield high-level water body in the central saddle to the north,

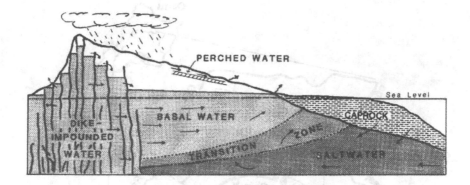

Figure 9.4. Schematic cross-section of typical ground-water bodies in Hawaii (after Hunt et al. [1988]).

and by the weathered contact with Waianae basalts to the west.

Natural recharge to the Pearl Harbor area aquifer occurs mainly in the upper-elevation areas as a result of direct infiltration, discharge from the central-saddle high-level water body, and discharge from the Koolau dike-impounded water body. Total recharge to the Pearl Harbor area is about 9×10^5 m^3/d (240 Mgal/d) [Mink, 1980]. Uncertainty in the actual quantities of recharge to the Pearl Harbor area aquifer is of critical importance in both modeling analysis and management of the aquifer. Natural discharge from the aquifer occurs along a line of springs at the inland boundary of the caprock near Pearl Harbor, and likely as diffuse leakage through the caprock to Pearl Harbor, the involuted water body located near the center of Oahu's south coast. Saltwater recharge likely occurs through offshore portions of the caprock, and saltwater discharge occurs in springs and likely in a diffuse manner through near-shore portions of the caprock (Figure 9.4).

Pumping from the aquifer began in the early 1880's initially for sugar cane irrigation. Rates of withdrawal increased continuously from the early 1900's to 1980 with resultant reductions in head and upward and landward movement of the TZ. Most wells for public and irrigation supply are located in a strip near the coast (about 1.5 km wide) that parallels the inland edge of the caprock (Figure 9.5). Some wells near the coast have been abandoned as a result of increasing salinity due to seawater intrusion.

Pre-development (pre-1880) heads in the Pearl Harbor aquifer

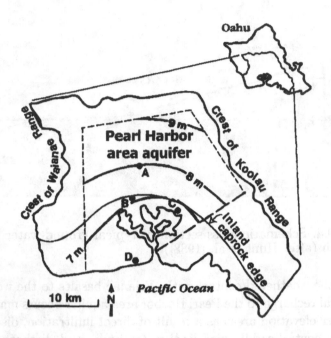

Figure 9.5. Map showing Pearl Harbor area aquifer with hydrogeologic boundaries (modified from Souza and Voss [1987]) and hydraulic heads in 1958 (modified from Mink [1980]). Observation wells: A, 2659-01 (Waipio); B, 2300-18 (Waipahu); C, 2356-51 (T67); D, 1959-05 (Ewa).

measured about 10 m near the spring discharge area, and are estimated to have been more than 12 m near the upstream boundary [Mink, 1980]. The pre-development freshwater lens in the Pearl Harbor area was thus 400 m to 500 m thick [Mink, 1980], assuming conditions of static equilibrium with seawater. Heads in 1958 are shown in Figure 9.5. In 1990, freshwater heads in the aquifer ranged from about 5 m to 7 m above sea level, and measured salinity profiles showed the freshwater lens to be only 200 m to 300 m thick [Voss and Souza, 1998]. Thus heads dropped 6 m from 1880 and the freshwater lens was reduced by half its thickness in 100 years of groundwater development in the aquifer. A cross-sectional view of the TZ based on recent data from deep observation wells in basalt, from spring salinity, and from a well penetrating the caprock is given in Figure 9.6. The TZ (shown to scale in Figure 9.7) is 100 m to 200 m thick at most points, but comprises less than 10 percent of assumed aquifer width, and may be considered narrow on the regional scale.

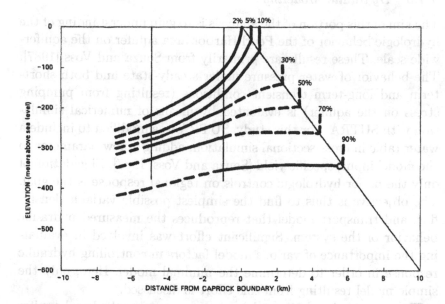

Figure 9.6. Actual transition zone in southern Oahu aquifer (1990) constructed from data in wells (A, B, C, D) and coastal springs (after Voss and Souza [1998]) . (Concentrations in percent seawater. Dashed lines hypothesized where data is missing.) Triangle at upper right is caprock area, remainder is layered basalts (compare with Figure 9.4). Inland well is observation well A (see Figure 9.10), and near-caprock well is combination of observation wells B and C. Circle at base of caprock is bottom-hole concentration in Well D (W. R. Souza, unpublished data from analysis of discrete water sample (1981)). Concentration in spring discharges above inland edge of caprock from Visher and Mink [1964].

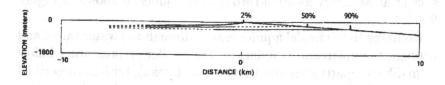

Figure 9.7. Actual transition zone in southern Oahu aquifer (1990) shown in true scale (after Voss and Souza [1998]). (Concentrations in percent seawater. Dashed lines hypothesized.)

9.7.3   Hydraulic Modeling

The aim of this portion of the analysis is to gain understanding of the hydrologic behavior of the Pearl Harbor area aquifer on the aquifer-wide scale. These results are primarily from Souza and Voss [1987]. The behavior of water pressure under steady-state and both short-term and long-term transient conditions (resulting from pumping stress on the aquifer) is evaluated by means of numerical simulation with SUTRA. For this study, SUTRA was modified to include a water table in cross-sectional simulation, adding a new parameter to the model input, specific yield [Souza and Voss, 1987]. Elucidation of only the major hydrologic controls on regional response is the goal. The objective is thus to find the simplest possible variable-density flow and transport model that reproduces the measured hydraulic behavior of the system. Significant effort was involved in evaluating the importance of various model factors in controlling hydraulic response in order to determine the minimal model. Here, only the simple model resulting from this effort is discussed.

The cross-sectional 2D-model domain represents the basal aquifer and caprock containing the basal freshwater lens, TZ, and deeper seawater as illustrated in Figure 9.4 to the right of the dike zone. A 2D representation is highly appropriate for evaluation of aquifer-wide response in this system for a few reasons:

1. Flow is approximately perpendicular to the coastal boundary.

2. Permeability anisotropy is primarily horizontal and vertical.

3. The caprock is parallel to the coast.

4. Sources and sinks are rather evenly spread along bands parallel to the coast.

The SUTRA mesh for hydraulic analysis, representing a 1 m vertical slice at an arbitrary location through the aquifer, is shown in Figure 9.8.

Recharge to the model aquifer enters through the water table, and through the upstream boundary of the aquifer (representing inflow from dike compartments and the central plateau), both at prescribed rates. Model response is not sensitive to the vertical distribution of the upstream recharge. Withdrawals occur near the coast in the modeled pumping center. Withdrawal and recharge quantity changes with time according to historical data (see Table 1 of Souza and Voss [1987]). The seaward model boundary is held at hydrostatic seawater

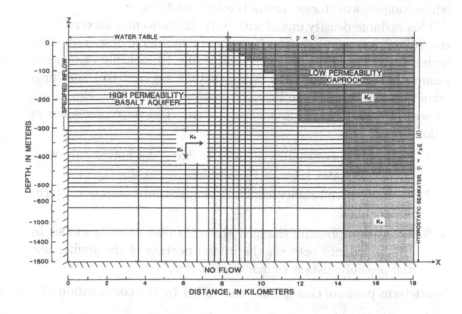

Figure 9.8. SUTRA finite-element mesh for cross-sectional hydraulic modeling of Pearl Harbor area aquifer (after Souza and Voss [1987]). Shows regions of constant permeability (k_C is caprock permeability, k_S is sea boundary permeability, k_h and k_v are horizontal and vertical permeabilities of layered basalt).

pressure and the top boundary above the caprock is held at a pressure of zero. The other boundaries are modeled as impermeable, with the bottom representing the location of pillow lavas at the base of the subaerial basalts and the rear representing the dike zone and central plateau boundary.

Permeability is defined for two regions, the basalt, which has different vertical and horizontal permeability, and the caprock, with equal vertical and horizontal values, although much lower than the basalt. In reality, the caprock is more complex (e.g. see Oki et al. [1998]) but this study focuses on water in the layered basalts and a bulk representation of the caprock is used. An additional region of permeability is defined along the vertical boundary below the caprock, with potentially lowered permeabilities, to represent the resistance to seawater flow of the portion of aquifer seaward of the boundary. Other hydraulic parameters are held constant throughout the model domain. A specific yield value is assigned to the water table only, while compressive storage exists throughout the section.

This variable-density model with only six constant-valued regional controlling parameters embodies system hydraulic response. This includes short-term hydraulic response due to monthly changes in pumping (such as during the 3-month 1958 sugar strike) and long-term response (such as the 100-year change in water-table elevation), as well as steady flow through the system. The six controls are as follows.

1. Horizontal permeability mainly controls the horizontal gradient of the water table in the aquifer.
2. Caprock permeability controls the absolute elevation of the water table.
3. Long-term pressure changes are controlled to some extent by the impedance between the inland portion of the aquifer and the possibly distant source of seawater.

Short-term pressure changes are controlled by the combination of:

4. specific yield,
5. aquifer matrix compressibility, and,
6. vertical permeability.

Table 9.1 gives values of fixed model parameters and the best-fit values of the six controlling parameters as determined by manual fitting of model response to the hydraulic data. The ability of this simple model to reproduce both short and long-term hydraulic responses is illustrated in Figure 9.9, which shows good response of the SUTRA model for 25 years of monthly-varying pumping in the aquifer.

9.7.4   Pearl Harbor Area Transition Zone

In many coastal aquifer systems where the aquifer is thin, seawater intrusion occurs only as a wedge along the coast so that the area of contact of saltwater with freshwater is relatively small. In the Pearl Harbor area of the southern Oahu aquifer, freshwater mixes with saltwater even beneath recharge areas and the fresh and mixed waters flow more than 10 km to the discharge areas. Because the aquifer is thick, the TZ between freshwater and saltwater occurs throughout the areal extent of the aquifer. The concentration distribution of solutes in this regional TZ thus provides an excellent large-scale

Parameter	Units	Base TZ Model	Range Tested (Best Model)
General			
Freshwater density	(kg/m^3)	1000	
Seawater density	(kg/m^3)	1024.99	
Water compressibility	$(1/Pa)$	4.47×10^{-10}	
Fluid viscosity	$(kg/m/s)$	10^{-3}	
Molecular diffusivity	(m^2/s)	1.5×10^{-9}	
Specific yield	(1)	0.04	$0.01 \rightarrow 0.15$ (0.04)
Aquifer matrix compressibility	$(1/Pa)$	2.5×10^{-9}	$10^{-11} \rightarrow 10^{-8}$ (2.5×10^{-9})
Caprock			
Hydraulic conductivity	(m/d)	0.0457	$0.01 \rightarrow 100$ (0.0457)
Longitudinal dispersivity α_L	(m)	250	
Transverse dispersivity α_T	(m)	0.25	

Table 9.1. Parameters for southern Oahu models. (Note: Base TZ model refers to TZ study, its base values are derived from best hydraulic model.)

natural tracer that can be used to study regional-scale dispersion processes.

This second portion of the analysis evaluates potential for seawater intrusion by study of the TZ. Concern about seawater intrusion in southern Oahu is motivated by the loss of about half of the freshwater lens thickness in only 100 years of aquifer development, and no evidence that the shrinkage has ended. The aim of this portion of the analysis is to gain understanding of TZ dynamics in the Pearl Harbor area aquifer on the aquifer-wide scale. This should be done with as simple a model as possible.

Using SUTRA for this analysis, aquifer parameters and dispersion processes that may control groundwater flow and solute transport are examined. Their effect is evaluated on both the steady-state character of the TZ and the regional movement and internal redistribution of solute concentrations in the TZ resulting from long-term and fluc-

Parameter	Units	Base TZ Model	Range Tested (Best Model)
Basalt			
Horizontal hydraulic conductivity	(m/d)	457	$150 \rightarrow 1500$
Anisotropy ratio			$1 \rightarrow 1000$
(K_H/K_V)	(1)		Hydraulic model
		200	$20 \rightarrow 500$
			TZ model
			(200)
Effective porosity	(1)	0.04	$0.02 \rightarrow 0.40$
			(0.04)
Leakance of		0.0125	4.57×10^{-5}
sea boundary	(1/d)		$\rightarrow 0.457$
			(0.457)
Horizontal longitudinal		250	$25 \rightarrow 2500$
dispersivity α_{LH}	(m)		(Model-A 250)
			(Model-I 100)
Vertical longitudinal		10	$0.5 \rightarrow 250$
dispersivity α_{LV}	(m)		(Model-A 50)
			(Model-I 100)
Horizontal transverse		0.25	$0 \rightarrow 10$
dispersivity α_{TH}	(m)		(0.25)
Vertical transverse		0.25	$0.25 \rightarrow 1000$
dispersivity α_{TV}	(m)		(arbitrary)

(Table 9.1 continued)

tuating short-term hydraulic stress. The TZ may be in a transient state of response to the major increases in withdrawals that occurred in the mid- to late-1900s, and responds regularly to yearly changes in stress. Unraveling the dynamics of the TZ in southern Oahu through measurements made today may thus be a difficult task.

Concentration data collected in deep monitor wells provide profiles characterizing the state of the TZ in the Pearl Harbor aquifer. As an example, the seawater concentration profile at about 5 km from the coast from Well A (Waipio, 2659-01) is shown in Figure 9.10 (see Figure 9.5 for location; this is also the inland well of Figure 9.6). At this location, the freshwater lens is about 250 m thick and the TZ

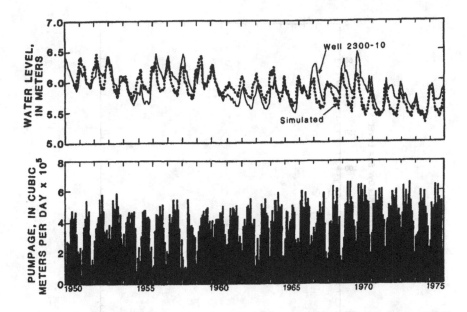

Figure 9.9. Comparison of simulated and actual monthly water level in well 2300-10 (near well B) in upper figure, and monthly pumpage in Pearl Harbor area aquifer (1950-1975) applied to SUTRA model (after Souza and Voss [1987]).

is about 100 m thick. The TZ, as observed in this open borehole, is characterized by a series of small sharp salinity changes. These discontinuities in the TZ profile are due to the heterogeneities in the aquifer fabric created by the more and less conductive sections in the lava flow stack. Groundwater flow occurs only in discrete layers. The quality of water in each of the conductive layers is distinctive and the water interacts with the open borehole in which measurements were made.

Most measurements of the TZ in Hawaii are made in such open boreholes, in which vertical flows within the holes can obscure the actual TZ in the formation. Measurements of vertical flow in such boreholes [Voss, unpublished data, 1989] indicate that sudden changes in the borehole fluid salinity occur where flows from above and below meet and exit the borehole in a conductive layer. Where flows enter the borehole, observed salinities change more gradually. Entries and exits occur every 10 to 20 m. Thus, while details of the TZ profile measured in an open borehole, such as the sharp changes noted

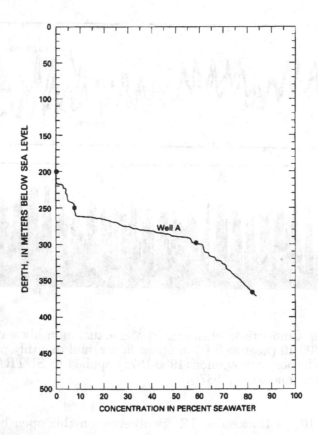

Figure 9.10. Concentration profile in Well A (from continuous fluid conductivity logs: Voss, unpublished data [1989]; and analyses of discrete water samples (circles): Voss, unpublished data [1989].

above, are not really indicative of the actual TZ in the formation, the general trend measured is a good characterization of the local TZ.

Observable at this location is another phenomenon occurring in all deep wells in the Pearl Harbor area. Concentrations at well bottoms are less than 100% seawater in terms of both total dissolved solids (TDS) and chloride. Although this may indicate only that wells are not deep enough to reach pure seawater, the approach to seawater concentration is also much more gradual than expected. In fact, concentration gradients in the lower TZ commonly decrease significantly with increasing depth. Aside from possible geochemical reactions, physical explanations for such 'tailing' of seawater concentration profiles in wells are possible. Tailing may be due to some hydraulic or dispersive behavior of the aquifer, or to the double-porosity nature of the aquifer fabric.

The dynamics of the TZ that are of note in studying the coastal aquifer system are: (1) the net rate of vertical movement of the entire TZ in response to freshwater withdrawal, (2) the regional movement of the potable water limit ($\sim$ 2% seawater) and its relation to solute distribution with depth through the TZ, and (3) the development of regional tailing of concentrations in the lower TZ. The shape of the vertical concentration profile as measured in observation wells is thus a diagnostic characteristic of transition-zone dynamics.

9.7.5 Modeling the Transition Zone

The model consists of a 2D cross-sectional representation of a 26-km-wide, three-dimensional block approximating the entire Pearl Harbor area of the southern Oahu aquifer system. The block extends from the upstream recharge boundary to an arbitrary line about 8 km offshore of the southern coast of Oahu. Vertically, the model encompasses the entire aquifer from the water table to a base at a depth of 1,800 m below sea level assumed to be impermeable. The overall length of the model is 20 km and the total vertical dimension is 1,800 m.

A coarse representation of the SUTRA finite-element mesh used for TZ modeling is shown in Figure 9.11. The horizontal discretization illustrated is that of the actual mesh. However, the vertical discretization is only schematically represented. The actual vertical discretization in the top 600 m of the mesh is 2 m; below a depth of 600 m, the vertical spacing in the bottom 40 rows of elements increases gradually with depth to a maximum of 57 m in the last row of elements. The actual mesh consists of 8,124 nodes and 7,760 elements. A constant time discretization of 2 months is used for all simulations.

In order to guarantee proper representation of the effects of small dispersion coefficients on transport, and because of the high vertical concentration gradients in the TZ, fine vertical discretization is required here. Spatial discretization in this mesh is therefore much finer than in the mesh used for the hydraulic analysis. Of note is that the upper discretization is on the same scale as the typical thickness

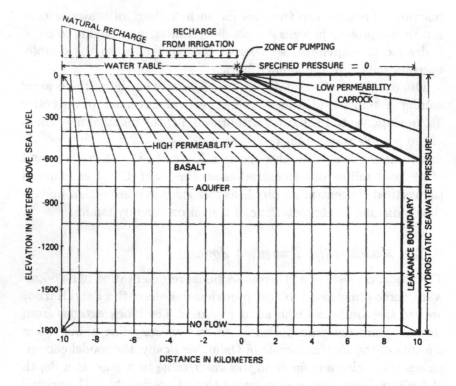

Figure 9.11. Schematic SUTRA finite-element mesh for transition zone modeling showing boundary conditions applied to model and regions of constant permeability (after Voss and Souza [1998]).

of lava flows that make up the aquifer fabric. This is an unusually fine vertical discretization compared to those used in other published coastal simulation studies at this scale.

The boundary conditions for the 2D cross-sectional model are shown schematically in Figure 9.11. Along the horizontal bottom and inland vertical edges of the domain, the boundary is closed to both flow and solute flux. Pressure is specified at seawater boundaries such that static equilibrium is maintained with overlying seawater. The TDS (total dissolved solids) concentration in fluid that flows into the model through specified-pressure boundaries is that of seawater (0.0357 kg-TDS/kg-fluid). The TDS concentration in fluid that flows out of the model at specified-pressure nodes or sink nodes is that of ambient aquifer fluid.

The inland and bottom model boundaries are impermeable, and

the vertical sea boundary is held at hydrostatic seawater pressure, allowing either inflow of seawater or outflow of aquifer water. The upper boundary is a water table with a specified amount of recharge inland of the caprock. It is held at a specified pressure of zero above the caprock to reflect the presence of the Pearl Harbor water and the ocean with concomitant inflow or outflow. Along the inland portion of the water table, time-independent natural recharge (from rain or dike-compartment spillage) is specified as increasing linearly towards the inland boundary. The natural recharge is assigned a seawater concentration (TDS) of zero. Between this area and the caprock edge, spatially constant recharge from irrigation is specified to occur on a time-varying schedule from 1880 to 1990 as determined from USGS records of pumping for irrigation. The irrigation recharge is arbitrarily assigned a low TDS concentration in order to track the movement of this water in the section. Near the caprock edge at a depth of 40 m below the water table in the basalt aquifer, the coastal strip of wells pumps water from the model aquifer at a specified time-varying rate as determined from USGS pumping records from 1880 to 1990.

The modeled distribution of the three water types (seawater, freshwater, and irrigation water) and the flow field for predevelopment conditions (1880) and for 1990 is shown in Figures 9.12 to 9.15 [Voss and Wood, 1994]. In the model, longitudinal dispersivities for horizontal flow and vertical flow are respectively, 250 m and 50 m. Transverse dispersivity is 0.25 m. This is the result of the best-fitting model resulting from the analysis of Voss and Souza [1998]. The irrigation water distribution is superposed on the seawater plot and is shown in percent irrigation water. It is the result of a steady-state transport simulation with the variable-density flow field of 1990. The shrinkage of the freshwater lens in 100 years of development is apparent, and even given constant recharge and pumping after 1990, the shrinkage would continue into the future [Souza and Voss, 1989]. In the 1990 simulation, the upstream monitor well (Well A–Waipio) intercepts all three water types separately, while further downstream (Well B–Waipahu), the irrigation-return water and the seawater overlap and mix across the freshwater lens (Figure 9.14).

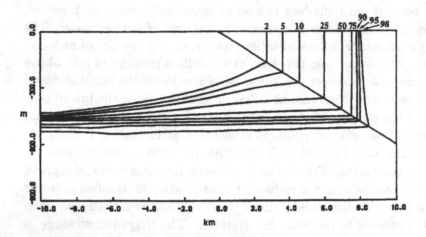

Figure 9.12. Modeled cross-section of Pearl Harbor area aquifer showing steady-state seawater distribution in 1880. (concentrations in percent seawater) (after Voss and Wood [1994]). Upper half of model section is shown.

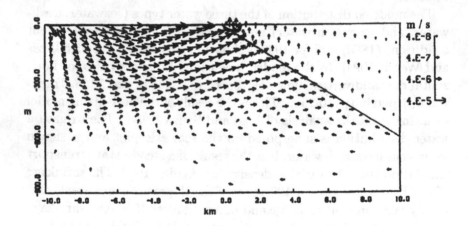

Figure 9.13. Modeled cross-section of Pearl Harbor area aquifer showing steady-state velocity distribution in 1880 (after Voss and Wood [1994]). Upper half of model section is shown.

9.7.6 Controls on Transition Zone Dynamics

A number of hydrologic processes are examined through simulation to determine their importance as controls on TZ dynamics under

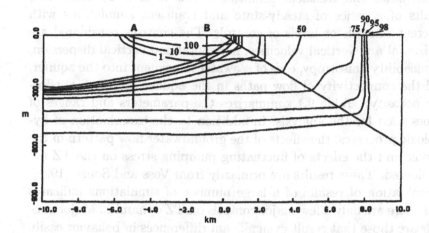

Figure 9.14. Modeled cross-section of Pearl Harbor area aquifer showing seawater distribution from transient simulation 1880 to 1990
(lower set of contours), and irrigation recharge water distribution
(upper set of contours) in 1990 in units of percent of irrigation water, and location of monitor wells (after Voss and Wood [1994]).
Contours are in percent seawater. Upper half of model section is
shown.

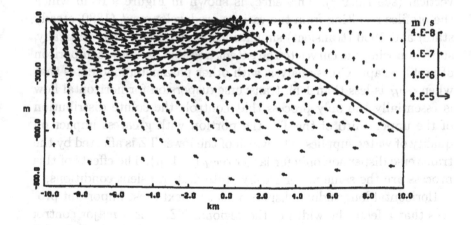

Figure 9.15. Modeled cross-section of Pearl Harbor area aquifer showing velocity distribution in 1990 from transient simulation 1880 to
1990 (after Voss and Wood [1994]). Upper half of model section is
shown.

steady-state and transient conditions. This is done by comparing results of a series of steady-state and transient simulations with different controls or levels of controls. The controls considered are horizontal and vertical velocities, horizontal and vertical dispersion, permeability anisotropy, ease of seawater movement into the aquifer, and the connectivity of flow paths in the aquifer matrix (i.e. effective porosity). Table 9.1 summarizes the parameters and ranges of values used for the analysis. In addition to the examination of hydrologic processes, the effects of the groundwater flow pattern in the aquifer and the effects of fluctuating pumping stress on the TZ are considered. These results are primarily from Voss and Souza [1998].

Evaluation of results of a large number of simulations indicates that there are only a few major controls on TZ dynamics. Major controls are those that result in significant differences in behavior easily observable in the field. Minor processes would be those with small effects on concentration that theoretically exist, but which would be difficult to observe in sparse data collected in a heterogeneous aquifer with a fluctuating flow field.

Analysis indicates that transverse dispersion in horizontal flow, with parameter, transverse dispersivity for horizontal flow, α_{TH}, is the main process controlling the width of the regional TZ. In this model, the 'max' and 'min' directions are respectively, horizontal and vertical (see Eq. 9.8). This effect is shown in Figure 9.16 in which the profiles resulting from the various predevelopment (1880, steady state) and 1990 (transient) simulations have been shifted vertically, superimposing the curves to the extent possible, to allow comparison of profile shape. Cross-analysis with other parameters indicates that when α_{TH} is less than 1 m, transverse dispersion for horizontal flow is essentially the only process that controls the solute distribution of the upper third of the TZ, the portion with greatest impact on quality of water supplies. The width of the lower TZ is affected by the transverse dispersion only for larger α_{TH} (> 1 m). The effects of this process are the same under steady-state and transient conditions.

Horizontal longitudinal dispersion is the next most important process that affects the width of the regional TZ. It is a major control on the concentration gradient through the central third of the TZ over the range of parameter, longitudinal dispersivity for horizontal flow, α_{LH} values considered (Figure 9.17, with vertically-shifted profiles). Its effect on the central third of the TZ is the same un-

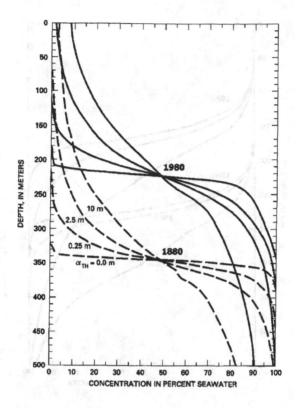

Figure 9.16. Effect of transverse dispersivity for horizontal flow (α_{TH}) on modeled transition zone profiles at -5 km (Well A) for steady-state predevelopment (1880) conditions (dashed lines) and transient conditions in 1980 (solid lines). Concentrations in percent seawater. Profiles are shifted vertically superposing 50% seawater points to allow comparison of shape (after Voss and Souza [1998]).

der steady-state and transient conditions. However, high values of α_{LH} additionally amplify the amount of lower TZ tailing exhibited at any given level of vertical longitudinal dispersion under transient conditions.

Vertical longitudinal dispersion is the main process that controls tailing in the lower third of the TZ when the freshwater lens shrinks due to long-term pumpage (Figure 9.18). In the figure the TZs are plotted at true depth without shifting, thus it is apparent that this process affects only the lower third of the TZ. The importance of the process in causing tailing is greater if there are seasonal or other short-term cyclic variations in pumping stress. The parameter con-

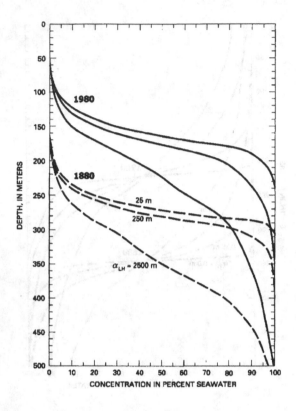

Figure 9.17. Effect of longitudinal dispersivity for horizontal flow (α_{LH}) on modeled transition zone profiles at -5 km (Well A) for steady-state predevelopment (1880) conditions (dashed lines) and transient conditions in 1980 (solid lines). Concentrations in percent seawater. Profiles are shifted vertically to allow comparison of shape (after Voss and Souza [1998]).

trolling tailing under any stress schedule is longitudinal dispersion for vertical flow, α_{LV}. This process is of little importance under steady-state conditions.

The effects of α_{TH} may be easily observed in field data and estimated from field measurement of the shape of the upper third of the TZ. When α_{TH} is low (as in southern Oahu), α_{LH} may be estimated from field measurements of the concentration profile in the central third of the TZ. Then α_{LV} may be estimated from profile data on the lower third of the TZ. Simulations show that sea boundary leakance, effective porosity, and anisotropy in permeability cannot be estimated independently from data on the depth of the TZ, nor can

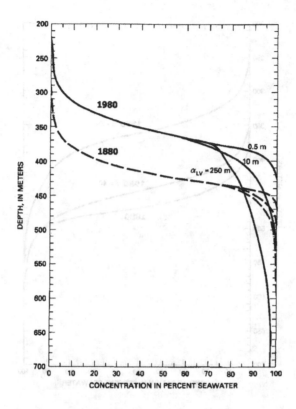

Figure 9.18. Effect of longitudinal dispersivity for vertical flow (α_{LV}) on modeled transition zone profiles at -5 km (Well A) showing steady-state predevelopment (1880) conditions (dashed lines) and transient conditions in 1980 (solid lines). Concentrations in percent seawater. Profiles are shifted vertically (less than 5 m) to superpose upper transition zone for shape comparison (after Voss and Souza [1998]).

anisotropy be estimated from TZ width independently of dispersion parameters. At high anisotropy (as in southern Oahu), however, differences in permeability anisotropy have only a minor effect on zone width, and other parameters may be estimated independently of its value [Voss and Souza, 1998].

Analysis indicates that the rate of net vertical and landward motion of the TZ, as induced by pumping stress, is controlled by: (1) the rate at which seawater moves into the aquifer at the seawater boundary, (2) the effective porosity of the aquifer (Figure 9.19 plotted at true depths), and (3) the TZ width. TZ movement tends to be

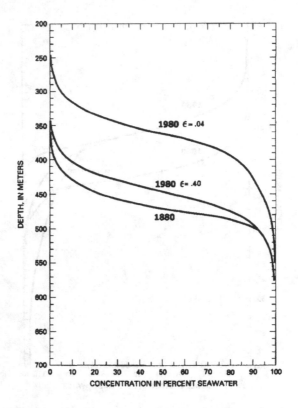

Figure 9.19. Effect of porosity (ε) on modeled transition zone profiles at -5 km (Well A). Transient conditions in 1980 compared with steady-state predevelopment (1880) conditions. Concentrations in percent seawater. Profiles shown at true depths (after Voss and Souza [1998]).

faster when seawater most freely enters the system, when effective porosity is lower, and when the TZ is narrower. A wide zone moves more slowly because buoyant driving forces depend on vertical fluid-density gradients that are lower. Thus, in one sense, wide TZs have an apparent advantage for water supply as they will rise more slowly due to pumping stress. However, while the net rise of a wide TZ is slower, the rise of the potable water boundary (2% seawater) is always faster for a wider zone negating the advantage of the net slower rise.

The analysis discussed to this point is based on a stress history for transient conditions using a long-term pumping schedule with no

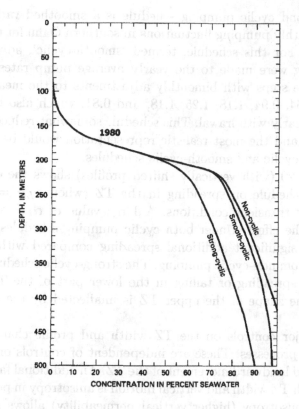

Figure 9.20. Effect of pumping schedule on modeled transition zone profiles at −5 km (Well A) for transient conditions in 1980. Concentrations in percent seawater. Profiles shifted vertically to superpose upper transition zone for shape comparison (after Voss and Souza [1998]).

stress variation within a given year ('non-cyclic' pumping). In reality, seasonal pumping for agricultural use exhibits an extreme range of pumping rates. Monthly variation may be more than 6.0×10^5 m^3/d. Effects of varying pumping rates on the TZ were investigated using two 'cyclic' pumping schedules.

The first cyclic schedule has a yearly period consisting of no pumping for six months, and twice the non-cyclic rate for six months resulting in the same total yearly withdrawal as non-cyclic pumping. Although an exaggeration of actual pumping practices, this schedule, referred to as 'strong-cyclic', is used to determine the maximum effect that pumping variability could have on the regional TZ.

The second cyclic pumping schedule is a smoothed variation of actual monthly pumping fluctuations in southern Oahu for the years 1946-1975. For this schedule, termed 'smooth-cyclic', adjustments in pumping were made to the yearly average pump rates in two-month time steps with bimonthly adjustments to the mean yearly rate of: 0.64, 0.94, 1.18, 1.25, 1.18, and 0.81, which also preserves the total yearly withdrawal. This schedule somewhat reduces actual variability and the most realistic representation would be between the strong-cyclic and smooth-cyclic schedules.

Figure 9.20 (with vertically shifted profiles) shows the effects of pumping schedule on spreading in the TZ (when $\alpha_{LH} = \alpha_{LV} = 250$ m) for transient conditions. A large value of α_{LV} is used to highlight the effect. Under both cyclic pumping schedules, the TZ undergoes significant additional spreading compared with the TZ resulting from non-cyclic pumping. The strong-cyclic schedule causes significant spreading or tailing in the lower part of the TZ, while notably, the shape of the upper TZ is unaffected by the pumping schedule.

The major controls on the TZ width and profile character are dispersion processes. These are independent of controls on the net vertical and landward movement of the TZ. An additional factor that affects both TZ width and vertical motion is anisotropy in permeability. Less anisotropy (higher vertical permeability) allows increased saltwater circulation velocities, with resultant dispersive widening of the TZ, and slower net vertical motion of the zone [Voss and Souza, 1998].

The processes identified are major controls on transition-zone dynamics because of the nature of the variable-density flow field. Horizontal flow dominates in the freshwater and in the upper half of the TZ (less than 50 percent seawater) where the ratio of horizontal to vertical velocity component is about 100:1. Thus, the main control on the TDS distribution in this region is transverse dispersion. Because the flow is almost, but not perfectly, parallel to concentration isopleths, longitudinal dispersion becomes an important process when the value of horizontal longitudinal dispersivity is large.

In the lower part of the TZ (greater than 50 percent seawater), horizontal velocity is much less than in the upper TZ. The ratio of horizontal to vertical velocity is less than 10:1, indicating that, in this part of the TZ, transport processes depend more equally on both components [Voss and Souza, 1998]. In the lower TZ, dispersive transport essentially consists of transverse dispersion for horizontal flow, longitudinal dispersion for vertical flow, and longitudinal dispersion for horizontal flow if horizontal longitudinal dispersivity is large. Under transient conditions, vertical velocity increases, making the importance of vertical longitudinal dispersion relatively greater than under steady conditions.

The models that best represent both the flow and solute transport processes in southern Oahu are defined by the aquifer parameters and system constants listed in Table 9.1. The only difference between the models is the value of longitudinal dispersivity parameters, α_{LH} and α_{LV}. In Table 9.1, Model A refers to the use of flow direction-dependent dispersivity and Model I refers to the use of flow direction-independent dispersivity. Using the smooth-cyclic pumping schedule, the simulated TZ in 1990 is compared to field salinity data. Figure 9.21 shows a cross-sectional comparison of Model I results and field data; the fit is very good. Considering that the model has only a small number of parameters, it captures the essence of the flow and transport behavior in this coastal aquifer. Elimination of the discrepancy in location between the simulated and field values at the highest concentrations (Figure 9.21) likely requires addition of other parameters to the model. This was not considered critical to understanding of TZ dynamics, and the model was not made more complex.

9.7.7 Ages of Water: Field Data vs. Model

To determine the age distribution of water in the modeled aquifer system, isochrones (lines of equal in-model residence time) may be calculated in the SUTRA model. Using the 1880 steady-state flow field, an extra simulation is run at steady state for an imaginary species. The concentration of this species will represent elapsed time in the model domain. The species has zero concentration at inflow, and undergoes zero-order production in the fluid (see Voss [1984a], for definition of this source) with a value of one per year. This numerical trick produces isochrones directly from the transport model (see Goode [1996]), indicating residence time in the model within regions having minimal mixing of fluids derived from different sources. The result (Figure 9.22) in both the freshwater lens and the deep saltwater gives in-model residence times [Voss and Wood, 1994]. Isochrones

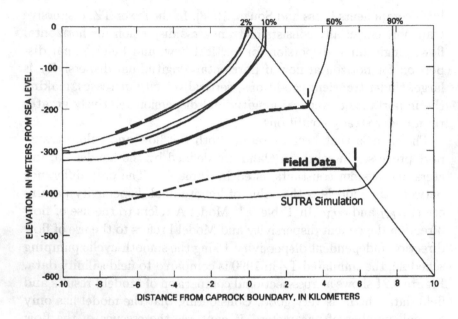

Figure 9.21. Comparison of actual transition zone data with transient conditions in 1980 simulated by 'best model I' with smooth-cyclic pumping stresses (after Voss and Souza [1998]). Concentrations in percent seawater. Best Model I has $\alpha_{LH} = \alpha_{LV} = 100$ m (see Table 9.1). Model I results (thin curves) and transition zone data (bold curves from Figure 9.6).

within the zone of mixing between freshwater and seawater are not plotted because the interpretation of mixing waters from different sources and with different ages is ambiguous.

An indication of actual ages and flow velocities of the three major water types found in the Pearl Harbor area aquifer in southern Oahu, Hawaii is given by Voss and Wood [1994]. Ages were based on analyses of samples collected in vertical profile in two wells along the regional groundwater flow direction, and subsequent geochemical and isotopic analysis of the data. These results may be compared objectively with the predictions of the above-described model, which was developed prior to and independently of the field-based geochemical analysis.

Voss and Wood [1994] analyzed water samples and found the uppermost water, consisting of recharge from irrigation, to be only a few tens of years old. It moves through the aquifer at a velocity of about

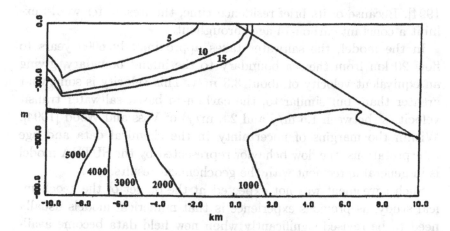

Figure 9.22. Modeled cross-section of Pearl Harbor area aquifer showing contours of equal travel time for conditions in 1880 (after Voss and Wood [1994]). Upper half of model section is shown. Travel times are in years. (Note: Times shown are not absolute water ages, but residence times within the model area.)

1 km/y. Incongruously, the freshwater core of the freshwater lens was found to have a great apparent age of about 1800 years. From hydrologic considerations, once this water enters the basal lens system from the inland compartments, it must move through the aquifer just as quickly as the irrigation recharge water, implying some inconsistency in these data. It may be possible to explain the great apparent freshwater age as being due to long storage in groundwater compartments in the recharge areas of the island. Simple volumetric calculations, however, show that inland storage can at most account for less than one-half of the apparent age. Alternatively, or in conjunction with the first explanation, the excess apparent age may be due to a possible contribution of old dissolved organic carbon added during recharge, which would tend to make the water appear older than it is. With regard to saltwater, Voss and Wood [1994] found the deepest water is intruded seawater. This water has an apparent age of about 6000 years to 9000 years near the inland edge of Pearl Harbor and older saltwater exists further inland.

In the model, total residence time of water in the freshwater lens is less than 20 years (Figure 9.22). This is in conformance with the tritium-based findings for irrigation-return water [Voss and Wood,

1994]. Because of its brief residence time, the freshwater would exhibit a constant carbon-14 age throughout.

In the model, the saltwater takes approximately 6000 years to flow 20 km from the sea boundary to the inland boundary, giving an equivalent velocity of about 3.3 m/y. This velocity is somewhat greater than, but similar to, the carbon-14-based saltwater transit velocities (between 1.6 m/y and 2.8 m/y) of Voss and Wood [1994]. Within the margins of uncertainty in the chemical data and age interpretations, the flow behavior represented by the SUTRA model is in general agreement with the geochemical results.

Such agreement was not expected at the outset of the geochemical study, as previous experience is that numerical models usually need to be revised significantly when new field data become available. Perhaps most interesting is the similarity in the velocity of saltwater moving inland as determined by carbon-14 dating (about 2 m/y) and that occurring in the numerical model (about 3 m/y). While this agreement may be only a coincidence, it may be the first time that a numerical variable-density groundwater model prediction of saltwater circulation in a coastal aquifer has been independently corroborated by geochemical field data [Voss and Wood, 1994].

The inland flow of saltwater in a steady-state system is driven by the amount of salt entrained in and subsequently discharged with the freshwater flow towards the coast. This, in turn, depends on two factors, the flux of freshwater and the dispersion process. To slow down the modeled saltwater inflow to better conform with the isotope-based velocity, requires either a decrease in the model's aquifer recharge by up to one half, or a decrease in the model's dispersion coefficients. Another factor, not considered in the model isochrone calculations, is that the effective porosity experienced by the slow-moving saltwater may be greater than the value of 0.04, which was assigned to the entire aquifer. The low value had been determined mainly by data on the freshwater zone. As discussed in 9.6.2, in slow-moving saltwater, the entire volumetric porosity, up to about 0.4, may participate in solute transport. This would occur by both solute diffusion and advection through blocks of basalt alongside the well-connected conductive rubbly beds for which the regional effective porosity is apparently only 0.04. In the case of an effective porosity of 0.4, the same saltwater flux as determined in the simulations can be obtained with a saltwater velocity lower by ten times. This provides ample flexibility to account for the discrepancy in modeled and isotope-determined saltwater velocities even with only a small portion of the less-conductive basalt porosity participating in saltwater zone solute transport.

9.7.8   Application of Model for Aquifer Management

When analyzing water resources in a coastal aquifer system where water quality may be affected by seawater intrusion, the movement of very low concentrations of solute are critical with respect to water supply. The limiting concentration of potable water is defined here as the equivalent of freshwater containing about 2 percent seawater ($\sim$ 700 ppm total dissolved solids). Tracking the boundary of potable water as delineated by this low concentration can be done only by use of a solute transport model with variable fluid density. However, this type of analysis is possible only with a numerically accurate code when used for simulations on a fine mesh.

The model for analyzing the TZ, described above, is used to predict the general evolution of water quality that will be drawn from the coastal pumping zone in the major aquifer system on the island of Oahu, Hawaii. The period 1980–2080 is considered, and the impact of several alternative pumping-recharge scenarios is evaluated [Souza and Voss, 1989]. Pumping is simulated by withdrawing water from the model area identified as the pumping zone in Figure 9.11. This pumping zone represents a strip of scattered wells along the coast about 1.5 km (5,000 ft) wide. In a practical sense, a regional-scale model cannot be used to predict the salinity of the water pumped from an individual well in the zone. Due to the large scale of processes modeled, only some average concentration in the zone may be considered when analyzing general salinity changes expected. Thus, meaningful salinity values of the pumped water are the mean concentrations of the three nodes that represent the pumping center. This represents average salinity of the water pumped from the areally extensive well field that exists in a strip across the Pearl Harbor area aquifer.

The modeled state of the aquifer in 1980 is the starting condition for the simulation of water quality evolution in the pumping center for the period 1980–2080 for eight scenarios. These scenarios have different withdrawal rates and/or different recharge rates (Table 9.2).

Historically, the recharge to the aquifer included both natural

Scenario	Recharge Rate		Withdrawal Rate		Net Discharge Rate
	Natural	Irrigation Water			
	$10^5 \, m^3/d$ (Mgal/d)	$10^5 \, m^3/d$ (Mgal/d)	$10^5 \, m^3/d$ (Mgal/d)	% of Total Recharge	$10^5 \, m^3/d$ (Mgal/d)
1	9.03 (225)	1.52 (37.8)	7.28 (182)	69%	3.27 (81.3)
2	9.03 (225)	0 (0)	5.76 (144)	64%	3.27 (81.3)
3	9.03 (225)	0 (0)	6.78 (169)	75%	2.25 (56.2)
4	9.03 (225)	0 (0)	7.28 (182)	81%	1.75 (43.5)
5	9.03 (225)	0 (0)	9.03 (225)	100%	0 (0)
6	6.77 (168)	1.52 (37.8)	7.28 (182)	69%	1.01 (21.1)
7	6.77 (168)	0 (0)	7.28 (182)	108%	-0.51 (-12.7)
8	6.77 (168)	1.52 (37.8)	5.03 (125)	61%	3.27 (81.3)

Table 9.2. Hypothetical water budgets for 100-year water management scenarios in the Pearl Habor area of southern Oahu, Hawaii.

recharge and recharge attributed to intensive furrow irrigation of sugar cane. It is estimated that the irrigation-return component increased total recharge by 15%. However, this source dwindled as agricultural practices changed after 1980 and sugar cane disappeared as a crop. Irrigation-return as a source of recharge is therefore eliminated for most of the scenarios. A reduction in total withdrawals has also occurred due to great reductions in pumping for sugar cane irrigation. Thus, some scenarios consider withdrawals lower than the 1980 rate.

The first five scenarios deal with long-term water-quality trends under various conditions. In these scenarios, all recharge and withdrawal rates are held constant in simulation from 1980 to 2080. The full quantity of natural recharge is applied in all five scenarios. The simulated changes in average salinity with time in the pumping center are compared in Figure 9.23. Salinity of water produced increases with time because of the general shrinkage of the freshwater lens resulting from withdrawals that decrease the head in the aquifer. The amount of salinity increase varies directly with pumping rates. Scenario 1 continues the recharge and withdrawal rates for the period 1970-1980 after 1980. It is the only scenario of the first five that includes recharge from canal irrigation of sugar cane. At the lower pumping rates (scenarios 1 through 4) salinity gradually rises from the 1980 level but reaches a near steady-state level by the end of the 100-year period. Scenario 4 continues 1980 pumping rates, scenario 3 has 93% of the 1980 withdrawals, and scenario 3 considers withdrawals at 79% of 1980 levels. Salinity eventually tends to stabilize in the pumping center at potable levels for lower net withdrawal rates, but the quality of water produced at the pumping center continually degrades at the higher rates. Scenario 5 is an extreme case where the long-term withdrawal equals 100% of the recharge (124% of 1980 withdrawal rate). This eventually has a catastrophic effect on water quality, wherein the average concentration of water withdrawn exceeds 2% seawater after only 12 years and continues to increase past 2080.

Some management strategies thus lead to sustainable potable water supplies, while others lead to significant seawater intrusion. Up to about 75% of the assumed recharge can be withdrawn before significant seawater intrusion affecting potability occurs in the modeled pumping center by 2080.

Three other scenarios deal with water quality changes that result from the significant reduction in recharge that occurs during a long drought. To simulate a drought, natural recharge is reduced by 25% for the first 10 years of the 100-year simulation. During this period, the withdrawal rate is held constant or adjusted downward as a strategy to control water quality. In scenarios 6 and 7, pumping is continued at the pre-drought level. In scenario 7, recharge from irrigation water is eliminated causing withdrawal to be greater than total recharge. The predicted average salinity changes in the pumping center for drought scenario 6 is shown in Figure 9.24 which also shows the evolution predicted by the corresponding scenario without

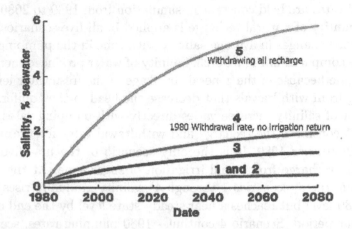

Figure 9.23. Evolution of average salinity of water produced in the Pearl Harbor area aquifer's zone of pumping, 1980 to 2080 for management scenarios 1 through 5 as predicted by SUTRA simulations (after Souza and Voss [1989]).

drought (scenario 1). Following the 10-year drought, salinity eventually drops toward the level simulated without the drought though the effects on the system remain at least twice as long as the drought duration.

The drought's impact on water quality is mitigated by the management strategy in scenario 8, where, for the conditions of drought scenario 6, pumping is decreased by an amount equal to the total reduction in recharge. The initial decrease in salinity that occurs (Figure 9.24) is unintentional and is caused by a near-well response to the decrease in the net withdrawal rate. These results show that the system responds quickly to damaging as well as to beneficial changes, and that careful management of system withdrawals during droughts can preserve water quality.

An analysis of controls on water-quality evolution indicates that the key control on long-term saltwater intrusion is the net system discharge (equal to the difference between total recharge and pumpage) [Souza and Voss, 1989]. The effect is illustrated in Figure 9.25 wherein the average concentration of seawater in water produced at the pumping center in the year, 2080, is a strong function of net system discharge. The reason for the importance of net system discharge is that it controls the ultimate thickness of the freshwater lens downstream

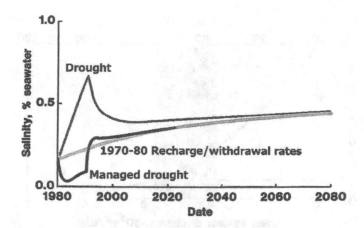

Figure 9.24. Evolution of average salinity of water produced in the Pearl Harbor area aquifer's zone of pumping, 1980 to 2080 for drought, managed drought, and non-drought scenarios as predicted by SUTRA simulations (scenarios 6, 8 and 1, respectively) (after Souza and Voss [1989]).

of the pumping center.

A range of "minimum safe discharge" from the Pearl Harbor area aquifer is shown in Figure 9.25. "Minimum safe discharge" is defined on the basis of simple interpretations of what the average salinity of water produced in the coastal pumping zone means in terms of a safe yield, as follows. A low safe yield (corresponding to a high net discharge) is selected by arbitrarily assuming that one-third of the wells will be pumping water containing 2% seawater by 2080. Two-thirds, then, would be pumping water containing no seawater, and the average concentration of pumped water is 0.67% of the salinity of seawater. Referring to Figure 9.25, this implies that the net system discharge cannot be allowed to drop below about 2.9×10^5 m^3/d (72 Mgal/d). In the case where total recharge is 9.0×10^5 m^3/d (225 Mgal/d), then no more than 68% of this amount can safely be pumped over the long term. The highest possible safe yield (corresponding to the lowest net discharge) is obtained by the assumption that all pumped water reaches the 2% seawater limit simultaneously. This is equivalent to the assumption that the average concentration of the water produced is 2% of the salinity of seawater. Then, the net discharge rate determined from Figure 9.25 is 1.7×10^5 m^3/d (43

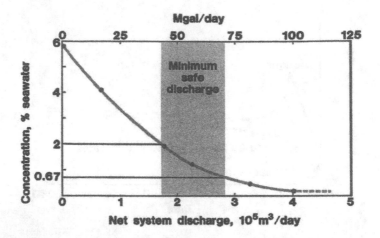

Figure 9.25. Dependence of average salinity of water produced in the Pearl Harbor area aquifer's zone of pumping after 100 years of pumping (1980–2080) upon the quantity of water that still discharges naturally from the aquifer after water is removed by pumping (called 'net system discharge'). The curve combines the results of a number of simulations. It is constructed from the circles which show average salinity in pumping zone in 2080 and the net discharge as simulated in various scenarios. The 'minimum safe discharge' indicates the *least* quantity of water that must be allowed to discharge naturally from the aquifer in order to guarantee production of potable water in 2080 from the zone of pumping.

Mgal/d), implying that the average pumping rate may not exceed 81% of the average recharge rate.

Thus, the long-term "maximum safe withdrawal" from this coastal aquifer based on the interpretations of these two safe yield limits is between 70% and 80% of the assumed recharge. The key parameter, however, controlling saltwater intrusion is the net system discharge, not the percentage of recharge pumped, and the maximum safe withdrawal will vary depending on the actual long-term recharge rate. This modeling indicates that if the long-term net system discharge rate drops to within the range of minimum safe discharge (1.7 to 2.9×10^5 m^3/d), then significant seawater intrusion in parts of the Pearl Harbor area pumping zone will occur by 2080.

At present, the Pearl Harbor area aquifer is still adjusting to historical increases in pumping. Adjustments are also occurring because of seasonal and yearly variations primarily due to variability in irrigation. In the past, development of new wells in this aquifer moved inland as water in coastal wells became salty. This is not a viable long-term solution. If the aquifer is being over-pumped, moving the pumping zone inland only acts to delay the inevitable arrival of the TZ. In the long term, the only possibility for effective management of seawater intrusion and maximizing total withdrawals from the aquifer is to control the difference between average rates of recharge and withdrawal in the system (the net system discharge). Finding the 'correct' value of net system discharge will depend on obtaining accurate measures of aquifer recharge.

The general approach to coastal seawater intrusion modeling presented here provides reasonable predictions of changes in the regional position of the potable-water boundary above the transition zone that result from changes in stress in a coastal aquifer. A strategy that controls withdrawals on the basis of (1) accurate estimates of recharge, and, (2) water-quality predictions from coastal aquifer modeling, should be a major component of a total management scheme for a coastal aquifer water resource.

Acknowledgments: The author thanks reviewers Alex Cheng and William Souza for their good thoughts which helped to improve this chapter.

Chapter 10

Three-Dimensional Model of Coupled Density-Dependent Flow and Miscible Salt Transport

G. Gambolati, M. Putti & C. Paniconi

10.1 Introduction

Saltwater intrusion in coastal aquifers is a very serious threat to subsurface water quality worldwide. This contamination of freshwater resources occurs, in a typical scenario, when the wide cone of depression formed by extensive groundwater pumping comes into contact with underlying or surrounding seawater. Twenty years ago Newport [1977] reported that at least twenty coastal areas in the United States were contaminated by saline water. Documented cases have since been reported for many other countries. Contamination by salt deteriorates water quality dramatically. A two to three percent mixing with seawater makes freshwater unsuitable for human consumption, and five percent mixing makes it unusable for irrigation as well [Custodio et al., 1987; Sherif and Singh, 1996].

Salt alters the water density in such a way as to induce important effects on the flow field, exerting a chemical control on these flows in addition to the hydrogeological, physical controls. Similar processes are observed in long-term salt dissolution and brine transport around saline diapirs such as those being considered for radioactive waste disposal, and in groundwater transport of leachates emanating from landfills and industrial waste sites. These anthropogenic disturbances generally occur in natural systems, which have been in hydrogeological equilibrium for very long periods of time. Depending on the scale of the phenomena, the transition from original equilibrium to a new equilibrated state could require decades in the case of

J. Bear et al. (eds.), Seawater Intrusion in Coastal Aquifers, 315–362.
© 1999 Kluwer Academic Publishers.

waste disposal, centuries for regional saltwater intrusion, and millennia for salt dissolution in deep formations [Frind, 1982a]. A proper evaluation of environmental impacts and economic effects associated with saltwater intrusion phenomena requires monitoring and analysis of the short, medium, and long-term response of the threatened system.

In a typical aquifer, seawater and freshwater are separated by an interface across which a mixing zone develops due to dispersive effects. Many modeling studies replace the mixing zone with a sharp front. If, in addition, one adopts the Dupuit assumption of predominantly horizontal flow, the problem of finding the interface may sometimes be solved in closed form [Cooper et al., 1964; Schmorak and Mercado, 1969; Collins and Gelhar, 1971; Mualem and Bear, 1974; Strack, 1976]. For the more general sharp interface problem, several two-dimensional numerical solutions have been published [Pinder and Page, 1977; Mercer et al., 1980b; Liu et al., 1981; Taigbenu et al., 1984; Wirojanagud and Charbeneau, 1985, Essaid, 1990b].

When vertical flow and/or dispersion become important, the sharp front approach may not be adequate, as was first recognized by Henry [1964]. As a result, many recent two and three-dimensional models of seawater intrusion handle both vertical flow and dispersion [Ségol et al., 1975; Frind, 1982a; Volker and Rushton, 1982; Sanford and Konikow, 1985; Huyakorn et al., 1985; Voss and Souza, 1987; Hill, 1988; Kakinuma et al., 1988; Diersch, 1988; Galeati et al., 1992]. None of these models, however, considers the unsaturated zone, important for treating cases such as ground sources of saltwater (e.g., salt mounds, contaminated irrigation water) and seawater intrusion in the vadose zone following significant lowering of the water table.

The most widely used numerical modeling approaches can be classified as either Eulerian or Eulerian-Lagrangian [Gray and Pinder, 1976; Kinzelbach, 1986] each with its advantages and limitations. Standard Eulerian methods are subject to numerical dispersion if the mesh resolution does not satisfy certain constraints that are dependent on the parameters of the particular problem to be solved. Eulerian-Lagrangian methods do not have this limitation in theory [Garder et al., 1964], but in practice some numerical dispersion is introduced via interpolation. They are also difficult to implement in the three-dimensional case, and they can be affected by problems of "demixing", depending on the particle tracking scheme being used [Baptista et al., 1984].

In this chapter we derive the mathematical and numerical formulation for a three-dimensional Eulerian finite element model that treats density-dependent variably saturated flow and miscible (dispersive) solute (salt) transport. Special consideration is given to how the coupling and nonlinearities are handled, and how the discrete linearized systems are solved. The formulation and procedures described form the basis of the CODESA-3D (COupled variable DEnsity and SAturation) model. Some representative examples are given to illustrate the application and features of the CODESA-3D model.

10.2   Mathematical Model

10.2.1   Governing Equations

The mathematical model of density-dependent flow and transport in groundwater is expressed here in terms of an equivalent freshwater head h, defined as [Huyakorn et al., 1987; Frind, 1982a; Gambolati et al., 1993]

$$h = \psi + z \qquad (10.1)$$

where $\psi = p/(\rho_o g)$ is the equivalent freshwater pressure head, p is the pressure, ρ_o is the freshwater density, g is the gravitational constant, and z is the vertical coordinate directed upward. The density ρ of the saltwater solution is written in terms of the reference density ρ_o and the normalized salt concentration c:

$$\rho = \rho_o(1 + \epsilon c) \qquad (10.2)$$

where $\epsilon = (\rho_s - \rho_o)/\rho_o$ is the density ratio, typically $\ll 1$, and ρ_s is the solution density at the maximum normalized concentration $c = 1$. Depending on the application, ρ_s can represent, for instance, the density of seawater, or of the solution nearest a surface salt mound or around an underground saline diapir. The dynamic viscosity μ of the saltwater mixture is also expressed as a function of c and of the reference viscosity μ_o as

$$\mu = \mu_o(1 + \epsilon' c) \qquad (10.3)$$

where $\epsilon' = (\mu_s - \mu_o)/\mu_o$ is the viscosity ratio and μ_s is the viscosity of the solution at $c = 1$. With these definitions, the coupled system of variably saturated flow and miscible salt transport equations is (see Appendix A)

$$\sigma \frac{\partial \psi}{\partial t} = \nabla \cdot \left[K_s \frac{1 + \epsilon c}{1 + \epsilon' c} K_r \left(\nabla \psi + (1 + \epsilon c) \eta_z \right) \right]$$
$$- \phi S_w \epsilon \frac{\partial c}{\partial t} + \frac{\rho}{\rho_o} q \tag{10.4}$$

$$\mathbf{v} = -K_s \frac{1 + \epsilon c}{1 + \epsilon' c} K_r \left(\nabla \psi + (1 + \epsilon c) \eta_z \right) \tag{10.5}$$

$$\phi \frac{\partial S_w c}{\partial t} = \nabla \cdot (D \nabla c) - \nabla \cdot (c \mathbf{v}) + q c^* + f \tag{10.6}$$

where ∇ is the gradient operator, K_s is the saturated hydraulic conductivity tensor at the reference density, $K_r(\psi)$ is the relative conductivity, η_z is a vector equal to zero in its x and y components and 1 in its z component, $\sigma(\psi, c)$ is the general storage term or overall storage coefficient, t is time, ϕ is the porosity, $S_w(\psi)$ is the water saturation, q is the injected (positive)/extracted (negative) volumetric flow rate, $\mathbf{v}$ is the Darcy velocity vector, D is the dispersion tensor, c^* is the normalized concentration of salt in the injected/extracted fluid, and f is the volumetric rate of injected (positive)/extracted (negative) solute that does not affect the velocity field.

Initial conditions and Dirichlet, Neumann, or Cauchy boundary conditions are added to complete the mathematical formulation of the flow and transport problem expressed in Eqs. 10.4–10.6. For the flow equation, these take the form

$$\psi(\mathbf{x}, 0) = \psi_o(\mathbf{x}) \tag{10.7a}$$
$$\psi(\mathbf{x}, t) = \psi_p(\mathbf{x}, t) \quad \text{on } \Gamma_1 \tag{10.7b}$$
$$\mathbf{v} \cdot \mathbf{n} = -q_n(\mathbf{x}, t) \quad \text{on } \Gamma_2 \tag{10.7c}$$

where $\mathbf{x} = (x, y, z)^T$ is the Cartesian spatial coordinate vector, superscript T is the transpose operator, ψ_o is the pressure head at time 0, ψ_p is the prescribed pressure head (Dirichlet condition) on boundary Γ_1, $\mathbf{n}$ is the outward normal unit vector, and q_n is the prescribed flux (Neumann condition) across boundary Γ_2. We use the sign convention of q_n positive for an inward flux and negative for an outward flux, consistent with the convention used for q and f in system Eqs. 10.4–10.6.

For the transport equation, the initial and boundary conditions are [Galeati and Gambolati, 1989]

$$c(\mathbf{x}, 0) = c_o(\mathbf{x}) \tag{10.8a}$$
$$c(\mathbf{x}, t) = c_p(\mathbf{x}, t) \qquad \text{on } \Gamma_3 \tag{10.8b}$$
$$D\nabla c \cdot \mathbf{n} = q_d(\mathbf{x}, t) \qquad \text{on } \Gamma_4 \tag{10.8c}$$
$$(\mathbf{v}c - D\nabla c) \cdot \mathbf{n} = -q_c(\mathbf{x}, t) \qquad \text{on } \Gamma_5 \tag{10.8d}$$

where c_o is the initial concentration, c_p is the prescribed concentration (Dirichlet condition) on boundary Γ_3, q_d is the prescribed dispersive flux (Neumann condition) across boundary Γ_4, and q_c is the prescribed total flux of solute (Cauchy condition) across boundary Γ_5. The sign convention for q_d and q_c is the same as for q_n, q, and f.

10.2.2   Coupling and Nonlinearity in the Model

Coupling in Eqs. 10.4–10.6 is due to the concentration terms in the flow equation 10.4 and the head terms that appear in the transport equation 10.6 via the Darcy velocities. In the simpler case of non-density-dependent flow and transport, the system is coupled only through the head terms in the transport equation. In this case there is physical coupling, but mathematically the system can be reduced ("decoupled") and solved sequentially, first the flow and then the transport equation, without iteration. For our density-dependent case, the system is irreducible and any sequential solution procedure requires iteration.

The nonlinearity of the coupled model, Eqs. 10.4–10.6, is due to the dependence of solution density on concentration that arises from relationship Eq. 10.2. For the flow equation 10.4, the saturated conductivity ($K_s(1+\epsilon c)/(1+\epsilon' c)$), total head, and time derivative terms are affected, as shown in Appendix A. As a consequence of the dependence on concentration in the flow equation, the transport equation 10.6 is also nonlinear, in its convective and dispersive flux terms.

An additional source of nonlinearity is introduced when the unsaturated zone is included in the saltwater intrusion model, as expressed through Eq. 10.4. This nonlinearity arises from strong pressure head dependencies in the relative hydraulic conductivity and general storage terms. These dependencies have been extensively studied and are expressed through semi-empirical constitutive relationships describing the soil hydraulic properties. Examples will be given in section 10.4.2. The nonlinearities connected to the coupling of the flow and transport equations, on the other hand, are not restricted to a few coefficients with well-known functional dependencies, so some analysis is needed to gain insight into the relative strengths of the many terms that contribute to the nonlinear coupling. For this purpose a heuristic analysis based on a simplified one-dimensional saturated version of the flow and transport system is presented below.

We consider the simple case of one-dimensional vertical flow with $S_s = 0$, $S_w = 1$, $D_o = 0$, $q = f = 0$, $\epsilon' = 0$, and constant coefficients. Assuming $v_z > 0$, substituting Eq. 10.5 into the transport equation, and collecting the ϵ-terms, we obtain

$$K_{sz}\frac{\partial^2 h}{\partial z^2} = \epsilon\left[\phi\frac{\partial c}{\partial t} - K_{sz}\left(1 + \frac{\partial h}{\partial z} + 2\epsilon c\right)\frac{\partial c}{\partial z} - K_{sz}c\frac{\partial^2 h}{\partial z^2}\right] \quad (10.9)$$

$$\alpha_L K_{sz}\frac{\partial}{\partial z}\left(\frac{\partial h}{\partial z}\frac{\partial c}{\partial z}\right) - K_{sz}\frac{\partial}{\partial z}\left(\frac{\partial h}{\partial z}c\right) + \phi\frac{\partial c}{\partial t}$$
$$= \epsilon\left\{K_{sz}\left[2c\frac{\partial c}{\partial z} + \frac{\partial}{\partial z}\left(c\frac{\partial h}{\partial z}\right) + 3\epsilon c^2\frac{\partial c}{\partial z}\right]\right.$$
$$\left. -\alpha_L K_{sz}\left[\frac{\partial}{\partial z}\left(c\frac{\partial c}{\partial z}\right) + \frac{\partial}{\partial z}\left(\frac{\partial h}{\partial z}\frac{\partial c}{\partial z}\right) + \epsilon\frac{\partial}{\partial z}\left(c\frac{\partial c}{\partial z}\right)\right]\right\} \quad (10.10)$$

For $\epsilon = 0$, Eqs. 10.9–10.10 can be decoupled by first solving the flow equation 10.9 for potential head, then using Eq. 10.5 to calculate the Darcy velocities, and finally solving the transport equation 10.10 for concentration. In this case the equations are linear.

If $\epsilon \neq 0$, the flow and transport equations are coupled, and nonlinear, and must be solved simultaneously for h and c. The coupling term is now given by

$$\epsilon\left[\phi\frac{\partial c}{\partial t} - K_{sz}\left(1 + \frac{\partial h}{\partial z} + 2\epsilon c\right)\frac{\partial c}{\partial z} - K_{sz}c\frac{\partial^2 h}{\partial z^2}\right] \quad (10.11)$$

If this term is zero the system is uncoupled, and the two equations can be solved separately. In this case the flow equation is linear, while the transport equation is nonlinear. If term 10.11 is nonzero, the degree of nonlinearity in the flow equation depends mainly on the magnitude of the $\epsilon K_{sz}(\partial h/\partial z)(\partial c/\partial z)$ and $\epsilon K_{sz}c(\partial^2 h/\partial z^2)$ terms. If the head and concentration gradients are small, we expect these nonlinearities to be weak. The transport equation contains two nonlinear components. The first, on the left hand side, is

$$\alpha_L K_{sz}\frac{\partial}{\partial z}\left(\frac{\partial h}{\partial z}\frac{\partial c}{\partial z}\right) - K_{sz}\frac{\partial}{\partial z}\left(\frac{\partial h}{\partial z}c\right) \tag{10.12}$$

The degree of nonlinearity of this component is controlled by the dependence of $\partial h/\partial z$ on term Eq. 10.11 via Eq. 10.9. When concentration and head gradients in space and time are small the coupling is weak, and consequently Eq. 10.12 is weakly nonlinear. The second nonlinear component, on the right hand side of Eq. 10.10, is

$$\epsilon\left\{K_{sz}\left[2c\frac{\partial c}{\partial z} + \frac{\partial}{\partial z}\left(c\frac{\partial h}{\partial z}\right) + 3\epsilon c^2\frac{\partial c}{\partial z}\right]\right.$$
$$\left. -\alpha_L K_{sz}\left[\frac{\partial}{\partial z}\left(c\frac{\partial c}{\partial z}\right) + \frac{\partial}{\partial z}\left(\frac{\partial h}{\partial z}\frac{\partial c}{\partial z}\right) + \epsilon\frac{\partial}{\partial z}\left(c\frac{\partial c}{\partial z}\right)\right]\right\} \tag{10.13}$$

Note that this component does not depend on h, but only on c, ϵ, and spatial concentration and head gradients. When transport is dispersion dominated, spatial and temporal concentration gradients are generally small. In this case the influence of the coupling and nonlinear terms 10.11 and 10.12, and to a lesser extent 10.13, is weak.

In summary, the saturated flow equation is not as strongly nonlinear as the transport equation, and we expect the importance of coupling, and the degree of nonlinearity in the transport equation, to decrease as ϵ decreases or as dispersion becomes dominant.

10.3   Numerical Discretization

10.3.1   Finite Element Discretization

The numerical model CODESA-3D is a standard finite element Galerkin scheme, with tetrahedral elements and linear basis functions, complemented by weighted finite differences for the discretization of the time derivatives. We present first the finite element discretization for the flow equation, followed by an analogous development for the transport equation. For an introduction to finite element techniques in engineering and groundwater applications, see Zienkiewicz [1986] and Huyakorn and Pinder [1983].

The finite element solution of the coupled system, Eqs. 10.4–10.6, approximates the exact solution (ψ, c) by $(\hat{\psi}, \hat{c})$ using linear basis functions $W(\mathbf{x})$ defined over a domain Ω discretized by E tetrahedral elements and N nodes:

$$\psi \approx \hat{\psi} = \sum_{j=1}^{N} \hat{\psi}_j(t) W_j(\mathbf{x})$$

$$c \approx \hat{c} = \sum_{j=1}^{N} \hat{c}_j(t) W_j(\mathbf{x}) \qquad (10.14)$$

where $\hat{\psi}_j$ and $\hat{c}_j$ are the components of the nodal solution vectors $\hat{\Psi}$ and $\hat{c}$.

10.3.1.1   Flow equation

Recasting Eq. 10.4 in operator notation

$$L(\psi, c) = \nabla \cdot \left[K_s \frac{1 + \epsilon c}{1 + \epsilon' c} K_r \left(\nabla \psi + (1 + \epsilon c) \eta_z \right) \right]$$
$$- \sigma \frac{\partial \psi}{\partial t} - \phi S_w \epsilon \frac{\partial c}{\partial t} + \frac{\rho}{\rho_o} q = 0 \qquad (10.15)$$

the error, or residual, represented by the finite element approximation Eq. 10.14 is given as $L(\hat{\psi}, \hat{c}) - L(\psi, c)$, or simply $L(\hat{\psi}, \hat{c})$. This error is minimized by imposing an orthogonality constraint between the residual and the basis functions, which yields the Galerkin integral

$$\int_{\Omega} L(\hat{\psi}, \hat{c}) W_i(\mathbf{x}) \, d\Omega = 0 \qquad i = 1, \ldots, N$$
$$(10.16)$$

We assume that the coordinate directions are parallel to the principal directions of hydraulic anisotropy, so that the off-diagonal components of the conductivity tensor K are zero. Expanding Eq. 10.16 and applying Green's lemma to the spatial derivative term we get, for $i = 1, \ldots, N$

$$-\int_\Omega K_r \left[K_s \frac{1+\epsilon\hat{c}}{1+\epsilon'\hat{c}} \left(\nabla\hat{\psi} + (1+\epsilon\hat{c})\eta_z \right) \cdot \nabla W_i \right] d\Omega$$

$$+\int_\Gamma K_r \left[K_s \frac{1+\epsilon\hat{c}}{1+\epsilon'\hat{c}} \left(\nabla\hat{\psi} + (1+\epsilon\hat{c})\eta_z \right) \cdot \mathbf{n} \right] W_i \, d\Gamma$$

$$-\int_\Omega \sigma \frac{\partial\hat{\psi}}{\partial t} W_i \, d\Omega - \int_\Omega \phi S_w \epsilon \frac{\partial\hat{c}}{\partial t} W_i \, d\Omega + \int_\Omega \frac{\rho}{\rho_o} q W_i \, d\Omega = 0$$

$$(10.17)$$

Substituting Eq. 10.14, changing sign, and making use of boundary condition Eq. 10.7c to replace the boundary integral term above, we obtain the following system of ordinary differential equations

$$H(\hat{\Psi}, \hat{c})\hat{\Psi} + P(\hat{\Psi}, \hat{c})\frac{d\hat{\Psi}}{dt} + \mathbf{q}^*(\hat{\Psi}, \hat{c}) = \mathbf{0} \qquad (10.18)$$

where

$$h_{ij} = \sum_{e=1}^{E} \int_{V^e} K_r^e \frac{1+\epsilon\bar{c}}{1+\epsilon'\bar{c}} \left(K_s^e \nabla W_j^e \cdot \nabla W_i^e \right) dV$$

$$p_{ij} = \sum_{e=1}^{E} \int_{V^e} \sigma^e W_j^e W_i^e \, dV$$

$$q_i^* = \sum_{e=1}^{E} \left[\int_{V^e} K_r^e K_{sz}^e \frac{(1+\epsilon\bar{c}^e)^2}{1+\epsilon'\bar{c}} \frac{\partial W_i^e}{\partial z} \, dV + \int_{V^e} \phi^e S_w^e \epsilon \frac{\partial\bar{c}^e}{\partial t} W_i^e \, dV \right.$$

$$\left. - \int_{V^e} \frac{\rho^e}{\rho_o} q^e W_i^e \, dV - \int_{\Gamma_2^e} q_n^e W_i^e \, d\Gamma \right]$$

$$(10.19)$$

In the above equations, $H = \{h_{ij}\}$ is the flow stiffness matrix, $P = \{p_{ij}\}$ is the flow mass (or capacity) matrix, $\mathbf{q}^* = \{q_i^*\}$ accounts for the prescribed boundary flux, the withdrawal or injection rate, the gravitational gradient term, and the term accounting for time variation of the concentration, K_{sz} is the vertical component of the saturated conductivity tensor, and $\bar{c}^e$ represents the average normalized concentration over the element volume V^e. Model parameters that are spatially dependent are considered constant for each element. Parameters that depend on pressure head or concentration are evaluated using ψ or c values averaged over each element and are also elementwise constant. Dirichlet boundary conditions are imposed after the discretized system has been completely assembled.

Equation 10.18 is integrated in time by the weighted finite difference scheme

$$\left(\nu_f H^{k+\nu_f} + \frac{P^{k+\nu_f}}{\Delta t_k} \right) \hat{\Psi}^{k+1} =$$

$$\left(\frac{P^{k+\nu_f}}{\Delta t_k} - (1 - \nu_f) H^{k+\nu_f} \right) \hat{\Psi}^k - \mathbf{q}^{*^{k+\nu_f}} \qquad (10.20)$$

where k and $k + 1$ denote the previous and current time levels, Δt_k is the time step size, and H, P, and $\mathbf{q}^*$ are evaluated at pressure head $\hat{\Psi}^{k+\nu_f} = \nu_f \hat{\Psi}^{k+1} + (1 - \nu_f) \hat{\Psi}^k$ and at concentration $\hat{c}^{k+\nu_f} = \nu_f \hat{c}^{k+1} + (1 - \nu_f) \hat{c}^k$. For numerical stability, parameter ν_f must satisfy the condition $0.5 \leq \nu_f \leq 1$.

The Darcy velocities are evaluated from Eq. 10.5 as

$$\mathbf{v}_\ell^{k+1} = U_\ell \hat{\Psi}^{k+1} + \mathbf{g}_\ell^{k+1} \qquad \ell = x, y, z \qquad (10.21)$$

where $\mathbf{v}_\ell$ is the vector of length E containing the component of the velocity in the ℓth direction for each tetrahedron, U_ℓ is an $E \times N$ matrix containing the head gradient component of Eq. 10.5, and $\mathbf{g}_\ell$ contains the gravity term of Eq. 10.5.

The evaluation of the matrices and vectors in Eqs. 10.19 and 10.21 is given in Appendix B.

10.3.1.2   Transport equation

Applying the same Galerkin procedure that was used for the flow equation, the analog to Eq. 10.17 for the transport model is

$$-\int_\Omega (D \, \nabla \hat{c}) \cdot \nabla W_i \, d\Omega - \int_\Omega \nabla \cdot (\mathbf{v} \hat{c}) W_i \, d\Omega + \int_\Gamma (D \, \nabla \hat{c}) \cdot \mathbf{n} W_i \, d\Gamma$$

$$-\int_\Omega \phi \frac{\partial S_w \hat{c}}{\partial t} W_i \, d\Omega + \int_\Omega (qc^* + f) W_i \, d\Omega = 0 \qquad i = 1, \ldots, N \qquad (10.22)$$

In the previous equation, Green's lemma has been applied only to the dispersive component of the transport equation, in order to avoid numerical instabilities that are typically introduced when the convective terms are also transformed [Huyakorn et al., 1985; Galeati and Gambolati, 1989].

Substituting the approximate solution Eq. 10.14 into Eq. 10.22, changing sign, and incorporating boundary conditions Eqs. 10.8c and

with $0.5 \leq \nu_c \leq 1$. A value close to 0.5 leads to accurate but possibly unstable solutions, while values close to 1 yield good stability but with larger numerical dispersion [Peyret and Taylor, 1983]. Another source of instability arises when convective fluxes dominate over dispersive fluxes. For these cases, alternative spatial discretization techniques are recommended, such as the ones described in section 10.3.2.3.

10.3.2 Alternative Discretization Techniques

10.3.2.1 Orthogonal subdomain collocation approach

Linear Galerkin finite element discretizations of the groundwater flow equation produce non-positive stiffness coefficients for internal element edges of two-dimensional Delaunay triangulations. This property, also called the positive transmissibility (PT) condition, ensures that the discrete flux is in the opposite direction of the head gradient [Prakash, 1987]. Violation of the PT condition means that, at least locally, nonphysical discrete fluxes can be generated. These may cause large approximation errors, especially for coupled systems of flow and transport (such as in multiphase flow or saltwater intrusion). For nonlinear problems, these errors may severely affect the behavior of the linearization scheme, and eventually lead to non-convergence.

It is known that the Galerkin approach on quadrilaterals, prisms, and tetrahedra, with bilinear, trilinear, and linear basis functions, respectively, does not always satisfy the PT condition [Forsyth, 1991]. The reason for this failure in the case of linear tetrahedral Galerkin schemes can be understood from the physical interpretation of the Galerkin stiffness coefficient. It is possible to show that the linear Galerkin stiffness coefficient in two-dimensional triangulations is the flux of the basis function gradient across a face midperpendicular to the element edges. However, in three-dimensional triangulations, the orthogonality of the face is lost, and thus the flux across this face is incorrectly evaluated. A modification of the three-dimensional Galerkin scheme called OSC (orthogonal subdomain collocation) [Putti and Cordes, 1998]. uses orthogonal subdomain boundaries and always satisfies the PT condition if a three-dimensional Delaunay triangulation is used.

10.3.2.2   Mixed finite element method for flow

In standard finite element simulations, groundwater velocities obtained element-wise from hydraulic head gradients introduce discontinuous fluxes, and thus divergence, at element edges. To avoid this problem, the mixed finite element (MFE) approach uses continuous fluxes across element edges as additional degrees of freedom. The MFE technique can be developed from an alternative formulation of the flow equation 10.4 using pressure head ψ and Darcy flux $\mathbf{v}$ as unknowns:

$$\mathbf{v} + K_s \frac{1+\epsilon c}{1+\epsilon' c} K_r \left(\nabla\psi + (1+\epsilon c)\eta_z\right) = 0 \qquad (10.26)$$

$$\sigma\frac{\partial\psi}{\partial t} + \nabla \cdot \mathbf{v} = \frac{\rho}{\rho_o}q - \phi S_w\epsilon\frac{\partial c}{\partial t} \qquad (10.27)$$

The finite element method is used to discretize these two equations by means of scalar and vector basis functions for ψ and $\mathbf{v}$, respectively. The typical scheme employed for this type of problem is the Raviart-Thomas (RT0) approach that uses piecewise constant basis functions for ψ and linear vector basis functions for $\mathbf{v}$. The pressure head is defined at the centroid while the fluxes are defined at each edge of the elements. This yields a system of equations that can be solved numerically for $\mathbf{v}$, giving a total number of unknowns equal to the number of edges in the computational mesh.

The mixed finite element method was introduced more than two decades ago for the discretization of groundwater flow equations [Meissner, 1972] (see also Douglas et al., [1983]), but has only recently been extensively studied and compared to more classical techniques [Brezzi and Fortin, 1991; Chavent and Roberts, 1991; Durlofski, 1993; Cordes and Putti, 1996] The method is extremely attractive for the solution of coupled flow and transport problems when used in conjunction with the finite volume method for the transport equation [Durlofski, 1993; Bergamaschi et al., 1995].

The main drawback of MFE is that the formulation generates about twice as many degrees of freedom as standard finite elements of the same order, and about 1.5 as many as finite volumes. However, the accuracy of the flow field attainable with MFE is superior to that of standard finite elements.

10.3.2.3   Finite volume method for transport

When transport is advection-dominated, sharp concentration fronts are usually present. Special numerical methods are then required to avoid spurious oscillations and to minimize the amount of artificial viscosity, which may considerably affect the resolution of sharp fronts. One such class of methods is upwind finite volumes (FV), developed over the last decades [Van Leer, 1977; Roe, 1986] and recently applied to groundwater contamination problems [Putti et al., 1990; Cox and Nishikawa, 1991; Unger et al., 1996]. FV methods are globally high-order accurate and nonoscillatory.

The basic idea behind these methods can be outlined as follows. First, the dependent variable is represented as a volume average in a control volume or cell. Its rate of change is determined by the fluxes across the cell interfaces. The flux at an interface is obtained from the "upwind" and "downwind" volume averaged states by solving the local physical problem, or so-called Riemann problem, exactly. This leads to a first-order accurate scheme. Note that, in solving the Riemann problem, the variables are assumed to be piecewise constant. Second order accuracy is achieved with the determination of the linear distribution of the dependent variable in the control volumes, using interpolation from neighboring cells. The crucial step is to do this without introducing numerical oscillations, i.e., no overshoots or undershoots with respect to neighboring cell averages should be created. In the vicinity of discontinuities, this requirement results in slopes that are smaller than those obtained by usual interpolation. This procedure is called "limiting". Once the piecewise linear distributions are determined, the fluxes are obtained from the solution of the local Riemann problem at cell interfaces. The scheme results in second order spatial accuracy on all points except where limiting occurs, where only first-order accuracy is attained. The role of the limiting procedure is fundamental in that it maintains the stability of the scheme, introducing at the same time numerical viscosity only in the neighborhood of the discontinuities.

10.4   Linearization

10.4.1   General Approaches for Nonlinear Systems

Techniques for solving large scale nonlinear problems, such as those that arise from multi-dimensional numerical discretization, are computationally intensive and require highly efficient and robust algorithms. Efficiency ensures optimal utilization of CPU and storage resources to attain a desired level of solution accuracy, while robustness implies that a given algorithm exhibits acceptable convergence behavior across a wide spectrum of simulation scenarios. Unfortunately, there are no good, general algorithms for solving systems of more than one nonlinear equation [Press et al., 1989]. The classical method around which most other iterative algorithms are developed, Newton's method, has excellent convergence properties, but only if a good initial solution estimate is provided.

Newton's method, also known as Newton-Raphson, can be derived for the general system $\mathbf{f}(\mathbf{x}) = \mathbf{0}$ by taking a Taylor series expansion of $\mathbf{f}$ about its solution $\mathbf{x}^*$.

$$0 = \mathbf{f}(\mathbf{x}^*) = \mathbf{f}(\mathbf{x}) + \mathbf{f}'(\mathbf{x})(\mathbf{x}^* - \mathbf{x}) + \cdots \qquad (10.28)$$

For $|\mathbf{x}^* - \mathbf{x}|$ small, we can define the following iteration:

$$\mathbf{x}^{m+1} = \mathbf{x}^m - \left(\mathbf{f}'(\mathbf{x}^m)\right)^{-1}\mathbf{f}(\mathbf{x}^m) \qquad (10.29)$$

where m is the iteration level. The matrix $\mathbf{f}'$ is called the Jacobian and is given by $J = \{\partial f_i / \partial x_j\}$.

Numerical discretization of nonlinear partial differential equations generally lead to quasi-linear systems of the type $\mathbf{f}(\mathbf{x}) = A(\mathbf{x})\mathbf{x} - \mathbf{b}(\mathbf{x}) = \mathbf{0}$. In this case the Jacobian is

$$J = A + A'\mathbf{x} - \mathbf{b}' \qquad (10.30)$$

and the Newton scheme is expressed as

$$\begin{aligned} J(\mathbf{x}^m)\mathbf{s}^m &= -\mathbf{f}(\mathbf{x}^m) \\ \mathbf{x}^{m+1} &= \mathbf{x}^m + \mathbf{s}^m \end{aligned} \qquad (10.31)$$

where vector $\mathbf{s}^m$ is called the Newton direction or the search direction.

In practice the Newton method's attractive convergence properties need to be weighed against the cost of computing the Jacobian. As there is no ideal, general-purpose nonlinear solver, many variants of the basic Newton method have been proposed, often based on introducing different approximations to the Jacobian. Some of the simplest ones include modified Newton, where the Jacobian is updated selectively rather than at every iteration, and the secant method, where the Jacobian is evaluated by finite differences.

One of the most commonly used variants is Picard iteration, also known as simple iteration, nonlinear Richardson iteration, or the method of successive substitution. It is a straightforward scheme that is easy to implement in a numerical code and is computationally inexpensive on a per iteration basis, but its theoretical local convergence speed is linear as compared to the Newton method's quadratic behavior. The Picard method is formulated as in Eq. 10.31, but with Jacobian $J = A$ instead of the Newton Jacobian given in Eq. 10.30.

Another class of methods, quasi-Newton, has been developed with the aim of both reducing the cost of the Jacobian evaluation step associated with Newton iteration, and maintaining good theoretical convergence properties. The basic idea behind quasi-Newton methods is to replace the Jacobian with a less costly approximation based on easily calculated recursive updates to the Jacobian or its inverse.

The initial solution estimate has a big effect on the behavior of iterative schemes. Unlike iterative techniques for linear systems, where theorems exist that guarantee global convergence under specified conditions, for the Newton-type methods described above only local convergence results are available, that is, provided the initial estimate is "close enough" to the solution. In practice, relaxed iteration, defined as

$$\mathbf{x}^{m+1} = \mathbf{x}^m + \lambda \mathbf{s}^m \qquad (10.32)$$

can often help achieve or accelerate convergence when a poor initial estimate is used. Line search algorithms systematically compute the relaxation parameter by finding the optimal step length to be taken along the search direction indicated by the iterative scheme.

The CODESA-3D model contains nonlinearities in the unsaturated flow equation and in the coupled flow and transport system. In the following two subsections we survey the solution methods that can be used to solve these different nonlinear problems. Although the general approaches described above can be used for both problems, much more investigative work has been reported for the solution of the variably saturated flow equation than for the coupled system. We discuss the flow equation first, detailing how the Newton, Picard, quasi-Newton, and line search relaxation schemes can be implemented for this equation. The Newton and Picard methods are then presented for the coupled system, as well as a new partial Newton scheme which avoids the computational complexities of full Newton for coupled systems.

The iterative solution procedure for the CODESA-3D model thus consists of an inner loop for the unsaturated flow equation within an outer loop for the coupled system. As will be described in section 10.5, iterative procedures are also used to solve the symmetric or nonsymmetric systems resulting from linearization of the flow and transport equations, thus adding a third, innermost loop in the CODESA-3D solution procedure. In its current implementation, each of these three levels is solved independently of the others. That is, for each coupled iteration, a Newton-type iterative method is applied to the nonlinear flow equation until convergence is achieved; for each of these inner iterations, a conjugate gradient-like iterative method is applied to the linearized flow equation until convergence is achieved. Some efficiency, and probably robustness, can be gained by linking these three iteration levels. For instance, for early iterations in the outer-coupled loop, it is probably wasteful to iterate until convergence is achieved in the inner flow loop.

10.4.2 Iterative Methods for Variably Saturated Flow

Equation 10.4 is highly nonlinear due to pressure head dependencies in the general storage and relative hydraulic conductivity terms. These terms can be modeled using various constitutive or characteristic relations describing the soil hydraulic properties.

The characteristic equations introduced in van Genuchten and Nielsen [1985] are commonly used. These can be written as

$$
\begin{aligned}
\theta(\psi) &= \theta_r + (\theta_s - \theta_r)[1 + \beta]^{-\gamma}; & \psi &< 0 \\
\theta(\psi) &= \theta_s; & \psi &\geq 0
\end{aligned}
\tag{10.33}
$$

$$
\begin{aligned}
K_r(\psi) &= (1 + \beta)^{-5\gamma/2}\,[(1 + \beta)^\gamma - \beta^\gamma]^2\,; & \psi &< 0 \\
K_r(\psi) &= 1; & \psi &\geq 0
\end{aligned}
\tag{10.34}
$$

where θ is the volumetric moisture content, θ_r is the residual moisture content, θ_s is the saturated moisture content (generally equal to the porosity ϕ), $\beta = (\psi/\psi_s)^n$, ψ_s is the capillary or air entry pressure head value, n is a constant, and $\gamma = 1 - 1/n$ for n approximately in the range $1.25 < n < 6$. The corresponding general storage term is (see Appendix A)

$$\sigma = S_w S_s (1 + \epsilon c) + \phi (1 + \epsilon c) \frac{dS_w}{d\psi} \tag{10.35}$$

where $S_w = \theta/\theta_s$ and S_s is the specific storage.

Another widely used set of characteristic relations expresses the water saturation S_w in terms of effective saturation S_e, in the form $S_w(\psi) = (1 - S_{wr})S_e(\psi) + S_{wr}$, where S_{wr} $(= \theta_r/\theta_s)$ is the residual water saturation. The set of equations is [Huyakorn et al., 1984]

$$
\begin{aligned}
S_e(\psi) &= \left[1 + \kappa^\beta (\psi_a - \psi)^\beta\right]^{-\gamma}; & \psi < \psi_a \\
S_e(\psi) &= 1; & \psi \geq \psi_a
\end{aligned} \tag{10.36}
$$

$$K_r(\psi) = K_r\left(S_e(\psi)\right) = 10^{G(S_e)} \tag{10.37}$$

where ψ_a is the air entry pressure, $G(S_e) = aS_e^2 + (b - 2a)S_e + a - b$, and κ, β, γ, a, and b are constants. The general storage term is again given by Eq. 10.35.

Expressing the discretized flow equation 10.20 as

$$\mathbf{g}\left(\hat{\Psi}^{k+1}, \hat{\mathbf{c}}^{k+1}\right) = H^{k+\nu_f}\hat{\Psi}^{k+\nu_f} + \frac{1}{\Delta t_k}P^{k+\nu_f}\left(\hat{\Psi}^{k+1} - \hat{\Psi}^k\right)$$
$$+ \mathbf{q}^{*^{k+\nu_f}} = 0 \tag{10.38}$$

Newton's method can be written as

$$J(\hat{\Psi}^{k+1,m}, \hat{\mathbf{c}}^{k+1,m})\mathbf{s}^m = -\mathbf{g}(\hat{\Psi}^{k+1,m}, \hat{\mathbf{c}}^{k+1,m}) \tag{10.39}$$

where $\mathbf{s}^m = \hat{\Psi}^{k+1,m+1} - \hat{\Psi}^{k+1,m}$ and

$$
\begin{aligned}
J_{ij} &= \nu_f H_{ij} + \frac{1}{\Delta t_k}P_{ij} + \sum_s \frac{\partial H_{is}}{\partial \hat{\psi}_j^{k+1}}\hat{\psi}_s^{k+\nu_f} \\
&\quad + \frac{1}{\Delta t_k}\sum_s \frac{\partial P_{is}}{\partial \hat{\psi}_j^{k+1}}(\hat{\psi}_s^{k+1} - \hat{\psi}_s^k) + \frac{\partial q_i^*}{\partial \hat{\psi}_j^{k+1}}
\end{aligned} \tag{10.40}
$$

The Picard method is usually arrived at by evaluating all nonlinear terms in Eq. 10.20 at the previous iteration level, m, and the linear terms at $m + 1$. This yields

$$\left(\nu_f H^{k+\nu_f,m} + \frac{1}{\Delta t_k} P^{k+\nu_f,m} \right) \hat{\Psi}^{k+1,m+1}$$
$$= \left(\frac{1}{\Delta t_k} P^{k+\nu_f,m} - (1 - \nu_f) H^{k+\nu_f,m} \right) \hat{\Psi}^k - \mathbf{q}^{*^{k+\nu_f,m}}$$

$$(10.41)$$

By simple algebraic manipulation, the above equation can be rearranged to give

$$\left(\nu_f H^{k+\nu_f,m} + \frac{1}{\Delta t_k} P^{k+\nu_f,m} \right) \mathbf{s}^m = -\mathbf{g}(\hat{\Psi}^{k+1,m}, \hat{c}^{k+1,m})$$

$$(10.42)$$

Comparing Eqs. 10.39 and 10.42, it is apparent that the Picard scheme can be viewed as an approximate Newton method. An important difference between the two schemes is that Newton linearization generates a nonsymmetric system matrix, whereas Picard preserves the symmetry of the original discretization of the flow equation.

It has been observed that the Newton method is more sensitive to the initial solution estimate than the Picard scheme. In order to exploit the best features of both methods, a mixed Picard-Newton approach was proposed in Paniconi and Putti, [1994], based on the idea of using Picard iteration to improve the initial solution estimate for the faster converging Newton method. In this mixed approach, Picard is used for the first few iterations, until a specified reduction in convergence error has been achieved, and Newton for the remaining iterations.

The quasi-Newton family is defined by substituting in the Newton equation 10.39 an approximation K to the Jacobian J, yielding

$$K(\hat{\Psi}^{k+1,m}, \hat{c}^{k+1,m}) \mathbf{s}^m = -\mathbf{g}(\hat{\Psi}^{k+1,m}, \hat{c}^{k+1,m})$$

$$(10.43)$$

Rather than calculate the Jacobian at each iteration, matrix K is updated using recursion formulae that require fewer operations than a full evaluation of J. In implementations of the quasi-Newton method, updates for K^{-1}, which we denote below as K_I, are often used instead of K updates, as these inverse updates avoid the need to solve a linear system at each iteration [Paniconi and Putti, 1996]. Several update formulae have been proposed in the literature, amongst the most popular being the Broyden and BFGS updates [Dennis and Moré, 1977; Fletcher, 1980]. The expression for the Broyden inverse update is

$$K_I^m = K_I^{m-1} + \frac{\left(\mathbf{s}^m - K_I^{m-1}\mathbf{y}^m\right)(\mathbf{s}^m)^T K_I^{m-1}}{(\mathbf{s}^m)^T K_I^{m-1}\mathbf{y}^m} \tag{10.44}$$

where $\mathbf{y}^m = \mathbf{g}^{m+1} - \mathbf{g}^m$, while the BFGS inverse update can be written as

$$K_I^m = \left(I - \omega^m \mathbf{s}^m(\mathbf{y}^m)^T\right) K_I^{m-1} \left(I - \omega^m \mathbf{y}^m(\mathbf{s}^m)^T\right)$$
$$+ \omega^m \mathbf{s}^m(\mathbf{s}^m)^T \tag{10.45}$$

where I is the identity matrix and $\omega^m = ((\mathbf{s}^m)^T\mathbf{y}^m)^{-1}$. In our implementation, we use the exact Jacobian J_0 for the initial update K_0. The inverse is never calculated explicitly. Instead, the LU factorization of K is used, and sparsity-preserving recursion expressions that calculate directly the solution vector $\mathbf{s}^{m+1}$ are developed.

Convergence of an iterative scheme can be enhanced by introducing a relaxation (or damping) parameter λ as described in Eq. 10.32. In applications of Picard and Newton iteration to Richards' equation, ad hoc relaxation methods, such as constant λ, have been used [Cooley, 1983; Huyakorn et al., 1986; Paniconi and Putti, 1994]. Line search algorithms, originally developed for optimization problems, provide a systematic procedure for determining λ. In this context, vector $\mathbf{s}^m$ can be viewed as a search direction in the N-dimensional vector space along which the new approximate solution will be sought. The relaxation parameter λ can be interpreted as the length of the step that will be taken along $\mathbf{s}^m$. The optimal value for λ can be chosen as the point that minimizes the objective function along the search direction. In our implementation, this minimum is found by solving, by simple bisection, the one-dimensional nonlinear problem [Dennis and Schnabel, 1983; Papadrakakis, 1993; Paniconi and Putti, 1996]

$$(\mathbf{s}^m)^T \mathbf{g}(\hat{\Psi}^{k+1,m} + \lambda \mathbf{s}^m) = 0 \tag{10.46}$$

10.4.2.1   Some performance results

Figure 10.1 compares the convergence profiles of the Newton, Picard, Broyden, and BFGS methods for a one-dimensional steady state flow

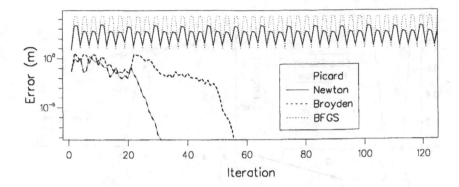

Figure 10.1. Convergence profiles for a 1-D steady state flow problem comparing the performance of Newton, Picard, and quasi-Newton methods.

problem. The parameters used for this problem are: uniform grid spacing $\Delta z = 0.1$ m, $\psi - 0$ at the base of the soil column $z = 0$, Darcy flux of 0.0001 m/h at the top of the column $z = 10$, $K_s = 0.01$ m/h, and characteristic equation 10.34 with $n = 5$ and $\psi_s = -3$ [Paniconi and Putti, 1996]. Both the Picard and Newton methods produced oscillations and failed to converge, while the Broyden and BFGS methods converged in 31 and 56 iterations, respectively.

The performance of the line search algorithm is shown in Figure 10.2 for the same test problem. We observe that the Newton scheme converged in as few as 34 iterations with a fixed relaxation parameter λ, and that when line search was applied only 12 iterations were required for convergence. The top left plot in Figure 10.2 illustrates how widely the "optimal" value of λ can vary from one iteration to the next, particularly in the first iterations, and thus the utility of a line search routine that can calculate dynamically the relaxation parameter.

Table 10.1 summarizes the convergence results obtained using Picard, Newton, and mixed Picard-Newton iteration for a problem involving steady state two-dimensional seepage through a square embankment [Cooley, 1983; Paniconi and Putti, 1994]. This problem was used to examine the effects of grid discretization and soil characteristics on convergence behavior. The performance of the Picard and Newton schemes deteriorated for the finer grid in all runs, whereas the mixed Picard-Newton scheme was successful at both grid dis-

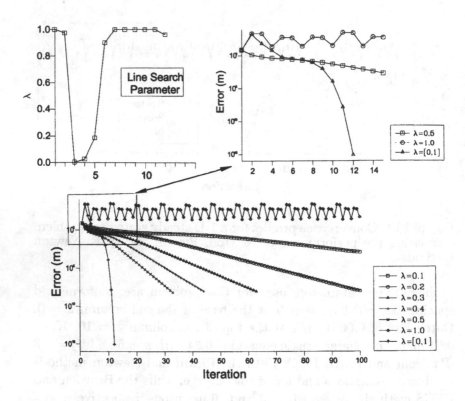

Figure 10.2. Convergence profiles for a 1-D steady state flow problem comparing the performance of the Newton scheme with constant relaxation parameter ($\lambda = 1$ is unrelaxed Newton) and with the relaxation parameter computed by line search ($\lambda = [0,1]$). The top right plot is a zoom on the first 15 iterations, and the top left plot shows the λ value computed by line search at each of these 15 iterations.

cretizations. In the same vein, the more strongly nonlinear $K_r(\psi)$ relationships caused convergence difficulties for the Picard and Newton schemes, while mixed Picard-Newton was very effective in all cases.

10.4.3 Iterative Methods for Coupled Flow and Transport

The most commonly used algorithm for the numerical solution of the nonlinearly coupled flow and transport system can be described as follows. At the $(m+1)$-st iteration the flow equation is solved for

n	ψ_s	$\Delta x = \Delta z$	Number of Nonlinear Iterations		
			Picard	Newton	Mixed Picard-Newton
5	−20	10	16	24	6 + 14 †
5	−20	2	20	29	5 + 19
5	−3	10	17	176	8 + 15
5	−3	2	74	failed	10 + 33
3	−3	10	24	50	8 + 21
3	−3	2	74	failed	8 + 30
1.5	−3	10	failed	44	9 + 29
1.5	−3	2	failed	failed	6 + 29

† Picard iterations + Newton iterations.

Table 10.1. Summary of convergence results for a 2-D steady state flow problem with $K_r(\psi)$ in Eq. 10.34 (Parameters n and ψ_s).

$\hat{\Psi}^{k+1,m+1}$ using, as initial guess, the values of pressure head, $\hat{\Psi}^{k+1,m}$, and concentration, $\hat{c}^{k+1,m}$, from the previous iteration. The velocity field Eq. 10.21 is then calculated using $\hat{\Psi}^{k+1,m+1}$ and $\hat{c}^{k+1,m}$. Using these updated values of velocity and pressure head, the transport equation becomes linear and is solved for $\hat{c}^{k+1,m+1}$. This three-step procedure is repeated until convergence is achieved. We refer to this iterative procedure as the Picard method.

The algorithm can be expressed mathematically as

$$\left(\nu_f H^{k+\nu_f,m} + \frac{1}{\Delta t_k} P^{k+\nu_f,m} \right) \hat{\Psi}^{k+1,m+1} =$$
$$\left(\frac{1}{\Delta t_k} P^{k+\nu_f,m} - (1 - \nu_f) H^{k+\nu_f,m} \right) \hat{\Psi}^k - q^{*^{k+\nu_f,m}}$$
$$(10.47)$$

$$v_\ell^{k+1,m} = U_\ell^{k+1,m} \hat{\Psi}^{k+1,m+1} + g_\ell^{k+1,m} \qquad ; \ell = x, y, z$$
$$(10.48)$$

$$\left[\nu_c (A + B + C)^{k+\nu_c,m} + \frac{1}{\Delta t_k} M^{k+1,m} \right] \hat{c}^{k+1,m+1} =$$
$$\left[\frac{1}{\Delta t_k} M^k - (1 - \nu_c)(A + B + C)^{k+\nu_c,m} \right] \hat{c}^k - r^{*^{k+\nu_c}}$$
$$(10.49)$$

Note that Eq. 10.47 is the Picard scheme of Eq. 10.41, and can be substituted by any other linearization of the variably saturated flow equation, such as Newton's method Eq. 10.39. Also note that in Eq. 10.48 we write $\mathbf{v}_\ell^{k+1,m}$ instead of $\mathbf{v}_\ell^{k+1,m+1}$ because, while the velocity field is calculated using the new estimate $\hat{\Psi}^{k+1,m+1}$ for pressure head, the previous estimate $\hat{c}^{k+1,m}$ is used for concentration. Analogously, the transport matrices are expressed at iteration m since they are evaluated using $\mathbf{v}_\ell^{k+1,m}$. In the Picard method two $N \times N$ linear systems are solved sequentially at each iteration, the flow system Eq. 10.47 and the transport system Eq. 10.49.

An alternative procedure for solving the coupled system of equations is to use the Newton technique. Let $\mathbf{f}$ be the vector-valued function

$$\mathbf{f}(\mathbf{u}) = \left[\begin{array}{c} \mathbf{f}_1(\mathbf{u}) \\ \mathbf{f}_2(\mathbf{u}) \end{array} \right] \tag{10.50}$$

where $\mathbf{u} = \left(\hat{\Psi}^{k+1}, \hat{c}^{k+1} \right)^T$ and

$$\mathbf{f}_1(\mathbf{u}) = \left(\nu_f H^{k+\nu_f} + \frac{P^{k+\nu_f}}{\Delta t_k} \right) \hat{\Psi}^{k+1} +$$

$$\left[(1 - \nu_f) H^{k+\nu_f} - \frac{P^{k+\nu_f}}{\Delta t_k} \right] \hat{\Psi}^k + \mathbf{q}^{*k+\nu_f}$$

$$\mathbf{f}_2(\mathbf{u}) = \left[\nu_c (A + B + C)^{k+\nu_c} + \frac{M^{k+1}}{\Delta t_k} \right] \hat{c}^{k+1}$$

$$+ \left[(1 - \nu_c)(A + B + C)^{k+\nu_c} - \frac{M^k}{\Delta t_k} \right] \hat{c}^k + \mathbf{r}^{*k+\nu_c} \tag{10.51}$$

We want to find the correction $\Delta \mathbf{u}$ defined as

$$\Delta \mathbf{u} = \left(\begin{array}{c} \Delta \hat{\Psi} \\ \Delta \hat{c} \end{array} \right) = \left(\begin{array}{c} \hat{\Psi}^{k+1,m+1} - \hat{\Psi}^{k+1,m} \\ \hat{c}^{k+1,m+1} - \hat{c}^{k+1,m} \end{array} \right) \tag{10.52}$$

The Jacobian matrix of $\mathbf{f}$ is

$$J = \mathbf{f}' = \left[\begin{array}{cc} \partial \mathbf{f}_1 / \partial \hat{\Psi}^{k+1} & \partial \mathbf{f}_1 / \partial \hat{c}^{k+1} \\ \partial \mathbf{f}_2 / \partial \hat{\Psi}^{k+1} & \partial \mathbf{f}_2 / \partial \hat{c}^{k+1} \end{array} \right] \tag{10.53}$$

and is a $2N \times 2N$ nonsymmetric matrix subdivided into four matrix blocks.

The assembly of the Jacobian matrix and the solution of the $2N \times 2N$ nonsymmetric system renders the Newton scheme computationally very demanding. It is therefore desirable to decouple the Newton system, taking into consideration the most important of the nonlinear derivative components of Eq. 10.53. Below we describe the partial Newton scheme, developed in Putti and Paniconi, [1995a 1995b], which produces a decoupled system, in the manner of the Picard scheme or other similar approaches used in petroleum reservoir simulation [Aziz and Settari, 1979], while retaining some of the best properties of the Newton technique. This scheme takes advantage of the fact that the coupling in the flow and transport equations is weak, as shown in section 10.2.2.

If we rewrite Eqs. 10.47–10.49 with the velocities expressed using $\hat{c}^{k+1,m+1}$ rather than $\hat{c}^{k+1,m}$, we obtain

$$\left(\nu_f H^{k+\nu_f,m} + \frac{P^{k+\nu_f,m}}{\Delta t_k} \right) \hat{\Psi}^{k+1,m+1} =$$
$$- \left[(1 - \nu_f) H^{k+\nu_f,m} - \frac{P^{k+\nu_f,m}}{\Delta t_k} \right] \hat{\Psi}^k - \mathbf{q}^{*^{k+\nu_f,m}}$$

$$(10.54)$$

$$\left[\nu_c (A + B + C)^{k+\nu_c,m+1} + \frac{M^{k+1,m+1}}{\Delta t_k} \right] \hat{c}^{k+1,m+1} =$$
$$- \left[(1 - \nu_c)(A + B + C)^{k+\nu_c,m+1} - \frac{M^k}{\Delta t_k} \right] \hat{c}^k - \mathbf{r}^{*^{k+\nu_c}}$$

$$(10.55)$$

where

$$\mathbf{v}_\ell^{k+1,m+1} = U_\ell \hat{\Psi}^{k+1,m+1} + \mathbf{g}_\ell^{k+1,m+1} \qquad (10.56)$$

The transport equation is now nonlinear and can be linearized using the Newton method.

The $N \times N$ Jacobian matrix for Eq. 10.55, i.e., the derivative of the transport equation with respect to concentration, yields

$$J\left(\hat{\Psi}^{k+1}, \hat{c}^{k+1} \right) = \nu_c (A + B + C)^{k+\nu_c}$$
$$+ \left[\frac{\partial}{\partial \hat{c}^{k+1}} (A + B + C)^{k+\nu_c} \right] \hat{c}^{k+\nu_c} + \frac{1}{\Delta t_k} M^{k+1}$$

$$(10.57)$$

This term is identical to the fourth block of the Jacobian matrix for the full Newton scheme, that is, $\partial \mathbf{f}_2 / \partial \hat{c}$ of Eq. 10.53. Evaluation of the different components of the derivative term in Eq. 10.57 for the saturated case is reported in [Putti and Paniconi, 1995a]. The Newton equation for Eq. 10.55 is therefore written as

$$J\left(\hat{\Psi}^{k+1,m}, \hat{c}^{k+1,m}\right) \Delta\hat{c} = -f_2\left(\hat{\Psi}^{k+1,m}, \hat{c}^{k+1,m}\right)$$
(10.58)

The solution procedure for one iteration of the partial Newton method is a sequence similar to the Picard scheme, and consists of the following two steps:

1. calculate $\hat{\Psi}^{k+1,m+1}$ by solving the Picard- or Newton-linearized flow equation using $\hat{c}^{k+1,m}$ and $\hat{\Psi}^{k+1,m}$ as initial values;

2. solve the Newton-linearized transport equation 10.58 for the new values of concentration $\hat{c}^{k+1,m+1}$.

Comparing Eq. 10.57 with the left hand side of Eq. 10.49, it can be easily shown that the additional cost of the partial Newton scheme relative to the Picard scheme amounts to two matrix-vector products and assembly of their corresponding elements. The remaining operations are formally the same, and the final system matrix is nonsymmetric for both schemes.

In the saturated flow case, the partial Newton scheme can also be derived directly from the full Newton procedure. Since the flow equation is weakly nonlinear in $\hat{c}$, we can neglect the $\partial f_1/\partial \hat{c}$ term of Eq. 10.53, obtaining the head difference solution

$$\Delta\hat{\Psi} = -\left[\frac{\partial f_1}{\partial \hat{\Psi}^{k+1}}\right]^{-1} f_1\left(\hat{\Psi}^{k+1,m}, \hat{c}^{k+1,m}\right)$$
(10.59)

The concentration difference can now be calculated solving the system

$$\left[\frac{\partial f_2}{\partial \hat{c}^{k+1}}\left(\hat{\Psi}^{k+1,m}, \hat{c}^{k+1,m}\right)\right] \Delta\hat{c} = -f_2\left(\hat{\Psi}^{k+1,m}, \hat{c}^{k+1,m}\right)$$
$$-\frac{\partial f_2}{\partial \hat{\Psi}^{k+1}}\left(\hat{\Psi}^{k+1,m}, \hat{c}^{k+1,m}\right) \Delta\hat{\Psi}$$
(10.60)

If the $(\partial f_2/\partial \hat{\Psi})\Delta\hat{\Psi}$ term is neglected, Eq. 10.60 reduces to Eq. 10.58 of the partial Newton scheme.

Because of the application of a quadratically convergent Newton linearization to the transport equation, faster convergence for the partial Newton scheme compared to the Picard technique is often observed. As noted previously, the degree of nonlinearity of the transport equation varies with the magnitude of the density ratio and dispersion coefficient. Thus the partial Newton scheme often converges successfully where Picard linearization may fail, that is, when ϵ is large or dispersion is small. For problems where ϵ is small or dispersion is dominant, on the other hand, the Picard scheme is usually adequate. Some performance results are given in Example 1 of section 10.6.

10.5   Projection Solvers for Linear Systems

The discretization and linearization procedures described in the previous two sections yield large sparse systems of linear equations that need to be repeatedly solved (at each nonlinear iteration and for every time step). These discrete linearized systems can be symmetric positive definite or nonsymmetric—they are symmetric in the case of Picard-linearized flow, and nonsymmetric for the transport equation and for quasi-Newton and Newton-linearized flow.

Projection methods relying on Krylov subspaces have recently been developed for large sparse symmetric and nonsymmetric systems, and several of these methods are implemented in the CODESA-3D model. We give a short review of the basic ideas underlying projection methods. For a more detailed and theoretical description, the reader is referred to Saad, [1981; 1990].

We express the linear or linearized equations to be solved in the general form

$$A\mathbf{x} = \mathbf{b} \tag{10.61}$$

where A represents the flow, transport, or Jacobian-type matrix of dimension $N \times N$. A projection method projects $A\mathbf{x} = \mathbf{b}$ onto subspaces (called Krylov subspaces) of increasing size ℓ, and solves the reduced system. The procedure converges to the correct solution $\mathbf{x}$ after N iterations in exact arithmetic. In practice, however, convergence long before the Krylov subspace dimension approaches this maximum value is sought. For this reason, and because of its repetitive nature, projection solvers are considered to be iterative methods.

A projection method starts from an initial (in general arbitrary) solution $\mathbf{x}_0$ to Eq. 10.61 and looks for a new approximation $\mathbf{x}_\ell$ from the subspace $\mathbf{x}_0 + K_\ell$, where K_ℓ is the Krylov subspace of size ℓ

$$K_\ell = \text{span}\{\mathbf{r}_0, A\mathbf{r}_0, A^2\mathbf{r}_0, \ldots, A^{\ell-1}\mathbf{r}_0\} \qquad (10.62)$$

with $\mathbf{r}_0 = \mathbf{b} - A\mathbf{x}_0$ being the initial residual associated to $\mathbf{x}_0$. The new solution $\mathbf{x}_\ell$ is obtained by prescribing the well-known Petrov-Galerkin condition of orthogonality between the new residual $\mathbf{r}_\ell$ and an auxiliary subspace L_ℓ with the same dimension as K_ℓ, i.e., $\mathbf{r}_\ell \perp L_\ell$.

Choosing different L_ℓ subspaces leads to different projection methods. In this respect there is a strong conceptual similarity between a projection method of the kind discussed above and a classical weighted residual approach, such as finite elements, used for solving an (initial) boundary value problem.

The most effective choices for L_ℓ are:

1. $L_\ell = K_\ell$. This is a Galerkin or orthogonal method. The classical conjugate gradient (CG) method [Hestenes and Stiefel, 1952] belongs to this class. The solution $\mathbf{x}_\ell$ is guaranteed to exist at each iteration ℓ (or Krylov subspace of dimension ℓ) if A is symmetric positive definite, i.e., in our case only for the Picard-linearized flow problem. Under these circumstances $\mathbf{x}_\ell$ can be shown to minimize the A^{-1} norm of $\mathbf{r}_\ell$ [Freund et al., 1992], hence the orthogonal or Galerkin method is also a minimal residual (MR) method with respect to a suitable $\mathbf{r}_\ell$ norm. On the other hand, if A is nonsymmetric, or is symmetric but indefinite, $\mathbf{x}_\ell$ does not satisfy any optimal property and is not guaranteed to exist at each iteration. In this case the Galerkin projection method may break down at some point during the iterative procedure. This is often the case when CG methods are applied with nonsymmetric or indefinite matrices.

2. $L_\ell = AK_\ell$. An orthogonal residual projection method in this class is also an MR method which minimizes the Euclidean residual norm $\|\mathbf{r}_\ell\|_2$ [Saad and Schultz, 1985]. A solution $\mathbf{x}_\ell$ exists at each step of the procedure, and $\|\mathbf{r}_\ell\|$ cannot grow with ℓ. The GMRES (generalized minimal residual) algorithm of Saad and Schultz [1986] for nonsymmetric matrices belongs to this class.

3. L_ℓ is formed using the transpose A^T. No optimization property is satisfied by a projection method is this class (unless $A = A^T$ and is positive definite, in which case the method reduces to a class 1 method). Again a solution $\mathbf{x}_\ell$ does not necessarily exist in each Krylov subspace, and the method may fail at some point. The most robust algorithms in this class are Bi-CGSTAB (biconjugate gradient stabilized, [van der Vorst, 1992] and TFQMR (transpose-free quasi-minimal residual, [Freund, 1993]).

In GMRES the computational work per iteration steadily increases with ℓ, so in practice a truncated or restarted version, GMRES(p), is used, where p stands for either the reduced Krylov subspace dimension in which $\mathbf{x}_\ell$ is currently sought or the maximum dimension of the Krylov subspace after which GMRES is restarted. It is only in the latter GMRES version that the property of $\|\mathbf{r}_\ell\|_2$ not increasing with ℓ is preserved.

Bi-CGSTAB and TFQMR are both variants of the biconjugate gradient (BCG) method of Lanczos [1952]. It should be mentioned that neither Bi-CGSTAB nor TFQMR addresses the problem of possible breakdown that may affect BCG. In fact in exact arithmetic Bi-CGSTAB and TFQMR fail whenever BCG fails. By contrast, convergence of the truncated GMRES depends on p but no blow-up can occur during the iterative procedure. This does not imply that GMRES(p) will always converge, as stagnation may affect the residual $\mathbf{r}_\ell$ after some ℓ.

In practical calculations all of the above projection algorithms (CG, GMRES, Bi-CGSTAB, TFQMR) must be preconditioned to enhance convergence. Effective and widely used preconditioners belong to the class of methods based on incomplete factorization of A. In the so-called incomplete Cholesky or Crout factorization, often referred to as ILU(0), the lower and upper triangular factors L and U of A ($U = L^T$ if $A = A^T$) are obtained by performing a standard triangularization and dropping all fill-in elements newly generated during the process [Meijerink and van der Vorst, 1977; Kershaw, 1978].

ILU(0) is the most inexpensive preconditioner based on the A factorization. A better, but more expensive, preconditioning is the one which allows for a partial (controlled) fill-in of L and U, thus ensuring faster convergence. The corresponding factorization is denoted by

ILUT(ρ, τ) where ρ and τ are user-specified parameters that control the fill-in process, i.e., the quality (and cost) of the preconditioner [Saad, 1994]. Parameter ρ specifies the maximum number of new elements that are to be preserved in each row of L and U, while τ sets a limit on the magnitude of any newly created L and U coefficient (relative to the corresponding row of A), below which this coefficient is dropped. With $\rho = N$ and $\tau = 0$, ILUT($N, 0$) provides the exact triangular factors of A, and the preconditioned projected method converges in just one iteration. This is a special case which is, of course, of no practical interest. Conversely, with $\rho = 0$ (and any τ) we obtain ILU(0).

Selecting the most convenient preconditioner, in terms of total computational cost of the projection methods at hand, will involve a trade-off between two opposing requirements for a preconditioning matrix: acceleration of asymptotic convergence and minimal cost of both computing L and U and performing each iteration. The final choice will be dependent to some extent on A, i.e., on the type of problem (flow, transport, Jacobian) to be solved.

All of the projection methods mentioned in this section, along with the preconditioners ILU(0) and ILUT(ρ, τ), are implemented in the CODESA-3D code. Extensive experience with both flow [Gambolati and Perdon, 1984] and transport [Pini and Putti, 1994; Gambolati et al., 1996] problems indicates that CG with ILU(0) performs extremely well for symmetric problems, and that Bi-CGSTAB is the most robust nonsymmetric projection method. GMRES(p) with $p = 20$ turns out to be the least robust. In most of the nonsymmetric examples dealt with, ILU(0) is not sufficient to provide acceptably fast convergence, and ILUT(ρ, τ) with appropriate ρ and τ values is needed. In terms of overall computational cost, Bi-CGSTAB, GMRES(20) and TFQMR are comparable when all converge, with Bi-CGSTAB being slightly more efficient than the other two methods.

10.6   Applications

10.6.1   Example 1: Two-Dimensional Homogeneous Flow and Transport in a Rectangular Domain

Our first example is used to validate the CODESA-3D code. It is a problem of steady state saturated flow and transport in a two-

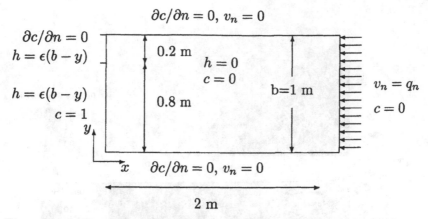

Figure 10.3. Two-dimensional domain with boundary and initial conditions for Example 1. In this figure n is the normal to the boundary.

dimensional domain, and is an adaptation of Henry's problem as described by Voss and Souza [1987]. The geometry and initial and boundary conditions are shown schematically in Figure 10.3. The incoming flux of freshwater on the right boundary is $q_n = 6.6 \times 10^{-5}$ m/s. The two-dimensional domain is discretized using a three-dimensional grid of unit width. The discretized domain contains 693 nodes and 2,400 tetrahedra.

The following parameter values are used for all runs: $K_{sx} = K_{sz} = 0.01$ m/s, $\phi = 0.35$, $D_o = 0$ m^2/s, $\epsilon' = 0$. To test the iterative procedures used to resolve the nonlinear coupling, four combinations of dispersivity and density ratio values are used: $\alpha_L = \alpha_T = 0.035$ m, $\epsilon = 0.0245$ (run 1); $\alpha_L = \alpha_T = 0.035$ m, $\epsilon = 0.05$ (run 2); $\alpha_L = \alpha_T = 0.05$ m, $\epsilon = 0.0245$ (run 3); $\alpha_L = \alpha_T = 0.05$ m, $\epsilon = 0.05$ (run 4). The solution to this problem is shown in Figure 10.4, where the head h, concentration, and velocity fields are shown for run 1. The effects of the density variation can be seen from the velocity field, where a flux inversion occurs near the left boundary.

Convergence plots for the four runs are shown in Figure 10.5. The convergence error plotted on the ordinate is the maximum absolute difference in concentration between the current and previous iterations. For these runs the convergence behavior of the coupled problem is dictated by the concentration errors. We observe that the partial Newton method is generally more robust than both the relaxed and unrelaxed Picard schemes, converging for all runs. On the other hand, the Picard method shows a smoother initial con-

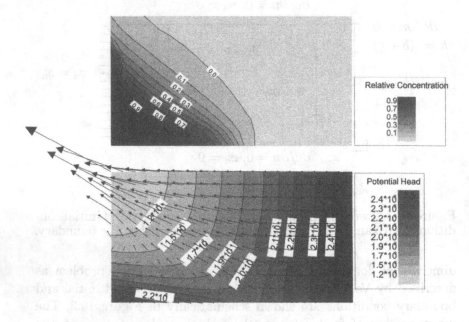

Figure 10.4. Steady state head, velocity, and concentration fields for run 1 of Example 1.

vergence behavior, but then stagnates, failing to converge except for the two runs with the lower density ratio value, when aided by relaxation. This situation suggests that an effective linearization strategy for strongly nonlinear problems may be to use the Picard method for the first few iterations and the partial Newton method for subsequent iterations, a strategy described earlier for the unsaturated flow equation.

The partial Newton scheme is more effective than the Picard method at high density ratios and lower dispersivity values. Convergence difficulties are expected at low dispersivity values, due to instabilities in the finite element discretization that are typical for advection dominated problems [Putti and Paniconi, 1995a].

10.6.2   Example 2: Heterogeneous Flow and Transport in a Coastal Aquifer with Recharge from a Channel

The second example is adapted from Sherif and Singh [1996] and considers a heterogeneous and anisotropic coastal aquifer with two

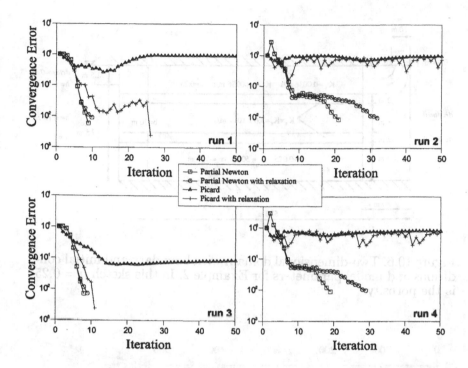

Figure 10.5. Convergence plots for runs 1, 2, 3, and 4 of Example 1.

impermeable layers. Infiltration of freshwater from a channel parallel to the coast creates a seaward flux that limits the extent of saltwater intrusion in the permeable layers. Figure 10.6 is a schematic description of the geometry and boundary conditions of the test problem, together with the values of the flow and transport model parameters (the molecular diffusion coefficient D_o and the viscosity ratio ϵ' are set to zero). The two-dimensional domain is discretized into a three-dimensional grid of unit width, with 6,642 nodes, 25,440 tetrahedra, and 40 vertical layers.

This test case illustrates the ability of the CODESA-3D model to capture the main features of flow and transport in heterogeneous formations. The results of the steady state simulation are shown in Figure 10.7. Recharge from the channel creates a seaward flux that acts to limit the extent of salt encroachment in the upper permeable layer. In the middle permeable layers freshwater recharge from the channel generates smaller fluxes, and the seawater is able to pene-

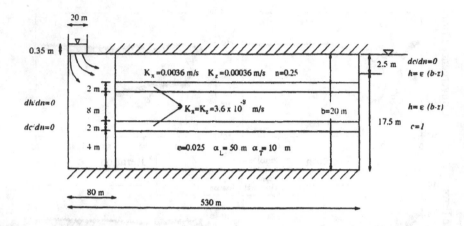

Figure 10.6. Two-dimensional domain with boundary and initial conditions and model parameters for Example 2. In this sketch $n = 0.25$ is the porosity.

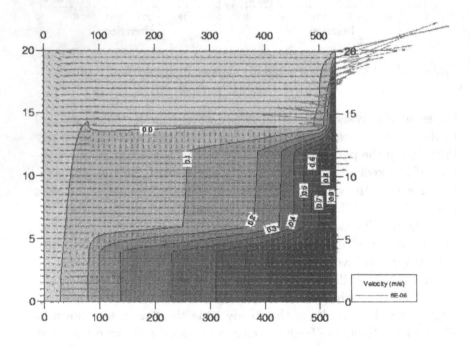

Figure 10.7. Steady state velocity field and salt concentration contours for Example 2.

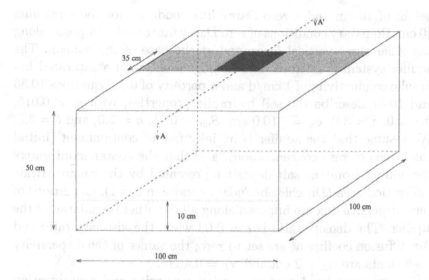

Figure 10.8. Three-dimensional domain for Example 3.

trate further inland, in particular along the boundary of the upper
semi-pervious layer. In the bottom layer the influence of the channel
is not appreciable, and the flow patterns are directed inland, causing
extensive salt intrusion.

10.6.3 Example 3: Contamination of a Ditch-Drained Aquifer by Trickle Infiltration from a Surficial Salt Deposit

In this example we present a three-dimensional problem of variably
saturated flow and transport in a ditch-drained aquifer with incident
steady rainfall and trickle infiltration of a salt contaminant. The test
case was described by Gureghian [1983], modified by Gambolati et
al., [1994a], and is further extended here to include density effects in
a three-dimensional scenario. The geometry is shown schematically
in Figure 10.8. The domain contains 2,009 nodes and 3,840 triangles
at the surface, and is discretized into 20 vertical layers, to yield
42,189 nodes and 230,400 tetrahedra for the 3-D grid.

The aquifer is initially completely saturated, with a uniform pressure head of zero throughout. A Darcy flux of 0.15 cm/d is applied
over the darkened square area, while the rest of the surface is sub-
jected to a Darcy flux of 0.1 cm/d. Along the front vertical face a
Dirichlet boundary condition of zero pressure head is applied to a

height of 10 cm and a zero Darcy flux condition for the remaining
40 cm. Boundary conditions of zero Darcy flux are also imposed along
the other three vertical faces, and at the base of the domain. The
aquifer system is isotropic and homogeneous, with a saturated hydraulic conductivity of 1 cm/d and a porosity of 0.3. Equations 10.36
and 10.37 describe the soil hydraulic properties, with $\kappa = 0.015$,
$\beta = 2.0$, $\gamma = 3.0$, $\psi_a = -10.0$ cm, $S_{wr} = 0.01$, $a = 2.0$, and $b = 3.5$.
We assume that the aquifer is initially free of contaminant (initial
conditions of zero concentration), and that the contaminant enters
the aquifer from the salt deposit represented by the square trickle
infiltration area (Dirichlet boundary condition $c = 1$). Conditions of
zero dispersive flux are imposed along all the other boundaries of the
aquifer. The density ratio is $\epsilon = 0.03$ while the viscosity ratio and
the diffusion coefficient are set to zero; the values of the dispersivity
coefficients are $\alpha_L = 2$ cm and $\alpha_T = 0.4$ cm.

The pressure head contours and the velocity and concentration
fields along cross section AA' after 60 days are shown in Figures 10.9
and 10.10. The results show how the aquifer drains through the front
vertical face at a rate faster than the recharge from the surface,
generating unsaturated conditions in the top portion of the aquifer.
The simulation mimics the behavior of a seepage face, with the water
table dropping until, by the time steady state is reached, it intersects
the front vertical face at the height of 10 cm imposed by the Dirichlet
condition (at time 60 days the position of this "exit point" is at about
17 cm, as can be seen from Figure 10.9). The salt plume is roughly
symmetric near its infiltration point at the surface, but begins to
preferentially migrate towards the seepage face in the lower, front
portion of the aquifer where the velocities are higher.

10.6.4   Example 4: Seawater Intrusion in a Southern Italian Coastal Aquifer

Our final example concerns steady state and transient simulations of
saltwater intrusion into a confined coastal aquifer in southern Italy
[Gambolati et al., 1994b]. The 36 km^2 study site is located on the
Tyrrhenian Sea, and is bounded by rivers on its northern and south-
ern borders, by the sea on the west, and by a high plain on the east
(Figure 10.11).

The construction of a channel system to discharge water into the
sea requires the lowering of the water table of the underlying aquifer

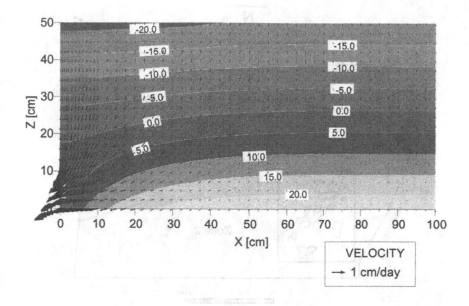

Figure 10.9. Example 3: pressure head contours and velocity fields along cross section AA' at time 60 days.

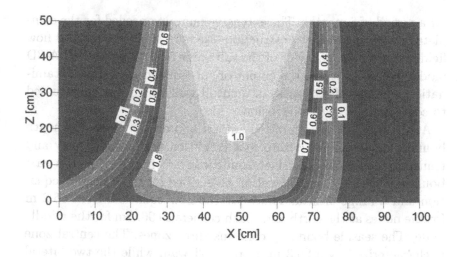

Figure 10.10. Example 3: salt concentration contours along cross section AA' at time 60 days.

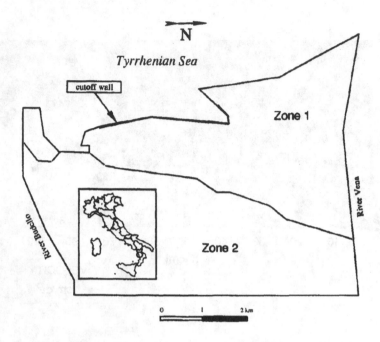

Figure 10.11. Location and geometry of the coastal aquifer for Example 4.

for a period of 6 months. The extensive pumping needed for the complete dewatering of the construction site may affect the natural flow field and alter the behavior of the saltwater front. The CODESA-3D model is used to assess the extent of subsequent saltwater contamination, and the effectiveness of a cutoff wall which was constructed to contain the seawater intrusion.

At the top and bottom of the aquifer and on the north and south boundaries we impose a zero flux condition for both the flow and transport equations. Dirichlet conditions are imposed on the east boundary, with zero concentration prescribed for the transport equation and a range of head values for the flow equation, from 66.2 m for the nodes at the north and south corners to 96.7 m for the middle node. The seaside boundary contains three zones. The central zone is characterized by a 26.3 m deep cutoff wall, while the two lateral zones are in direct contact with the sea. In these lateral zones, hydrostatic equivalent freshwater head is prescribed, along with zero concentration flux for the top 10.9 m and unitary (seawater) concen-

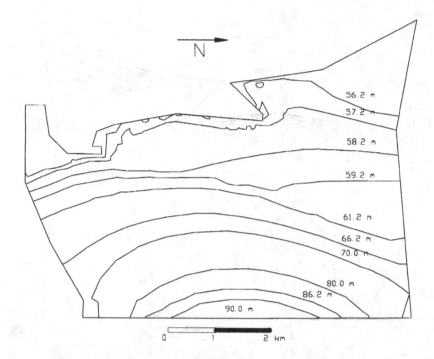

Figure 10.12. Example 4: steady state total head contours at the
bottom of the aquifer (after Gambolati et al. [1994]).

tration in the lower part. In the central zone, zero flux is prescribed
for the flow and transport equations for the top 26.3 m, while for the
lower part hydrostatic head and unitary concentration are imposed.

The surface mesh contains 2,829 triangles and 1,459 nodes, and
is duplicated vertically to form 10 layers of depths 3.7, 3.7, 1.5, 2.0,
7.9, 7.5, 7.5, 7.5, 7.5, and 7.5 m from top to bottom, for a total
aquifer depth of 56.3 m. The three-dimensional grid contains 84,870
tetrahedra and 16,049 nodes.

The aquifer contains two distinct hydrogeological zones (Figure
10.11). Zone 1 is stratified, with hydraulic conductivity values of
$K_{sx} = K_{sy} = 1.8 \times 10^{-3}$ m/s and $K_{sz} = 3.0 \times 10^{-4}$ m/s in the
top 7.4 m, and $K_{sx} = K_{sy} = K_{sz} = 6.0 \times 10^{-4}$ m/s in the bottom
48.9 m. In zone 2 we have $K_{sx} = K_{sy} = K_{sz} = 8.0 \times 10^{-5}$ m/s. For
the entire aquifer $D_o = 0.0$, $\phi = 0.22$, $\epsilon = 0.03$, $\epsilon' = 0$, $\alpha_L = 10$ m,
and $\alpha_T = 1$ m.

Figures 10.12 and 10.13 show the steady state head contour lines

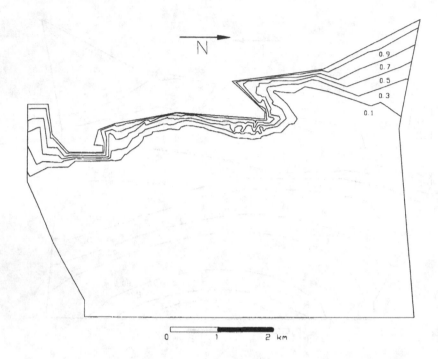

Figure 10.13. Example 4: steady state equiconcentration lines at the bottom of the aquifer (after Gambolati et al. [1994]).

and the steady state equiconcentration lines at the bottom of the aquifer before pumping. The steady state solution is attained by running a transient simulation until variations in head and concentration are negligible, which occurred after 1342 days of simulation, using time step sizes that varied from 1 to 25 days.

A second simulation is run using the steady state conditions from the first simulation as initial conditions. In this second simulation, pumping at a rate of 2.189 m^3/day is applied to the aquifer. A period of 185 days (which is the expected duration of the channel construction phase) is simulated. Figure 10.14 shows the head drawdown at the end of the pumping period, while Figure 10.15 shows the saltwater equiconcentration lines for the same period. Both figures are referred to a horizontal section at the bottom of the aquifer. The pumping area is evident in Figures 10.14 and 10.15 from the concentration contour lines and flow field that converge towards the extraction wells. The results of this simulation show that water withdrawal

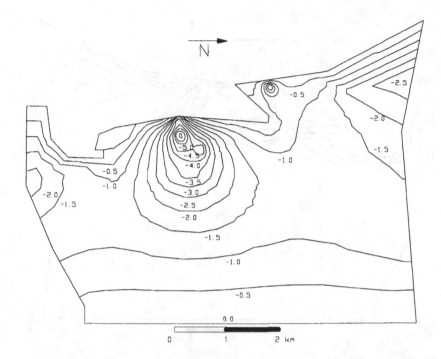

Figure 10.14. Example 4: drawdown contour lines (m) at the bottom of the aquifer after 185 days of pumping (after Gambolati et al. [1994]).

from pumping causes only a limited increase in saltwater intrusion.

Appendix A. Derivation of the Mathematical Model

The mass conservation equation for the saltwater mixture can be written as

$$\frac{\partial(\phi S_w \rho)}{\partial t} = -\nabla \cdot (\rho \mathbf{v}) + \rho q \qquad (10.63)$$

Recalling that

$$\frac{\partial \phi}{\partial t} = \frac{S_s}{\rho_o g} \frac{\partial p}{\partial t} \qquad (10.64)$$

where $S_s = \rho_o g(\alpha + \phi \beta)$ is the elastic storage coefficient as defined in Gambolati [1973] and α and β are the compressibility coefficients of

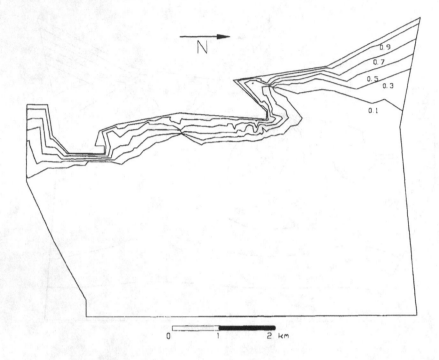

Figure 10.15. Example 4: equiconcentration lines at the bottom of the aquifer after 185 days of pumping (after Gambolati et al. [1994]).

the porous matrix and the freshwater, respectively, we can apply the chain rule and relation Eq. 10.2 to the accumulation term to obtain

$$\frac{\partial(\phi S_w \rho)}{\partial t} = \left[\frac{S_w S_s}{g}(1 + \epsilon c) + \phi \rho_o (1 + \epsilon c)\frac{\partial S_w}{\partial p} \right] \frac{\partial p}{\partial t}$$
$$+ \phi S_w \rho_o \epsilon \frac{\partial c}{\partial t} \tag{10.65}$$

Substituting the definition of equivalent pressure head $\psi = p/(\rho_o g)$, we obtain

$$\frac{\partial(\phi S_w \rho)}{\partial t} = \left[S_w S_s(1 + \epsilon c) + \phi \rho_o g(1 + \epsilon c)\frac{\partial S_w}{\partial p} \right] \rho_o \frac{\partial \psi}{\partial t}$$
$$+ \phi S_w \rho_o \epsilon \frac{\partial c}{\partial t} \tag{10.66}$$

Darcy's velocity can be expressed in the case of variable density as [Bear, 1979]

$$\mathbf{v} = -\frac{kK_r}{\mu}(\nabla p + \rho g \eta_z) \tag{10.67}$$

where k is the intrinsic permeability of the porous medium and μ is the dynamic viscosity, given by Eq. 10.3. The spatial conservation term can then be written as

$$\nabla \cdot (\rho \mathbf{v}) = -\nabla \cdot \left[\frac{\rho k K_r}{\mu} \left(\nabla p + \rho_0 g (1 + \epsilon c) \eta_z \right) \right]$$

(10.68)

which, using once again the definition of equivalent pressure head, becomes

$$\nabla \cdot (\rho \mathbf{v}) = -\rho_0 \nabla \cdot \left[\frac{\rho g k K_r}{\mu} \left(\nabla \psi + (1 + \epsilon c) \eta_z \right) \right]$$

(10.69)

Assembling Eqs. 10.66 and 10.69 and dividing through by ρ_0, the flow equation for the saltwater mixture can be written as

$$\sigma \frac{\partial \psi}{\partial t} = \nabla \cdot \left[K_s \frac{1 + \epsilon c}{1 + \epsilon' c} K_r \left(\nabla \psi + (1 + \epsilon c) \eta_z \right) \right]$$
$$- \phi S_w \epsilon \frac{\partial c}{\partial t} + \frac{\rho}{\rho_0} q$$

(10.70)

where $K_s = \rho_0 g k / \mu_0$ is the saturated hydraulic conductivity tensor of freshwater and

$$\sigma = \left[S_w S_s (1 + \epsilon c) + \phi \rho_0 g (1 + \epsilon c) \frac{\partial S_w}{\partial p} \right]$$

(10.71)

is the general storage term.

To derive the transport equation we start from the expression for the solute mass conservation:

$$\frac{\partial (\phi S_w \tilde{c})}{\partial t} = -\nabla \cdot (\tilde{c} \mathbf{v} - D \nabla \tilde{c}) + q \tilde{c}^* + \tilde{f}$$

(10.72)

where $\tilde{c}$ is the mass of solute per unit volume of the mixture. Assuming that ϕ does not depend on concentration and dividing the previous equation by the maximum concentration c_s, we obtain the transport equation in terms of normalized concentration $c = \tilde{c}/c_s$ as

$$\phi \frac{\partial S_w c}{\partial t} = \nabla \cdot (D \nabla c) - \nabla \cdot (c \mathbf{v}) + q c^* + f$$

(10.73)

The dispersion tensor $D = \phi S_w \tilde{D}$, where $\tilde{D}$ is defined as in Bear [1979], is given by

$$D_{ij} = \phi S_w \tilde{D}_{ij} = \alpha_T \mid \mathbf{v} \mid \delta_{ij} + (\alpha_L - \alpha_T)\frac{v_i v_j}{\mid \mathbf{v} \mid}$$

$$+\phi S_w D_o \tau \delta_{ij}; \qquad i, j = x, y, z \qquad (10.74)$$

where α_L and α_T are the longitudinal and transverse dispersivity coefficients, respectively, $\mid \mathbf{v} \mid = \sqrt{v_x^2 + v_y^2 + v_z^2}$, δ_{ij} is the Kronecker delta, D_o is the molecular diffusion coefficient, and τ is the tortuosity ($\tau = 1$ is usually assumed).

Other formulations of the coupled saturated flow and transport equations have been used in the literature [Frind, 1982a; Huyakorn et al., 1987, Galeati et al., 1992]. These formulations vary depending on the different assumptions that are made during the derivation process. For example, Galeati et al., [1992] assume that $\nabla \rho \approx 0$, and thus obtain the following saturated flow equation:

$$S_s \frac{\partial h}{\partial t} = \nabla \cdot [K_s (\nabla h + \epsilon c \eta_z)] - \phi \epsilon \frac{\rho_o}{\rho} \frac{\partial c}{\partial t} + q$$

$$(10.75)$$

Appendix B. Finite Element Matrices

In this appendix we write the expressions for the finite element matrices of the flow and transport equations, using linear basis functions on tetrahedral elements. The basis function W_i^e for a generic tetrahedron e with vertices i, j, k, and m is $W_i^e = (\varrho_i + \varsigma_i x + \zeta_i y + \xi_i z)/6V^e$ where the volume of the element is given by

$$V^e = \frac{1}{6} \begin{vmatrix} 1 & x_i & y_i & z_i \\ 1 & x_j & y_j & z_j \\ 1 & x_k & y_k & z_k \\ 1 & x_m & y_m & z_m \end{vmatrix} \qquad (10.76)$$

and

$$\varrho_i = \begin{vmatrix} x_j & y_j & z_j \\ x_k & y_k & z_k \\ x_m & y_m & z_m \end{vmatrix} \qquad \varsigma_i = - \begin{vmatrix} 1 & y_j & z_j \\ 1 & y_k & z_k \\ 1 & y_m & z_m \end{vmatrix}$$

$$\zeta_i = \begin{vmatrix} 1 & x_j & z_j \\ 1 & x_k & z_k \\ 1 & x_m & z_m \end{vmatrix} \qquad \xi_i = - \begin{vmatrix} 1 & x_j & y_j \\ 1 & x_k & y_k \\ 1 & x_m & y_m \end{vmatrix} \quad (10.77)$$

B.1 Flow Equation

In this appendix we evaluate matrices H and P and vector $\mathbf{q}^*$ of Eq. 10.19. The ijth element of matrix H is given by $h_{ij} = \sum_{e=1}^{E} h_{ij}^e$ where

$$
\begin{aligned}
h_{ij}^e &= \int_{V^e} K_r^e \left(K_s^e \frac{1+\epsilon\bar{c}}{1+\epsilon'\bar{c}} \nabla W_j^e \cdot \nabla W_i^e \right) dV \\
&= \int_{V^e} K_r^e \frac{1+\epsilon\bar{c}}{1+\epsilon'\bar{c}} \left(K_{sx}^e \frac{\varsigma_j}{6V^e} \frac{\varsigma_i}{6V^e} + K_{sy}^e \frac{\zeta_j}{6V^e} \frac{\zeta_i}{6V^e} \right. \\
&\qquad \left. + K_{sz}^e \frac{\xi_j}{6V^e} \frac{\xi_i}{6V^e} \right) dV \\
&= \frac{K_r^e}{36 \, |V^e|} \frac{1+\epsilon\bar{c}}{1+\epsilon'\bar{c}} (K_{sx}^e \varsigma_j \varsigma_i + K_{sy}^e \zeta_j \zeta_i + K_{sz}^e \xi_j \xi_i) \qquad (10.78)
\end{aligned}
$$

where K_{sx}, K_{sy}, and K_{sz} are the diagonal components of the saturated conductivity tensor, and the superscript e indicates that the quantity is averaged over the element. The nonlinear coefficient K_r^e is evaluated using the average value of $\hat{\Psi}$ at the centroid of each tetrahedron. This treatment of nonlinear coefficients is also used for all the other integral terms.

The ijth element of P is $p_{ij} = \sum_{e=1}^{E} p_{ij}^e$ where

$$
p_{ij}^e = \int_{V^e} \sigma^e W_j^e W_i^e \, dV = \sigma^e \frac{|V^e|}{20} \cdot \begin{cases} 2 & \text{if } i = j \\ 1 & \text{if } i \neq j \end{cases} \qquad (10.79)
$$

Finally, $\mathbf{q}^* = \mathbf{g}_z + \mathbf{b}_f + \mathbf{q}_f$ where the gravity vector $\mathbf{g}_\ell$, $\ell = x, y, z$ is such that $\mathbf{g}_x = \mathbf{g}_y = 0$ and $\mathbf{g}_z = \{g_{z_i}\} = \sum_{e=1}^{E} G_i^e$, $\mathbf{b}_f = \{b_{f_i}\} = \sum_{e=1}^{E} F_i^e$, $\mathbf{q}_f = \{q_{f_i}\} = \sum_{e=1}^{E} L_i^e$. The components of these three vectors are given by

$$
\begin{aligned}
G_i^e &= \int_{V^e} K_r^e K_{sz}^e \frac{(1+\epsilon\bar{c})^2}{1+\epsilon'\bar{c}} \frac{\partial W_i^e}{\partial z} \, dV = \frac{|V^e|}{6V^e} K_r^e K_{sz}^e \frac{(1+\epsilon\bar{c})^2}{1+\epsilon'\bar{c}} \xi_i \\
F_i^e &= \int_{V^e} \left(\phi^e S_w^e \epsilon \frac{\partial \bar{c}}{\partial t} - \frac{\rho}{\rho_o} q^e \right) W_i^e \, dV = \left(\phi^e S_w^e \epsilon \frac{\partial \bar{c}}{\partial t} - \frac{\rho}{\rho_o} q^e \right) \frac{|V^e|}{4} \\
L_i^e &= - \int_{\Gamma_2^e} q_n^e W_i^e \, d\Gamma = -q_n^e \frac{|\Delta^e|}{3} \qquad (10.80)
\end{aligned}
$$

The quantity $|\Delta^e|$ denotes the area of the triangular face of the tetrahedron where the boundary condition is imposed, while the time derivative of $\bar{c}$ is computed using a backward finite difference formula.

B.2 Velocity Calculation

In linear tetrahedral finite elements the Darcy velocity vector is constant in each element. The components of the vector $\mathbf{v}^e$ for a tetrahedron with vertices i, j, k, and m are

$$
\begin{aligned}
v_x^e &= -K_r^e K_{sx}^e \frac{1 + \epsilon \bar{c}}{1 + \epsilon' \bar{c}} \frac{\partial \hat{\psi}^e}{\partial x} \\
v_y^e &= -K_r^e K_{sy}^e \frac{1 + \epsilon \bar{c}}{1 + \epsilon' \bar{c}} \frac{\partial \hat{\psi}^e}{\partial y} \\
v_z^e &= -K_r^e K_{sz}^e \frac{1 + \epsilon \bar{c}}{1 + \epsilon' \bar{c}} \left(\frac{\partial \hat{\psi}^e}{\partial z} + 1 + \epsilon \bar{c} \right)
\end{aligned}
\tag{10.81}
$$

where

$$
\hat{\psi}^e = \hat{\psi}_i W_i^e + \hat{\psi}_j W_j^e + \hat{\psi}_k W_k^e + \hat{\psi}_m W_m^e
\tag{10.82}
$$

The velocity vector $\mathbf{v}_\ell$, where $\ell = x, y, z$, can then be written in matrix form as

$$
\mathbf{v}_\ell = U_\ell \hat{\Psi} + \mathbf{g}_\ell
\tag{10.83}
$$

in which $\mathbf{g}_\ell$ is the gravity vector given previously and the rows of the $E \times N$-dimensional matrices U_ℓ, $\ell = x, y, z$ are given by the N-dimensional vectors

$$
\begin{aligned}
-\frac{K_r^e}{6V^e} K_{sx}^e \frac{1 + \epsilon \bar{c}}{1 + \epsilon' \bar{c}} (\varsigma_i \mathbf{e}_i + \varsigma_j \mathbf{e}_j + \varsigma_k \mathbf{e}_k + \varsigma_m \mathbf{e}_m) \quad &\text{for } \ell = x \\
-\frac{K_r^e}{6V^e} K_{sy}^e \frac{1 + \epsilon \bar{c}}{1 + \epsilon' \bar{c}} (\zeta_i \mathbf{e}_i + \zeta_j \mathbf{e}_j + \zeta_k \mathbf{e}_k + \zeta_m \mathbf{e}_m) \quad &\text{for } \ell = y \\
-\frac{K_r^e}{6V^e} K_{sz}^e \frac{1 + \epsilon \bar{c}}{1 + \epsilon' \bar{c}} (\xi_i \mathbf{e}_i + \xi_j \mathbf{e}_j + \xi_k \mathbf{e}_k + \xi_m \mathbf{e}_m) \quad &\text{for } \ell = z
\end{aligned}
\tag{10.84}
$$

In the above, the components of the N-dimensional vector $\mathbf{e}_i$ are zero except for the ith component which is equal to one.

B.3 Transport Equation

We evaluate A, B, C, M, and $\mathbf{r}^*$ of Eq. 10.23. The coefficients of matrix A are computed by performing a rotation of axes so as to align the new coordinate direction x' with the direction of the Darcy velocity. The new coordinate system is defined as $x' = (v_x x + v_y y + v_z z)/ |\mathbf{v}|$, $y' = (v_y x - v_x y)/\delta$, and $z' = (v_x v_z x + v_y v_z y - \delta^2 z)/(\delta |\mathbf{v}|)$, where $\delta = \sqrt{v_x^2 + v_y^2}$. Thus the elements of the dispersion tensor become

$$
\begin{aligned}
D_{x'} &= \alpha_L |\mathbf{v}| + \phi S_w D_o \tau \\
D_{y'} &= \alpha_T |\mathbf{v}| + \phi S_w D_o \tau \\
D_{z'} &= D_{y'}
\end{aligned}
\tag{10.85}
$$

The ij^{th} element of matrix A is given by $a_{ij} = \sum_{e=1}^{E} a_{ij}^e$ where

$$
\begin{aligned}
a_{ij}^e &= \int_{V^e} \left(D_{x'}^e \frac{\partial W_j^e}{\partial x'} \frac{\partial W_i^e}{\partial x'} + D_{y'}^e \frac{\partial W_j^e}{\partial y'} \frac{\partial W_i^e}{\partial y'} + D_{z'}^e \frac{\partial W_j^e}{\partial z'} \frac{\partial W_i^e}{\partial z'} \right) dV \\
&= \int_{V^e} \left(D_{x'}^e \frac{\varsigma_j'}{6V^e} \frac{\varsigma_i'}{6V^e} + D_{y'}^e \frac{\zeta_j'}{6V^e} \frac{\zeta_i'}{6V^e} + D_{z'}^e \frac{\xi_j'}{6V^e} \frac{\xi_i'}{6V^e} \right) dV \\
&= \frac{1}{36 |V^e|} \left(D_{x'}^e \varsigma_j' \varsigma_i' + D_{y'}^e \zeta_j' \zeta_i' + D_{z'}^e \xi_j' \xi_i' \right)
\end{aligned}
\tag{10.86}
$$

where ς', ζ', and ξ' are computed in the local (rotated) reference frame x', y', and z'. The ijth element of B is $b_{ij} = \sum_{e=1}^{E} b_{ij}^e$ where

$$
\begin{aligned}
b_{ij}^e &= \int_{V^e} \left(v_x^e \frac{\partial W_j^e}{\partial x} + v_y^e \frac{\partial W_j^e}{\partial y} + v_z^e \frac{\partial W_j^e}{\partial z} \right) W_i^e \, dV \\
&= \int_{V^e} \left(v_x^e \frac{\varsigma_j}{6V^e} + v_y^e \frac{\zeta_j}{6V^e} + v_z^e \frac{\xi_j}{6V^e} \right) W_i^e \, dV \\
&= \frac{|V^e|}{24V^e} (v_x^e \varsigma_j + v_y^e \zeta_j + v_z^e \xi_j)
\end{aligned}
\tag{10.87}
$$

For $c_{ij} = \sum_{e=1}^{E} c_{ij}^e$,

$$
\begin{aligned}
c_{ij}^e &= - \int_{\Gamma_5^e} (v_x^e n_x + v_y^e n_y + v_z^e n_z) W_j^e W_i^e \, d\Gamma \\
&= -(v_x^e n_x + v_y^e n_y + v_z^e n_z) \frac{|\Delta_5^e|}{12} \cdot \begin{cases} 2 & \text{if } i = j \\ 1 & \text{if } i \neq j \end{cases}
\end{aligned}
\tag{10.88}
$$

where $\mathbf{n} = (n_x, n_y, n_z)^T$ is the unit outward normal vector. If we lump all the extradiagonal contributions onto the main diagonal, we get

$$
\begin{aligned}
c_{ii}^e &= -(v_x^e n_x + v_y^e n_y + v_z^e n_z) \frac{|\Delta_5^e|}{3} \\
c_{ij}^e &= 0 \quad \text{if } i \neq j
\end{aligned}
\tag{10.89}
$$

The elements of matrix M are given by $m_{ij} = \sum_{e=1}^{E} m_{ij}^e$, where

$$m_{ij}^e = \int_{V^e} \phi^e S_w^e W_j^e W_i^e \, dV = \phi^e S_w^e \frac{|V^e|}{20} \cdot \begin{cases} 2 & \text{if } i = j \\ 1 & \text{if } i \neq j \end{cases} \tag{10.90}$$

Finally, $r_i^* = G_i^1 + G_i^2 + G_i^3$, where

$$G_i^1 = \sum_{e=1}^{E} G_i^{1e}; \qquad G_i^{1e} = -\int_{V^e} (q^e c^* + f^e) W_i^e \, dV$$

$$= -(q^e c^* + f^e) \frac{|V^e|}{4}$$

$$G_i^2 = \sum_{e=1}^{E} G_i^{2e}; \qquad G_i^{2e} = -\int_{\Gamma_4^e} q_d^e W_i^e \, d\Gamma = -q_d^e \frac{|\Delta_4^e|}{3}$$

$$G_i^3 = \sum_{e=1}^{E} G_i^{3e}; \qquad G_i^{3e} = -\int_{\Gamma_5^e} q_c^e W_i^e \, d\Gamma = -q_c^e \frac{|\Delta_5^e|}{3} \tag{10.91}$$

In the expressions for c_{ij}^e, G_i^{2e}, and G_i^{3e}, $|\Delta_4^e|$ and $|\Delta_5^e|$ are the lengths of the boundary segments Γ_4^e and Γ_5^e, respectively.

Acknowledgments: This work has been partially supported by the European Commission through grants AVI-CT93-2-073 and IC15-CT96-0211, and by the Sardinia Regional Authorities. The authors wish to thank Anna Chiara Bixio (University of Padua) and Giuditta Lecca (CRS4) for the simulations described in Examples 2 and 3.

Chapter 11

Modified Eulerian Lagrangian Method for Density Dependent Miscible Transport

S. Sorek, V. Borisov & A. Yakirevich

11.1 Introduction

The lithologic sequence of the Israeli coastal aquifer is composed of calcareous sandstone of the Pleistocene age, silt and intercalation of clay and loam which can appear as lenses. The basis of this aquifer is built on sea clays and shales from the Neogene age. Within a zone of approximately five km. from the sea, these clay layers separate the aquifer into three major sub-aquifers. Mostly in the south, the central one is further subdivided into additional three sub-aquifers. Each of these sub-aquifers constitutes its own hydrologic characteristics. This type of aquifers is highly heterogeneous due to the influences of the following factors [Correns, 1950; McBrride, 1977; Curtis, 1961]:

1. Lithologic (i.e. high spatial variability of thickness of the silt, clay and sandstone layers),

2. Diagenesis (i.e. cement re-sedimentation, high spatial variability of the secondary porosity), and

3. Post sedimentation tectogenesis, that may be reflected as fractures of the cemented sandstones.

This type of heterogeneity may introduce steep gradients of conductivity and dispersion coefficients. This, in return, may highly affect the accuracy of the obtained numerical results concerning flow and transport problems.

It is an accepted argument that the transport equation may conform to become hyperbolic dominant when the advective flux is dominant over the dispersive one. In what follows we shall show that this

J. Bear et al. (eds.), Seawater Intrusion in Coastal Aquifers, 363–398.
© 1999 *Kluwer Academic Publishers.*

equation can also have hyperbolic features for very small fluid velocities but with steep gradients of the dispersion coefficient. In fact
for the parabolic flow equation, the hyperbolic characteristic can be
evident by steep gradients of the conductivity coefficient [Bear et al.,
1997]. The numerical solution of advective dominant equations may
result in problems of numerical dispersion, spurious oscillations and
over/under shooting errors.

The Eulerian-Lagrangian (EL) methods have been developed as
special techniques for solving the advection dominated transport
problems. Neuman [1981, 1984], Neuman and Sorek [1982], Sorek
and Braester [1988], and Sorek [1985a, 1985b, 1988] decompose the
dependent variable into advective and residual parts. The resulting
advection problem is solved by tracking particles along their characteristic pathlines while the "dispersive" problem is solved on a fixed
grid. For the EL methods, particles are conveyed by fluid velocity.
With the MEL method, this velocity is modified to account also for
the heterogeneity of the medium.

Here in what follows the MEL formulation is developed for both
flow and solute migration, in reference to the saltwater intrusion
problem. In the case of the flow equation, particle's velocity is associated with the gradient of the hydraulic conductivity while for the
transport equation it accounts also for gradient of the hydrodynamic
dispersion. The numerical scheme is based on the decomposition of
the differential operator and the dependent variables.

11.2   Mathematical Statement

11.2.1   The General Case

Consider the transport of an extensive quantity, E, of a fluid phase
occupying part of the void space of a porous medium domain. Let S
and e denote the saturation of the phase and the density of E within
it, respectively.

The macroscopic Eulerian balance equation of E of a phase, within
a porous medium domain, can be written in the form

$$\frac{\partial}{\partial t}(\phi S e) = -\nabla \cdot \left[\phi S \left(e\mathbf{V} - \mathbf{D}_h^E \cdot \nabla e\right)\right] - f_{\alpha \to \beta}^E + \phi S \rho \Gamma^E \tag{11.1}$$

where ϕ denotes the porosity, $\mathbf{V}$ denotes the mass weighted velocity of the phase, $f_{\alpha \to \beta}^E$ denotes the transfer of E from the considered α phase to all other (β) phases across common (microscopic) interphase boundaries, ρ denotes the mass density of the phase, and Γ^E denotes the rate of growth (= source) of E per unit mass of the phase. We often combine, $\mathbf{J}^{*E}$, the dispersive flux of E (due to variations in e and in $\mathbf{V}$ at the microscopic level) and, $\mathbf{J}^E$, the diffusive flux of E, (relative to the mass averaged velocity) by introducing the hydrodynamic dispersive flux, $\mathbf{J}_h^E$ ($= \mathbf{J}^{*E} + \mathbf{J}^E$). Let us also assume that the flux of hydrodynamic dispersion of E can be expressed as a *Fickian-type* expression, i.e.,

$$\mathbf{J}_h^{*E} = -\mathbf{D}_h^E \cdot \nabla e, \quad \text{with} \quad \mathbf{D}_h^E = \mathbf{D}^E + \mathcal{D}\mathcal{D}^E$$

$$(11.2)$$

in which $\mathbf{D}^E$ denotes the coefficient of dispersion of E, and $\mathcal{D}^E$ denotes the coefficient of diffusion of E.

The *MEL* formulation of Eq. 11.1 reads

$$\frac{1}{\phi}\frac{D\mathbf{v}}{Dt}(\phi) + \frac{1}{S}\frac{D\mathbf{v}}{Dt}(S) + \frac{1}{e}\frac{D\widetilde{\mathbf{v}}}{Dt}(e) =$$

$$-\nabla \cdot \mathbf{V} + \frac{\mathbf{D}_h^E}{e} : (\nabla\nabla e) - \frac{1}{\phi S e}f_{\alpha \to \beta}^E + \frac{\rho}{e}\Gamma^E \quad (11.3)$$

where

$$\frac{D\mathbf{v}}{Dt}(..) \equiv \frac{\partial}{\partial t}(..) + \mathbf{V} \cdot \nabla(..) \qquad (11.4)$$

denotes the *material derivative* of $(..)$ associated with a particle carrying $(..)$, moving at a velocity $\mathbf{V}$ along its *characteristic curve*. Since $\phi S e$ expresses the quantity of E per unit volume of porous medium, the l.h.s. of Eq. 11.3 describes how this quantity varies along the curve that refers to the movement of the particle.

The velocity that serves as the *characteristic velocity* for the propagation of e reads

$$\widetilde{\mathbf{V}} \equiv \mathbf{V} - \frac{1}{\phi S}\nabla(\phi S \mathbf{D}_h^E) \qquad (11.5)$$

This is referred to as a *modified velocity* due to spatial variations in ϕ, S, and $\mathbf{D}_h^E$. We note that Eq. 11.5 accounts for the fluid velocity and an apparent velocity term associated with the heterogeneity

control coef.	Eulerian		EL		MEL	
	Flow	Transp.	Flow	Transp.	Flow	Transp.
f	1	1	—	0	1	0
φ	0	0	—	0	1	1
λ	0	1	—	0	0	0

Table 11.1. Choice of control coefficients.

characteristics of the medium. The first is along a gradient from a higher energy level to a lower one, the latter along a gradient from high hydrodynamic dispersion to a lower one.

Following the Lagrangian concept, e will propagate along a pathline, say carried with a particle, defined by

$$\frac{D\widetilde{\mathbf{V}}}{Dt}(\mathbf{r}) = \widetilde{\mathbf{V}} \tag{11.6}$$

where $\mathbf{r}$ denotes a spatial position vector.

We note that in the absence of the velocity (i.e. $\mathbf{V} = 0$), Eq. 11.1 will describe the Eulerian form of the flow equation while Eq. 11.3 will provide its MEL formulation. Actually, the Eulerian, EL and MEL forms for the flow and the transport equations can be combined to read

$$\frac{1}{\phi}\frac{D_{\mathbf{V}^*}}{Dt}(\phi) + \frac{1}{S}\frac{D_{\mathbf{V}^*}}{Dt}(S) + \frac{1}{e}\frac{D_{\widetilde{\mathbf{V}}^*}}{Dt}(e) =$$

$$-(1 - f - \lambda)\nabla \cdot \mathbf{V} + \varphi\frac{\mathbf{D}_h^E}{e}:(\nabla\nabla e)$$

$$+\frac{1 - \varphi}{\phi Se}\nabla \cdot \left[\phi S\left(\mathbf{D}_h^E \cdot \nabla e\right)\right]$$

$$-\frac{\lambda}{e}\mathbf{V} \cdot \nabla e - \frac{1}{\phi Se}f_{\alpha \to \beta}^E + \frac{\rho}{e}\Gamma^E \tag{11.7}$$

where

$$\widetilde{\mathbf{V}^*} \equiv \mathbf{V}^* - \frac{\varphi}{\phi S}\nabla(\phi S\mathbf{D}_h^E); \quad \mathbf{V}^* \equiv (1 - f)\mathbf{V} \tag{11.8}$$

The various formulations that can be obtained are described in Table 11.1.

11.2.2   Why MEL

The numerical schemes of the Eulerian and the EL methods do not specifically address the heterogeneity of the medium which may be presented in the form of steep gradients of the permeability and/or the dispersion coefficients. This for the *MEL* method is specifically accounted for as additional information which is included in the velocity, Eq. 11.5, of the particle and thus affects the solution process.

The *EL* formulation is reported to be effective in overcoming *numerical dispersion* phenomena when simulating problems which are dominated by advection. In view of Eq. 11.5, it is argued that the second term on the r.h.s. of Eq. 11.5 can also be significant in dictating a situation similar to the advection dominant characteristics. Note that this may occur for governing equations with no reference to $\mathbf{V}$, or even when $\mathbf{V} = 0$. Accordingly, as described in Table 11.1, the *EL* formulation can not be implemented to flow problems as in the case of the *MEL* formulation. We thus suggest that it can be preferable to simulate advection dominated problems by, Eq. 11.3, the *MEL* formulation rather then the *EL* formulation.

We suggest that in view of Eq. 11.1 for a transport equation (similar arguments hold upon referring to the flow equation), when considering the flux $\phi S(\mathbf{D}_h^E \cdot \nabla e)$ for a heterogeneous medium (with steep gradients of $\phi S\mathbf{D}_h^E$), it may be that $\left| \nabla(\phi S\mathbf{D}_h^E) \cdot \nabla e \right| \gg \left| \phi S\mathbf{D}_h^E : \nabla \nabla e \right|$. This will conform Eq. 11.1 to become hyperbolic dominant. The validity of this argument can be demonstrated, for the sake of simplicity and with no affect on the general conclusions, in a 1-D case.

The purely hyperbolic, advective , transport problem can be described by

$$\frac{\partial e}{\partial t} + V\frac{\partial e}{\partial x} = f \qquad (11.9)$$

in which f denotes a source term. On the basis of Eq. 11.9 and when accounting also for a diffusion-dispersion process, we refer to a parabolic transport equation in the form

$$\frac{\partial e}{\partial t} + V\frac{\partial e}{\partial x} = D\frac{\partial^2 e}{\partial x^2} + f \qquad (11.10)$$

in which D denotes a generalized diffusion-dispersion coefficient. Notice that in terms of the mass flux, $D\,\partial e/\partial x$, Eq. 11.10 can be rewritten to read

$$\frac{\partial e}{\partial t} + C\frac{\partial e}{\partial x} = \frac{\partial}{\partial x}\left(D\frac{\partial e}{\partial x}\right) + f \quad \text{with} \quad C \equiv V + \frac{\partial D}{\partial x} \qquad (11.11)$$

in which C is regarded as an apparent velocity. This velocity will conform to a similar form as in Eq. 11.11 if the first r.h.s. of Eq. 11.10 will be replaced by $\partial(D \, \partial e / \partial x) / \partial x$, the divergence of the mass flux.

The extent to which an advection-dispersion equation is considered to be advection dominant, is commonly related to P_e, the Peclet number. In view of Eq. 11.11, we suggest to expand the definition for the Peclet number to read

$$P_e \equiv \frac{L_c}{D} \left| V + \frac{\partial D}{\partial x} \right| \qquad (11.12)$$

in which L_c denotes a characteristic length. By virtue of Eq. 11.12 we note that a high Peclet number ($P_e \gg 1$), which is associated with an advective dominant equation, can be resulting from steep gradients of D also when $V = 0$. The latter can be referred to the case of the flow equation for which D is regarded as the permeability coefficient. Hence, Eq. 11.12 can be used for both, a flow and/or a transport equation. Moreover, the advective-dispersive transport equation can be prescribed with no reference to the fluid velocity. For such a case, the characteristic of the transport equation can be investigated only by Eq. 11.12. To demonstrate that such a form of a transport equation may exist, let us multiply Eq. 11.11 by $\Lambda(x) \neq 0$, and assume that,

$$\Lambda \frac{\partial}{\partial x} \left(D \frac{\partial e}{\partial x} \right) - \Lambda C \frac{\partial e}{\partial x} = \frac{\partial}{\partial x} \left(D \Lambda \frac{\partial e}{\partial x} \right) \qquad (11.13)$$

The equality between both sides of Eq. 11.13 can be ensured if

$$\Lambda C = -D \frac{\partial \Lambda}{\partial x} \qquad (11.14)$$

which is solved by,

$$\Lambda = \Lambda_0 \exp \left[-\int_0^x \frac{C(\eta)}{D(\eta)} \, d\eta \right] \qquad (11.15)$$

By virtue of Eq. 11.15, we can rewrite Eq. 11.11 in the form

$$\Lambda \frac{\partial e}{\partial t} = \frac{\partial}{\partial x} \left(D \Lambda \frac{\partial e}{\partial x} \right) + \Lambda f \qquad (11.16)$$

We note that Eqs. 11.11 and 11.16 are equivalent although the latter does not contain a velocity term.

11.2.3   Numerical Analysis

When considering a system of coupled PDE's, we can not guarantee a Eulerian monotone (i.e. obeys to the *maximum principle* and does not yield spurious oscillations) numerical scheme. Here in what follows we will prove that the *MEL* scheme is unconditional monotonic and is thus specifically suited for the numerical solution of, say, the flow and transport equations simulating saltwater intrusion problems.

Let us consider, without lose of generality, a 1-D system of PDE's. The general Eulerian formulation can be transformed to read

$$\mathbf{A}\frac{\partial \mathbf{e}}{\partial t} + \mathcal{V}\frac{\partial \mathbf{e}}{\partial x} = \frac{\partial}{\partial x}\left(\mathbf{D}\frac{\partial \mathbf{e}}{\partial x}\right) + \mathbf{F} \qquad (11.17)$$

where $\mathbf{e}$ (e.g. $\mathbf{e}^T \equiv [p, c, T]$ pressure, concentration and temperature, respectively) denotes the vector of the unknowns and $\mathbf{A}$ is given by

$$A_{ij} = \begin{cases} 0 & \text{if } i \neq j \\ 0 \text{ or } 1 & \text{if } i = j \end{cases} \qquad (11.18)$$

The nonconservative form of Eq. 11.17 reads

$$\mathbf{A}\frac{\partial \mathbf{e}}{\partial t} + \mathbf{B}\frac{\partial \mathbf{e}}{\partial x} = \mathbf{D}\frac{\partial^2 \mathbf{e}}{\partial x^2} + \mathbf{F} \quad \text{where} \quad \mathbf{B} \equiv \mathcal{V} - \frac{\partial \mathbf{D}}{\partial x} \qquad (11.19)$$

The *MEL* formulation of Eq. 11.19 can be written in the form

$$\sum_m \left[G_{\ell m}\frac{\partial e_m}{\partial t} + \frac{d_{\ell m}}{dt}(e_m) \right] = \sum_m D_{\ell m}\frac{\partial^2 e_m}{\partial x^2} + F_\ell; \quad G_{\ell m} \equiv A_{\ell m} - 1 \qquad (11.20)$$

in which

$$\frac{d_{\ell m}}{dt}(e_m) \equiv \frac{\partial e_m}{\partial t} + B_{\ell m}\frac{\partial e_m}{\partial x} \qquad (11.21)$$

and the pathline equations read

$$\frac{d}{dt}(x_{\ell m}) = B_{\ell m} \qquad (11.22)$$

For the numerical approximation of Eq. 11.17 or Eq. 11.20, let us consider uniform steps in time, $\Delta t \, (\equiv t^{k+1} - t^k)$, $(t^k = k\Delta t, \ k =$

$0, 1, \dots$), and uniform spatial increments, $\Delta x (\equiv x_{i+1} - x_i)$, $(x_i = i\Delta x$, $i = 0, 1, \dots, N)$.

The general *implicit* difference scheme to approximate Eq. 11.17 reads

$$\frac{\mathbf{A} e_i^{k+1} - \mathbf{H}_i^{k+1} e_i^k}{\Delta t} = \mathbf{R}_i^{k+1} e_{i-1}^{k+1} - \mathbf{Q}_i^{k+1} e_i^{k+1} + \mathbf{S}_i^{k+1} e_{i+1}^{k+1} + \mathbf{F}_i^{k+1}$$

(11.23)

in which $\mathbf{H}_i^{k+1}$ denotes, possibly, nonlinear operators, and the approximation of spatial derivatives and source terms was considered on the $\{x_{i-1}, x_i, x_{i+1}\}$ template. In view of Eq. 11.23, if we take $\mathbf{H}_i^{k+1} = \mathbf{A}$, we obtain a possible scheme to approximate Eq. 11.17. Hence a Eulerian difference scheme for Eq. 11.17 reads

$$\mathbf{A}\frac{e_i^{k+1} - e_i^k}{\Delta t} = \mathbf{R}_i^{k+1} e_{i-1}^{k+1} - \mathbf{Q}_i^{k+1} e_i^{k+1} + \mathbf{S}_i^{k+1} e_{i+1}^{k+1} + \mathbf{F}_i^{k+1}$$

(11.24)

The *MEL implicit* difference scheme to approximate Eq. 11.20 on the $\{x_{i-1}, x_i, x_{i+1}\}$ template is given by

$$\sum_m \left(G_{\ell mi}^{k+1} \frac{e_{mi}^{k+1} - e_{mi}^k}{\Delta t} + \frac{e_{mi}^{k+1} - {}^k e_{\ell mi}}{\Delta t} \right) =$$
$$\sum_m D_{\ell mi}^{k+1} \frac{e_{m,i-1}^{k+1} - 2 e_{mi}^{k+1} + e_{m,i+1}^{k+1}}{(\Delta x)^2} + F_{\ell i}^{k+1} \quad (11.25)$$

in which ${}^k(\ )_i$ denotes a value at the i node associated with the t^k time level, from a backward position defined by Eq. 11.6. In view of Eq. 11.20 we can rewrite Eq. 11.25 in a form

$$\sum_m \left(A_{\ell m} \frac{e_{mi}^{k+1} - e_{mi}^k}{\Delta t} + \frac{e_{mi}^k - {}^k e_{\ell mi}}{\Delta t} \right) =$$
$$\sum_m \left(R_{\ell mi}^{k+1} e_{m,i-1}^{k+1} - Q_{\ell mi}^{k+1} e_{mi}^{k+1} + S_{\ell mi}^{k+1} e_{m,i+1}^{k+1} \right) + F_{\ell i}^{k+1} (11.26)$$

in which

$$\mathbf{R}_i^{k+1} = \frac{\mathbf{D}_i^{k+1}}{(\Delta x)^2}; \quad \mathbf{Q}_i^{k+1} = 2\frac{\mathbf{D}_i^{k+1}}{(\Delta x)^2}; \quad \mathbf{S}_i^{k+1} = \frac{\mathbf{D}_i^{k+1}}{(\Delta x)^2}$$

(11.27)

Comparing Eq. 11.23 with Eq. 11.26, we note that the $\mathbf{H}_i^{k+1}$ operators are obtained by

$$\sum_m H_{\ell mi}^{k+1} e_{mi}^k = \sum_m \left[A_{\ell mi} \, e_{mi}^k - \left(e_{mi}^k - {}^k e_{\ell mi} \right) \right]$$

(11.28)

These operators can be nonlinear, depending on the interpolation implemented to obtain ${}^k e_{\ell mi}$ at its backward position.

We will now investigate the conditions that will prevent the occurrence of spurious oscillations at the t^{k+1} time level, assuming that these did not exist at the previous time level t^k. Hence, without affecting the general implications, let us set the initial conditions (at each t^k time level and for all i nodal points) to be

$$\mathbf{e}_i^k = 0$$

(11.29)

To investigate the "pure" performance of the scheme, let us also omit the influence of the conditions prescribed at the $\{x_0, x_N\}$ (say, $x_0 = 0$ and $x_N = 1$) boundaries. Hence, we will set the boundary conditions to be

$$\begin{aligned}
\mathbf{e}(0, t^{k+1}) &= 0 \\
\mathbf{e}(1, t^{k+1}) &= 0
\end{aligned}$$

(11.30)

Actually, any general initial and boundary conditions can by simple transformation be set to conditions Eqs. 11.29 and 11.30, namely

$$\left. \begin{aligned}
\mathbf{e} &= \varphi(x) & \text{at} \quad & 0 < x < 1, & t &= t^k \\
\mathbf{e} &= \gamma_0(t) & \text{at} \quad & x = 0, & t &\geq t^k \\
\mathbf{e} &= \gamma_0(t) & \text{at} \quad & x = 1, & t &\geq t^k
\end{aligned} \right\} \Longrightarrow$$

$$\left\{ \begin{aligned}
\boldsymbol{\rho} &= \mathbf{e} - \varphi(x) & \text{at} \quad & 0 < x < 1, & t &= t^k \\
\boldsymbol{\rho} &= \mathbf{e} - \gamma_0(t) & \text{at} \quad & x = 0, & t &\geq t^k \\
\boldsymbol{\rho} &= \mathbf{e} - \gamma_0(t) & \text{at} \quad & x = 1, & t &\geq t^k
\end{aligned} \right.$$

(11.31)

We thus solve Eq. 11.17 with $\boldsymbol{\rho}$ as its vector of unknowns, subject to homogeneous boundary and initial conditions.

Focusing on the characteristic of the numerical schemes for Eq. 11.17, let us disregard its source terms, i.e., let $\mathbf{F} = 0$. Hence by substituting Eqs. 11.29 and 11.30 into Eq. 11.24 (for the Eulerian numerical scheme) or into Eq. 11.26 (for the *MEL* numerical scheme), we obtain

$$\mathbf{R}_i^{k+1}\mathbf{e}_{i-1}^{k+1} - \mathbf{C}_i^{k+1}\mathbf{e}_i^{k+1} + \mathbf{S}_i^{k+1}\mathbf{e}_{i+1}^{k+1} = 0; \quad \mathbf{e}_0^{k+1} = 0; \quad \mathbf{e}_N^{k+1} = 0$$

(11.32)

in which

$$\mathbf{C}_i^{k+1} \equiv \frac{\mathbf{A}}{\Delta t} + \mathbf{Q}_i^{k+1}$$

(11.33)

In what follows we will show that the conditions that guarantee the absent of spurious oscillations for Eq. 11.32 are given by

$$\| (\mathbf{C}_i^{k+1})^{-1}\mathbf{R}_i^{k+1} \| + \| (\mathbf{C}_i^{k+1})^{-1}\mathbf{S}_i^{k+1} \| \leq 1; \quad i = 0, 1, \dots, N$$

(11.34)

Samarsky and Nikolaev [1978] prove that the algorithm to solve for the tridiagonal system of Eq. 11.32 is given by

$$\mathbf{e}_i^{k+1} = \boldsymbol{\alpha}_{i+1}^{k+1}\mathbf{e}_{i+1}^{k+1} + \boldsymbol{\beta}_{i+1}^{k+1}$$

(11.35)

in which

$$\boldsymbol{\alpha}_{i+1}^{k+1} = (VC_i^{k+1} - VR_i^{k+1}\alpha_i^{k+1})^{-1}VS_i^{k+1};$$
$$i = 1, 2, \dots, N-1; \quad \alpha_1^{k+1} = 0 \quad (11.36)$$
$$\boldsymbol{\beta}_{i+1}^{k+1} = (VC_i^{k+1} - VR_i^{k+1}\alpha_i^{k+1})^{-1}VR_i^{k+1}\beta_i^{k+1};$$
$$i = 1, 2, \dots, N-1; \quad \beta_1^{k+1} = 0 \quad (11.37)$$

Accounting in Eq. 11.35 for the homogeneous boundary condition Eq. 11.30 (i.e. $\mathbf{e}_0^{k+1} = 0$), we note in Eqs. 11.36 and 11.37 the zero values at node $i = 1$. Furthermore, in view of Eq. 11.37 we obtain

$$\boldsymbol{\beta}_i^{k+1} = 0; \quad i = 0, 1, \dots, N$$

(11.38)

If conditions Eq. 11.34 is fulfilled, then in view of Eq. 11.36 [Samarsky and Nikolaev, 1978] we have

$$\| \boldsymbol{\alpha}_i^{k+1} \| \leq 1; \quad i = 0, 1, \dots, N$$

(11.39)

By virtue of Eqs. 11.35, 11.38 and 11.39 we write

$$\| \mathbf{e}_i^{k+1} \| = \| \boldsymbol{\alpha}_{i+1}^{k+1}\mathbf{e}_{i+1}^{k+1} \| \leq \| \boldsymbol{\alpha}_{i+1}^{k+1} \| \, \| \mathbf{e}_{i+1}^{k+1} \| \leq \| \mathbf{e}_{i+1}^{k+1} \|$$

(11.40)

Hence, by virtue of Eq. 11.40, we prove that the scheme Eq. 11.32 is monotonous. Moreover, due to the homogeneous boundary conditions, Eq. 11.40 implies that the scheme Eq. 11.32 will also yield a unique solution in the form

$$\mathbf{e}_i^{k+1} = 0; \qquad i = 0, 1, \ldots, N \tag{11.41}$$

Actually, since conditions Eq. 11.34 can be implemented for any arbitrary time step (e.g. $\Delta t \gg 1$), we can refer, instead of Eq. 11.34, to conditions in the form

$$\| (\mathbf{Q}_i^{k+1})^{-1} \mathbf{R}_i^{k+1} \| + \| (\mathbf{Q}_i^{k+1})^{-1} \mathbf{S}_i^{k+1} \| \leq 1 , \qquad i = 0, 1, \ldots, N \tag{11.42}$$

In view of Eq. 11.27 and by virtue of Eq. 11.42, we note that the *MEL* implicit difference equations, Eq. 11.26, guarantee a monotonous scheme. This, however, is not granted with the Eulerian implicit numerical schemes.

11.3   Formulation of Saltwater Intrusion Problem

11.3.1   Flow, Transport and Heat Equations in 3-D Space

The mathematical model is based on the balance equations of momentum and mass for the liquid, mass balance equation of a solute in the liquid and on the solid phases, and energy balance equation in the porous media.

11.3.1.1   The Eulerian Forms

The Eulerian water flow, solute and heat transport equations are given in the form:
The liquid mass balance equation,

$$\frac{\partial \phi \rho}{\partial t} + \nabla \cdot (\phi \rho \mathbf{V}) = \rho_R Q_R - \rho Q_P \tag{11.43}$$

where $\phi(p)$ denotes the pressure (p) dependent porosity, $\rho(c, T)$ denotes the liquid density being a function of solute's concentration c (relative to fluid's mass) and fluid's temperature T, and ρ_R denotes the density of a recharged liquid with Q_R as its flux and Q_P as the pumping discharge.

Constitutive relations and definitions,

$$\rho = \rho_0 \exp[\chi_c^\rho (c - c_0) + \chi_T^\rho (T - T_0)]; \quad \chi_c^\rho \equiv \frac{1}{\rho}\frac{\partial \rho}{\partial c}, \quad \chi_T^\rho \equiv \frac{1}{\rho}\frac{\partial \rho}{\partial T}$$

$$\phi = 1 - (1 - \phi_0)\exp[-\chi_p^\phi(p - p_0)]; \quad \chi_p^\phi \equiv \frac{1}{1-\phi}\frac{d\phi}{dp} \quad (11.44)$$

where ρ_0 , c_0 , T_0 , χ_c^ρ , χ_T^ρ , χ_p^ϕ , ϕ_0 and p_0 are prescribed at the same reference.

Liquid's linear (Darcy) momentum,

$$\phi\mathbf{V} = -\frac{\mathbf{k}}{\mu} \cdot (\nabla p - \rho\mathbf{g})$$

$$= -\mathbf{K} \cdot \left(\nabla H + \frac{\rho - \rho_0}{\rho_0}\nabla Z\right) \quad (11.45)$$

where $\mathbf{k}$ denotes the intrinsic permeability tensor, $\mathbf{g}$ denotes the vector of gravity acceleration, Z denotes elevation parallel to $\mathbf{g}$, μ denotes the fluid viscosity, $\mathbf{K}$ $(= \mathbf{k}g\rho_0/\mu)$ denotes the hydraulic conductivity and H $(= p/g\rho_0 + Z)$ denotes an hydraulic head with reference to ρ_0.

Assuming decay with η as the decay coefficient, and adsorption (governed by linear equilibrium isotherm) on the solid matrix, we write:

Solute mass balance equation for the porous medium,

$$\frac{\partial}{\partial t}[\phi\rho c + (1 - \phi)\rho_s c_s] = -\nabla \cdot [\phi\rho(c\mathbf{V} - \mathbf{D}_h \cdot \nabla c)]$$

$$- \eta\phi\rho c + c_R\rho_R Q_R - c\rho Q_P \quad (11.46)$$

in which $\mathbf{D}_h$ denotes the dispersion tensor, c_s $(= \kappa_d\rho c)$ denotes the concentration on the solid matrix, ρ_s denotes the constant solid density, κ_d denotes the partitioning coefficient (= volume of fluid per unit mass of solid) and c_R denotes the concentration of a solute associated with an injected liquid. An other form, replacing Eq. 11.46, can be obtained if multiplying Eq. 11.43 by c (for $c \neq 0$) and subtracting this product from Eq. 11.46. We thus obtain the solute mass balance equation in the form

$$\phi\rho\frac{\partial c}{\partial t} + \frac{\partial}{\partial t}[(1 - \phi)\rho_s c_s] = \nabla \cdot (\phi\rho\mathbf{D}_h \cdot \nabla c) - \phi\rho\mathbf{V} \cdot \nabla c$$

$$- \eta\phi\rho c + (c_R - c)\rho_R Q_R \quad (11.47)$$

In writing the energy balance equation for the porous medium we assume linear thermodynamics. We assume small deformations so that we neglect the energy associated with heat dissipation and with change of volume. Further more, the liquid specific heat at constant volume, C_f, and the solid specific heat at constant strain, C_s, are assumed constant. We assume that liquid's and solid's temperatures are practically equal, solid's velocity and its dispersive heat flux are much smaller than that of the liquid. Hence, following Bear and Bachmat [1990], we write,

the porous medium energy balance equation,

$$\frac{\partial}{\partial t}\left\{[\phi\rho C_f + (1-\phi)\rho_s C_s]\,T\right\} =$$
$$-\nabla\cdot\left\{\phi\left[(\rho C_f T)\,\mathbf{V} - \mathbf{D}^{*H}\cdot\nabla(\rho C_f T) - \boldsymbol{\lambda}^*\cdot\nabla T\right]\right\}$$
$$+\rho_R C_f T_R Q_R - \rho C_f T Q_P \qquad (11.48)$$

where $\mathbf{D}^{*H}$ denotes the liquid thermal dispersion coefficient as suggested by Nikolaevskij [1990] and $\boldsymbol{\lambda}^*$ (for isotropic medium, $\lambda^*_{ij} \equiv \lambda\delta_{ij}$) denotes the thermal conductivity coefficient of the porous medium.

11.3.1.2   The MEL Forms

By differentiating Eqs. 11.43, 11.46 and 11.48, and in view of Eqs. 11.44 and 11.45, we can define a matrix form describing the flow, transport and heat transfer equations. This reads

$$\mathcal{A}_{ij}\frac{\partial \mathcal{Y}_j}{\partial t} + \boldsymbol{\mathcal{B}}_{ij}\cdot\nabla\mathcal{Y}_j = \mathcal{C}_{ij}:\nabla\nabla\mathcal{Y}_j + \mathcal{E}_i \qquad (11.49)$$

where

$$\mathcal{C} = \begin{bmatrix} (\mathbf{k}/\mu) & 0 & 0 \\ c(\mathbf{k}/\mu) & \phi\mathbf{D}_h & 0 \\ T(\mathbf{k}/\mu) & T\chi_c^\rho\,\phi\mathbf{D}^{*H} & (1/\rho C_f)\phi\boldsymbol{\lambda}^* + (1+\chi_T^\rho T)\phi\mathbf{D}^{*H} \end{bmatrix}$$
$$(11.50)$$

$$\mathcal{A}_{11} = (1-\phi)\chi_p^\phi; \qquad \mathcal{A}_{12} = \phi\chi_c^\rho$$

$$\mathcal{A}_{21} = c(1-\phi)\chi_p^\phi(1-\rho_s\kappa_d)$$

$$\mathcal{A}_{22} = (1+c\chi_c^\rho)[\phi+(1-\phi)\rho_s\kappa_d]$$

$$\mathcal{A}_{31} = (1-\phi)\chi_p^\phi T\left(1-\frac{\rho_sC_s}{\rho C_f}\right)$$

$$\mathcal{A}_{32} = \phi\chi_c^\rho T; \qquad \mathcal{A}_{13} = \phi\chi_T^\rho$$

$$\mathcal{A}_{23} = c\chi_T^\rho[\phi+(1-\phi)\rho_s\kappa_d]$$

$$\mathcal{A}_{33} = \phi(1+\chi_T^\rho T)+(1-\phi)\frac{\rho_sC_s}{\rho C_f} \qquad (11.51)$$

$$\mathcal{Y} = \begin{bmatrix} p \\ c \\ T \end{bmatrix}$$

$$\mathcal{E} = \begin{bmatrix} \rho\nabla\cdot(\mathbf{k}/\mu)\cdot\mathbf{g}+(\rho_R/\rho)Q_R-Q_P \\ \rho c\nabla\cdot(\mathbf{k}/\mu)\cdot\mathbf{g}-\eta\phi c+c_R(\rho_R/\rho)Q_R-cQ_P \\ \rho c\nabla\cdot(\mathbf{k}/\mu)\cdot\mathbf{g}+c_RT_R(\rho_R/\rho)Q_R-TQ_P \end{bmatrix} (11.52)$$

$$\mathcal{B}_{11} = -\nabla\cdot(\mathbf{k}/\mu)$$

$$\mathcal{B}_{21} = -(\mathbf{k}/\mu)\cdot[(1+c\chi_c^\rho)\nabla c+c\chi_T^\rho\nabla T]-c\nabla\cdot(\mathbf{k}/\mu)$$

$$\mathcal{B}_{31} = -(\mathbf{k}/\mu)\cdot[\chi_c^\rho T\nabla c+(1+T\chi_T^\rho)\nabla T]-T\nabla\cdot(\mathbf{k}/\mu)$$

$$\mathcal{B}_{12} = -\chi_c^\rho(\mathbf{k}/\mu)\cdot(\nabla p+2\rho\mathbf{g})$$

$$\mathcal{B}_{22} = -\phi\mathbf{D}_h\cdot(\chi_c^\rho\nabla c+\chi_T^\rho\nabla T)-\nabla\cdot(\phi\mathbf{D}_h)$$
$$\qquad -(1+2c\chi_c^\rho)\rho(\mathbf{k}/\mu)\cdot\mathbf{g}$$

$$\mathcal{B}_{32} = -(1+2c\chi_c^\rho)\rho(\mathbf{k}/\mu)\cdot\mathbf{g}-\chi_c^\rho T\nabla\cdot(\phi\mathbf{D}^{*H})$$

$$\mathcal{B}_{13} = -\chi_T^\rho(\mathbf{k}/\mu)\cdot(\nabla p+2\rho\mathbf{g})$$

$$\mathcal{B}_{23} = -2\rho c\chi_T^\rho(\mathbf{k}/\mu)\cdot\mathbf{g}$$

$$\mathcal{B}_{33} = -2\rho c\chi_T^\rho(\mathbf{k}/\mu)\cdot\mathbf{g}-\phi\mathbf{D}^{*H}\cdot(\chi_c^\rho\nabla c+2\chi_T^\rho\nabla T)$$
$$\qquad -\nabla\cdot(\phi\mathbf{D}^{*H})-(1/\rho C_f)\nabla\cdot(\phi\boldsymbol{\lambda}^*) \qquad (11.53)$$

On the basis of Eqs. 11.49 to 11.53, we can write the *MEL* formulation for the flow, transport and heat transfer problems. This reads,

$$\mathcal{G}_{ij}\frac{\partial y_j}{\partial t}+\frac{D_{\mathcal{B}_{ij}}}{Dt}(\mathcal{Y}_j)=\mathcal{C}_{ij}:\nabla\nabla y_j+\mathcal{E}_i \qquad (11.54)$$

in which

$$\mathcal{G}_{ij} \equiv \mathcal{A}_{ij} - 1, \quad \text{and} \quad \frac{D\mathcal{B}_{ij}}{Dt} \equiv \frac{\partial}{\partial t} + \mathcal{B}_{ij} \cdot \nabla$$
$$(11.55)$$

The solution of Eq. 11.54 can be viewed in terms of particles moving along characteristic pathlines, carrying pressure, $(\ )_p$, solute's concentration, $(\ )_c$, and temperature, $(\ )_T$. Such are the particles which are associated with the flow equation, $(\ )^\ell$, with velocities $\mathbf{V}_p^\ell$, $\mathbf{V}_c^\ell$ and $\mathbf{V}_T^\ell$. Particles associated with the solute mass balance equation, $(\ )^s$, have the velocities $\mathbf{V}_p^s$, $\mathbf{V}_c^s$ and $\mathbf{V}_T^s$. Particles associated with the heat transfer equation, $(\ )^H$, have the velocities $\mathbf{V}_p^H$, $\mathbf{V}_c^H$ and $\mathbf{V}_T^H$. In view of Eq. 11.55, we note that these velocities are obtained from the components of $\mathcal{B}$ namely

$$\mathcal{B} \equiv \begin{bmatrix} \mathbf{V}_p^\ell & \mathbf{V}_c^\ell & \mathbf{V}_T^\ell \\ \mathbf{V}_p^s & \mathbf{V}_c^s & \mathbf{V}_T^s \\ \mathbf{V}_p^H & \mathbf{V}_c^H & \mathbf{V}_T^H \end{bmatrix} \qquad (11.56)$$

Moreover, we note that based on the grid size $\triangle x_\ell$ $(\ell = 1, 2, 3)$, we can define the analogous Peclet numbers, Pe_{ij}, and the analogous Courant numbers, Cr_{ij}. In view of Eqs. 11.50 and 11.56 these are given by

$$Pe_{ij} \equiv |(\mathcal{B}/\mathcal{C})_{ij}| \triangle x_\ell; \qquad \forall \ \mathcal{C}_{ij} \neq 0 \qquad (11.57)$$

and

$$Cr_{ij} \equiv \frac{|\mathcal{B}_{ij}|\triangle t}{\triangle x_\ell} \qquad (11.58)$$

11.3.2   Flow, Transport and Heat Equations in 2-D Horizontal Plane

Actually, the 3-D balance equations will describe the saltwater intrusion problem in its full spatial extent. Numerical simulations of this may, however, consume significant computation time. Yet, controlling the migration of a salt plume and studying the effect of different scenarios of stressing the aquifer, can rely on 2-D horizontal plane simulations for regional groundwater management. It is for these above mentioned reasons that we will develop the formulations for simulating the saltwater intrusion problem in 2-D horizontal plane regions.

11.3.2.1   The Eulerian Balance Equations

Consider a *phreatic aquifer* and let h [$= h(x, y, t)$, i.e. function of Cartesian plane coordinates and time, respectively] and b [$= b(x, y)$], denote the upper (i.e. at the point where $p = 0$) and lower elevations, respectively, of a phreatic aquifer. Flow, transport and heat transfer equations in a 2-D horizontal plan space, can be obtained by integrating (from b to h) all of the terms in Eqs. 11.43, 11.45, 11.46 and 11.48. A vertical averaged term (i.e. along the, z, vertical coordinate) will thus be in the form

$$\overline{(..)} \equiv \frac{1}{h - b} \int_b^h (..) \, d\zeta \qquad (11.59)$$

and the average over a product of terms will be assumed to be equal to the product of the averaged terms.

We shall make use of Leibnitz rules for any vector Π and scalar π in the forms:

$$\int_b^h \nabla \cdot \Pi \, dz = \nabla_{xy} \cdot \int_b^h \tilde{\Pi} \, dz + \Pi_h \cdot \nabla (z - h) - \Pi_b \cdot \nabla (z - b) \qquad (11.60)$$

$$\int_b^h \frac{\partial \pi}{\partial t} \, dt = \frac{\partial}{\partial t} \int_b^h \pi \, d\tau + \pi_h \frac{\partial h}{\partial t} - \pi_b \frac{\partial b}{\partial t} \qquad (11.61)$$

where $\Pi^T = [\Pi_x, \Pi_y, \Pi_z]$ and $\tilde{\Pi}^T = [\Pi_x, \Pi_y]$, $(\ )_h$ and $(\ )_b$ denote the values at the elevations h and b, respectively. Integrating Eq. 11.61 and assuming that

$$\pi = \beta(\tau)(t - \tau), \qquad h \equiv t, \qquad \text{and} \qquad \frac{\partial b}{\partial t} = 0 \qquad (11.62)$$

we obtain, for any scalar β

$$\int_b^t \left(\int_b^\tau \beta \, d\theta \right) d\tau = \int_b^t \beta(\tau)(t - \tau) \, d\tau \qquad (11.63)$$

The vertical component of the fluid velocity will vary mostly at the transition between the fresh and the saline water zones. Bearing this in mind, we will postulate that the vertical average form of the specific flux can be obtained by investigating two possibilities. The first by averaging the flow equation when assuming that the vertical pressure is hydrostatic and thus the vertical velocity is zero. The second, by referring to a sharp interface model for which the vertical fluid velocity is zero at the fresh and saline water zones and infinity at the interface.

Accordingly, introducing the notations

$$\Psi \equiv \frac{p}{\rho_0 \, g} = H - Z \,, \qquad \rho^* \equiv \frac{\rho}{\rho_0} \qquad (11.64)$$

we write, in the case of hydrostatic pressure, Ψ_z (e.g. say at $z = b$) in the form

$$\Psi_z = \int_z^h \rho^* \, d\zeta \qquad (11.65)$$

Applying Eqs. 11.60, 11.61 and 11.63 to the liquid mass, Eq. 11.43, and momentum, Eq. 11.45, balance equations together with its constitutive, Eq. 11.44, relations yields a specific discharge $\mathbf{u}$ in the form

$$\mathbf{u} = -\overline{\mathbf{K}} \left[\overline{\rho^*} \nabla_{xy} h + \frac{1}{2} \, (h - b) \, \nabla_{xy} \overline{\rho^*} \right] \qquad (11.66)$$

Actually, Eq. 11.66 can be considered as a good approximation in regions where the vertical flow component is indeed small. However, following Bear [1972] and Ségol [1993] for saltwater intrusion problems, this assumption is not always justified for the flow in the transition zone between fresh and saline water. Let us obtain an estimation of the specific discharge by using the sharp interface model which accounts separately for saltwater, $(\ )_S$, and freshwater, $(\ )_F$, zones. For seawater at $20°C$ it can be shown [Wolf et al., 1980] that

$$\rho_S = 1.025 \, \rho_F \qquad \text{and} \qquad \mu_S = 1.067 \, \mu_F$$
$$(11.67)$$

Since viscosity is in general a function of density and temperature, we can assume that viscosity of water increases with density. Thus we obtain

$$\mathbf{K}_S \approx 0.96 \, \mathbf{K}_F \qquad \Rightarrow \qquad \mathbf{K}_S \approx \mathbf{K}_F \approx \overline{\mathbf{K}}$$
$$(11.68)$$

Furthermore, we can relate the sharp interface elevations (m) with ρ^* which is associated with the 2-D plane model. This, in terms of fluid mass balance relation, is given by

$$\rho_S^* (m - b) + \rho_F^* (h - m) = \overline{\rho^*} (h - b) \qquad (11.69)$$

from which, we deduce that

$$m = \omega h + (1 - \omega) b; \qquad \omega \equiv \frac{\overline{\rho^*} - \rho_F^*}{\rho_S^* - \rho_F^*} \qquad (11.70)$$

In view of Eq. 11.69, we can in a similar way obtain the average specific discharge $\overline{\mathbf{u}}$ based on a sharp interface model. We thus write

$$\overline{\mathbf{u}} (h - b) = \mathbf{u}_S (m - b) + \mathbf{u}_F (h - m) \quad \text{or}$$
$$\overline{\mathbf{u}} = \omega \mathbf{u}_S + (1 - \omega) \mathbf{u}_F \qquad (11.71)$$

After additional considerations in which $\mathbf{u}_S$ and $\mathbf{u}_F$ are evaluated on the basis of ideas following a sharp interface model, we can rewrite Eq. 11.71 in the form

$$\overline{\mathbf{u}} = -\overline{\mathbf{K}} \cdot \left[\nabla_{xy} h + \omega (h - b) \nabla_{xy} \left(\frac{\overline{\rho^*}}{\rho_S^*} \right) \right] \qquad (11.72)$$

Realizing that Eqs. 11.66 and 11.72 are two realizations based on almost two opposite premises, we postulate that, $\overline{\mathbf{q}}$, the average specific discharge is a linear combination of both. Hence we write

$$\overline{\mathbf{q}} = -\overline{\mathbf{K}} \cdot [\nabla_{xy} h + \Omega (h - b) (\chi_c^\rho \nabla_{xy} \overline{c} + \chi_T^\rho \nabla_{xy} \overline{T})] ;$$
$$\text{with} \quad \Omega = 0.5\theta + (1 - \theta) \omega \qquad (11.73)$$

where θ $(0 \leq \theta \leq 1)$ is a parameter of the process being investigated. We note that for $\theta = 0$ we obtain Eq. 11.72, and with $\theta = 1$ we have a form which is almost identical to Eq. 11.66. Really, Eq. 11.73 suggests that the pressure distribution along the vertical is based on a linear hybrid between the hydrodynamic model that assumes hydrostatic pressure and the one that assumes a sharp interface. Moreover, associated with the 2-D plane flow, we can also use Eq. 11.70 to estimate the spatial distribution of the depth of the saltwater body.

In averaging, Eq. 11.46, the transport equation, we need to account for the variation of the horizontal velocities along the vertical direction. To estimate, $\overline{c\phi\tilde{\mathbf{V}}}$, the average advective term, we refer to a sharp interface model. Analogously to Eq. 11.69 for the solute mass balance equation within the fluid phase reads

$$(h-b)\overline{c\phi\tilde{\mathbf{V}}} = \mathbf{u}_S \int_b^m c\,d\zeta + \mathbf{u}_F \int_m^h c\,d\zeta \qquad (11.74)$$

in which we change $\mathbf{u}_S$ and $\mathbf{u}_F$ with relations based on the average discharge given by Eq. 11.73. In view of Eq. 11.74 it can be shown that an additional dispersion term should be introduced to account for variations in horizontal fluid velocities along the vertical direction. The factor, ξ, of this additional dispersion is in the form

$$\xi = \frac{\alpha_A}{\alpha_A + \sqrt{\alpha_L\,\alpha_T}} \qquad (11.75)$$

where α_L and α_T denote the longitudinal and transversal dispersivities, respectively, and α_A denotes the additional apparent dispersivity. The average energy balance equation is obtained in a similar procedure.

Finally, the Eulerian forms of the balance equations to simulate the saltwater intrusion in a 2-D plane will read:
The flow equation

$$\alpha_{11}\frac{\partial h}{\partial t} + \alpha_{12}\frac{\partial \bar{c}}{\partial t} + \alpha_{13}\frac{\partial \tilde{T}}{\partial t} = -\nabla_{xy} \cdot [(h-b)\bar{\rho}\bar{\mathbf{q}}] + \rho_h q_h + \rho_b q_b$$
$$+ (h-b)\left(\overline{\rho_R Q_R} - \overline{\rho Q_P}\right) \qquad (11.76)$$

where

$$\begin{aligned}
\alpha_{11} &\equiv \rho_h S_y + \bar{\rho}^2(1-\bar{\phi}_0)\chi_p^\phi\,(h-b) + \bar{\phi}_0\,(\bar{\rho}-\rho_h) \\
\alpha_{12} &\equiv \bar{\phi}_0\bar{\rho}\chi_c^\rho\,(h-b) \\
\alpha_{13} &\equiv \bar{\phi}_0\bar{\rho}\chi_T^\rho\,(h-b)
\end{aligned} \qquad (11.77)$$

in which q_h, q_b and ρ_h, ρ_b denote, respectively, prescribed flux terms and densities at the upper, $(\ )_h$, and lower, $(\ )_b$, elevations and S_y denotes the specific yield
The transport equation

$$\alpha_{21}\frac{\partial h}{\partial t} + \alpha_{22}\frac{\partial \bar{c}}{\partial t} + \alpha_{23}\frac{\partial \tilde{T}}{\partial t} =$$
$$-\nabla_{xy} \cdot [(h-b)\left(\bar{\rho}\bar{c}\bar{\mathbf{q}} - \mathbf{D}_{2c}\cdot\nabla_{xy}\bar{c}\right)]$$
$$+\rho_h c_h q_h + \rho_b c_b q_b - (h-b)\,\eta\bar{\phi}_0\bar{\rho}\bar{c}$$
$$+ (h-b)\left(\overline{\rho_R c_R Q_R} - \overline{\rho c Q_P}\right) \qquad (11.78)$$

where

$$\begin{aligned}
\alpha_{21} &\equiv \rho_h c_h S_y + \bar{\rho}^2 \bar{c}(1 - \bar{\phi}_0)\chi_p^\phi (h - b) R_d \\
&\quad + \bar{\phi}_0 \left[\bar{\rho}(\bar{c} - c_h) + \bar{c}(\bar{\rho} - \rho_h) \right] R_d \\
\alpha_{22} &\equiv \bar{\rho}\bar{\phi}_0(1 + \bar{c}\chi_c^\rho) (h - b) R_d \\
\alpha_{23} &\equiv \bar{c}\bar{\rho}\bar{\phi}_0\chi_T^\rho (h - b) R_d
\end{aligned} \tag{11.79}$$

$$\mathbf{D}_{2c} \equiv \bar{\rho}\bar{\phi}_0\bar{\mathbf{D}}_h + \xi(1 - \omega)(h - b)\bar{\rho}\bar{c}\chi_c^\rho\bar{\mathbf{K}} \tag{11.80}$$

in which c_h and c_b denote prescribed solute concentrations at the upper and lower elevations, respectively, and R_d denotes the constant retardation factor.

The heat transfer equation

$$\begin{aligned}
\alpha_{31}\frac{\partial h}{\partial t} &+ \alpha_{32}\frac{\partial \bar{c}}{\partial t} + \alpha_{33}\frac{\partial \bar{T}}{\partial t} = \\
&- \nabla_{xy} \cdot \left[(h - b) \left(C_f \bar{\rho}\bar{T}\bar{\mathbf{q}} - \mathbf{D}_{3T} \cdot \nabla_{xy}\bar{T} \right) \right] \\
&+ C_f \rho_h T_h q_h + C_f \rho_b T_b q_b \\
&+ (h - b) \left(\overline{C_f \rho_R T_R Q_R} - \overline{C_f \rho T Q_P} \right)
\end{aligned} \tag{11.81}$$

where

$$\begin{aligned}
\alpha_{31} &\equiv \rho_h C_f T_h S_y + (\bar{\rho}C_f - \rho_s C_s)\bar{\rho}\bar{T}\chi_c^\rho(1 - \bar{\phi}_0)(h - b) \\
&\quad + \left[\bar{\rho}\bar{\phi}_0 C_f + (1 - \bar{\phi}_0)\rho_s C_s \right] (\bar{T} - T_h) + \bar{\phi}_0\bar{T}\bar{\rho}C_f(\bar{\rho} - \rho_h) \\
\alpha_{32} &\equiv \bar{\rho}\bar{\phi}_0 C_f \chi_c^\rho \bar{T}(h - b) \\
\alpha_{33} &\equiv \left[\bar{\rho}\bar{\phi}_0 C_f(1 + \chi_T^\rho \bar{T}) + (1 - \bar{\phi}_0)\rho_s C_s \right](h - b)
\end{aligned} \tag{11.82}$$

$$\begin{aligned}
\mathbf{D}_{3T} &\equiv \bar{\phi}_0\bar{\boldsymbol{\lambda}}^* + \bar{\rho}\bar{\phi}_0(1 + \bar{T}\chi_T^\rho)\bar{\mathbf{D}}^{*H} \\
&\quad + \xi\bar{\rho}C_f\bar{T}(1 - \omega)(h - b)\chi_T^\rho\bar{\mathbf{K}}
\end{aligned} \tag{11.83}$$

in which T_h and T_b denote prescribed temperature values at the upper and lower elevations, respectively.

11.3.2.2   The MEL Balance Equations

We now rewrite Eqs. 11.76, 11.78 and 11.81 to obtain, respectively, the 2-D horizontal plane forms of the *MEL* flow, transport and energy equations in terms of the h, $\bar{c}$ and $\bar{T}$ variables.

These forms are obtained in a similar procedure as described in Eqs. 11.49 to 11.56. Accordingly, we will obtain the 2-D plane velocities associated with particles moving along their characteristic pathlines carrying the groundwater elevation, $(\ )_h$, solute concentration, $(\ )_c$, and, $(\ )_T$, temperature. The ones associated with the 2-D plane *MEL* flow equation will be denoted by $\bar{\mathbf{V}}_h^\ell$, $\bar{\mathbf{V}}_c^\ell$ and $\bar{\mathbf{V}}_T^\ell$. The ones associated with the 2-D plane *MEL* transport equation will be denoted by $\bar{\mathbf{V}}_h^s$, $\bar{\mathbf{V}}_c^s$ and $\bar{\mathbf{V}}_T^s$. The ones associated with the 2-D plane *MEL* heat transfer equation will be denoted by $\bar{\mathbf{V}}_h^H$, $\bar{\mathbf{V}}_c^H$ and $\bar{\mathbf{V}}_T^H$.

11.4 Numerical Examples

Hereinafter we will examine the numerical performance of the *MEL* scheme. In the 1st example, we do this in terms of deviation from an analytical solution and compare to the results obtained by the Eulerian and the *EL* schemes. In the 2nd example we check the validity of the *MEL* 2-D horizontal plane code to solve saltwater intrusion problems. In the 3rd example, we demonstrate the use of the developed code to solve for regional problems for which field observations are distributed in a horizontal plane.

11.4.1 1st Example

Consider a fluid with constant density flowing through an horizontal 1-D saturated porous matrix. The flow and transport of a single solute read, respectively

$$g\rho\chi_p^\phi(1-\phi)\frac{\partial H}{\partial t} = \frac{\partial}{\partial x}\left(K\frac{\partial H}{\partial x}\right) \qquad (11.84)$$

$$\frac{\partial}{\partial t}(\phi c) = \frac{\partial}{\partial x}\left[\phi\left(D\frac{\partial c}{\partial x} - Vc\right)\right] \qquad (11.85)$$

To solve for h and c, we can consider several possibilities.

Case A: A single Eulerian transport equation

In this case we note that Eq. 11.84 is decoupled from Eq. 11.85. The fluid velocity, V, can be solved separately by

$$\phi V = -K\frac{\partial H}{\partial x} \qquad (11.86)$$

10.8d, the following system of ordinary differential equations is obtained:

$$\left[A(\hat{\Psi}, \hat{c}) + B(\hat{\Psi}, \hat{c}) + C(\hat{\Psi}, \hat{c})\right] \hat{c} + \frac{d}{dt}\left[M(\hat{\Psi})\hat{c}\right] + \mathbf{r}^* = \mathbf{0}$$
(10.23)

where matrices A, B, C represent the dispersive, advective, and Cauchy boundary components, respectively, of the transport stiffness matrix, M is the transport mass matrix, and $\mathbf{r}^*$ accounts for sources and sink terms, and for the dispersive component of the Neumann and Cauchy boundary conditions. The coefficients of these arrays are

$$a_{ij} = \sum_{e=1}^{E} \int_{V^e} (D^e \nabla W_j^e) \cdot \nabla W_i^e \, dV$$

$$b_{ij} = \sum_{e=1}^{E} \int_{V^e} \nabla \cdot \left(\tilde{v}^e W_j^e\right) W_i^e \, dV$$

$$c_{ij} = -\sum_{e=1}^{E} \int_{\Gamma_5^e} \tilde{v}^e \cdot \mathbf{n} W_j^e W_i^e \, d\Gamma$$

$$m_{ij} = \sum_{e=1}^{E} \int_{V^e} \phi^e S_w^e W_j^e W_i^e \, dV$$

$$r_i^* = \sum_{e=1}^{E} \left[-\int_{V^e} (q^e c^* + f^e) W_i^e \, dV - \int_{\Gamma_4^e} q_d^e W_i^e \, d\Gamma \right.$$

$$\left. -\int_{\Gamma_5^e} q_c^e W_i^e \, d\Gamma \right]$$
(10.24)

where $\tilde{v}^e = \left(v_x^e, v_y^e, v_z^e\right)^T$. These arrays are evaluated in Appendix B. As in the discretized flow equation 10.18, Dirichlet boundary conditions are imposed after the system has been completely assembled.

Integration in time of Eq. 10.23 is again performed using a weighted finite difference scheme:

$$\left[\nu_c(A + B + C)^{k+\nu_c} + \frac{M^{k+1}}{\Delta t_k}\right] \hat{c}^{k+1} =$$

$$\left[\frac{M^k}{\Delta t_k} - (1 - \nu_c)(A + B + C)^{k+\nu_c}\right] \hat{c}^k - \mathbf{r}^{*k+\nu_c}$$
(10.25)

and is then introduced, as a prescribed value, into Eq. 11.85. The single transport equation reads

$$\phi \frac{\partial c}{\partial t} = \frac{\partial}{\partial x}\left(\phi D \frac{\partial c}{\partial x}\right) - \phi V \frac{\partial c}{\partial x} \tag{11.87}$$

Case B: A single EL transport equation

Again as in *Case A* we follow Eq. 11.86, however we rewrite Eq. 11.87 in the *EL* form

$$\phi \frac{D\mathbf{v}}{Dt}(c) = \frac{\partial}{\partial x}\left(\phi D \frac{\partial c}{\partial x}\right) \tag{11.88}$$

Case C: A system of the Eulerian flow and transport equations

We now introduce Eq. 11.86 into Eq. 11.85 and thus together with Eq. 11.84 we obtain a system of equations in the form

$$g\rho \chi_p^\phi (1 - \phi)\frac{\partial H}{\partial t} = \frac{\partial}{\partial x}\left(K \frac{\partial H}{\partial x}\right)$$

$$\frac{\partial}{\partial t}(\phi c) = \frac{\partial}{\partial x}\left(\phi D \frac{\partial c}{\partial x} + Kc \frac{\partial H}{\partial x}\right) \tag{11.89}$$

We note that Eq. 11.89 is a coupled system of nonlinear PDE's for the solution of the H and c variables.

Case D: A system of the MEL flow and transport equations

Let us rewrite the flow equation, Eq. 11.84, in the form

$$g\rho \chi_p^\phi (1 - \phi)\frac{\partial H}{\partial t}] - \frac{\partial K}{\partial x} \cdot \frac{\partial H}{\partial x} = K \frac{\partial^2 H}{\partial x^2} \tag{11.90}$$

while the transport equation, Eq. 11.89, can be rewritten in the form

$$-\frac{\partial}{\partial x}(Kc) \cdot \frac{\partial H}{\partial x} + \left[\frac{\partial}{\partial t}(\phi c) - \frac{\partial}{\partial x}(\phi D) \cdot \frac{\partial c}{\partial x}\right] = Kc \frac{\partial^2 H}{\partial x^2} + \phi D \frac{\partial^2 c}{\partial x^2} \tag{11.91}$$

The *MEL* form of Eqs. 11.90 and 11.91, reads

$$\left[g\rho \chi_p^\phi (1 - \phi) - 1\right]\frac{\partial H}{\partial t} + \frac{D\mathbf{v}_H^\ell}{Dt}(H) = K \frac{\partial^2 H}{\partial x^2}$$

$$-\frac{\partial H}{\partial t} + \frac{\partial}{\partial t}(\phi c) + \frac{D\mathbf{v}_H^s}{Dt}(H) + \frac{D\mathbf{v}_c^s}{Dt}(c) = Kc \frac{\partial^2 H}{\partial x^2}$$

$$+ \phi D \frac{\partial^2 c}{\partial x^2} \tag{11.92}$$

in which

$$V_H^\ell \equiv -\frac{\partial K}{\partial x}; \quad V_H^s \equiv -\frac{\partial}{\partial x}(Kc); \quad V_c^s \equiv -\frac{\partial}{\partial x}(\phi D)$$
(11.93)

Let us further assume that $\chi_p^\phi = 0$ (i.e. incompressible matrix and flow with fluid velocity constant in space), K and D are constant values so that Eqs. 11.87 and 11.90 become linear PDE's, while Eqs. 11.89 and 11.91 are nonlinear.

We note that Eqs. 11.84 and 11.85 can be solved analytically. Moreover, *Case A* and *Case B* will obviously require less numerical approximations (say, in terms of interpolations) to obtain the solution for H and c. However, in general, we may need to solve a problem consisting of a nonlinear (coupled) set of PDE's such as, e.g. the saltwater intrusion problem. We will hence compare the numerical solution obtained in *cases C and D*.

Let us also note that the grid Peclet, Pe, and Courant, Cr, numbers read

$$Pe \equiv \frac{|V|\Delta x}{D}; \quad Cr \equiv \frac{|V|\Delta t}{\Delta x}$$
(11.94)

and are valid for *cases A and B*. In *case D*, in view of Eqs. 11.57, 11.58 and 11.93, we obtain the analogous Peclet and Courant numbers, in the form

$$Pe_H^\ell = 0; \quad Pe_H^s = \frac{(\partial c/\partial c)\Delta x}{c}; \quad Pe_c^s = 0$$
(11.95)

and

$$Cr_H^\ell = 0; \quad Cr_H^s = \frac{K(\partial c/\partial c)\Delta t}{\Delta x}; \quad Cr_c^s = 0$$
(11.96)

Comparisons between an analytical solution and the numerical solution for the above mentioned four cases, is depicted in Figure 11.1. We note that the *EL* and *MEL* (*Cases B* and *D*, respectively) yield identical results. However, since the latter is solving a set of nonlinear coupled PDE's, we can thus consider it to be more suitable for the solution of saltwater intrusion problems. The Eulerian formulation for the coupled PDE's is written in terms of nonlinear parabolic (diffusion) equations. Yet, its numerical scheme produces spurious oscillations and over/under shooting errors.

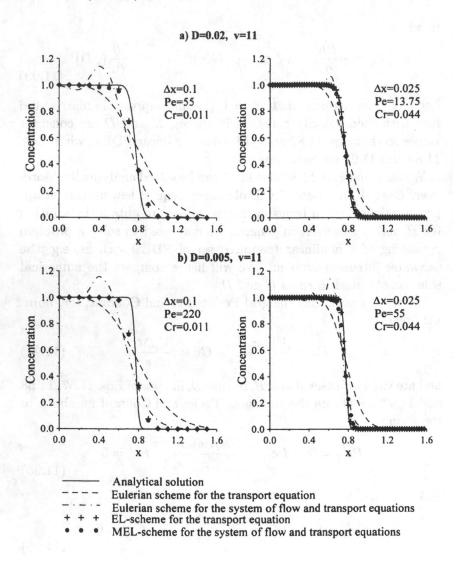

a) D=0.02, v=11

$\Delta x=0.1$
Pe=55
Cr=0.011

$\Delta x=0.025$
Pe=13.75
Cr=0.044

b) D=0.005, v=11

$\Delta x=0.1$
Pe=220
Cr=0.011

$\Delta x=0.025$
Pe=55
Cr=0.044

_____ Analytical solution
- - - - Eulerian scheme for the transport equation
- · - · Eulerian scheme for the system of flow and transport equations
+ + + EL-scheme for the transport equation
• • • MEL-scheme for the system of flow and transport equations

Figure 11.1. Comparison with the 1-D analytical solution of the 1st example, $t = 0.07$.

11.4.2   2nd Example

Let us verify the *MEL* 2-D horizontal plane formulation with another saltwater intrusion computer code. Consider an isothermal phreatic aquifer. Let the initial groundwater level and concentration values be equal to zero [$h(x, y, 0) = 0.0$, $\bar{c}(x, y, 0) = 0.0$], depth of the aquifer

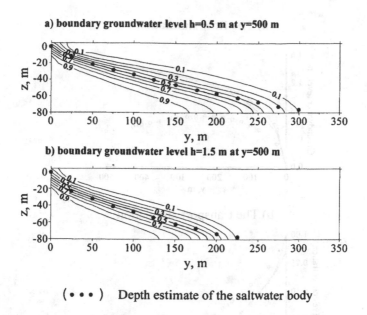

a) boundary groundwater level h=0.5 m at y=500 m

b) boundary groundwater level h=1.5 m at y=500 m

(• • •)   Depth estimate of the saltwater body

Figure 11.2. Concentration profiles obtained by SUTRA in a vertical cross section after 50 years and the MEL 2-D horizontal plane model (MEL2DSLT) estimate of the saltwater depth.

bottom $b = -80.0$ m, the hydraulic conductivity $K = 1.0$ m/day, for the average discharge we chose $\theta = 0.2$ (see Eq. 11.73), longitudinal and transversal dispersivities $\alpha_L = 10.0$ and $\alpha_T = 1.0$ m, respectively, the apparent dispersivity was chosen as $\alpha_A = 2.86$ m (see Eq. 11.75), and the specific yield $S_y = 0.3$. Along the sea boundary, let $h(x, 0, t) = 0.0$ and $(\bar{c}(x, 0, t)/c_s) = 1.0$ (relative concentration) while along the boundary $y = 500$, we assigned $\bar{c}(x, 500, t) = 0.0$. At that boundary we had assigned two different groundwater levels $h(x, 500, t) = 0.5$ and $h(x, 500, t) = 1.5$, for two separate simulations. Solute concentration for seawater was $c_s = 35.7\,g/kg$. Impervious boundary conditions were prescribed for flow and transport along $x = 0.0$ m and $x = 200.0$ m. These flow conditions generated a 1-D flow and transport problem. Simulations were performed using constant spatial increments of $\triangle x = 100.0$ and $\triangle y = 25.0$ m.

Simulations were compared to those obtained by SUTRA [Voss, 1984a], in which flow and transport are evaluated through a saturated-unsaturated, vertical cross section (i.e. averaged along the horizontal, coastal line, direction), of a porous medium. Hence to enable such

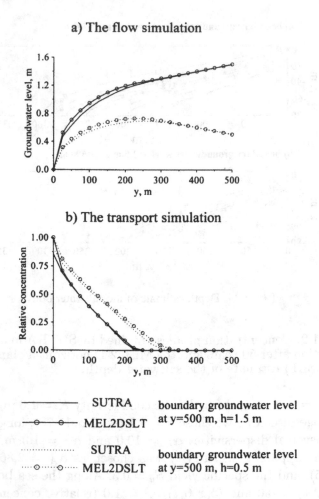

a) The flow simulation

b) The transport simulation

Figure 11.3. Comparison between simulations with SUTRA and with MEL2DSLT, after the time period of 50 years.

a comparison with the developed *MEL* 2-D horizontal plane model (MEL2DSLT), we averaged the concentrations obtained by SUTRA over the thickness of the saturated zone. Such an averaging will generate a 1-D solute distribution.

Using SUTRA, we had approximated the prescribed boundary (at $y = 0.0$) concentrations by Cauchy conditions. As demonstrated by the solution obtained with SUTRA (Figure 11.2) for a vertical cross section, a significant gradient of concentration is noticed. We also note that following ideas of a sharp interface model, Eq. 11.70, the

MEL 2-D horizontal plane code (MEL2DSLT) produced an estimate of the saltwater depth. We note (Figure 11.2) that this follows the 0.5 isoconcentration line. Figure 11.3a demonstrates a good agreement between the SUTRA and the MEL2DSLT simulation of the flow problem. The averaged concentration profiles produced by SUTRA, also prove to be in very small deviation from those obtained by MEL2DSLT (Figure 11.3b).

Numerical experiments comparing between the solutions obtained by SUTRA and MEL2DSLT proved that the extent of change of, α_A, the apparent dispersivity is significantly less than that of α_L and α_T, the longitudinal and transversal dispersivities, respectively. Hence, e.g., with $\alpha_L = 25.0$ m and $\alpha_T = 2.5$ m we found that $\alpha_A = 3.0$ m produced almost no deviation between the solution of SUTRA and that of MEL2DSLT.

11.4.3 3rd Example

Let us consider the implementation of the MEL2DSLT code to a field problem. We will investigate the saltwater intrusion problem into the multi-layered coastal aquifer of Gaza Strip. A general view and a typical hydrogeological cross section of the Gaza strip region is described in Figure 11.4. Our study concerning the flow and transport processes, is focused on the Khan Yunis region. The spatial distribution of the pumping wells at the Khan Yunis area is delineated in Figure 11.5.

Actually, Figure 11.4b and Figure 11.5 demonstrate characteristics for which we developed the MEL2DSLT code. In the case of a multi-layered aquifer, we address the possible discontinuities in fluid velocities (and steep gradients of the aquifer parameters) by the *MEL* algorithm. Note, however, that with the MEL2DSLT code, such discontinuities are accounted for only through the 2-D horizontal plane. To enable regional assessment of different stress scenarios together with comparably quick resolution, we developed the 2-D horizontal plane model. We chose to calibrate the model using information obtained from a rectangular area, limited from the North-West by the coastal line from strips 83 to 86 (Figure 11.4) and stretching 8 km inland North-East. This area (8 km × 8 km) was divided by rectangular cells 200 m × 200 m each, altogether $41 \times 41 = 1681$ nodes. Information regarding concentration values was scarce and not reliable. We therefore calibrated the model using mainly information about

a) Location map

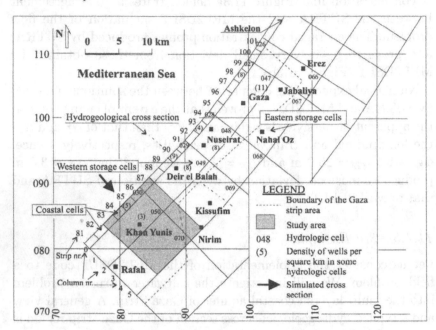

b) Hydrogeological cross section of the Gaza strip region

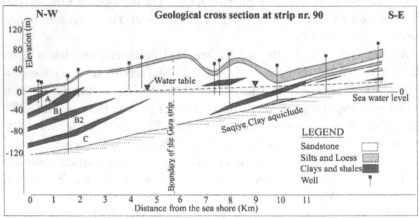

Figure 11.4. General view and a typical hydrogeological cross section (after Melloul and Collin [1994]).

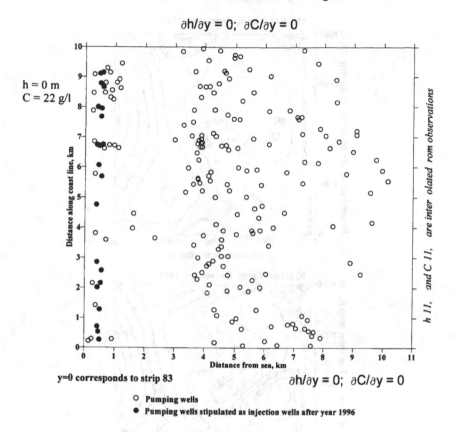

Figure 11.5. Stipulated management and boundary conditions at the Khan Yunis region.

the measured groundwater levels. Such, exists for the years 1985 to 1991, as well as values of pumping and natural replenishment. These data was not provided for all wells and were not observed at the same time. To overcome these, we chose a time step of one month to represent the diversity of observations at different time. Altogether 12 month× 7 year= 84 files were completed. These can be considered as "slices" of a random field of groundwater levels in 84 time points. Information concerning the random field of the groundwater levels was assembled into 84 files, each represents the distribution of the groundwater levels in a particular month from the years pe-

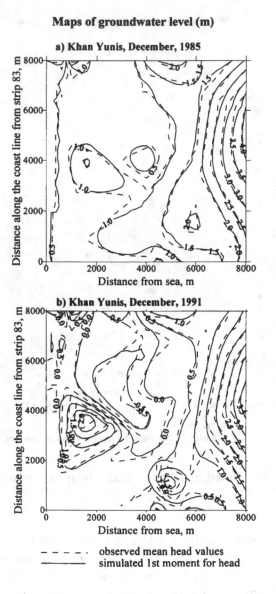

Figure 11.6. Samples of calibration maps of groundwater levels.

riod of 1985 to 1991. We had eliminated an inherent spatial plane distribution of a linear time trend in the random distribution of the groundwater levels (e.g., due the anthropogenic effects prevailing in the study area). We then evaluated the spatial distributions of the

Map of chloride concentration (g/kg)

Khan Yunis, year 1991

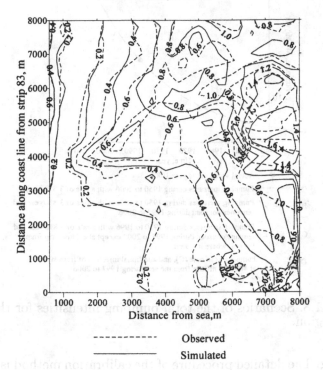

--------- Observed

———————— Simulated

Figure 11.7. Observed distribution of concentration values compared with the simulated ones.

mean of the initial groundwater levels and those of the Cl (chloride) concentrations, at the beginning of the year 1985. Dirichlet conditions for the hydraulic head and for the Cl concentration, were assigned at the boundaries. These were the evaluated mean values based on observation in time along the boundaries.

For the sake of obtaining the estimated groundwater levels, we used in our simulation the following values: specific yield $S_y = 0.3$, porosity $\phi = 0.35$, for the averaged specific discharge we used $\theta = 0.22$, longitudinal dispersivity $\alpha_L = 25$ m, transversal dispersivity $\alpha_T = 0.5$ m and apparent dispersivity $\alpha_A = 2.7$ m. We had also accounted for an observed regional average annual rate of the natural replenishment with Cl concentration in the infiltrating water of $C_h = $

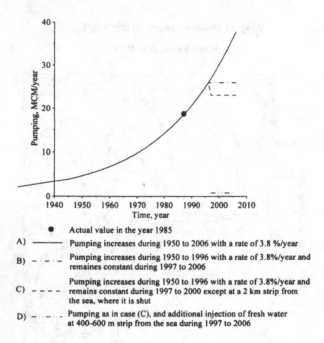

Figure 11.8. Scenarios of the total pumping intensities for the Khan Yunis region.

110 g/kg. The detailed procedure of the calibration method is beyond the scope of the present report and will, thus, not be discussed herein.

The comparison between the observed groundwater level values and the simulated ones is depicted in Figure 11.6. The resulted concentration distribution in compare to the observed ones (the latter is situated only from 800 m inland) is delineated in Figure 11.7.

After calibrating for the aquifer parameters and the boundary conditions, we investigated the code predictions resulting from possible stress patterns. To obtain the initial distribution of groundwater levels and concentrations, we run a steady state simulation without any source/sink stresses. Using these as initial conditions, we had now implemented various pumping scenarios using the actual pumping intensity from the year 1985 and extrapolating on the basis of 3.8% annual population growth. These pumping scheme is delineated in Figure 11.8, while the actual spatial wells distribution at the study area, is presented in Figure 11.5.

The predictions of the regional distributions resulting from the

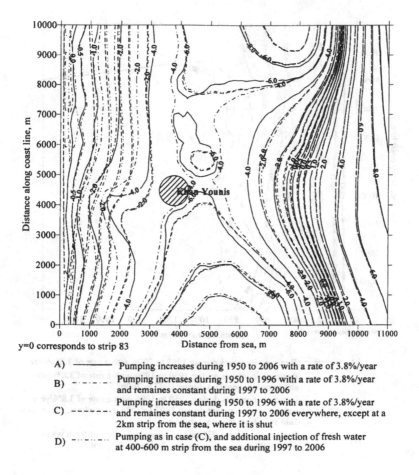

A) ———— Pumping increases during 1950 to 2006 with a rate of 3.8%/year

B) – · – · – Pumping increases during 1950 to 1996 with a rate of 3.8%/year
and remaines constant during 1997 to 2006

C) – – – – – Pumping increases during 1950 to 1996 with a rate of 3.8%/year
and remaines constant during 1997 to 2006 everywhere, except at a
2km strip from the sea, where it is shut

D) – · · – · · – Pumping as in case (C), and additional injection of fresh water
at 400-600 m strip from the sea during 1997 to 2006

Figure 11.9. Predicted groundwater levels resulting from different pumping scenarios.

stipulated pumping scenarios is described in Figure 11.9, for the simulated groundwater levels and in Figure 11.10 for the simulated chloride migration.

By virtue of Eq. 11.70, we can also assess the depth of the fresh-water body (Figure 11.11).

11.5 Conclusion

The saltwater intrusion problem is described for 3-D and 2-D, horizontal plane, spatial coordinates. The coupled flow, transport and

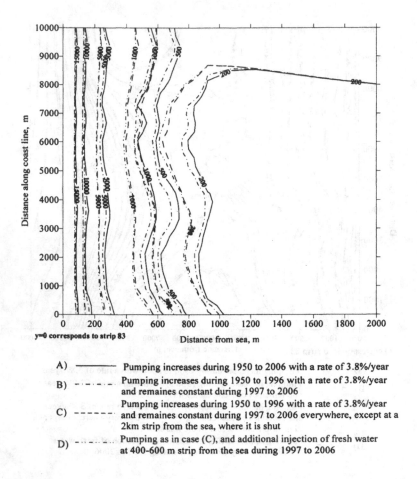

y=0 corresponds to strip 83

A) ———— Pumping increases during 1950 to 2006 with a rate of 3.8%/year

B) – · – · – · Pumping increases during 1950 to 1996 with a rate of 3.8%/year
and remaines constant during 1997 to 2006

C) – – – – – Pumping increases during 1950 to 1996 with a rate of 3.8%/year
and remaines constant during 1997 to 2006 everywhere, except at a
2km strip from the sea, where it is shut

D) – ·· – ·· – Pumping as in case (C), and additional injection of fresh water
at 400-600 m strip from the sea during 1997 to 2006

Figure 11.10. Predicted chloride concentration (mg/kg) resulting
from different pumping scenarios.

heat transfer equations are formulated using Eulerian and Modified
Eulerian Lagrangian (*MEL*) concepts. The latter is also accounting
for the gradients of the permeability, hydrodynamic dispersion and
heat dispersion coefficients. Accordingly, every balance equation is
associated with three groups of particles. Each particle travels with
its own velocity along its characteristic pathline.

Rigorous discussion is provided on the mathematical and numerical analysis associated with the *MEL* formulation proving that:

- The *MEL* scheme accounts for additional information, in between
  grid nodes, concerning gradients of the permeability, hydrody-

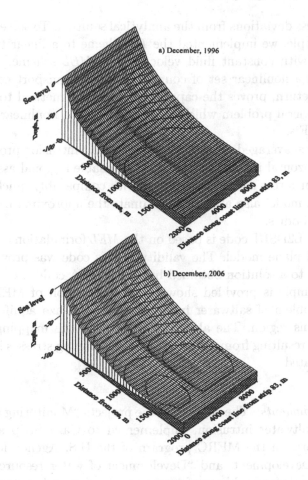

a) December, 1996

b) December, 2006

Figure 11.11. Estimate of the saltwater depth following pumping scenario A.

namic dispersion and heat dispersion coefficients. A generalized Pe number is developed in which such gradients are introduced.

- The *MEL* scheme (unlike the *EL* scheme) can also be implemented to balance equations (e.g. the flow equation) that do not refer explicitly to the phase velocity.

- The *MEL* scheme (unlike the Eulerian scheme), is unconditionally free of spurious oscillations for the solution of a system of coupled PDE's.

Numerical comparisons between the Eulerian scheme, the *EL* scheme and the *MEL* one, demonstrate that the two latter yield significant less deviations from the analytical solution. To solve the specific example, we implemented the *EL* scheme to a linear transport equation with constant fluid velocity. The *MEL* scheme, however, addressed a nonlinear set of coupled flow and transport equations. This, in return, proves the capacity of the *MEL* method to reliably solve a general problem which is formulated by a nonlinear coupled set of PDE's.

A vertical averaged model for simulating salt intrusion problems in a 2-D horizontal plane was developed to enable regional assessment of different stress scenarios together with comparably quick resolution. This model also allows the estimate the approximate depth of saltwater bodies.

The MEL2DSLT code is based on the *MEL* formulation of the 2-D horizontal plane model. The validity of this code was proven when compared to a solution obtained by the SUTRA code.

An example is provided showing the calibration of MEL2DSLT to the problem of saltwater intrusion into the Gaza aquifer at the Khan Yunis region. The ability to investigate the mapping of salt migration resulting from different regional pumping stresses is clearly demonstrated.

Acknowledgments: The support of the projects "Monitoring and modeling of saltwater intrusion, implemented to Gaza Strip and Morocco," through the MERC program of the U.S. Agency for International Development, and "Development of water resources management tools for problems of seawater intrusion and contamination of fresh-water resources in coastal aquifers" through the European Community AVICENNE program is acknowledged with appreciation.

Chapter 12

Survey of Computer Codes and Case Histories

S. Sorek & G. F. Pinder

12.1 General

To assemble the pertinent research in this field we have provided
sections explaining the essence of each contribution. The contributors
were asked to provide a series of short summaries introducing the
attributes of their models and, where appropriate, case histories.
Each contributor was asked to make a series of brief statements that
would fit into a table (some are in a condensed form). These were:

A. Physical Concepts

1. Advective-dispersive or sharp-interface transport;

2. Saturated or unsaturated flow;

3. Multiple or single phases;

4. Adsorption or conservative transport;

5. Transient or steady-flow;

6. Transient or steady-transport;

7. Isothermal or non-isothermal conditions;

8. Multiple or single component (species) transport;

9. Chemical reactions or non-reactive chemistry.

B. Computational Concepts

1. Dimensionality (i.e. one dimensional, vertical plane, horizontal
   plane, quasi two-dimensional, three dimensional, etc.);

J. Bear et al. (eds.), Seawater Intrusion in Coastal Aquifers, 399–461.
© 1999 *Kluwer Academic Publishers.*

2. Numerical method (i.e. finite volumes, analytical elements, stochastic elements, boundary elements, spectral elements, finite elements, finite differences, lumped parametric approach etc.);

3. Boundary conditions (i.e. can accommodate first, second or third-type boundaries);

4. Specified internal conditions (i.e. can specify a known flux or potential value on the interior of the mesh);

5. Stresses (i.e. pumping/recharge, evapotranspiration, rainfall, etc.);

6. Parameter input (i.e. can accommodate heterogeneity, random fields, etc.);

7. Kind of equation solvers employed (i.e. Gauss elimination, GMRES, etc.);

8. If the system is non-linear, how are the nonlinearities handled (i.e. Picard, Newton-Raphson etc.);

C. Other Issues

1. Applicable graphical interfaces if they exist;

2. Platforms and operating systems on which code will run;

3. Method of verification (i.e. against analytical solutions, other numerical solutions, mass balance accuracy etc.);

4. Perceived strengths and weaknesses of the model.

Some of our contributors have developed models, some have applied models and some have done both. Accordingly, contributors were invited to provide one case history that includes the following information:

D. Case Histories

1. Site location;

2. Purpose and scope of investigation;

3. Hydrogeological setting;

4. Model employed;

5. Qualitative description of parametric input, imposed stresses, and boundary conditions;

6. Results of analysis.

We did not review the documentation of the various models nor did we test the codes in any way. Thus we do not attest to the suitability, accuracy, or functionality of the models. Indeed we provide in this chapter only information provided to us by the code developers and users.

We first present information referring to the developed models (i.e. the above A, B and C sections). The essence of these remarks are then compiled into tables.

12.2   3DFEMFAT/2DFEMFAT

George Yeh (gty@darcy.psu.edu)

12.2.1   Description

The 3DFEMFAT/2DFEMFAT codes are based on a Galerkin finite element discretization of the variably saturated flow equation and a hybrid Eulerian-Lagrangian finite-element discretization of the advective-dispersive or pure advective solute-transport equation. The codes are applicable to either three (3DFEMFAT) or two (2DFEMFAT) dimensional systems. The transport is isothermal and either transient or steady state. The model applies to either transient or steady-state, single-phase saturated-unsaturated flow and considers adsorption. First order chemical and biodegradation reactions are incorporated in this code. Non-linear terms are approximated using Picard iteration and the resulting algebraic equations are solved using a choice of several solvers. Standard boundary conditions and first-type internal conditions can be accommodated. Point and areal stresses can be handled as well as heterogeneous, anisotropic aquifers. The code runs under UNIX and DOS operating systems. The use of a hybrid Eulerian-Lagrangian discretization scheme allows for coarser time and grid discretization. Adaptive local grid refinement and a multigrid solver provides improved efficiency for this code. As with other models, when the density difference between fresh and salt water becomes large, the model encounters difficulties in achieving non-linear convergence.

12.2.2 Physical Concepts

1. Simulation is via the advective-dispersive transport equations or the purely advective sharp interface equations.

2. Variably saturated and slightly deformable media are assumed [Cheng et al., 1998].

3. The model considers single-phase flow.

4. Reversible adsorption is included in transport.

5. Transient and steady-flow conditions are considered.

6. Transient and steady-state transport is accommodated.

7. Isothermal conditions are assumed.

8. Single-component transport is permitted.

9. First-order chemical kinetic reactions or biodegradation reactions are allowed.

12.2.3 Computational Concepts

1. The models are two-dimensional cross sectional for 2DFEM-FAT or three dimensional (3DFEMFAT).

2. The numerical procedure used is Eulerian-Lagrangian Galerkin finite element.

3. First, second and third type boundary conditions can be used.

4. First-type conditions can also be applied on the interior of the domain.

5. Point and area sources can be specified.

6. Parameter input permits heterogeneity and anisotropy.

7. Algebraic equations are solved using direct Gauss elimination, block iteration, point iteration, preconditioned conjugate gradient method (PCG) with polynomial, incomplete Cholesky decomposition, modified incomplete Cholesky decomposition, or symmetrical successive over relaxation as the preconditioner.

8. When a non-linear set of equations is encountered, Picard iteration is used.

12.2.4 Other Issues

1. The GMS-FEMWATER graphical interface is used for 3DFEM-FAT.

2. The model runs under UNIX and DOS operating systems.

3. Verification includes comparison against analytical solutions, other numerical solutions, and mass tracking.

4. The primary advantage of this method is the use of the Eulerian-Lagrangian method of numerical approximation which effectively handles convection dominated transport.

5. The use of adaptive local grid refinement and multigrid methods has greatly improved the efficiency and convergence of the model.

6. The major disadvantage is the difficulty occasionally encountered in convergence when the density difference between fresh and salt water is large. When this occurs the adjustment of iteration parameters to overcome the problem requires the attention of expert users.

12.2.5 Case History

1. Wilmington Harbor groundwater study.

2. The purpose of the study was to determine any significant effects on the aquifer system due to deepening the 26 miles of shipping channel.

3. The study area is underlain by eastward thickening wedges of sediments and sedimentary rocks of Cretaceous to Quaternary age. They consist of interbedded sands, calcareous sands, silts, clays and fossiliferous limestones. Aquifers recognized within the study area include the surficial, Castle Hayne, and the Pcedee.

4. The model employed was 3DFEMFAT.

5. Dirichlet boundary conditions are assigned along the coastal shore and the Cape River. Rainfall as recharge is specified over the study area. Many pumping wells are located in the study area.

6. Preliminary results indicate no significant effects on the aquifer system due to channel deepening will occur.

12.3   CODESA-3D

COupled variable **DE**nsity and **SA**turation **3D** model

Giuseppe Gambolati (gambo@dmsa.unipd.it)
Claudio Paniconi (cxpanico@crs4.it)
Mario Putti (putti@dmsa.unipd.it)

12.3.1   Description

CODESA-3D is a finite-element model that can simulate the salt-water interface by solving the convective-dispersive transport equation. The model handles saturated and variably saturated porous media. It is based on discretization into tetrahedral elements that can be generated automatically from a two-dimensional surface grid. Picard or Newton iteration methods are used to resolve the non-linearities in the unsaturated flow case, and Picard or partial-Newton methods are employed for the coupled flow-transport system. Preconditioned conjugate gradient-type algorithms are used to solve the resulting symmetric or non-symmetric linear systems. Time stepping is adaptive and based on the convergence behavior of the iterative schemes. The code handles temporally and spatially variable boundary conditions of Dirichlet, Neumann, or Cauchy type, including seepage faces and atmospheric (rainfall/evaporation) inputs which are checked at each time step for type-switching. The model also accepts heterogeneous material and solute properties including hydraulic characteristics.

12.3.2   Physical Concepts

1. Simulation is via the advective-dispersive transport equation.
2. Variably saturated aquifers are assumed.
3. The model is restricted to single phase flow.
4. Linear instantaneous sorption is used.
5. Steady or transient flow is considered.
6. Transient and steady-state transport is accommodated.
7. Isothermal conditions are assumed.
8. Single-component transport is permitted.
9. The model does not allow for chemical reactions.

12.3.3   Computational Concepts

1. The model is fully three-dimensional.

2. The numerical procedure used is finite elements.

3. First, second and third-type boundary conditions can be used.

4. First and second-type conditions can also be applied on the interior of the domain.

5. Point and area sources can be specified.

6. Hydraulic conductivity is heterogeneous and anisotropic.

7. Algebraic equations are solved using PCG (Preconditioned Conjugate Gradient) for symmetric systems and Bi-CGSTAB (BiConjugate Gradient STABilized), GMRES (Generalized Minimal RESidual) or TFQMR (Transpose-Free Quasi-Minimal Residual) for non-symmetric systems, with different choices of preconditioners.

8. Picard and Newton methods are used to solve the non-linear flow equation and Picard and partial-Newton methods are used for the coupled flow and transport equations.

9. An adaptive time-stepping procedure can recover from non convergence of the Picard and Newton method by reducing the time step-size appropriately.

12.3.4   Other Issues

1. A graphical interface is not available.

2. The model is written in FORTRAN 77 and is portable across different architectures.

3. The model has been verified using published numerical solutions.

4. The model uses fully three-dimensional tetrahedral elements defined automatically from a triangular two-dimensional grid and employs several functional relationships to describe the saturated-unsaturated flow behavior.

12.3.5   Case History

1. The CODESA-3D code was applied to Gioia Tauro and Calabria in southern Italy.

2. During the construction of a channel system to discharge water into the sea, it was necessary to lower the water table of the underlying aquifer for a period of six months. It was feared that the extensive pumping needed for the complete dewatering of the construction site could impact the natural flow field and alter the behavior of the saltwater front. The model was used to assess the extent of subsequent saltwater contamination, and the effectiveness of a cut-off wall which was constructed to contain the seawater intrusion.

3. The hydrogeological setting is a confined 36 km^2 coastal aquifer in southern Italy located on the Tyrrhenian Sea. It is bounded by rivers on its northern and southern borders, by the sea on the west, and by a high plain on the east.

4. An early version of CODESA-3D for saturated porous media was used.

5. At the top and bottom of the aquifer and on the northern and southern boundaries a zero flux condition for both the flow and transport equations was imposed. Dirichlet conditions were imposed on the east boundary, with zero concentration prescribed for the transport equation and a range of head values for the flow equation.

The sea-side boundary contained three zones. The central zone was characterized by a 26.3 m deep cutoff wall, while the two lateral zones were in direct contact with the sea. In these lateral zones, hydrostatic equivalent freshwater head was prescribed, along with zero concentration flux for the top 10.9 m and a unitary (seawater) concentration specified in the lower part. In the central zone, zero flux was prescribed for the flow and transport equations for the top 26.3 m, while for the lower part hydrostatic head and unitary concentration were imposed.

The surface mesh contained 2829 triangles and 1459 nodes. This array was duplicated vertically to form 10 layers over the total aquifer depth of 56.3 m. The three-dimensional grid contained 84,870 tetrahedra and 16,049 nodes.

The aquifer contained two distinct hydrogeological zones, one stratified and anisotropic and the other homogeneous and isotropic. The density ratio was 0.03 and the longitudinal and transverse dispersivity values were 10 m and 1 m, respectively.

6. A steady-state solution was attained by running a transient simulation until variations in head and concentration were negligible. This occurred after 1,342 days of simulation, using time step sizes that varied from 1 to 25 days. A second set of simulations was run using the steady-state solution as an initial condition. In this second set, pumping at a rate of 2,189 m^3/day was applied. A period of 185 days (which was the expected duration of the channel-construction phase) was simulated. The simulation results showed that water withdrawal from pumping would cause only a limited increase in saltwater intrusion. The total simulation (1342 + 185 days) took less than one hour of CPU on an IBM RS6000/390 workstation.

12.4 DSTRAM

Density-dependent **S**ubsurface **T**ransport **A**nalysis **M**odel

Peter Huyakorn (psh@hgl.com)
Sorab Panday

12.4.1 Description

The DSTRAM code is based upon a finite-element discretization of the three-dimensional advective-dispersive transport equations. It provides a transient or steady-state simulation of non-isothermal flow in a saturated aquifer with linear decay. The resulting set of algebraic equations can be solved with a choice of two solvers and the non-linear terms are approximated using Picard iteration. Standard boundary conditions are accommodated and the code runs under Windows and UNIX. Up to 500 thousand node problems have been solved. It has been observed that the one-step steady-state solution option may fail to converge for complex physical systems.

12.4.2 Physical Concepts

1. Simulation is via the advective-dispersive transport equation.
2. Saturated aquifers are assumed.
3. The model is restricted to single-phase flow.
4. Linear adsorption is utilized.
5. Transient and steady flow is considered.

6. Transient transport is accommodated.

7. Non-isothermal conditions are simulated.

8. Single-component transport is permitted.

9. The model allows for linear decay of components.

12.4.3   Computational Concepts

1. The model is fully three dimensional.

2. The numerical procedure used is finite elements.

3. First, second and third-type boundary conditions can be used.

4. First, second and third-type conditions can also be applied on the interior of the domain.

5. Point, line and area sources can be specified.

6. Parameter input is zone-wise constant.

7. Algebraic equations are solved using conjugate-gradient or an ORTHOMIN solver with an ILU preconditioner. Red-black ordering is used.

8. When a non-linear set of equations is encountered, Picard's method is used.

12.4.4   Other Issues

1. A graphical interface is not currently available.

2. The model runs under DOS, Windows 3.1, Windows 95, Windows NT and UNIX operating systmes.

3. The model has been verified by comparison to analytical solutions and against the sharp-interface solution using the Ghyben-Herzberg relationship.

4. The model has been used for problems up to 500 thousand nodes and has been field tested.

5. It has been observed that the one-step steady-state solution option may fail under very complex physical situations.

12.5   FAST-C(2D/3D)

Fast **A**lgorithm for **S**aline or **T**hermal **C**onvection in porous media

Ekkehard Holzbecher (holzbecher@igb-berlin.de)

12.5.1 Description

The FAST codes are based on finite difference/finite volume discretizations of partial-differential equations. FAST-C(2D) can be
used for modeling density-driven flow in vertical cross-sections. Variable density and viscosity can be considered. Viscosity and density
dependencies can be linear or nonlinear. The FAST codes do not
account for flow in the unsaturated zone and do not consider matrix deformations. The code has been tested extensively in several
benchmarks. The recent version of the program includes a graphical interface (GUI) and on-line-help. FAST-C (3D) can be used for
modeling density-driven flow in three dimensions. Pressure is used
as a flow variable. FAST-C(3D) has been tested for a few cases only.

12.5.2 Physical Concepts

1. Solute transport is governed by the advection–dispersion equation.
2. Solute transport is estimated for a saturated confined aquifer.
3. Only one fluid (liquid) phase is assumed to be mobile.
4. Solute is assumed to be retarded.
5. Transient and steady-state conditions are employed for the simulation of the flow problem.
6. Transient and steady-state conditions are employed for simulating transport.
7. Isothermal conditions are implemented.
8. The model is based on the transport of one solute species.
9. Generation of solute mass due to chemical reactions is not considered.

12.5.3 Computational Concepts

1. Simulations are done for a two-dimensional vertical plane or
   for a three dimensional space.
2. The finite-difference method is employed. The transient flow
   can, optionally, be simulated by either explicit, implicit or
   Crank-Nicolson schemes.
3. Boundary conditions of the first and second type can be accommodated for simulating the flow and the transport problems.

5. Dirichlet and Neumann boundary conditions are accommodated for flow and transport. The FAST-C(3D) code also accommodates Neumann conditions.

6. Heterogeneity and anisotropy are accommodated in terms of parameter input. The FAST codes also account for the generation of random fields.

7. Linear solvers are: CG for the flow equation and CG, CGS, BiCG and BiCGSTAB for the transport equation. All algorithms can be run with or without preconditioning. The user has the choice of specifying an over-relaxation parameter in the case of SOR preconditioning.

8. Picard's iteration method is used to handle the nonlinear system.

12.5.4   Other Issues

1. The graphical interface GEO-Shell is incorporated in both the FAST-C(2D) and FAST-C(3D) codes.

2. The FAST-programs are written in standard FORTRAN code. The routines for graphical output cannot be transferred from one machine to the other. On the PC version there are several options to enable the use of the SURFER graphics software for post-processing. The Geo-Shell user interfaces for FAST are written in C language for WINDOWS and work on PCs only.

3. FAST-C(2D) was verified against analytical and numerical solutions obtained by other codes. The cases compared were saline and thermal problems. Thermal problems: steady and oscillatory convection, the Elder experiment (HYDROCOIN test case 2.2), Yusas model for geothermal flow and horizontal heat transfer. Saline problems: saltwater intrusion, saltwater upconing, flow above a salt-dome (HYDROCOIN test case 1.5) and flow in desert sedimentary basins.

4. In FAST-C(2D) flow needs to be specified in terms of stream functions. FAST-C(3D) is based on a pressure formulation. Additional testing is required.

12.5.5   Case History

1. The FAST-C(2D) code was applied to the Nile delta region covering an area of more than 23,000 km^2 which is a major

supplier for the increasing food demand in Egypt.

2. The Nile delta aquifer was investigated in terms of becoming an alternate water supply source. One main aspect of the study was to assess saltwater intrusion as an outcome of increased pumping.

3. The Nile Delta's Pleistocene aquifer has a north-south extension of more than 100 km, limited on both sides by the desert. The only source of water into the region is the Nile river, which bifurcates into two delta arms, Damietta and Rosetta, 20 km north of Cairo. The longest east-west extension, at the coast, is more than 150 km. With distance from the sea, the depth of the aquifer decreases gradually from a maximum of 1000 m.

   The permeability of the aquifer has been obtained on the basis of numerous pumping tests reported in former publications. The conductivity is 6×10^{-4} m/s. Some semi-permeable lenses with a conductivity of 1×10^{-7} m/s can be found locally. In most parts the aquifer is confined by a clay cap with a conductivity of 2.8×10^{-9} m/s. The clay cap originates from the Nile river sediments in the Holocene and its depth reaches from a few meters in the south up to 75 meters below the coastline. In preceding publications, the transition zone from freshwater to saltwater has been calculated to lie between 80 and 130 km inland from the Mediterranean shoreline at the base of the aquifer.

   Field measurements down to a depth of 330 m indicate seawater concentrations (35,000 ppm) only along the coastline, while the transition zone reaches no more than 40 km inland. In the middle delta, freshwater concentrations (less than 1000 ppm) were found through electrical conductivity measurements.

4. The field measurements have been confirmed by the FAST-C(2D) code. This was implemented along the symmetry axis of the delta. The vertical cross-section in the north-south axis through the aquifer reaches an extent of 100 km from the Mediterranean. The depth of the model cross-section varies and takes its maximum of 1 km at the seaside.

   An extension of the FAST-C(2D) was implemented to use cylindrical coordinates and an arbitrary opening angle of the segment. Steady-state conditions were assumed. Grids with different spacing were used to study the effect of grid-dependent numerical dispersion. A typical run for the fine grid model took 16 minutes CPU and 12 Picard iterations on a SUN workstation to obtain a relative accuracy of 1×10^{-5}. This can also be implemented on common PC-computers.

5. The flow boundary conditions at the surface and at the in-flow boundary in the south were determined on the basis of the hydrological balance of the Nile delta region referring to the groundwater recharge (2.17 km^3/yr) and withdrawal (1.82 km^3/yr). At the bottom a no-flow condition was required. Along the vertical seaside boundary, a condition of no vertical component of the velocity was ascribed. Dirichlet conditions of high salinity were set on the vertical boundary below the seashore and of low salinity on the vertical inflow boundary. A Neumann condition of no salt flow was prescribed at the horizontal boundaries. The type of conditions are identical to those given for Henry's classical test case of saltwater intrusion. An effective dispersivity of 4×10^{-3} m^2/s was assumed to govern the dispersion in the delta aquifer.

6. Seawater was found to migrate to an extent of 40 to 50 km south of the coast. Recorded in-situ measurements correspond with simulations. Grids with 50, 200, 400 and 800 equally spaced blocks have been used to investigate the significance of the numerical dispersion. The maximum grid Peclet numbers were 0.88, 0.24, 0.12 and 0.065 respectively. Numerical dispersion decreases with grid refinement. Results for the 400 and 800 blocks show almost identical outputs, so that the solution for these grids can be considered as grid-independent. Grid-Peclet numbers are much smaller than prescribed by the classical grid-Peclet criterion. The numerical results obtained with the FAST-C(2D) agree well with the measurements of salt concentrations within the project. Predictions show that although the saltwater intrusion is most likely not advancing, and the groundwater balance does not yield significant negative change in storage, yet groundwater extraction should not be accelerated. Otherwise, the reservoir will be overexploited.

12.6 FEFLOW

A Finite-Element simulation package for 3D and 2D density-dependent **FLOW**, mass and heat transport processes in groundwater

Hans J. G. Diersch (H.Diersch@wasy.de)

12.6.1 Description

FEFLOW is a finite-element package for simulating 3D and 2D fluid density-coupled flow, contaminant mass (salinity) and heat transport in the subsurface. It is capable of computing:

- Groundwater systems with and without free surfaces (phreatic aquifers, perched water tables, moving meshes);
- Problems in saturated-unsaturated zones;
- Both salinity-dependent and temperature-dependent transport phenomena (thermohaline flows);
- Complex geometric and parametric situations.

The package is fully graphics-based and interactive. Pre-, main- and post-processing are integrated. There is a data interface to GIS (Geographic Information System) and a programming interface. The implemented numerical features allow the solution of large problems. Adaptive techniques are incorporated.

12.6.2 Physical Concepts

1. Advection-dispersion and heat transfer equations are solved for two or three spatial dimensions.
2. Solute migration can be obtained for an unsaturated and/or saturated medium. For unsaturated conditions, the nonlinear Richards equation is solved for optional parametric models such as van Genuchten-Mualem, Brooks-Corey and Haverkamp. Alternatively, for 3D regional unconfined aquifers a multiple free surface (perched water table) approach is implemented with moving meshes.
3. Only one fluid (liquid) phase is assumed to be mobile.
4. Adsorption effects can be modeled via Henry, Freundlich or Langmuir isotherms.

5. The flow problem can be simulated either for transient or steady-state conditions.

6. The transport problem can by solved for either transient or steady-state conditions.

7. The transport process can be considered as isothermal or non-isothermal. Viscosity and fluid density effects due to salinity and temperature can be considered. Boussinesq and extended Boussinesq approximations are selectable. The latter is appropriate for strong density coupling (large density contrasts). Furthermore, a variable density expansion for temperature effects permits the solution of problems with large temperature ranges and allows one to tackle the 4°C anomaly (cold groundwater, also under variable saturation). Salinity and heat can be simultaneously coupled as given for thermohaline flow (double diffusion convection DDC).

8. The model is based on the transport of one solute.

9. A decay rate and a zero order reaction term can be specified. Non-reactive processes represent a special case within this general approach.

12.6.3 Computational Concepts

1. FEFLOW is capable of solving full 3D and 2D problems. Among 2D problems vertical plane, horizontal plane and axisymmetric domains can be accommodated.

2. A Galerkin-based finite-element method for unstructured meshes is used. For convection-dominant transport problems upwind techniques are available, such as streamline upwinding and shock capturing. For 3D problems hexahedral and pentahedral elements using trilinear and triparabolic approximation are available. For 2D problems quadrilateral and triangular elements of bilinear and biparabolic accuracy are available. Different time marching schemes can be chosen. For example, one can use automatic time stepping schemes based on the Gresho-Lee-Sani (GLS) predictor-corrector time integrator for second order (Adams-Bashforth/trapezoidal rule) or, alternatively first order (forward Euler/backward Euler), Crank-Nicolson, or fully implicit schemes at fixed (user-defined) time increments. A fully adaptive approach is provided for 2D that

is based on mesh refinement and derefinement (AMR) techniques required to enhance the reliability of the numerical simulation. An entropy-based a-posteriori error estimator controls the spatial discretization by using optimality criteria given by Zienkiewicz & Zhu and Onate & Bugeda. Consistent and lumped mass matrices are optional (the latter is recommended for unsaturated media).

3. The 2D and 3D codes can accommodate Dirichlet (1st type), Neumann (2nd type) and Cauchy (3rd type) boundary conditions that are specified for flow, mass, and heat. A so-called 4th type boundary condition exists for singular (pumping or injection) wells. All boundary conditions can be specified either as steady-state or as transient and can account for constraint formulations.

4. The boundary conditions are node-related while material parameters are element-related. Both nodal and elemental quantities can be transient if desired.

5. Rainfall or evaporation (areal groundwater recharge) are normally modeled by sink/source formulations. They may also be time-dependent. Pumping or injection conditions of singular wells (occurring at selected nodal points) are described by 4th type boundary conditions. Multi-layer pumping or injection wells are also accommodated in 3D.

6. All parameters are handled on an element level, i.e., they can differ from element to element and it is possible to consider them as steady-state or transient quantities (e.g., conductivity can be changed with time). The input and assignment of the parameter (including such distributed parameters as capillary pressure or relative conductivity curves with all related fitting coefficients) is done via a graphical problem attribute editor. A general programming interface allows the user to implement his/her own rules for parameter assignment and generation (e.g., random fields, alternative kriging techniques).

7. For symmetric equations the preconditioned conjugate gradient (PCG) is provided. Standard preconditioners such as the incomplete factorization (IF) technique and alternatively a modified incomplete factorization (MIF) technique based on the Gustafsson algorithm are used. Different alternatives are

available for the CG-like solution of the unsymmetric transport equations: a restartable ORTHOMIN (orthogonalization-minimization) method, a restartable GMRES (generalized minimal residual) technique and Lanczos-type methods such as CGS (conjugate gradient square), BiCGSTAB (bi-conjugate gradient stable) and BiCGSTABP (post-conditioned bi-conjugate gradient stable). For preconditioning, the incomplete Crout decomposition scheme is applied. For small (2D) or ill-posed problems direct Gaussian elimination techniques for symmetric and unsymmetric matrices are available. Here, the Reverse Cuthill-McKee (RCM) nodal reordering technique is used to minimize fill-in entries of the matrices.

8. Temporal discretization (predictor-corrector versus marching with fixed predefined time steps), Newton or Picard iterative techniques are employed. The full Newton method is preferred as a first choice for the transport equations when adaptive techniques are applied. Picard iteration is used in combination with the relaxation of the primary variables, fixed time stepping, or steady-state approaches.

12.6.4   Other Issues

1. FEFLOW contains interactive graphical editors and mesh generators. It provides an open data interface for importing, exporting (GIS interface) and programming (interface manager IFM) as well as encompasses many graphical tools in the post-processing analysis of the results. There is a layer configurator to generate, refine and reposition stratigraphic and layer subdivisions as well as to admit data.

2. The code is written in C and C^{++} and currently encompasses more than one million code statements. FEFLOW is available for UNIX and PC.

3. With regard to saltwater intrusion and density-dependent problems, FEFLOW has been verified against cellular-free convection problems (Elder problem, Benard problem) for solute, and thermal and thermohaline processes in 2D and 3D, the salt-dome problem (HYDROCOIN case 5 level 1) and Henry's problem (saltwater encroachment), saltwater upconing below pumping wells (comparison with sharp-interface model results and experimental data), development of interception techniques

(so-called "saltwater prevention wells"), and flushing of saltwater horizons (comparison with analytical solutions).

4. FEFLOW incorporates an advanced interactive graphical working environment. It integrates powerful and modern numerical techniques together with tools highly useful in 'daily' handling of data and computational results. FEFLOW is not a small program. It requires more resources and computational power from the hardware to take advantage of the implemented features of the software. Nevertheless, the design of the package follows the standard capabilities of graphics workstations and PCs that are available today and is not necessarily dependent on high-end or super-computing technology.

12.6.5 Case History

P. Perrochet (pierre.perrochet@epfl.ch)
L. Tacher

1. FEFLOW was applied to the nuclear testing site of Mururoa Atoll (French Polynesia).

2. Computational investigations were performed to assess the hydrogeological effects of underground nuclear explosions focusing on the evolution of density-driven advective-dispersive processes through the Atoll. A variety of local and regional scale, short and long term 2D and 3D models were designed in order to gain some insights into the significance of test-induced perturbations.

3. The Atoll consists of a volcanic basement above the ocean crust and is covered by carbonate formations with a relatively high permeability. The thickness of the carbonates can reach a few hundred meters with, at the top, an emerging coral rim separating the ocean from a 50 m deep lagoon. The porous structure of the Atoll is fully saturated with seawater, except below the rim where a freshwater lens develops at depth. Groundwater circulation in the Atoll is essentially governed by the buoyancy forces arising from the combination of salinity with the geothermal flux heating the system from below. Under virtually equal and constant ocean and lagoon free surface elevations, cold and denser ocean waters penetrate from the flanks

of the Atoll, flow toward the central warmer regions, gradually heat up and move upward toward the lagoon. This phenomenon is particularly marked in the carbonate formations where the high permeabilities allow important, almost horizontal centripetal fluxes of cold water. These fluxes generate thermal anomalies and can occur over large distances until sufficient heat exchanges at the volcanic interface drive the water back to the surface. In principle, nuclear devices were detonated in the low permeability volcanic formations at sufficient depths to ensure containment of the radionuclides produced. Induced fracturing in a large volume around, and subsequent creation of a rubble-filled collapse chimney above the location of each explosion are the major hydrogeological concerns. With full water saturation and the large amount of heat generated by the explosion, highly convective circulations develop in the chimney and its fractured vicinity. Hence, the risk is one of relatively fast migration paths for radionuclides intersecting high permeability carbonate layers, and from there one of transport to the lagoon or the ocean.

4. A mathematical model including higher order salinity-temperature-dependent density and viscosity functions was used. Dense and heterogeneous meshes (up to a few hundreds of kilo-nodes) were designed to account for the natural geological settings as well as the damaged zones due to the explosions.

5. Pre- and post-test rock permeabilities were calibrated on the basis of observed temperature profiles and re-filling times of collapse chimneys. Other unknown parameters were given standard values and submitted to sensitivity analysis. Boundary conditions consisted of imposed geothermal flux at the bottom (-2000 m) of the model, imposed salt concentrations, temperatures and density-corrected pressures on the flanks of the Atoll and at the bottom of the lagoon, and infiltration at the rim surface.

6. At Atoll scale, centripetal buoyancy-driven circulations were demonstrated. Calibrated, unperturbed steady-state Darcy fluxes range from a few mm/yr in the volcanics to a few m/yr in the carbonates. Maximum groundwater ages in the volcanic are of the order of 100,000 yr. Horizontal inflows in karstic horizons can reach 10 to 20 m/yr. At the scale of a single explosion,

Darcy fluxes are temporarily vertical above the collapse chimney and can reach a few tens of m/yr during the cooling period (a few tens of years). Total steady-state groundwater discharge in the lagoon is of the order of 60,000 m^3/day, and is increased by less than 1% when all the damaged zones are included in the model.

12.7   HST3D

Heat and Solute Transport in 3 Dimensional groundwater systems

Ken Kipp (klkipp@usgs.gov)

12.7.1   Description

The HST3D code is based upon a finite-volume discretization on a rectangular mesh of the three-dimensional advective-dispersive transport equations.

It provides a transient simulation of non-isothermal flow in a saturated aquifer with transport of a single solute with possible linear adsorption and linear decay. Steady-state results may be obtained by marching out in time. The resulting set of algebraic equations can be solved with a choice of two solvers and the non-linear terms are approximated using Picard iteration. Standard boundary conditions are accommodated and first and second type conditions are applicable on the interior. The code runs on a UNIX workstation and on WINTEL PC machines. Several input-output options exist with spatial data being input by coordinate location. As with most finite-difference formulations, numerical dispersion or oscillations are possible under certain mesh and time-step conditions. It has been observed that the Picard iteration can perform poorly when there are wide density variations.

12.7.2   Physical Concepts

1. Simulation is via the advective-dispersive transport equation.
2. Saturated aquifers are assumed.
3. The model is restricted to single-phase flow.
4. Linear adsorption is accommodated.
5. Transient and steady flow are considered.

6. Transient transport is simulated.

7. Steady-state may be achieved by marching out in time.

8. Non-isothermal conditions are accommodated.

9. Single-component transport is permitted.

10. The model allows for linear decay of the solute component.

12.7.3   Computational Concepts

1. The model is fully three dimensional.

2. The numerical procedure used is finite volumes.

3. First, second and third type boundary conditions can be used.

4. First and second-type conditions can also be applied on the interior of the domain.

5. Point and area sources can be specified.

6. Leakage, evapotranspiration, and aquifer influence function boundary conditions are available.

7. Well bore and well riser models are available.

8. Parameter input allows for heterogeneous aquifers.

9. Algebraic equations are solved using generalized conjugate gradient (with Schur complement) or direct elimination.

10. Picard iteration is used for the non-linear set of equations.

12.7.4   Other Issues

1. The ARGUS-ONE graphical interface may be used for input and output.

2. The model runs under UNIX and Windows.

3. The code has been verified by comparison to analytical and numerical solutions, and by examination of the mass balance.

4. The organization of input and output, especially using the ARGUS graphical interface is very effective.

5. The code allows for variable viscosity.

6. Because the finite-volume method is a variant on finite differences, numerical dispersion and oscillatory solutions are possible under certain conditions.

7. When wide density variations are encountered, the Picard method of linearization can perform poorly.

12.7.5   Case History

1. Bonneville Salt Flats, Western Utah: the Bonneville Salt Flats study area is located in the western part of the Great Salt Lake Desert of northwestern Utah.

2. A hydrologic investigation of the Bonneville Salt Flats was done over three years by the U.S. Geological Survey Water Resources Division, in cooperation with the Bureau of Land Management, U.S. Department of Interior. A loss of salt crust had been documented over a 28-year period, but no definitive data had shown what processes were involved. The Bureau of Land Management needed to know the dynamics of the salt loss to make management decisions. The purpose of this investigation was to provide them with data and interpretations concerning the movement of salt through the present hydrologic system.

   The specific objectives of this investigation were to define the hydrology of the brine groundwater system and the natural and human processes causing salt loss and, where feasible, quantify these processes. Seasonal and annual variability in salt transport precluded any rigorous quantification. Lack of long-term data prevented assessment of climatic trends and rates of mineral production.

   The primary goals of the ground-water flow and solute transport modeling were to:

   (a) develop a fluid and solute balance for the ground-water system in the study region;

   (b) evaluate the effect on the salt crust of the brine production from the ditches; and

   (c) identify the major and minor solute fluxes to or from the system.

3. The groundwater system in the Bonneville Salt Flats study area consists of an alluvial-fan aquifer located along the margin of the basin, a deep basin-fill aquifer, and a shallow-brine aquifer. The shallow-brine aquifer is composed of a halite and gypsum crust at the surface surrounded by carbonate mud, and contains the most valuable concentrated brines. Evaporation of

these brines contributes to the perennial salt crust. Rainfall on the playa surfaces during summer and winter results in the dissolution of salt and subsequent seepage into the subsurface.

Water and brine enter the system from:

(a) recharge of precipitation; and

(b) groundwater inflow from adjacent aquifer areas to the north, east, and west.

Brine leaves the system by:

(a) evaporation leaving a salt crust deposit;

(b) groundwater outflow at the south boundary to the adjacent aquifer area;

(c) surface-water outflow by the production ditch;

(d) groundwater outflow to the adjacent alluvial-fan aquifer along the flank of the Silver Island Mountains; and

(e) leakage to the underlying basin-fill aquifer.

Ponding occurs over the lower areas of the salt crust in response to seasonal precipitation and reduction in evaporation rate during the winter months.

4. The three-dimensional Heat and Solute Transport (HST3D) computer code [Kipp, 1987, 1997] of the U.S. Geological Survey was the simulator used to model the shallow-brine aquifer underlying the Bonneville Salt Flats. Effects of temperature variations on the density of the brine were assumed to be negligible for study purposes.

5. The shallow-brine aquifer is a fractured medium with fissures up to 1 inch in width extending down to 25 ft with fissure spacing in polygonal patterns of 3 to 200 ft across. The assumption of an equivalent porous medium was made because the density of fractures was high enough to be represented as distributed porosity and permeability for the purposes of this model.

The available data showed considerable annual variation of precipitation, evaporation, ditch production, and other hydrologic conditions. Model calibration by history matching was not done because insufficient data were available to define initial conditions, and boundary conditions show transient variations on a much shorter time scale (days to weeks) than the time scale (years to decades) of the transport mechanisms pertinent to this study.

For simulation purposes, a synthetic seasonal sequence was formulated representing a typical year split into a six-month summer-fall season of production and evaporation followed by a six-month winter-spring season of recovery and recharge. A periodic steady state was simulated representing a typical climatic year with average brine production volume from the collection ditch.

6. Results of the simulations from the calibrated model included water-table elevations, density fields, and cumulative fluid volume and solute mass amounts through the boundary surfaces. The following balances were computed:

   (a) a flow balance on the shallow-brine aquifer under the Bonneville Salt Flats and

   (b) a mass balance on the total crystalline salt deposit.

Primary fluid recharge to the shallow-brine aquifer is by infiltration of rainfall through the playa surface which dissolves salt as it infiltrates. Major flow rates are through the land surface, production ditch, and south boundary. A small quantity of subsurface inflow from the east and north boundaries contributes an insignificant quantity of salt to the shallow brine aquifer. Ponding is an important hydrological feature which had to be included in the model. The high transmissivity values at the south boundary interpreted by a previous study appear inconsistent with the equivalent transmissivity values of the calibrated model, which are about 1/6 of the former. The flow system does not reach steady state during the production or recovery seasons and the model is very sensitive to net precipitation during the recovery season and net evaporation during the production season.

This model shows the effects of brine production on the solute and salt crust balance of this system, and supports the concept that brine production is a major cause of salt-crust loss. Eliminating brine production from Federal leases would decrease but not eliminate the disappearance of salt crust from the Bonneville Salt Flats.

12.8 MEL2DSLT

Modified Eulerian–Lagrangian, method for **2-D** horizontal plane simulation of **SaLT** water intrusion problems

Shaul Sorek (shaul@hydro.boker.bgu.ac.il)
Viacheslav Borisov
Alex Yakirevich

12.8.1 Description

The MEL2DSLT code is based on the Modified Eulerian–Lagrangian (MEL) formulation implemented on the 2-D horizontal (i.e. averaging along the vertical depth) equations of flow, transport and heat transfer of a phreatic aquifer. It is specifically aimed at simulating the temporal and regional variations of the water level, solute concentration and the fluid temperature within a heterogeneous medium. The temporal and regional variations of the vertical depth of the saltwater body is also estimated.

The following principal assumptions constitute the conceptual model:

- The flow is limited to the saturated zone of the phreatic aquifer.
- The water carries a non-volatile contaminant that can be adsorbed on the solid matrix as well as be subject to decay reactions.
- The matrix can undergo small deformations.
- Liquid density and viscosity depend on the solute concentration and temperature. Solid density remains practically constant.
- Solid velocity is negligible in comparison to that of the liquid.
- Liquid and solid temperatures are practically equal.
- The flow is essentially horizontal.

12.8.2 Physical Concepts

1. Solute transport is governed by the advection–dispersion equation.

2. Fluid flow, solute transport and heat transfer are estimated for the saturated zone of a phreatic aquifer.

3. Only one fluid (liquid) phase is assumed to be mobile.

4. Solute is assumed to be adsorbed on the solid matrix. Adsorption is governed by a linear equilibrium isotherm.

5. Transient conditions are employed for the simulation of the flow problem.

6. Transient conditions are employed for simulating transport and heat transfer.

7. Isothermal and non-isothermal conditions can be implemented.

8. The model is based on the transport of one solute.

9. Generation of solute mass due to chemical reaction between different components is not considered. The radioactive-decay process is accommodated.

12.8.3 Computational Concepts

1. Simulations are done for the two-dimensional horizontal plane. In addition, the vertical depth of the saltwater body can be estimated on the basis of the averaged sharp interface concepts.

2. An explicit finite difference scheme is employed.

3. Boundary conditions of the first, second and third type can be accommodated for simulating flow, transport and heat transfer problems.

4. Flux, groundwater levels, concentration and temperature values can be prescribed as internal conditions at specified nodes on the interior of the mesh.

5. Stresses such as rainfall, evapotranspiration and pumping/ recharge can be accommodated.

6. Heterogeneity can be specified in terms of the saturated, anisotropic, hydraulic and heat conductivities, specific yield, porosity and the longitudinal and transverse dispersivities.

7. The set of linear algebraic equations resulting from the coupled flow, transport and heat transfer equations, are solved by the LU-decomposition method.

8. The explicit scheme dictates conditioned time steps thus there is no need for any iterations involving methods such as Picard, Newton-Raphson etc.

12.8.4   Other Issues

1. There is no graphical interface incorporated in the MEL2DSLT code.

2. The computer code uses the FORTRAN compiler and can be run under PC or UNIX operating systems.

3. The code was verified against analytical solutions. The model was also tested against a numerical solution for a cross-sectional problem of saltwater intrusion simulated by the SUTRA code. The solute concentrations obtained by the SUTRA code were averaged over the depth and the pressure distribution was taken in the phreatic zone (namely below the boundary along which pressure equals zero). The comparison was done on the basis of a 1-D flow and transport problem.

4. In contrast to localized, vertical-section simulations, the MEL2DSLT code addresses 2-D horizontal, regional, simulations. In addition, it also provides estimates of the vertical distribution of saltwater bodies. The incorporated MEL method accounts specifically for spatial heterogeneities in the aquifer. The 2-D horizontal model requires the estimate of two additional parameters (although, practically, almost constant) to accommodate for the variations of the vertical water velocities along the aquifer depth.

5. The MEL method is employed for the flow, transport and heat transfer equations.

12.8.5   Case History

1. The code was implemented in a case study of saltwater intrusion into the Gaza aquifer, focusing on the Khan Yunis area.

2. The Gaza strip coastal phreatic aquifer is under severe hydrological stress due to over exploitation and anthropogenic land-usage. An excess of pumpage over the past decades in the Gaza region has created a significant lowering of groundwater level, altering in some regions the normal washing of salt into the sea and the groundwater flow. The model was used to assess the extent of subsequent saltwater intrusion under an assumption that the pumpage rate would increase in accordance with a population growth of about 3.8% a year.

3. The coastal aquifer of the Gaza Strip region is a Pleistocene granular aquifer. The aquifer is composed of different layers of dune sand, sandstone, calcareous sandstone, silt, and intercalations of clays and loams which can appear as lenses. Most of the lenses begin at the coast and stretch to approximately 4 km from the sea, separating the aquifer into various subaquifers. The aquifer is built upon a base of sea clays and shales. Up to 5 km from the seashore, the exploitation of the aquifer is mainly limited to the upper subaquifer where there is still saltwater with a relatively low rate of salinity increase.

4. The investigated area is densely populated with pumping wells. To allow the prediction of different pumping/injection scenarios on the region, we implemented the 2-D horizontal MEL2DSLT code.

5. Observations regarding concentration values were scarce and not reliable. Data was not provided for all the wells at the same time. The model was, hence, calibrated using groundwater level data from the years 1985–1991. This provided 84 files (12 months × 7 years) of "slices" representing a random field of groundwater levels at 84 time points. The study area (8km×8km) was divided into rectangular blocks 200m×200m, altogether 41 × 41 = 1681 nodes. We used a trial and error process to match the simulated mean of the groundwater level and the measured average one, corrected using an estimate of the variance of the random field of these observations. The observed distributions of the average values of the groundwater levels and the chloride concentrations (at the beginning of 1985) were assigned as initial conditions. Dirichlet boundary conditions for the groundwater levels and chloride concentrations were assigned at the boundaries. The aquifer was subject to natural replenishment and pumping.

6. At first we ran a transient simulation for a period of about 200 years until variations in groundwater levels and concentration were negligible. The results of this run were used as the initial conditions. Following this a second set of simulations was run using the calibrated values of the pumping rate for the year 1985. Pumping was extrapolated on the same basis as the

annual population growth (approximately equal to 3.8%). Simulation was carried out for the period from the years 1950 to 2006. Results indicate a considerable decrease in groundwater levels during the last and especially the next ten years. In the year 2006, the simulated groundwater level in the Khan Yunis city (3–4 km from sea shore) is predicted to be 4–6 m below sea level. Expected encroachment of saline water with concentrations higher than 0.5 g/kg will be about 800 m inland.

12.9 MLAEM/VD

Multi-Layer Analytic Element Method Variable Density code

Otto D. L. Strack (Strack.Inc@Worldnet.att.net)
Wim J. de Lange (W.dLange@RIZA.RWS.minvenw.nl)

12.9.1 Description

MLAEM/VD is the variable density version of the analytic element model for multiple-layers, which is based on the superposition of analytic functions, each representing a particular geohydrologic feature in an infinite aquifer. Analytic functions are used to meet boundary conditions at points, along straight and curved lines, and inside areas. Apart from the classical function for a well, functions exist for almost any geohydrologic feature such as inhomogeneities, areal infiltration and leakage across resistance layers, narrow obstructions to flow, rivers, canals and polders. The analytic-element method can handle extremely different scales within one model, enables flawless coupling of models and allows refinement, redefinition and reshaping of all elements in any area within an existing model. MLAEM/VD simulates fully three-dimensional density-driven flow within the aquifers of an incompressible quasi-three-dimensional model for saturated groundwater flow. The changes over time of the density distribution can be simulated by numerical integration through time. MLAEM/VD has been tested against analytic solutions and has been extensively used for the well-calibrated models of the complex coastal area of the Netherlands.

12.9.2   Physical Concepts

1. Solute transport is governed by the advection equation.
2. Fluid flow and solute transport are estimated within the saturated zone.
3. Only one fluid (liquid) phase is assumed to be mobile.
4. Solute is assumed to be adsorbed and conservative.
5. Flow is assumed incompressible and (semi-)confined in the aquifers in which density-driven flow is computed.
6. Solute transport is assumed steady under successive steady state conditions.
7. Simulation is based on isothermal conditions.
8. The model is based on the transport of one solute.
9. Modification of solute mass is considered in terms of radioactive decay. Chemical reaction between different components is not considered.

12.9.3   Computational Concepts

1. Simulations are done for quasi three-dimensional, including fully three-dimensional continuous density-driven flow.
2. Computation follows the Analytic Element Method (AEM). This is based on the superposition of (non-standard) analytic solutions for 2-D flow (i.e. analytic elements). One solves for the unknown strengths (fluxes, dipole strengths). There is no element mesh, but the position of any element is exactly determined based on the geohydrological feature to be represented.
3. Boundary conditions of the first, second and third type are accommodated for simulating flow and transport problems. The top of the system is defined by area elements with constant or space-variable flux rates (first and third type including a stepwise-linear relationship between potential and flux). Horizontal boundaries can be defined using line-elements for the three boundary types.
4. The three types of boundary conditions can be applied at internal, prescribed, locations. In aquifers: points and lines with potential, flux, and a linear relation between flux and resistance. In aquitards: area elements similar to those used on the top of the system as well as the third type elements for the leakage between aquifers.

5. The code accommodates stresses in terms of changes in flux, head values, groundwater recharge and surface water level.

6. Input consists of point-wise geohydrological parameters which are interpolated using the multi-quadric method, which is a radial basis interpolator.

7. Equations are solved by the LDU-decomposition.

8. Non-linear systems are linearized using a Newton Raphson like approach.

12.9.4   Other Issues

1. The MLAEM/VD code uses a proprietary graphical interface which includes import and export facilities in AUTOCAD-type file format and enables the use of SURFER for post-processing.

2. Platforms and operating systems include PCs (WINDOWS 95/NT) and UNIX.

3. The code was verified against analytical benchmarks.

4. The code yields the exact values of potentials and their derivatives at any spatial point using analytic expressions, water balance is exact for any volume and the 3-D density dependent flow is included in the exact mathematical expression.

12.9.5   Case History

1. The MLAEM/VD code was especially, but not exclusively, developed for the NAtional GROundwater Model (NAGROM) of the Netherlands program, which consists of nine sub-models, four of which accommodate density variation and five that do not. Along the coastal part $(250 \times 25 \text{ km}^2)$ of the Netherlands variation in density due to variation in chloride content significantly influences groundwater flow. In this area a three-dimensional continuous density distribution was obtained based on interpolation of measured data.

2. The major objective was to analyze the national water-management strategies (such as the reduction of the area with a desiccated nature or agriculture) by means of a combination of the redistribution of drinking water supply, surface water level changes, changes of vegetation (land use) and changes in drainage systems. The effects of climate changes on the seepage

of salty groundwater has been computed for the entire country. In a region in the northern part of the country, the expected changes in the salt distribution of the groundwater was investigated for a period of a century.

3. The coastal aquifer is formed of thick Pleistocene sand layers above which different aquitards and aquifers are present only in parts of the area. The top of the system consists of (generally thick) clay and peat layers in which artificial drainage systems maintain the (surface and ground) water levels (polder areas).

5. Three aquifers and aquitards were modeled, including a continuous three-dimensional density distribution based on chloride contents in more than 10,000 (3-d distributed) measurement points. The interpolated distribution was directly included in the MLAEM/VD code and resulted in a three-dimensional continuous distribution of heads and fluxes. A third type boundary condition was used to describe the confining layers and the water levels. Variations in the transmissivities of the aquifers and the resistances of the aquitards were included as well as hundreds of wells. The 1–2 km wide dunal area, in which artificial infiltration for drinking water supply occurs, has been lumped, due to the focus on national-scale problems. Calibration was done against more than 7000 head measurements averaged over a period of six weeks in the wet season of 1990.

6. For more than 90% of the points the model is within 1 m of the measured values and in more than 75% within 0.5 m.

12.10 MOCDENSE

Ward E. Sanford
Leonard F. Konikow (lkonikow@wrdmail.er.usgs.gov)

12.10.1 Description

The MOCDENSE code is based upon a finite-difference representation of flow and a method-of-characteristics formulation for transport. It accommodates variable density advective transport in a two-dimensional cross section. Fluid flow is isothermal and steady or transient and transport is transient. The model applies to saturated flow and does not incorporate adsorption. Transport considers two components which are non-reactive. Several methods are available to solve the approximating equations. The algorithm is non-iterative for non-linear conditions but there is a user-specified tolerance for relative concentration change that will assure that the pressures are recalculated for the subsequent time increment if the density field has changed significantly. The code is considered more efficient than standard finite-difference approaches because it can accommodate a coarser mesh. The numerical approach provides challenges for highly divergent or convergent flow and does not guarantee an exact mass balance.

12.10.2 Physical Concepts

1. Simulation is via the advective-dispersive transport equation.
2. Saturated and unsaturated aquifers are accommodated.
3. The model is restricted to single-phase flow.
4. Transport is conservative.
5. Steady or transient flow is assumed.
6. Transient transport is accommodated.
7. Isothermal conditions are assumed.
8. Two-component transport is permitted.
9. The model does not allow for chemical reactions.

12.10.3 Computational Concepts

1. The model assumes a two-dimensional cross section.
2. The numerical procedure used is finite difference for flow and method of characteristics for transport.
3. First, second and third type boundary conditions can be used for flow and first type for transport.
4. First and second type-conditions can also be applied on the interior of the domain for flow and first for transport.
5. Point, line and area sources are accommodated.
6. Parameter input is nonhomogeneous.
7. Algebraic equations are solved using SIP or a D4 direct solver for flow, method of characteristics for advection and an explicit finite difference method for dispersion.

8. When a non-linear set of equations is encountered a non-iterative approach is used (see description above).

12.10.4 Other Issues

1. A text-based interface called PREDENSE, which is a menu-driven preprocessor for input data, is available.

2. The model runs under either a DOS or UNIX operating system.

3. The code has been verified against the Henry analytic solution and other numerical codes.

4. The primary advantage of this method is that it is more efficient than standard finite-difference codes for advection-dispersion because of accommodation of a coarser mesh. The challenges involve simulation of highly divergent or convergent flow and the fact that an exact mass balance is not guaranteed.

12.10.5 Case History

1. Donana National Park, Southwestern Spain.

2. The objectives were to develop a better understanding of:

   (a) transport processes governing the chemical composition of groundwater in a low-permeability confining layer, and

   (b) interactions between the confining layer and the regional groundwater-flow system.

3. The hydrological setting is a wetlands area near the mouth of the Guadalquiver River. The area is underlain by fluvial, marine, and estuarine sands, silts, and clays. Nearly impervious marls underlie the aquifer system.

4. The model employed was MOCDENSE.

5. The model was set up as a one-dimensional vertical grid, using specified pressure boundary conditions at the top and bottom of the column, and assuming as initial conditions that the confining layer was filled with either freshwater or marine saltwater. The sensitivity of the solution to variations in hydraulic conductivity and effective diffusion coefficient were evaluated. The evolution of the salinity profile under various conditions was simulated over tens of thousands of years.

6. Results indicated that at present there is probably an upward flow of the order of 1 mm per year to 1 cm per year and a delicate balance in the solute flux through the confining layer between upward advection (yielding evaporative salt deposits at the land surface) and downward diffusion (into the underlying freshwater aquifer). A small change in boundary conditions may drive the system in one direction or the other, and small changes in the average salinity in the confining layer may cause the flow direction to reverse. Such a flow reversal probably occurred in the past in response to sea-level changes or other long-term environmental stresses.

12.11   MOC-DENSITY/MOCDENS3D

(MOC-DENSITY)
Luc C. Lebbe (Luc.Lebbe@rug.ac.be)

(MOC-DENSITY and MOCDENS3D)
Gualbert H. P. Oude Essink (g.oude.essink@geof.ruu.nl)

12.11.1   Description

The original MOC code (also called USGS 2-D TRANSPORT) has been modified to simulate density-dependent groundwater flow in vertical profiles. The 3-D code MOC3D, which is interconnected with the modified MODFLOW (for solving the flow problem) module, is adapted for density differences and is called MOCDENS3D. In both codes, differences in so-called freshwater heads instead of pressures are used to determine groundwater flow.

12.11.2   Physical Concepts

1. Solute transport is governed by the advection–dispersion equation.

2. Solute transport is estimated for a saturated confined aquifer.

3. Only one fluid (liquid) phase is assumed to be mobile.

4. Solute is assumed to be adsorbed and conservative.

5. Transient conditions are employed to simulate the flow problem.

6. Transient conditions are also employed to simulate transport.

7. Isothermal conditions are implemented.

8. The models are based on the transport of one solute.

9. Generation of solute mass due to chemical reaction is considered in terms of radioactive decay and sorption (for MOC-DENSITY also ion exchange).

12.11.3 Computational Concepts

1. The MOC-DENSITY code enables simulations of 2-D vertical cross-sections. The MOCDENS3D enables simulations for 3-D spatial coordinates.

2. The finite-difference method is applied to the flow equation. The MOCDENS3D code is interconnected with an adapted version of MODFLOW. Solute transport is solved by the method of characteristics for the advection part (by means of a particle tracking procedure) and by finite difference for the dispersive transport.

3. Boundary conditions of all three types can be accommodated for simulating flow and transport problems.

4. Flux and potential values can be prescribed as internal conditions at specified nodes on the interior of the mesh.

5. Stresses such as groundwater recharge, pumping/infiltration can be accommodated. The MOCDENS3D code can also accommodate rivers, drains and evapotranspiration stresses.

6. Heterogeneity can be specified in terms of the saturated, anisotropic, hydraulic conductivity.

7. Solution is obtained for the MOC-DENSITY code by the Iterative Alternating-Direction Implicit procedure (ADI) for groundwater flow. The MOCDENS3D code applies the Strongly Implicit Procedure (SIP), Slice-Successive Overrelaxation (SSOR) and the Preconditioned Conjugate-Gradient 2 (PCG2) procedures for groundwater flow. After each time step the groundwater flow is re-adjusted taking into account the obtained distribution of fresh and salt water. For both codes, solute transport is solved by the method of characteristics.

8. The system is linear thus no iterations are needed.

12.11.4   Other Issues

1. MOCGRAPH 1.1 is a graphical interface developed specifically to display MOC-DENSITY and MOCDENS3D output.

2. The computer codes can be run under PC or UNIX operating systems.

3. MOC-DENSITY is verified with Henry's problem, whereas both codes are verified with the analytical solution of a sharp interface problem. A comparison is made between the results of MOC-DENSITY, SUTRA and SWICHA for a case in a sand-dune area.

4. Numerical errors are small in the case of high Peclet numbers. MOC-DENSITY accommodates only constant spatial values of porosities and dispersivities, whereas in MOCDENS3D spatial values of porosities and dispersivities can vary between layers. Under certain conditions, the particle tracking procedure causes significant errors in the solute mass balance.

12.11.5   Case History 1

Luc C. Lebbe (Luc.Lebbe@rug.ac.be)

1. Salt-fresh water flow under the shore at the French-Belgian border.

2. The first version of MOC-DENSITY was implemented to a case study of a particular salt-fresh water distribution in a phreatic aquifer under the dunes, the shore and the sea at the French-Belgian border. The purpose was to simulate the natural evolution of a large saltwater inversion under a shore with semi-diurnal tides as well as to study the over-exploitation of the freshwater lens in the dune area at the French-Belgian border.

3. At the studied cross section, the thickness of the aquifer ranges between 25 and 30 m and is rather homogeneous. It consists mainly of sands of medium size, at the base, which change gradually towards the top from well-sorted medium to fine sands. The substratum of the phreatic aquifer consists of a thick heavy clay of more than a 100 m thick. The central part of the studied cross section, is situated under a sandy runnel and ridge shore. The difference in level between the high-high water and the

low-low water line, is about 5 m and the horizontal distance between the high-high water line and the low-low water line is about 420 m. The second largest part is situated under the dune area and the smallest part under the sea.

4. To explain and to simulate this particular saltwater inversion under the shore, the original code of Konikow and Bredehoeft [1978] was converted to simulate density driven flow in a two-dimensional vertical cross section.

5. The lower boundary of the aquifer and the vertical boundary that coincides with the water divide line in the dune area, were impervious. The upper boundary under the dunes was a constant vertical flow boundary where freshwater infiltrates with a rate of 7.39×10^{-4} m/day. A constant freshwater head was prescribed at the upper boundary under the shore and the sea. The values of the freshwater heads under the sea correspond with the mean sea level. Under the shore the freshwater heads increase from the sea in the landward direction. The prescribed freshwater heads under the shore were derived by interpolation of the observed fluctuation means in five wells with screens which are only a few meters under the ground surface. The saltwater percentage of the infiltrated water on the shore is 100%. The seaward vertical boundary was also taken as a constant hydraulic head boundary. The prescribed freshwater head values correspond with the mean sea level and no vertical flow of pure saltwater was assumed. The saltwater percentage of the water which flows through this vertical boundary into the model was also 100%. To simulate the growth of the freshwater lens under the dunes and the saltwater above the freshwater tongue under the shore, it was supposed that the aquifer is filled with saltwater. This was based on the fact that under the eolian dune all sediments of the phreatic aquifer are tidal flat deposits.

6. Calibration was done by a trial and error procedure. Calculations were continued until the salt-fresh water distribution, which reaches a dynamic equilibrium after 500 years, corresponded at first glance with the one that was derived from the borehole resistivity measurements. The optimal values were 10.1 and 5.62 m/day for the horizontal and the vertical conductivity, respectively, 7.01 and 1.59 mm for the longitudinal and

transverse dispersivity, respectively, and 0.363 for the effective porosity. After 500 years of simulation, the freshwater head configuration and the saltwater percentage distribution under the dunes, shore and sea, did not change significantly and was assumed as the initial condition. The lower part of the phreatic aquifer under the shore was filled by freshwater and a saltwater lens occurred above it. This lens was due to the infiltration of saltwater on the upper part of the shore during the high tides. It flows rather quickly in a shallow cycle to the lower part of the shore. There, an important outflow of saltwater occurred especially during the low tides. The freshwater, which flows from the dunes towards the sea, left the aquifer at the seaward side of the low-low water line. On this basis, simulation was done to predict the saltwater intrusion due to the overexploitation of the freshwater lens under the dunes. When this started the saltwater distribution changed significantly. Under the high-water line, the saltwater that infiltrated on the upper part of the shore, started to flow in the landward direction in the upper part of the aquifer. Then this saltwater intrusion sank towards the lower part of the aquifer due to the higher density of the saltwater and due to the low horizontal flow velocity at that place. Finally, the saltwater intrusion advanced in the lower part of the aquifer towards the catchment. Hence, the main threat of saltwater intrusion came from the high waters at the high-water line and not from the mean sea level at the mean-water line. Under the shore, an isolated brackish lens was formed in the vicinity of the low-water line. After a lapse of time this brackish lens became more and more salty and moved in the direction of the sea. This explained the existence of the brackish lens under the lower part of the shore before the urban area of De Panne. This isolated brackish lens was also observed with the aid of borehole resistivity measurements.

12.11.6   Case History 2

1. The MOC-DENSITY code was implemented to study saltwater intrusion into the coastal aquifer of Holland due to sea level rise. (See also Chapter 14.)

2. The purpose was to investigate through numerical modeling the possible impacts of sea level rise and human activities on

vulnerable coastal aquifers in the Netherlands for the next millennium.

3. The hydrogeological setting consisted of eight profiles perpendicular to the coast with several aquifers varying from 30 to 280 m thick, separated by clayey and loamy aquitards. Directly inland of the coast, sand-dune areas comprise freshwater lenses of some 40 to 90 m thickness. Further on, low-lying polders occur with phreatic groundwater levels of up to 6.5 meters below mean sea level.

4. Simulations were done subject to boundary conditions of sea level rise at the seaside boundary (several scenarios of sea level rise from −0.6 m/c up to 1.5 m/c), pumping wells in the sand-dune areas and constant phreatic groundwater level at polders. Stresses were in terms of groundwater extractions, deep-well infiltrations, land-reclamation, subsidence (lowering of the phreatic groundwater level), physical barriers and changes in natural groundwater recharge in the sand-dune area.

5. Results of the analysis indicated that no state of dynamic equilibrium is reached. The present salinization process is already severe and is generated by previous human activities during past centuries. It is to be expected that sea level rise significantly accelerates the salinization. Eventually, many hydrogeologic systems will become completely saline due to: (1) the future sea level rise and (2) the present difference in polder level and sea level.

12.12   SALTFRES

A mixed finite element model for 3-D **SALT**water intrusion problems

David G. Zeitoun (optimod@inter.net.il)
George F. Pinder

12.12.1   Description

The SALTFRES code is based upon a three-dimensional coupled finite element, finite difference model. It simulates saltwater intrusion in single and multiple coastal aquifer systems with either a confined or phreatic uppermost aquifer. The code is an extension of the PTC

(Princeton Transport Code) software developed at Princeton University. The numerical algorithm involves discretizing the domain into approximately parallel horizontal layers using finite elements. The layers are connected vertically by a finite-difference discretization. This hybrid coupling of the finite element and finite difference methods provides the opportunity to apply a splitting algorithm. During a given time iteration, all the computations are divided into two steps. In the first step, each layer of the horizontal two-dimensional finite element equations is solved separately. In the second step, the one-dimensional vertical equations which connect the layers, are solved using a multi-level finite-difference algorithm.

12.12.2   Physical Concepts

1. Solute transport is governed by the advection–dispersion equation.
2. Flow and solute transport are limited to the saturated zones.
3. Only one fluid (liquid) phase is assumed to be mobile. The matrix can undergo small deformations but this matrix velocity is considered negligible in comparison to that of the liquid.
4. The water carries a non-volatile contaminant that can be adsorbed on the solid matrix. Grain density remains practically constant.
5. Transient conditions are employed for the simulation of the flow problem.
6. Transient conditions are also assumed for simulating transport and heat transfer. Liquid and solid temperatures are assumed practically equal.
7. Isothermal and non-isothermal conditions can be implemented.
8. The model is based on the transport of one solute.
9. Chemical transformation of solute mass due to chemical reaction between different components is not considered. Radioactive decay processes are accommodated.

12.12.3   Computational Concepts

1. Simulations are done in three-dimensional space.
2. The numerical method is based on a combination of finite elements in the horizontal layers and finite differences in the vertical.

3. First, second and third-type boundary conditions can be accommodated.

4. Prescribed flux and potential values can be specified as internal conditions on the interior of the mesh.

5. Stresses such as rainfall, evapotranspiration, pumping/recharge can be introduced.

6. Principal direction anisotropy can be accommodated via parameter input.

7. Gauss elimination and SSOR are employed as the equation solvers.

8. The nonlinearities are handled by a special Picard iteration procedure combined with an operator-splitting representation of the partial-differential equations.

12.12.4   Other Issues

1. The code uses the ArgusONE graphical interface. ArgusONE is a general purpose finite-element and finite-difference pre and post-processor. It supports various mesh and grid topologies and CAD-GIS like layers. It allows one to create relational information data bases in order to store geographical and physical data. It has also a programming language facility and a graphical user interface compatible with Visual C++.

2. The code has been compiled on PC Windows 95 platforms and UNIX operating systems.

3. The code has been verified against classical analytical solutions such as Henry's problem.

12.12.5   Case History

1. The code was employed in several case studies focusing on the Israel coastal aquifer.

2. One of the studies involved the modelling of the design of a coastal collector near the Yavne Artificial Recharge area.

3. One of the three main groundwater resources of Israel's water system is the Pleistocene coastal aquifer. The average total production from this aquifer for the last decade has amounted to 370–400 Mm^3/yr, while the natural recharge is estimated

to be 300 Mm³/yr. This aquifer supplies about 30% of the water consumption in Israel and is intended to serve as the major long-term storage reservoir. The aquifer is phreatic, and the depth of the water level varies from a few meters (above sea level) near the coast up to 80–90 m in the south. The Mediterranean Sea is the natural outlet of the aquifer system.

One of the major difficulties of long term water storage is the fact that the natural flow direction is from East to West. Thus, an important quantity of freshwater flows through the coastal aquifer to the sea. Moreover, the internal flow to the sea is increased near the regions where the recharge facilities exist. In the coastal aquifer, one of the main artificial recharge areas is the Yavne recharge area.

To intercept a portion of this residual outflow to the sea, coastal collectors (i.e. shallow pumping wells collecting freshwater, situated near the sea) have been extensively used in Israel.

4. The model employed was SALTFRES.

5. To investigate this strategy, three aquifers and aquitards were modeled using continuous three-dimensional measurements of chloride content from more than 2000 spatial points. These were directly interpolated into a three-dimensional continuous distribution of head and flow. The coastal aquifer was modeled as a multilayer leaky aquifer formed by four subaquifers. The transmissivities of the subaquifers and resistances of the aquitards were first estimated by results of pumping tests. Pumping and recharge wells were modeled in the 1–2 km wide dunal area, where pumping and artificial infiltration for drinking water supply occurs. Existing spreading field near the Shafdan area were modeled as recharge area.

A third type boundary condition was used to describe the leaky confining layers and the overlying water levels. Calibration was done against more than 350 head measurements averaged over a period of six months from 1995 to 1996.

6. In more than 90% of the points the model is within 1 m of the measured values and in more than 75% within 0.5 m.

12.13   SALTHERM/3D

Paul Chin
John Molson (molson@uwaterloo.ca)
Emil Frind

12.13.1   Description

The SALTHERM/3D code is based upon a Galerkin finite-element formulation for single species mass and heat transport. It accommodates transient, variable-density advective and dispersive transport in a fully three-dimensional system. Fluid flow is non-isothermal. The model applies to single-phase saturated flow, considers linear adsorption and does not consider chemical reactions, beyond first-order decay processes. The non-linear algebraic equations are accommodated using Picard iteration and the preconditioned conjugate gradient method is used to solve the equations. The model handles the standard suite of boundary and internally imposed conditions. A graphical user interface is not yet available and the code runs under either a DOS or UNIX based operating system. This model uses a deformable grid and can handle discretely fractured porous media. It uses the extended Boussinesq approximation and employs only block prism elements.

12.13.2   Physical Concepts

1. Simulation is via the advective-dispersive transport equation.

2. Saturated aquifers are assumed.

3. The model is restricted to single-phase flow.

4. Transport allows for adsorption and linear decay.

5. Transient or steady-state flow is considered.

6. Transient transport is accommodated.

7. Non-isothermal conditions can be simulated.

8. Single component transport is permitted.

9. The model does not allow for chemical reactions but includes first-order decay.

12.13.3 Computational Concepts

1. The model is fully three dimensional (one and two dimensions are also possible).
2. The numerical procedure uses Galerkin finite elements.
3. First, second and third type boundary conditions can be employed.
4. First, and second type conditions can also be applied on the interior of the domain.
5. Point and area sources can be specified.
6. Heterogeneous and random-field parameter input is possible.
7. Algebraic equations are solved using preconditioned conjugate gradients.
8. When a non-linear set of equations is encountered, a Picard iteration method is used.

12.13.4 Other Issues

1. Output files are compatible with existing graphical user interfaces, e.g. Tecplot.
2. The model runs under DOS and UNIX operating systems.
3. The code has been verified against Henry's semi-analytic solution and published test problems.
4. The primary advantage of this method is that it employs an iterative water-table search using a deformable grid and can accommodate discretely fractured or fractured porous media.
6. The primary limitations of the code are the use of the extended Boussinesq approximation and block prism elements.

12.13.5 Case History

1. The site location is a proposed nuclear waste repository within the Canadian Shield.
2. The purpose of this research is to assess the influence of deep brines on the evolution of radionuclide transport from buried nuclear fuel wastes. By coupling the processes of density-dependent flow and advective-dispersive heat and mass transport, the flow dynamics within the large-scale regions surrounding the proposed vaults can be investigated. The influence of fractures, and hydraulic overpressuring due to glaciation will also be assessed.

3. The hydrogeological setting is within a granitic pluton of the Canadian Shield. The two-dimensional cross-sectional domain measures $21,000$ m$\times 2,000$ m in the horizontal and vertical dimensions respectively. Host-rock hydraulic permeabilities are typically on the order of 10^{-12} to 10^{-15} m/s. Fractures have been found within the area and, in the scenario described here, are represented with apertures of 1.0×10^{-4} m. The proposed vault is located at a depth of 500 m.

   The salt concentrations at the sites being investigated for nuclear waste repositories in the Canadian Shield range from brackish-type waters ($c < 1$ g/l) to brines at depths of 800–1000 m where concentrations reach more than 200 g/l. Also, so-called "moose licks" occur at ground surface with concentrations of up to 9 g/l.

4. The model used is SALTHERM/3D.

5. A regional hydraulic gradient of 0.002 was imposed across the top of the domain. A steady-state solution was first obtained which was then used as the initial condition for the transient run.

   The host rock is conceptualized as a three-layer system with hydraulic conductivities varying from 1.0×10^{-8} m/s at ground surface to 1.0×10^{-13} m/s at depth. The domain is intersected by three horizontal and six vertical fractures and is discretized with 11,000 rectangular elements. The initial condition for the brine is assumed to be horizontally stratified, varying from a maximum concentration of 200 g/l at a depth of 1000 m and below to 0 g/l at ground surface. Heat flux from the vault was also considered in several of the simulations.

6. Saltherm/3D simulations of density-dependent mass transport in deep, regional flow systems have shown the important role of fractures in determining the behavior of transport away from the proposed repositories. Within the porous matrix blocks the excess head resulting from past glaciation was found to dissipate over a time period on the order of 3500 years and significantly increased the rate of mass removal from the vault.

   Heat transport was found to have only minor effects on transport rates.

12.14   SHARP and SWIP

Michael Merritt (mmerritt@usgs.gov)

12.14.1 Description

SHARP and SWIP are sharp-interface and convective-dispersive transport finite-difference codes respectively. SHARP was prepared by the U.S. Geological Survey [Essaid, 1990a], and SWIP was adapted from a petroleum industry code by a private firm under contract to the U.S. Geological Survey [INTERA, 1979]. SWIP has been substantially modified by U.S. Geological Survey personnel. SHARP is quasi three dimensional and SWIP is fully three dimensional. Both models simulate transient single-phase saturated density-dependent flow in confined or unconfined aquifers and, where applicable, transport conditions. SWIP is also capable of non-isothermal solute transport with linear adsorption and first-order chemical reactions and heat transport. SWIP is a two-fluid model in that it assumes resident and source fluids to be mixtures of two fluids whose characteristics are defined by input data and the solution variable is the local fraction of one of the two fluids in the mixture. SHARP uses a SIP equation solver and SWIP employs either L2SOR or a direct solution algorithm. Standard boundary conditions and sources and sinks can be accommodated by each code. The codes can be run on either the Windows 95 or UNIX operating systems. SHARP can be more computationally efficient than SWIP when an equilibrium solution is required, but, of the two model codes, only SWIP represents the transition zone between salt and fresh water.

12.14.2 Physical Concepts

1. Simulation is via the sharp-interface assumption for SHARP and the advective-dispersive transport equation for SWIP.

2. Saturated aquifers are the solution domain.

3. The models are restricted to single-phase flow.

4. Linear adsorption during transport is simulated by SWIP.

5. Transient flow (pressure) or coupled flow and transport solutions are computed by SWIP, and a steady-state flow solution may be obtained that requires several iterations.

6. The transient interface solution computed by SHARP is not separable into flow and transport components.

7. Non-isothermal conditions are assumed for SWIP.

8. SWIP is a two-fluid model and assumes local resident and source fluids to be mixtures of the two fluids defined by input specifications.

9. SWIP allows for first-order chemical reactions.

12.14.3 Computational Concepts

1. The SHARP model is quasi three-dimensional and SWIP is fully three dimensional and has a cylindrical-coordinates option.

2. The numerical procedure used in each code is finite differences, and centered and backward differencing schemes are available as options in SWIP.

3. First, second and third type boundary conditions can be used for SHARP and in SWIP first type can be used for pressure and second type for concentration and temperature.

4. First, second, and third-type conditions can also be applied on the interior of the domain with SHARP and first and second type can be used in SWIP.

5. Point and areal sources can be accommodated by both codes.

6. Parameter input allows for heterogeneous material properties and anisotropy in both codes.

7. Algebraic equations are solved using SIP in the case of SHARP and either L2SOR or a direct solver for SWIP.

8. When a non-linear set of equations is encountered by SWIP, Picard iteration is used.

9. Both models use block-centered, variably-spaced rectangular discretizations and SWIP allows cell-wise redefinition of vertical cell dimension and placement.

10. SWIP optionally provides for time-step length to be a function of the maximum computed concentration change.

11. SHARP solutions may be accelerated by adjusting storage and porosity parameters.

12.14.4   Other Issues

1. A graphical interface is not available, although line-printer plots are provided by SWIP.

2. The models run under Windows 95 and UNIX operating systems, but, as FORTRAN codes, are not system or machine dependent.

3. SHARP can be computationally more efficient than an advective-dispersive model when a steady-state analysis is needed. On the other hand SWIP, the advective-dispersive code, accommodates the transition zone although considerable computational effort is involved.

4. It has been observed that a fine mesh and small time steps are required to avoid numerical instability in simulations with SWIP.

5. Given the boundary condition constraints imposed by SWIP, it is not evident how one specifies other than no-flow along a boundary that transects the interface. Otherwise, if a difference in density occurs at this boundary during a simulation, large boundary fluxes are simulated that are not likely to be realistic.

12.14.5   Case History

1. Broward County, Florida.

2. In Broward County, in southeastern Florida, many cases in which public water supplies have been affected by saltwater intrusion from the Atlantic Ocean have occurred. The construction of major regional drainage canals, three of which transect Broward County, caused a lowering of the water table and a gradual inland movement of the coastal saltwater front toward well fields. The construction of long tidal canals extending inland from, or paralleling the coast, emplaced to provide ocean access to boat owners, further lowered the water table and allowed saltwater to enter the aquifer as direct seepage from canals in inland areas. In recent years, pumping has had to be curtailed at most wells used for supply by several local municipalities.

3. Southeastern Broward County, the "study area" of this case history, is an area of generally flat terrain. The annual aver-

age rainfall in Fort Lauderdale is about 61 inches, of which about 70 percent falls during the months of May through October. Evapotranspiration rates in the study area are high and tend to vary with the amount of solar radiation, which implies a strongly seasonal variation (higher in summer, lower in winter). Based on an examination of stage and water-table altitude records from stations in similar or analogous surface soil environments in still-natural areas within Dade County to the south, it is surmised that the average pre-development water table in parts of southeastern Broward County was as much as 2.5 ft higher than the present-day water table.

The highly-permeable and unconfined Biscayne aquifer occurs at about 35 to 50 ft below land surface in southeastern Broward County. The transmissivity of the Biscayne aquifer has been estimated to be greater than $1,000,000$ ft^2/day where sections of highly permeable coralline limestone are present. The average hydraulic conductivity of the Biscayne aquifer in the study area is probably between 7,500 and 10,000 ft/day. The base of the Biscayne aquifer occurs at a depth of between 80 and 120 ft below land surface more than 5 mi inland of the present-day coast. Closer to the coast, the Biscayne aquifer thickens appreciably and extends to a greater depth. Data describing the composition and vertical extent of the aquifer offshore are lacking.

Overlying the Biscayne aquifer throughout the study area are less permeable layers of sandstone and silty, limy sand that do not hydraulically confine the Biscayne aquifer. The materials underlying the Biscayne aquifer are clastics of relatively low permeability. The chloride concentrations of water samples collected between 1939 and 1994 from various monitoring and exploratory wells were used in this study to delineate the approximate position of the saltwater front and its movement over time. Results indicated that the front has moved as much as 0.5 mi inland in parts of the study area. Water samples from the large number of monitor wells east (seaward) of the Hallandale Well Field showed a clear pattern of saltwater intrusion toward the pumping wells since the first wells were drilled in 1969. In 1988, pumping from the main part of the well field was curtailed. The numerical simulation of the saltwater intrusion process was considered a worthwhile objective because such techniques have the potential to:

(a) help determine which of the various hydrologic and water-management parameters characterizing the coastal sub-surface environment and its use by man is dominant or significant in causing saltwater intrusion to occur, and

(b) provide a simulator that would show what effect changes in system management could have on the saltwater intrusion process.

4. Two model codes were applied in the course of the study. These codes were the SHARP code, which simulates the position of the saltwater front as a sharp interface without any transition zone (a zone in which a gradational change between freshwater and saltwater occurs), and the SWIP code, which simulates a two-fluid variable-density system with a convective-diffusion approach that contains a representation of the transition zone. The SHARP code was used for estimates of the likely interface position under pre-development conditions and under recent conditions as affected by water-management control and as affected by the Hallandale Well Field.

5. In order to estimate the location of the pre-development interface, it was assumed that natural flow was uniformly in the eastward direction throughout the area modeled. The annual average head along the western boundary of the model was specified to be 5 ft above sea level. The specified head on the eastern boundary was the average tidal stage at the Broward County coastline (0.75 ft above sea level).

The specified lateral aquifer hydraulic conductivity was 10,000 ft/day. The pre-development sharp interface was simulated as passing below the main well-field pumping zone at about 127 ft, which is inconsistent with the observation that water samples from a test hole drilled in the center of the well field were fresh to 300 ft after decades of water management and well-field pumping.

The simulation results were found to be insensitive to a major change in the bottom slope of the aquifer, or to a uniform

25 percent reduction in the hydraulic conductivity value, and only moderately sensitive to a 150-percent increase in net atmospheric recharge.

Because the SHARP pre-development simulation constrained flow to be uniform in the easterly direction, an analogous simulation could be accomplished with a cross-sectional application of the SWIP code in order to obtain a comparison of the sharp interface and diffuse interface methods and to make an assessment of the effect of dispersion on the equilibrium position of the interface.

Boundaries of the 11-layer cross section corresponded to the east and west boundaries of the SHARP grid; aquifer thickness, ocean bottom depths, and aquifer properties were described similarly. Simulation time periods ranged to 40 years.

6. All lines of equal concentration within the transition zone converged to an equilibrium position. The 50 percent saltwater concentration line of the diffuse interface simulated by SWIP converged to a position similar to the interface position simulated by SHARP when a low value of longitudinal dispersivity (4 ft) was specified in SWIP or when revised dispersion algorithms were used in SWIP that greatly reduced the degree of simulated vertical dispersion.

When a high longitudinal dispersivity value (400 ft) was used in SWIP together with the standard dispersion algorithms, a broad, diffuse, and nearly vertical transition zone was simulated, and the 50 percent saltwater concentration line was located thousands of feet seaward of the position simulated by SHARP. The simulated transition zone was much narrower when a longitudinal dispersivity value of 40 ft was specified than when a value of 400 ft was specified, regardless of the dispersion algorithm used. The position and degree of dispersal of the transition zone were insensitive to moderate variations of the low transverse dispersivity value.

A 20-layer cross-sectional grid, in which layers had uniform thickness, was designed for additional analyses with SWIP. Parts of layers of the grid were specified as having lower permeability than the Biscayne aquifer to represent the lower permeability of the overlying leaky sand layer (250 ft/day) and inland sections of the underlying layers of clastics (100 ft/day). The simulated equilibrium position of the 50 percent saltwater concentration line was somewhat sensitive to variations in the specified spatial distribution of permeability within the Biscayne aquifer layers. The position of the 50-percent saltwater concentration line was not appreciably sensitive to the specification of seasonally varying boundary conditions and net-recharge rate or to a physically realistic degree of variation in the density of saltwater.

Another cross-sectional grid with 28 layers and hypothetical stresses (arbitrary specifications for boundary heads, rainfall, and evapotranspiration rates) was used for additional sensitivity analyses with SWIP. Treating thin layers of widely contrasting hydraulic conductivity as one or several layers with thickness-weighted average hydraulic conductivities did not appreciably change the simulated equilibrium position of the 50 percent concentration line. When the layers were explicitly represented, the interface tended to be more horizontal in layers of high hydraulic conductivity and more vertical in layers of lower hydraulic conductivity. Other tests with the 28-layer model further demonstrated the strong sensitivity of the position and dispersal of the simulated transition zone to the choice of an algorithm for computing vertical dispersion and to the related choice of longitudinal dispersivity value.

A study of velocities simulated by the 20-layer model indicates a saltwater convection cell that extends seaward in the aquifer from the offshore freshwater discharge location. A slight degree of inland convection of saltwater beneath land was simulated. Freshwater discharge was upward through the leaky sand layer at the shoreline rather than at the presumed outcrop of the aquifer at the edge of the shelf when the leaky sand layer was assigned a hydraulic conductivity value of 250 ft/day everywhere. When the layer was assigned a hydraulic conductivity value of 0.05 ft/day offshore, discharge was at the outcrop location, but the inland extent of the saltwater zone was similar to that previously simulated.

In all cross-sectional simulations, the equilibrium interface position approximately coincided with the interface position that would be estimated using the Ghyben-Herzberg relation and simulated equilibrium head values, and a near-equilibrium position was established in zones of high hydraulic conductivity within one to several years. In contrast, the average head measured near the Hallandale Well Field since 1970 was less than half the values simulated in various cross-sectional model runs.

When pre-equilibrium conditions were simulated, the simulated head at the well field quickly rose to a value higher than the equilibrium value, and then decreased gradually as the equilibrium position of the interface was established. The simulations, therefore, failed to demonstrate that the low measured heads were a pre-equilibrium condition. The results of the cross-sectional simulations suggest that the conceptual model of freshwater flow eastward to a line sink of off-shore locations of discharge to tidewater should be reexamined.

12.15   T3DVAP.F/MOR3D.F

Bryan Travis (bjt@vega.lanl.gov)

12.15.1   Description

The T3DVAP.F/MOR3D.F codes are based upon an integrated finite-difference formulation. They accommodate variable-density advective transport in a fully three-dimensional system. Transport is transient and isothermal (T3DVAP.F) or nonisothermal (MOR3D.F). Transport allows linear/nonlinear equilibrium or rate-limited adsorption. Multiple species and a small set of chemical reactions are incorporated. Evaporation and resulting concentration changes are allowed (T3DVAP.F). Particle tracking and interface tracking submodules are also included. Non-linear terms are approximated using a Newton-Raphson formulation and the resulting algebraic equations are solved using an incomplete factorization preconditioner with GM-RES. Standard boundary conditions and internal conditions can be accommodated. Point and areal stresses can be handled as well as heterogeneous, anisotropic and random fields. The code runs under various operating systems. The model accounts for changes in permeability due to precipitation-dissolution.

12.15.2 Physical Concepts

1. Simulation is via the advective-dispersive transport equation, particle tracking, or sharp interface equations.

2. Saturated and unsaturated aquifers are assumed.

3. The models consider multiphase flow.

4. Adsorption is included in transport.

5. Transient and steady-flow conditions are considered.

6. Transient and steady transport is accommodated.

7. Isothermal and nonisothermal conditions are assumed.

8. Multiple-component transport is permitted.

9. The models allow for chemical reactions, including feedback to porosity and permeability changes.

12.15.3 Computational Concepts

1. The models are fully three dimensional.

2. The numerical procedure used is integrated finite differences.

3. First, second and third type boundary conditions can be used.

4. First and second-type conditions can also be applied on the interior of the domain.

5. Point and area sources can be specified.

6. Parameter input permits heterogeneity, anisotropy and random fields.

7. Algebraic equations are solved using GMRES with an incomplete factorization preconditioner.

8. When a non-linear set of equations is encountered, a Newton-Raphson method is used.

12.15.4 Other Issues

1. A graphical interface is not available, commercial graphics packages are used.

2. The model runs on UNIX or PC systems.

3. Verification includes comparison against analytic solutions, other numerical solutions, experimental data and mass and energy tracking.

4. The primary advantage of this method is the consideration of precipitation-dissolution effects on permeability and the coupled two-phase flow. T3DVAP.F is very robust under numerically difficult conditions of low water saturation. The major disadvantage is the limited number of chemical reactions.

12.15.5 Case History

1. (1) Los Alamos, NM; (2) Mid-Ocean Ridges of the Atlantic.

2. The objectives were to develop a better understanding of:

   (a) the effects of air flow through fractured systems, and

   (b) the effect of brine and precipitation/dissolution on hydrothermal convection at mid-ocean ridges.

3. Site 1: The mesas surrounding Los Alamos, New Mexico, experience a cool, dry environment, with wet summers and occasional snow cover in winter. The underlying soil is thin, covering fractured volcanic rock. Site 2: The hydrologic setting at mid-ocean ridges is very different: saturated conditions at very high temperatures (200–500°C), high salt concentration, and strongly varying water properties. The underlying rock is fractured and subject to chemical attack.

4. T3DVAP.F was employed at site 1, and MOR3D.F at site 2.

5. For site 1, a two-dimensional grid was used to describe a fracture with surrounding permeable, partially saturated rock. Air flows through the fracture, evaporating pore water as it passes. Two seasonal conditions were considered: low humidity air and frequent, episodic water infiltration (summer), and higher humidity air with very slow infiltration from snow melt (winter). For site 2, a three-dimensional geometry was specified, discretizing a region extending from the sea floor to a depth of six kilometers, with a mid-ocean ridge on one side. Seawater can flow into the top of the domain, while hot water can exit through the top as well. A strong thermal energy source is located on the bottom surface below the ridge. Various initial distributions of brine are considered.

6. Results at site 1 indicated that air flow through fractures can have a major impact on the water saturation profile normal to a fracture face. Dry air flow produces very low saturation at

a fracture face, inducing water flow towards the fracture face. Net water infiltration is greatest in winter even though more precipitation occurs in summer, due to the low humidity of summer air. Results for site 2 indicate that brine can complicate the picture of hydrothermal convection, creating unsteady convection at lower Rayleigh number than for a simple fluid. A brine layer above a fresh layer, e.g., can lead to a period of gradually weakening episodic convection. Changes in permeability due to precipitation/dissolution reactions also produce unsteady convection at lower Rayleigh numbers than when the rock matrix is inert. Under strong temperature variations, real water properties can result in a stable fluid layer below a convecting regime even with bottom-heating conditions. Chemical reactions coupled to convective motion show distinctly different mineral assemblages will develop over the course of an intrusive event when hydrothermal convection can occur as compared to the case of purely thermal diffusion. Model simulations match observed effluent water temperatures at ridges when measured rock permeabilities are used.

12.16   TVD-2D/TVD-3D

Transport with Variable Density for vertical 2-D plane or 3-D space

Philippe Ackerer (ackerer@imf.u-strasbg.fr)
Robert Mose
Anis Younes

12.16.1   Description

TVD-2D and TVD-3D are two codes which can simulate mass transport in porous media in 2D or 3D with variable density and viscosity. Based on mixed hybrid finite elements and discontinuous finite elements, the code can handle unstructured meshes based on quadrilateral elements (2D) or hexahedral elements (3D). All parameters are described at the element level. Anisotropy is taken into account. Flow is limited to the saturated zone. Phreatic aquifers are handled by a moving mesh. Boundary conditions and stresses (recharge, pumping) can be time dependent. Due to the splitting of the transport equation, the description of the boundary conditions are very flexible, they can be of finite-difference type (average concentration in a cell), or of finite-element type (concentration prescribed at a node), for convection or dispersion only or for both. The code has been tested extensively in several benchmarks and some laboratory experiments.

12.16.2   Physical Concepts

1. Solute transport is governed by the advection–dispersion equation.
2. Solute transport is simulated for saturated, free or confined aquifers.
3. Only one fluid (liquid) phase is assumed to be mobile.
4. Solute is assumed to be adsorbed or conservative.
5. Transient conditions are employed to simulate the flow problem.
6. Transient conditions are also employed to simulate transport.
7. Isothermal conditions are implemented.
8. The model is based on the transport of one solute.
9. Generation of solute mass due to chemical reaction is not considered.

12.16.3   Computational Concepts

1. Simulations are done for a two-dimensional vertical plane or for a three-dimensional space.
2. Mixed hybrid finite elements is implemented for the solution of the flow equation. Discontinuous finite elements are used for the advection part of the transport equation and mixed hybrid finite elements for the dispersion part.
3. Boundary conditions of the first and second type can be accommodated for simulating the flow and the transport problems.
5. Stresses such as pumping, infiltration and recharge can be accommodated.
6. Heterogeneity and anisotropy are introduced in terms of parameter input.
7. The preconditioned conjugate gradient method is employed as the equation solver.
8. Picard iteration is used to handle the nonlinear system.

12.16.4 Other Issues

1. There are no graphical interfaces incorporated in the TVD-2D/3D code.

2. The computer code uses the FORTRAN compiler and can be run under PC or UNIX operating systems.

3. The code was verified against standard benchmarks (Henry, Elder, Salt Dome) and laboratory experiments (3D heterogeneous media, 3D homogeneous media with variable bedrock, 2D layered media with viscosity contrasts).

4. The model presents accurate new numerical techniques. The mixed finite element method enables the accurate evaluation of velocities while the discontinuous finite element method can handle any grid Peclet number (we tested from $Pe = 0.1$ to 1000). Transport is handled by an explicit scheme.

12.17 Summary

12.17.1 Physical Concepts

See Table 12.1.

Legends:

AD–**A**dvective-**D**ispersive model
SI–**S**harp **I**nterface model
SU–**S**aturated/**U**nsaturated (**PH**reatic)
St–**S**aturated (confined)
MP–**M**ulti-**P**hase
MC–**M**ulti-**C**omponent
Ad–**Ad**sorption
CR–**C**hemical **R**eactions
SF–**S**teady **F**low
TF–**T**ransient **F**low
ST–**S**teady **T**ransport
TT–**T**ransient **T**ransport
Is–**Is**othermal
NIs–**N**on**I**sothermal

Notes:

1. Free surface optional without moving mesh.

2. Variable saturation and free-surface optional with/without moving mesh.

3. Free-surface with moving mesh.

12.17.2 Computational Concepts

See Table 12.2.

Legends

Spatial distribution

1D–One Dimension
2V–Vertical Plane
2H–Horizontal Plane
Q3–Quasi Three Dimensions
3D–Three Dimensions

Numerical Method

FE–Finite Elements
VE–Volume Elements
AE–Analytical Elements
FD–Finite Difference

Notes:

1. Also for axisymmetric problems

2. Temporal and regional variations of the vertical depth of the saltwater body is also estimated

3. Dupuit-Forchheimer flow in combination with 3-D density driven flow

4. Advective transport is solved by the method of characteristics

5. Mixed hybrid finite element for flow and dispersion, discontinuous finite element for convection

Code	AD	SI	SU	St	MP	MC	Ad	CR
3D(2D)FEMFAT	⋆		⋆				⋆	⋆
CODESA-3D	⋆		⋆	⋆			⋆	
DSTRAM	⋆			⋆			⋆	⋆
FAST-C(2D)	⋆			⋆			⋆	
FAST-C(3D)	⋆			⋆			⋆	
FEFLOW	⋆		⋆	⋆			⋆	
HST3D	⋆			⋆			⋆	⋆
MEL2DSLT	⋆		PH				⋆	
MLAEM/VD	⋆		⋆	⋆			⋆	
MOCDENSE	⋆			⋆		⋆		
MOC-DENSITY	⋆			⋆			⋆	
MOCDENS3D	⋆			⋆			⋆	
SALTFRES	⋆		PH	⋆			⋆	
SALTHERM/3D	⋆			⋆			⋆	
SHARP		⋆		⋆				
SWIP	⋆			⋆			⋆	⋆
T3DVAP.F/MOR3D.F	⋆		⋆	⋆	⋆	⋆	⋆	⋆
TVD-2D/3D	⋆		⋆	⋆			⋆	
Code	SF	TF	ST	TT	Is	NIs	Note	
3D(2D)FEMFAT	⋆	⋆	⋆	⋆	⋆			
CODESA-3D	⋆	⋆	⋆	⋆	⋆			
DSTRAM	⋆	⋆	⋆	⋆		⋆		
FAST-C(2D)	⋆		⋆	⋆	⋆			
FAST-C(3D)	⋆		⋆	⋆	⋆			
FEFLOW	⋆	⋆	⋆	⋆	⋆	⋆	(1)	
HST3D	⋆	⋆		⋆	⋆	⋆	(2)	
MEL2DSLT		⋆		⋆	⋆	⋆		
MLAEM/VD			⋆		⋆			
MOCDENSE	⋆	⋆		⋆	⋆			
MOC-DENSITY		⋆		⋆	⋆			
MOCDENS3D		⋆		⋆	⋆			
SALTFRES	⋆	⋆	⋆	⋆	⋆			
SALTHERM/3D	⋆	⋆	⋆	⋆		⋆		
SHARP		⋆		⋆	⋆			
SWIP	⋆	⋆		⋆		⋆		
T3DVAP.F/MOR3D.F	⋆	⋆	⋆	⋆	⋆			
TVD-2D/3D		⋆		⋆	⋆		(3)	

Table 12.1. Physical Concepts.

Code	1D	2V	2H	Q3	3D
3D(2D)FEMFAT					
CODESA-3D		⋆			⋆
DSTRAM					
FAST-C(2D)		⋆			
FAST-C(3D)					⋆
FEFLOW		⋆	⋆		⋆
HST3D					⋆
MEL2DSLT			⋆	⋆	
MLAEM/VD				⋆	
MOCDENSE		⋆			
MOC-DENSITY		⋆			
MOCDENS3D		⋆			⋆
SALTFRES		⋆	⋆	⋆	⋆
SALTHERM/3D	⋆				
SHARP			⋆	⋆	
SWIP		⋆	⋆		⋆
T3DVAP.F/MOR3D.F	⋆	⋆	⋆		⋆
TVD-2D/3D		⋆	⋆		⋆
Code	FE	VE	AE	FD	Note
3D(2D)FEMFAT					
CODESA-3D	⋆				
DSTRAM				⋆	(1)
FAST-C(2D)				⋆	
FAST-C(3D)		⋆		⋆	
FEFLOW	⋆				(1)
HST3D		⋆		⋆	(1)
MEL2DSLT				⋆	(2)
MLAEM/VD			⋆		(3)
MOCDENSE			⋆		(4)
MOC-DENSITY				⋆	(4)
MOCDENS3D				⋆	(4)
SALTFRES					
SALTHERM/3D					
SHARP				⋆	
SWIP				⋆	
T3DVAP.F/MOR3D.F		⋆			(3)
TVD-2D/3D	⋆				(5)

Table 12.2. Computational concepts.

Chapter 13

Seawater Intrusion in the United States

L. F. Konikow & T. E. Reilly

13.1   Introduction

The growth of industry and population in the United States during
most of its history has been concentrated in coastal areas having
good access to ports. Thus, those coastal areas that also had abun-
dant groundwater resources were the areas where seawater intrusion
became a serious concern or problem. These areas included much
of the Atlantic and Gulf Coastal Plains, the carbonate aquifers of
coastal Florida, alluvial basins along the California coast, and the
volcanic islands of Hawaii. These coastal zones combined represent
a total distance of approximately 5,000 km along which seawater is
potentially intruding into productive aquifers (see Figure 13.1). The
potential problem of seawater intrusion was recognized in the United
States as soon as the first U.S. water well was drilled in 1824 in New
Jersey [Back and Freeze, 1983, p. 90]. In reviewing the occurrence
and geochemical significance of saltwater, Back and Freeze [1983]
further note that the problem was recognized as early as 1854 on
Long Island, New York.

Over time, a variety of approaches have evolved for identifying,
monitoring, studying, and managing seawater intrusion in the United
States. Previous overviews of the situation in the United States were
presented by Krieger et al. [1957] and Task Committee on Saltwater
Intrusion [1969]. This chapter reviews a representative range of these
approaches, mostly through the description of a number of selected
case histories. The case histories also provide an illustration of how

J. Bear et al. (eds.), Seawater Intrusion in Coastal Aquifers, 463–506.

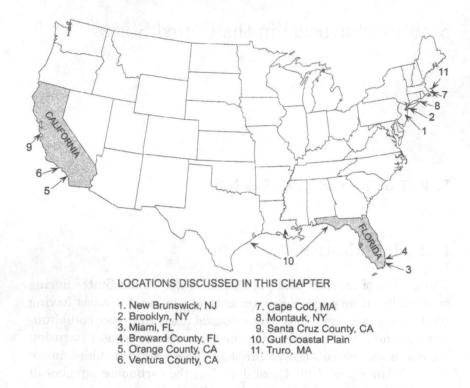

LOCATIONS DISCUSSED IN THIS CHAPTER

1. New Brunswick, NJ
2. Brooklyn, NY
3. Miami, FL
4. Broward County, FL
5. Orange County, CA
6. Ventura County, CA
7. Cape Cod, MA
8. Montauk, NY
9. Santa Cruz County, CA
10. Gulf Coastal Plain
11. Truro, MA

Figure 13.1. Map of the United States showing locations of selected areas referred to in the text.

fast seawater intrusion has progressed in selected environments in response to groundwater withdrawals and (or) changes in regional drainage that lower the water table. In this chapter, we attempt to integrate a historical perspective with technical highlights, and emphasize alternative modeling approaches to help analyze problems, as well as engineering approaches to manage or control seawater intrusion.

13.2   Early History

Carlston [1963] describes an early American statement of the principle of static equilibrium between freshwater and saltwater. This early work and publication [Du Commun, 1828] actually predates the commonly cited work of Badon Ghyben and Herzberg, but is not widely cited or recognized. According to Carlston [1963], Du Commun was an American of French birth who was teaching at the West Point Military Academy. The first water well drilled in the United States was completed in 1824 in the Coastal Plain sediments of New Brunswick, New Jersey. It was uncertain why water levels in this freshwater well were above sea level and also fluctuated in phase with the local tides. Du Common offered an explanation based on a density balance of static fluids in columns of freshwater and saltwater that are connected at depth—an explanation that is equivalent to the Ghyben-Herzberg principle developed about a half century later.

Back and Freeze [1983] note that 100 years after the first American well had been drilled, seawater intrusion had become widespread and a matter of public concern. Brown [1925] studied coastal groundwater problems and summarized the experience in the United States and Europe as of that time. Back and Freeze [1983] state by the time of Brown's paper, "... many of the principles required for management of coastal aquifers were well understood." For example, Brown [1925] correctly characterizes the nature of the contact between fresh and saline groundwater along coasts as a transition zone in which "there is a continual diffusion of marine salts from the seawater into the fresh groundwater of the land and an actual movement of fresh groundwater from the land seaward." He also describes correctly the process of upconing, and recognizes that in some cases a supply well located near the coast can be ruined by saltwater rising from below much sooner than seawater can migrate inland from the coast.

Brown [1925] drew several conclusions from his work, which are still appropriate today. These form the basis of modern groundwater management in coastal areas (although most water managers and hydrologists are probably unaware that these management principles were clearly stated in the United States so long ago). Brown's work shows that it is necessary to integrate all the data and processes at the scale of individual wells and to view the overall problem from the perspective or scale of a basin or watershed. Brown states [1925, p. 39], "All the evidence regarding the effect of pumping on wells near the sea tends to one conclusion—that if the quantity of freshwater removed is greater than that supplied from the contributory area the wells become salty." As will be discussed later in this chapter, this simple water-balance approach is the basis for management actions today in parts of California.

Another important conclusion is derived from an analysis of the relation between pumpage and salinity. Brown [1925] compared pumpage and salinity in a well field near Brooklyn, NY over the period 1897–1905 (Figure 13.2). The well field consisted of 12 production wells that were withdrawing water from a confined sand aquifer at depths between 125 and 180 ft. The chloride content of water being pumped at the well field began to increase significantly after about one year of operation. The pumping rates were reduced, more or less steadily, over the next several years. The chloride content decreased only gradually, however, and did not decrease in direct relation to the decrease in pumpage. The well field was abandoned in 1905. Brown [1925, p. 47] highlights an important lesson, which is that "... the freshness of the groundwater in the sands is not restored at once by shutting down the plant and permitting the groundwater to rise to its original level. When the seawater once reaches a system of wells the only remedy is to abandon them. Probably only in the course of many years will the saltwater be entirely washed from the sands by the slowly moving freshwater escaping into the sea." The lesson is that prevention is the best cure, and this philosophy too is a driving force in modern management of coastal groundwater systems, as will be discussed later in this chapter.

13.3   Case Studies

We have selected our primary discussion of case studies from examples in Florida and California. These are areas where groundwater use is high and seawater intrusion has been recognized and analyzed for many years, providing an excellent record in the literature upon which to draw generalizations.

As of 1990, Florida was the U.S. state having the second largest population served by groundwater supplies (about 90 percent of 11.2 million people) [Solley et al., 1993]. It is also second largest as measured by total groundwater withdrawals for public supply (6.4×10^6 m^3/d). Florida has five regional water management agencies whose mission is to manage water and related resources for the benefit of the public, including water supply and water-quality protection.

California is the U.S. state having the largest public water supply derived from groundwater sources; groundwater withdrawals in

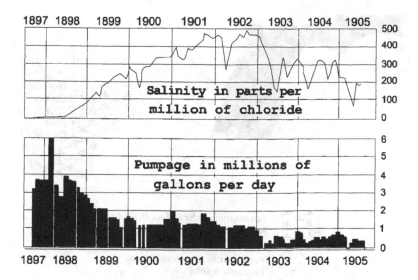

Figure 13.2. Comparison of pumpage and salinity at well field near Brooklyn, NY (modified from Brown [1925] and Spear [1912]).

1990 were approximately 12.3×10^6 m^3/d [Solley et al., 1993]. California also has the largest population of any state that is served by groundwater sources (13.6 million, which is more than half the population of the state). Seawater intrusion has been recognized in many coastal areas of California. In California, the protection of groundwater basins from seawater intrusion is considered to be the responsibility of local government [Smith, 1989]. Smith [1989, p. 57] also notes that if local officials fail to take steps to control a serious problem, the State Water Resources Control Board can file suit to restrict pumping or to impose other methods to help preserve the quality of the groundwater resource.

13.3.1 Seawater Intrusion in South Florida

13.3.1.1 Hydrogeologic Setting

The climate in south Florida is subtropical and humid. The average annual rainfall varies from a maximum of more than 60 inches (1520 mm) per year along the east coast, near Miami, to less than 50 inches (1270 mm) per year along the Florida Keys and in interior areas in the vicinity of Lake Okeechobee [Parker et al., 1955].

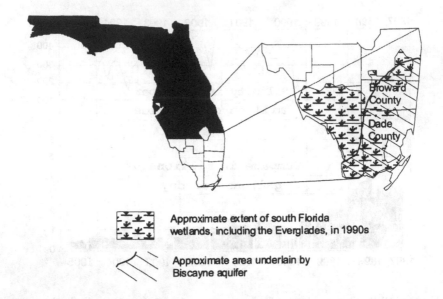

Approximate extent of south Florida wetlands, including the Everglades, in 1990s

Approximate area underlain by Biscayne aquifer

Figure 13.3. Location map of south Florida showing approximate extent of wetlands areas and area underlain by Biscayne aquifer.

Florida is underlain by two notably high-yielding aquifers. The Floridan aquifer is a confined regional carbonate system that underlies most of the state and extends northward into adjacent states. The Biscayne aquifer is an unconfined carbonate aquifer system that is present only in southernmost Florida, but it is the sole source of water supply for the city of Miami and adjacent areas (see Figure 13.3). Seawater intrusion is a problem throughout coastal Florida, but is perhaps most advanced, and especially well documented, in parts of the Biscayne aquifer. The regional groundwater systems and their relation to seawater intrusion are discussed in detail by Bush and Johnston [1988], Meyer [1989], and Miller [1990].

From 1909 through the 1930s, a network of drainage canals was constructed westward from the coastal areas in the Miami (Dade County) area. The canals were mostly uncontrolled through the 1930s and caused a continuous diversion to the ocean of large quantities of freshwater from the Everglades and drainage of groundwater from the Biscayne aquifer [Klein and Waller, 1985]. The resulting drainage lowered water levels about 2 m in the Everglades area [Kohout, 1965]. This, combined with drawdowns from several coastal well fields tapping the Biscayne aquifer, caused seawater to advance progressively inland. Although management actions were taken in the 1940s to bring the problem under control, several municipal well fields were abandoned prior to 1945 because of increased salinity levels. Sonenshein [1997] summarizes the history of seawater intrusion in Dade County. He notes that salinity-control structures had been installed by 1946 in all primary canals. These structures were located as close as possible to the coast and were effective in (1) preventing saltwater from moving upstream in the canals due to tidal and density effects, and (2) backing up the freshwater to maintain higher water levels in the Biscayne aquifer near the coast. Sonenshein [1997] states that seawater intrusion was slowed or reversed (in a few areas) by the effects of the higher water levels resulting from these control structures. The network of canals and control structures relative to the extent of seawater intrusion is shown in Figure 13.4.

The lowering of the water table in the Everglades and adjacent inland areas affects the equilibrium between fresh and salty groundwater in the coastal zone, and causes seawater intrusion to progress until a new equilibrium is attained between the saltwater and the water table conditions that prevailed after drainage. Parker [1945] estimated that for an approximately 20-km reach of coastline around Miami, during the 25-yr period up to 1943, the rate of saltwater encroachment averaged about 70 m/yr. During the next few years, however, a lengthy drought caused water levels to decline to record lows, and the leading edge of the saltwater wedge advanced inland at a faster rate of about 270 m/yr [Parker et al., 1955].

The base of the Biscayne aquifer near the Miami coastline lies at an elevation of approximately 100 ft (30.5 m) below mean sea level. On the basis of the Ghyben-Herzberg relation, Parker et al. [1944, p. 18] predicted that the saltwater at the base of the aquifer in this area would continue to advance inland until eventually it would stabilize in equilibrium with freshwater where the average annual water-table elevation had been lowered to 2.5 ft (0.76 m). Cooper et al. [1964, p. iii], however, noted that by the early 1950s, the advancement of the saltwater appeared to cease, although the front of the wedge was still as much as 8 miles seaward of the predicted ultimate position. Consideration of this phenomenon led Cooper and his colleagues to the concept that the position of the saltwater front can be in dynamic equilibrium between flowing groundwater and non-static (flowing) saltwater. Thus, seawater in coastal aquifers would continually cir-

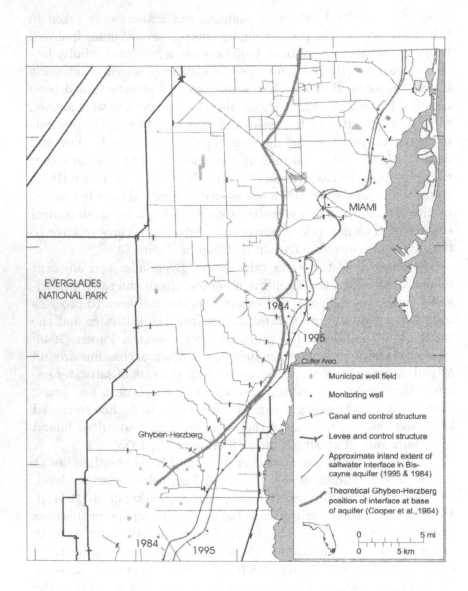

Figure 13.4. Map of eastern part of Dade County, FL, showing extensive network of canals, control structures, wells, and the approximate position of saltwater/freshwater interface in Biscayne aquifer (modified from Sonenshein [1997]). For comparison, the theoretical Ghyben-Herzberg position of the saltwater interface at the base of the Biscayne aquifer, as mapped by Kohout [in Cooper et al., 1964], is also shown.

culate and flow slowly inland, but the front (or interface) would be continually eroded by seaward flow of mixed water from the zone of diffusion. Carrier [1958] came to an equivalent conclusion (expressed from an opposite perspective) that the seaward discharge of salty water from the zone of diffusion must be balanced by an inflow of seawater. These evolving ideas led to a long-term study by the U.S. Geological Survey of this phenomenon [Cooper et al., 1964, p. v]. One major component was a field investigation, which aimed to document and define in detail the patterns of flow, head, and salt concentration in the Biscayne aquifer near Miami, Florida.

13.3.1.2   Detailed Field Definition of Seawater in a Coastal Aquifer

Kohout conducted a series of classic investigations in the coastal part of the Biscayne aquifer [Kohout, 1960; Cooper et al., 1964; Kohout, 1965]. In this area, the position of the saltwater front appeared to stabilize as much as 8 miles seaward of the position computed on the basis of either the Ghyben-Herzberg principle or the theory of dynamic equilibrium (with static saltwater) (see Figure 13.4). Kohout [in Cooper et al., 1964] attributes the discrepancy to the fact that two assumptions inherent in the two stated hypotheses are not met in the Biscayne aquifer. These are (1) that a sharp interface exists between freshwater and saltwater and (2) that the saltwater is static.

A number of fully cased wells, open only at the bottom, were drilled to various depths at locations close to and just beyond the shoreline along a line of cross section in the Cutler area. These wells were used to collect water samples and to measure fluid pressures at discrete points in the aquifer. The chloride data clearly defined the existence of a zone of diffusion (see Figure 13.5). The width of the zone of diffusion at the base of the aquifer is about 2,000 ft (600 m), although in the nearby Silver Bluff study area, the width of the zone of diffusion appeared to be much greater (more than 12,000 ft or 3,650 m). For comparison, the theoretical position of a sharp interface based on the Ghyben-Herzberg principle is also shown in Figure 13.5.

The hypothesis presented by Cooper et al. [1964] indicated that the circulation of saltwater is induced by dispersion of salts related primarily to the reciprocative motion of the saltwater front caused by tidal cycles in adjacent Biscayne Bay and Atlantic Ocean. The dispersive flux is mostly due to mechanical dispersion, which is pro-

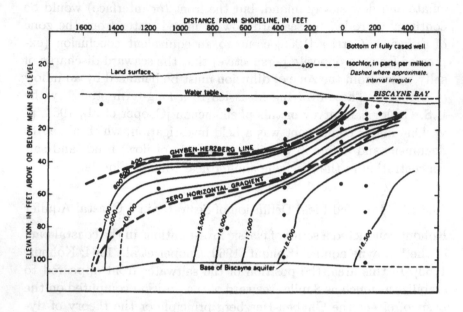

Figure 13.5. Cross section through the Cutler area, near Miami, FL, showing the zone of diffusion as defined by chloride concentrations on September 18, 1958 (from Cooper et al. [1964]). Also shown for comparison are the theoretical Ghyben-Herzberg position and the line of zero horizontal hydraulic gradient.

portional to the velocity of the fluid. Support for this argument requires a demonstration that there is indeed a significant velocity in the saltwater zone. Kohout [in Cooper et al., 1964] analyzed temporal and spatial changes in the fluid pressures in the observation well network. By converting all fluid pressure measurements to equivalent freshwater head, Kohout was able to map freshwater potential to provide a reasonably close approximation to the nature of the flow field throughout the entire system (see Figure 13.6). Equivalent freshwater head represents the hypothetical water level in a cased well if the well were filled with freshwater, instead of the water of variable salinity actually present. Equivalent freshwater head can be used to determine the direction of flow along a horizontal plane, but it is not necessarily useful in the determination of the direction of the vertical movement of the water. These potentials clearly indicate the seaward movement of freshwater in the upper part of the aquifer. Most of the discharge occurs into the floor of the sea. As supporting

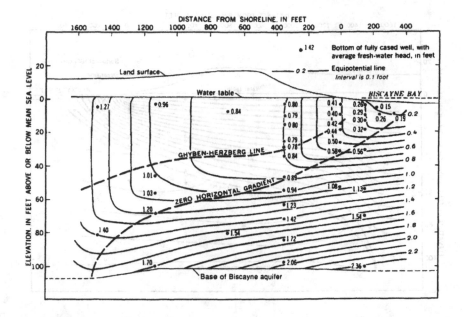

Figure 13.6. Cross section through the Cutler area, near Miami, FL, showing lines of equal freshwater potential, averaged for September 18, 1958, for inland conditions of relatively low water-table elevations (from Cooper et al. [1964]).

evidence, it was observed by Cooper [in Cooper et al., 1964, p. 6] that the shallow wells open to "... the zone of diffusion beneath the floor of the sea tap salty water under sufficient head to rise above sea level."

Kohout [in Cooper et al., 1964] also presented (but not shown here) maps of freshwater potential for times of both low tide and high tide on the same day. At high tide, the equipotential lines slope more strongly inland than shown in Figure 13.6, while at low tide the equipotential lines slope seaward to about the same degree that Figure 13.6 shows a slope inland for the daily averaged tidal conditions. These analyses indicate that the seawater is not static, but must be flowing in the direction of the slope. The daily average condition (Figure 13.6) indicates that the daily variations in flow directions average out to yield a net inland motion of saltwater. Kohout also presents equivalent data for May 29, 1958, when the water-table elevations were relatively high following a period of high recharge. On

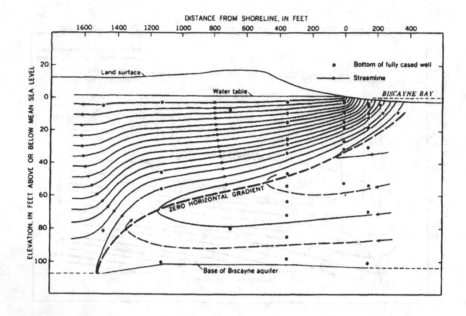

Figure 13.7. Cross section through the Cutler area, near Miami, FL, showing the pattern of flow of fresh and salty groundwater for September 18, 1958, based on potentials shown in Figure 13.6 (from Cooper et al. [1964]).

this day, all water in the aquifer moved seaward.

Kohout [in Cooper et al., 1964, p. 29] used the equipotential pattern in Figure 13.6 "... as a guide for separating the region of seaward-flowing water from that of the inland-flowing water. Such a separation is formed by a line passed through the points of horizontality of the individual equipotential lines." Along this line (shown in Figures 13.5 and 13.6) there is no horizontal flow, and water is flowing upward. The position of this line in Figure 13.5 indicates that some high-salinity water (greater than 1,000 ppm chloride) above this line is probably flowing seaward.

Kohout [in Cooper et al., 1964] also used these data to construct a flow net for the head conditions represented in Figure 13.6, also accounting for heterogeneous permeability in the aquifer. This resulted in the streamline pattern shown in Figure 13.7, which qualitatively reflects slow inland flow in the deep saltwater zone and faster seaward flow in the shallow freshwater part of the system. Most of the discharge is into the floor of Biscayne Bay. About one-eighth of the total discharge represents a return of seawater that entered the aquifer at locations further offshore. Kohout [in Cooper et al., 1964] notes two significant features in Figure 13.7, including (1) the zero-horizontal-gradient line, which separates the seaward and landward horizontal components of flow, and (2) the streamline that separates the groundwater into two regions according to whether it originated as freshwater or as seawater. These two lines intersect at the base of the aquifer [Cooper et al., 1964, p. 32]. This perspective is consistent with an analysis based on advective transport only.

13.3.1.3   Management Approaches

The South Florida Water Management District (SFWMD) is one of the five regional water management agencies in the state. The SFWMD includes an area of almost 18,000 mi^2 (47,000 km^2) that relies almost solely on groundwater as a source of supply. Hence, it is natural that a major focus would be on understanding, monitoring, and managing seawater intrusion.

As part of its operation, SFWMD controls and operates the extensive network of drainage and flood control canals that are distributed throughout the region. Most of these have control structures (or headgates) near their outlets that can be raised or lowered to control the water levels in the canals. Many canals are remotely monitored and many control structures are centrally operated and automated from the SFWMD headquarters to allow rapid responses to changing conditions. The canal and levee system is the primary means of controlling water levels in the Biscayne aquifer, and therefore represent a major management approach to controlling seawater intrusion. Keeping canal levels high for seawater intrusion control, however, sometimes conflicts with their flood control functions. More research is needed to define the intrusion risks and damage associated with brief periods (days to months) that canal levels might have to be lowered to minimize flooding potential. This could provide the basis for a model to help optimize canal operation to meet two sometimes conflicting goals while minimizing overall risks.

SFWMD also has the regulatory power to issue permits to construct water wells. This authority to grant or deny a permit represents one means to control or influence the distribution and magnitude of groundwater pumping. The impact of new wells on seawater intrusion is considered a major factor by SFWMD in making permitting decisions. SFWMD requires applicants to demonstrate conclusively that the new withdrawals will not cause seawater intrusion. If some risk is present, they may grant the permit but require the installation of saltwater monitoring wells. Permits can be granted for specific time periods, such as 5 years, and reevaluated on the basis of the monitoring outcome. If unanticipated intrusion and damage to the aquifer is detected, permitted withdrawals can be reduced or well fields shut down.

During periods of drought, SFWMD can impose water use restrictions throughout the District to conserve water. This minimizes pumping requirements, and reduces drawdown relative to what would otherwise occur. This, in turn, helps to minimize intrusion during periods of minimal aquifer recharge (although conservation clearly serves other purposes too).

Another management tool is surveillance of the aquifer system. A monitoring well network is used to observe changes in groundwater levels and chloride content. The monitoring network is an important component of the management strategy and provides (1) an early warning signal that water-supply operations may need to be modified, and (2) feedback on the effectiveness of prior management strategies. Monitoring efforts in south Florida are discussed in more detail below.

In the extreme event that seawater intrudes into the intake zones of a well or well field, and renders the pumped water too saline for use, by that time it may already be too late to rectify the problem through simple adjustments of pumping rates or pumping schedules. In this case, the only practical near-term solution may be to relocate the well field at a site further inland (or purchase water from another municipal agency that has wells further inland). This solution is costly, as it involves drilling new wells and (or) modifying the distribution system. This has already been required in parts of south Florida, but clearly represents a management approach taken more out of necessity than out of foresight.

In some cases, water in wells inland of the saltwater front may become degraded by increasing salinity. This is probably due to upconing rather than to lateral intrusion. For water-supply wells that are close to the position of the front or close to the coast, it may be difficult to determine whether a salinity increase is caused by upconing or lateral intrusion unless data are available from additional observation wells in the area. This distinction is important because different management actions would be warranted in the two different situations.

An innovative approach that is being operationally evaluated in this region involves the subsurface storage of injected freshwater in brackish or saline zones of the Floridan aquifer system. Freshwater is injected during periods of water surplus, or wet seasons, for recovery during future drought periods. The impact on seawater intrusion would only be indirect, to the extent that the recovery and use of injected water reduces the need to pump the freshwater aquifers during times of drought. The analysis of such systems, however, requires the use of the same types of variable-density flow and transport models as would be applied directly to simulate seawater intrusion (see, for example, Merritt [1997]).

Artificial recharge near coastal areas is being encouraged as a means to build up the water-table elevations and thereby control seawater intrusion. The many golf courses in the area are encouraged to irrigate with reclaimed waste water, which has the added benefit of reducing demand for freshwater, thereby also helping to control seawater intrusion. The shallow injection of treated waste water is also being considered as a means to build a mound in the water table exactly where needed to control seawater intrusion most effectively. However, unlike water for sprinkling, spraying, or irrigation, which recharges the shallow aquifer by infiltration, artificially recharged water that is injected directly into the aquifer must have undergone more expensive tertiary treatment.

Restoration of the Everglades and other wetlands has recently become a priority objective in south Florida. These efforts may require significant shifting of water-use patterns in the region, particularly with respect to irrigation and drainage. It remains to be seen if any of these efforts will have any impacts on seawater intrusion.

13.3.1.4   Monitoring

The U.S. Geological Survey, in cooperation with the South Florida Water Management District and with county and municipal agencies, has been operating an extensive network of monitoring wells in south Florida to help delineate changes in time and space in the extent of seawater intrusion. For operational purposes, the extent of seawater intrusion is defined as the landward limit of groundwater having a chloride concentration of 1,000 mg/l [Koszalka, 1995]. Because of density effects, this limit will be furthest inland at the base of the aquifer. Water samples collected at the end of the dry season were taken as the indicator.

The monitoring approach in eastern Broward County is described by Koszalka [1995] and offers an illustrative example. For the period from 1980-90, water levels and chloride content were measured in 63 monitoring wells in eastern Broward County. Many of the monitoring wells are near major water-supply well fields, and thus serve as an early-warning system for those specific water sources. To define flow and solute transport in a three-dimensional framework, the fluid pressure, density, and elevation must be known at a number of points. Thus, monitoring wells are intended to act as point samplers in the aquifer, so that any mixing or dilution during the sampling procedure is eliminated or minimized. To achieve this as closely as possible, the wells are constructed with a 1- to 10-ft open-hole section beneath the casing.

On the basis of data collected during 1980-90, Koszalka [1995] concluded that seawater intrusion has continued and progressed, so that in 1990 the 1,000 mg/l chloride line was more than 5 km from the coast near one well field south of Fort Lauderdale. Two individual monitoring wells located between individual well fields and the coast showed increases in chloride from less than 100 mg/l prior to 1980 to more than 5,000 mg/l in 1991.

Sonenshein and Koszalka [1996] describe trends in seawater intrusion in Dade County for a 20-year period (1974-93). During this period, observation wells were sampled regularly for changes in chloride concentration, and water-level fluctuations were measured continuously at as many as 73 locations at a time and intermittently at many additional sites. Although Sonenshein and Koszalka [1996] report that the data demonstrate that the saltwater front has continued to move inland in some parts of Dade County, there is also evidence that it has retreated and moved seaward in some areas near two of the well fields that are monitored. In one of these well fields, pumpage was decreased significantly during 1984-92, and the other well field had been shut down during 1977-82.

13.3.1.5   Natural Long-Term Trends

Where monitoring detects progressive seawater intrusion, it is normally attributed to the effects of drainage, pumping, and other water-supply development projects. However, some part of the change may actually be attributable to long-term background trends in response to geological processes, such as sea-level rise. Meyer [1989] analyzed in some detail the effects of rising sea level in south Florida on groundwater movement and seawater intrusion. During the past 150,000 yrs, sea-level fluctuations ranged from about 7 m above present-day sea level at about 140,000 yrs ago to about 100 m below present-day sea level about 18,000 yrs ago. Meyer [1989] noted that the last rise in sea level began about 18,000 years ago and is continuing, although at a much slower rate.

On the basis of geochemical, isotopic, and paleohydrologic evidence, Meyer [1989] concluded that sea-level rise would likely be accompanied by rising water tables, even if rainfall remained about the same. This is because sea level represents a base level to which groundwater flows, and an increase in this boundary elevation would lead to increased hydraulic heads throughout the system, which would partly counteract the induced seawater intrusion. Meyer [1989] also estimated that seawater moves inland about 1 mile for each 1-ft rise in sea level. The rate of encroachment of the saltwater front depends on aquifer properties, and would be relatively fast in the highly transmissive Biscayne aquifer.

Because the most recent eustatic sea-level change was a rise, which is probably continuing, the position of the seawater front in most coastal aquifers is probably moving slowly inland. This background trend may have to be accounted for in some areas in order to accurately determine or predict the effect of water-supply development on seawater intrusion. In other areas, the impact of pumping or other development activities may simply overwhelm the natural background trend of intrusion. In developing models of coastal systems, however, the slow response to prior sea-level changes may explain disparities between the present-day or predevelopment position of an interface and the postulated equilibrium position.

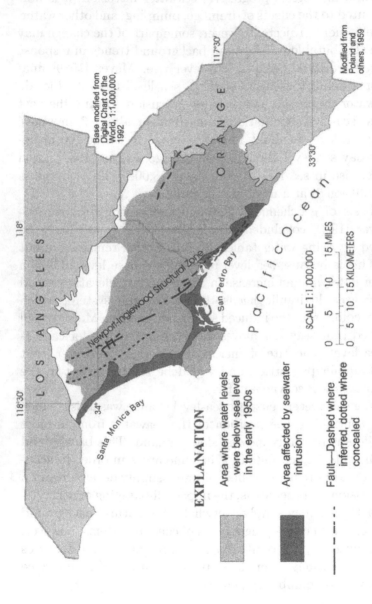

Map of Los Angeles-Orange County coastal plain basin. Withdrawals of groundwater in excess of natural recharge have resulted in declining water levels. By the 1950s, water levels had declined below sea level throughout most of the basin. The reversal of the hydraulic gradient from seaward to landward allowed seawater intrusion along most of the coast. (Modified from Planert and Williams [1995])

Figure 13.8.

13.3.2 Seawater Intrusion in Orange County, California

Orange County, California, is south of Los Angeles and has a semi-arid Mediterranean climate. Rainfall averages about 15 inches (380 mm) per year, and nearly all of it falls from late autumn to early spring. The Los Angeles-Orange County coastal plain basin (see Figure 13.8) is bounded on the west by the Pacific Ocean and on the north, east, and south by hills and mountains that are underlain by consolidated rocks of igneous, metamorphic, and marine-sedimentary origin [Planert and Williams, 1995]. These rocks surround and underlie thick unconsolidated alluvial deposits, which include several aquifer systems. Because the metropolitan Los Angeles area is one of the largest population centers in the world, the demand for water is great and the predominant use is for public supply [Planert and Williams, 1995]. Water sources include local groundwater and water imported from the Colorado River and from other parts of California. Groundwater withdrawals to meet the high demands exceed natural replenishment in the basin.

13.3.2.1 Hydrogeology

As described by Planert and Williams [1995], the coastal plain aquifer system is made up of as many as 11 aquifers. Each consists of a distinct layer of permeable sand and gravel, usually separated from other water-yielding units by clay and silt confining layers. The basin is crossed by a prominent structural feature—the Newport-Inglewood structural zone, which trends northwestward over a distance of about 40 mi (64 km). It represents a composite belt of anticlinal folds and subsidiary faults [Poland et al., 1956]. Structural activity in this zone has uplifted low-permeability marine sediments in places, and the fault zone itself constitutes a low-permeability barrier in most places. In Orange County, the Newport-Inglewood structural zone is located 1 to 3 mi (1.6 to 4.8 km) inland from (and parallel to) the coast (Figure 13.8). The sediments that form the uplifted mesas along the coast have low permeabilities; however, erosion formed gaps in the mesas, and the gaps were subsequently filled with more permeable alluvial deposits [Planert and Williams, 1995]. These gaps were the principal conduits for groundwater discharge under natural conditions, but also became the principal pathways for seawater intrusion after pumping reversed the groundwater flow directions near the coast.

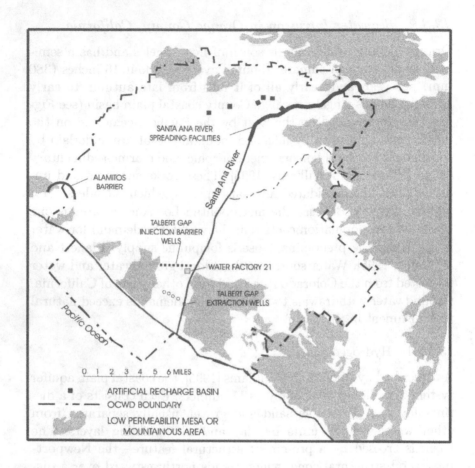

Figure 13.9. Map of Orange County Water District (OCWD) area, showing location of recharge facilities in coastal plain basin and relation to mesas and gaps (modified from OCWD data and maps).

The Santa Ana River discharges to the ocean through one of these gaps, referred to as either Talbert Gap or Santa Ana Gap in different publications (Figure 13.9). Poland [1959] used hydraulic and geochemical evidence to conclude that within the Santa Ana Gap the faults of the Newport-Inglewood structural zone do not constitute a complete barrier to groundwater flow and do not form a discontinuity in the hydraulic conductivity of the full sequence of permeable sediments. This is reflected in the cross-section shown in Figure 13.10. The continuity of the permeability in the shallow alluvial Talbert aquifer is not affected by the faults of the Newport-

Inglewood structural zone, although hydraulic continuity in deeper stratigraphic units is disrupted.

13.3.2.2   Water Management: Artificial Recharge

The Orange County Water District (OCWD) was formed in 1933 and given a mandate to protect the groundwater basin from depletion and irreparable damage [OCWD, 1993, p. 7] in an area of about 350 mi^2 (900 km^2) in the northern half of the county. One recognized source of damage resulting from depletion is seawater intrusion. A long-term drought that began in 1945 was accompanied by excessive groundwater withdrawals and drawdowns, which reduced average groundwater levels in the coastal basin to 5 m below sea level. Some coastal wells had to be abandoned because they began producing brackish water. This led to a more active water-management role by OCWD.

In 1948, OCWD started using imported water for artificial recharge. A law enacted in 1953 required those pumping groundwater to report semi-annually the amount of water they extracted. The cost of the imported water was then apportioned to all pumpers in the District; in 1954 pumpers were assessed \$3.50 per acre-foot (1.2×10^3 m^3) of groundwater withdrawn from the basin. This approach had the dual benefit of placing most of the costs of counteracting the depletion on those who were causing it and also of providing an added economic incentive for individual users to minimize their withdrawals and to eliminate any inefficient water use.

By 1956, Orange County's population was growing at an unprecedented rate, but groundwater levels had declined to an average of about 7 m below sea level, and saltwater was detected in aquifers as far as 3.5 mi (5.6 km) from the ocean [OCWD, 1993]. By the mid-1960s, increases in artificial recharge operations (see Figure 13.9) and shifting of some water use from groundwater to imported supplies had led to significant recovery from previous water-level declines. By the late-1960s, however, it became evident to OCWD that their recharge operations could not completely control seawater intrusion, and they decided to modify their basin management strategy and adopt a conjunctive-use policy [OCWD, 1993]. This approach tried to optimize between the use of imported water and groundwater, and continue the recharge program when water was available. While these management approaches were beneficial and worthwhile, they

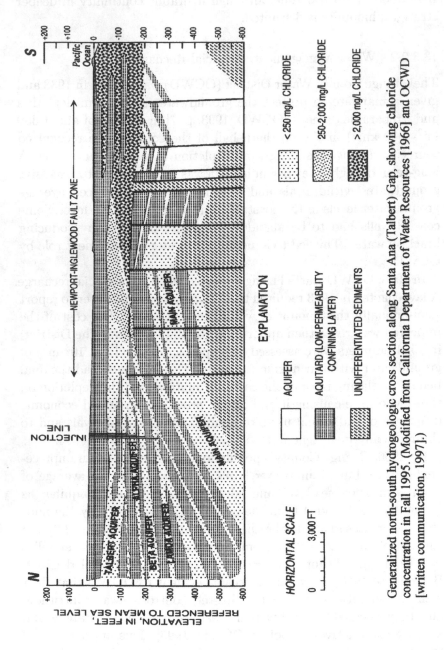

Figure 13.10.

Generalized north-south hydrogeologic cross section along Santa Ana (Talbert) Gap, showing chloride concentration in Fall 1995. (Modified from California Department of Water Resources [1966] and OCWD [written communication, 1997].)

were not sufficient to control seawater intrusion.

13.3.2.3   Hydraulic Barrier to Seawater Intrusion

In the Huntington Beach area, seawater migrated as far as 5 mi (8 km) inland through the buried valley of the Santa Ana (Talbert) Gap. As seawater intruded inland, there was a progressive increase in the abandonment of water wells [Calif. Dept. Wat. Res., 1966]. In 1965, OCWD began a pilot project to stop the intrusion in this area by injecting secondary treated wastewater. The injection causes the head to build up in the area to create a sufficiently high local groundwater divide and ensure that the hydraulic gradient would force flow from the injection area towards the coast. Thus, the injection wells would create a hydraulic barrier to seawater intrusion.

After six years of operating the pilot project, the California Department of Health Services (CDHS) approved a full-scale operation [OCWD, 1993]. However, they did not allow injection of only treated wastewater, but required it to be mixed with water from non-waste water sources that also met or exceeded drinking water standards. A new facility, called Water Factory 21, was constructed to provide advanced tertiary treatment, including granular activated carbon (GAC) adsorption and reverse osmosis (RO), of the wastewater prior to injection; the treatment plant had a capacity of 15 million gallons per day (MGD) (5.7×10^4 m^3/d). The RO plant had a capacity of 5 MGD (1.9×10^4 m^3/d). The minimum 33 percent non-waste water blend required by CDHS would be derived from four deep wells drilled into confined aquifers below those subject to seawater intrusion. The final mix that provided the water source for injection was derived in approximately equal proportions from the GAC process, the RO plant, and the deep wells. Aquifer recharge was implemented through 23 injection wells, which were spaced out along a 2.2-mi (3.5 km) east-west line located about 4 mi (6.4 km) from the Pacific Ocean (see Figures 13.9 and 13.10). About half-way between the injection wells and the coast, OCWD drilled seven withdrawal wells to extract saltwater that had already intruded into the aquifer and return it directly to the sea. Full-scale operation of the intrusion control project began in 1976.

The injection wells were constructed using multiple casings to allow recharge to flow directly into each of the four shallow aquifers subject to intrusion (see relation between injection wells and stratig-

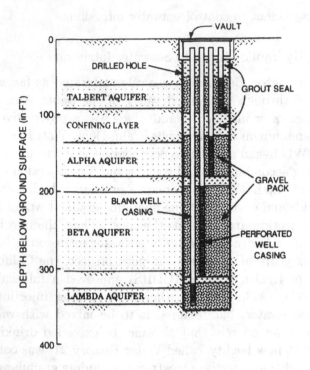

Figure 13.11. Schematic of typical injection well showing multiple screened intervals (modified from Wesner and Herndon [1990]).

raphy in Figure 13.10). A detailed schematic of a typical injection well is shown in Figure 13.11. Injection wells typically become clogged after some time, and require maintenance and redevelopment to restore them to their original capacity. Wesner and Herndon [1990] indicate that routine maintenance and redevelopment of the injection wells should occur at approximately 5-year intervals. Given the periodic maintenance, the operating efficiency of the injection wells has not deteriorated during more than 20 years of operations. A well rehabilitation program, however, is now underway to repair wells that were found to have corrosion holes in their casings and erosional piping around the borehole seals that allowed injection water to rise to the ground surface [R. Herndon, oral commun., 1997].

In 1991, the CDHS granted OCWD a permit to inject 100 percent recycled wastewater, without blending [OCWD, 1993]. The permit required that the operation continue as a research and demonstration project and that the injected water meet all drinking water standards. In order to assure compliance with stricter water quality standards of the new permit, however, it is expected that the capacity of the RO plant will have to be tripled, so that 75 to 90 percent of the flow will have been treated by RO.

OCWD also maintains an extensive monitoring well network as part of the injection barrier project. The primary purpose of the network is to assure that the injected water meets all quality standards. The secondary purpose of the network is to evaluate the effectiveness or efficiency of the operation relative to controlling seawater intrusion.

The OCWD was also evaluating the potential for desalting the brackish water produced in the extraction wells that are seaward of the injection barrier [OCWD, 1994]. This would provide additional potable water for the District. Toward that end, therefore, they were also simultaneously evaluating the possibility of adding additional extraction wells. The pumping from the expanded extraction system would help maintain a trough in the head distribution, which would act as an additional barrier to seawater intrusion. It could, however, also accelerate intrusion in the area between the coast and the line of extraction wells. This creates a complex problem requiring optimization of many factors, as well as related cost-benefit analyses. By 1997, the extraction wells had not been used for several years. Instead, recent planning in 1997 will evaluate the construction of an additional line of injection wells parallel to the existing line, but about 1 km closer to the coast [R. Herndon, oral commun., 1997].

OCWD, in cooperation with the Los Angeles County Flood Control District, operates another hydraulic barrier project (the Alamitos Barrier—see Figure 13.9) using a line of injection wells on the border between the two counties. This project started in the 1960s. By the early 1990s, continued seawater intrusion at the east end of the barrier and leakage through the central part have required the construction of five additional monitoring wells and four new injection wells to attain more effective control [OCWD, 1994]. Although injection barriers are complex and costly ventures to maintain, similar operations have been undertaken, or are being considered, by other jurisdictions elsewhere in coastal California.

13.3.2.4   Economic Benefit of Controlling Seawater Intrusion

The value of averting seawater intrusion in Orange County was assessed by the National Research Council [1997]. The underlying premise is that the loss of the groundwater basin to seawater intrusion is a long-term one (if not an irreversible one). This would cause the OCWD to rely more heavily on imported water, which is more expensive. It would also preclude the use of the aquifer as a component of the water storage and distribution system. Protecting the basin from intrusion will allow the water supply from this aquifer system to continue to be available for use.

Local groundwater is generally less expensive than imported water, primarily because of the development and transmission costs of the imported water [National Research Council, 1997, p. 147]. By projecting the value of Orange County's groundwater over a 20-year period, and comparing it to a similar projection for the value of imported water in the absence of fresh groundwater, the National Research Council [1997, p. 147] estimated that the present value difference between the two alternatives is about $3.41 billion. The economic benefit of controlling seawater intrusion thus appears to be significantly greater than the costs of constructing and operating the artificial recharge and seawater intrusion barrier systems. This did not include the added value of having an emergency water supply and distribution system available should the surface distribution system be disrupted or unusable or if the source of imported water were to be cut off or contaminated.

13.3.3   Oxnard Plain, Ventura County, California

The Oxnard Plain is a coastal basin located about 60 mi (100 km) northwest of Los Angeles and includes an area of about 120 mi^2 (330 km^2). It is underlain by a complex system of aquifers more than 1,400 ft (450 m) thick. Izbicki [1996] notes that the aquifer system can be divided into an upper and lower system. Seawater intrusion was first observed in the early 1930s and became a serious problem in the mid-1950s [Calif. Dept. Wat. Res., 1965]. This discussion of the Oxnard Plain is, to a large extent, extracted directly from Izbicki [1996].

13.3.3.1   Hydrogeology

As described by Izbicki [1996], the upper aquifer system consists of alluvial deposits about 400 ft (120 m) thick. This system contains two aquifers that have been developed for water supply—the Oxnard and Mugu aquifers. The Oxnard aquifer, about 180 ft (55 m) below land surface, is the primary water-yielding zone. It is underlain by the Mugu aquifer and overlain by a thick, areally extensive clay deposit, which separates it from a shallow unconfined aquifer. The Oxnard and Mugu aquifers crop out less than one-quarter mile offshore in Hueneme and Mugu submarine canyons.

Native water in the Oxnard and Mugu aquifers, which are part of the upper aquifer system, is generally fresh and has chloride concentrations of about 40 mg/l. In some areas, however, interbedded fine-grained deposits contain saline water. Prior to the onset of seawater intrusion, these two aquifers were extensively pumped for water supply. The shallow unconfined aquifer contains both fresh and saline water, but it is not used for water supply. Saline water in the shallow aquifer results from a combination of (1) seawater that recharged the aquifer through offshore outcrops or infiltrated into the aquifer through coastal wetlands or coastal flooding, (2) concentration of dissolved minerals from evaporative discharge of groundwater, and (3) infiltration of irrigation return water.

The lower aquifer system consists of alternating layers of alluvial sand and clay. These deposits grade to sediments of marine origin near the coast and overlie fine-grained marine sands; the sands are separated by marine silt and clay interbeds as much as 50 ft (15 m) thick. Native water in the lower aquifer system is fresh and has chloride concentrations that range from 40 to 100 mg/l. The system is surrounded and underlain by partly consolidated marine and volcanic rocks that contain saline water.

13.3.3.2   Seawater Intrusion and Monitoring

Groundwater pumping caused water levels in parts of the upper aquifer system to decline below sea level and below the water levels in the shallow aquifer. Seawater entered the aquifers through outcrop areas in Hueneme and Mugu submarine canyons in the mid-1950s and advanced inland over time, as shown in Figure 13.12. Historically, local agencies responsible for the management of groundwater used a criterion of 100 mg/l chloride to define the leading edge of the seawater front. It was assumed that all high-chloride water from wells behind the front originated from seawater that entered the aquifers' outcrop areas in submarine canyons.

By 1989, about 23 mi^2 (60 km^2) of the upper aquifer system was believed to be intruded by seawater (Figure 13.12). Because of increasing chloride concentrations, pumping was shifted from the upper system to the lower system. Subsequently, water levels in the lower aquifer system declined to below sea level. Increasing chloride concentrations were observed in the lower aquifer system near Mugu submarine canyon as early as 1985, and high-chloride water was discovered near Hueneme canyon in 1989 [Izbicki, 1991]. After 1993, a combination of groundwater management strategies and increased availability of surface water for groundwater recharge caused water levels in upper aquifer wells near the coast to rise above sea level. Water levels in parts of the lower aquifer system also rose above sea level, but they were still below sea level near Mugu canyon as late as 1996.

Prior to 1989, most wells used to monitor chloride concentrations were abandoned (unused) agricultural supply wells, and many of these were screened in more than one aquifer. Recently, however, 32 new monitoring wells were installed and screened in individual aquifers. Results of sample analyses showed that water from abandoned agricultural wells was not representative of groundwater in the Oxnard or Mugu aquifers. This is because these wells have (1) large diameter casings and are difficult to purge of stagnant water prior to sample collection, (2) screened intervals that may be open to more than one aquifer (hence, chloride in individual aquifers cannot be defined), and (3) steel casings that are subject to corrosion and failure—especially where water in the shallow aquifer is saline.

Corrosion and failure of abandoned wells may result in leakage of water from the shallow aquifer into underlying aquifers. Izbicki [1996] showed that this was the case at several sites where the shallow groundwater had very high chloride content and water levels higher than in the underlying systems, allowing downward leakage of saline water. This was confirmed using velocity logs in the suspect boreholes. The outcome is that the saline leakage gave false indications of the extent of seawater intrusion. Results of surface-resistivity surveys confirmed that the areal extent of seawater intrusion in the upper aquifer system was smaller than previously believed. The geo-

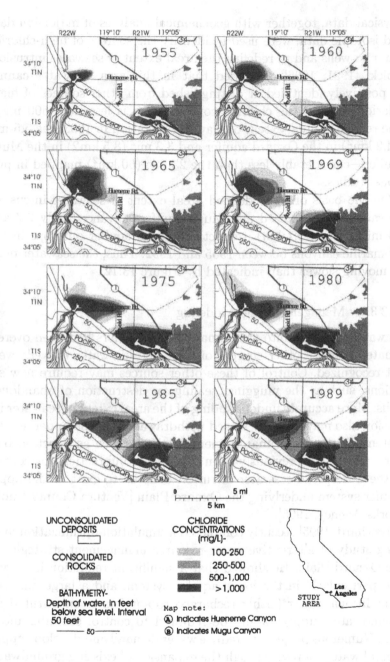

Figure 13.12. Chloride concentrations in water from wells in the upper aquifer system in the Oxnard Plain, 1955-89 (from Izbicki [1996], based on data from California Dept. of Water Resources and County of Ventura Public Works Agency.)

physical data, together with geochemical analyses of major-ion data and isotopic data, were used to identify the sources of high-chloride water to wells and to redefine the areal extent of seawater intrusion. Izbicki [1991; 1996] concluded that in this area, seawater cannot be positively identified or distinguished from other sources of high-chloride water if the chloride concentration is less than 500 mg/l. The redefined areal extent of seawater intrusion in 1993 was 5.5 mi^2 (14.2 km^2) in the Oxnard aquifer and 3.3 mi^2 (8.5 km2) in the Mugu aquifer—considerably less than the 23 mi^2 (60 km^2) mapped in previous studies.

On the basis of the redefined areal extent of seawater intrusion, seawater in the Oxnard and Mugu aquifers advanced only 2.7 and 1.9 mi (4.3 and 3.1 km), respectively, from aquifer outcrops in the submarine canyons between 1955 and 1992. Thus, the seawater must be moving slower than indicated by Figure 13.12.

13.3.3.3   Management and Modeling

Seawater intrusion into the Oxnard Plain aquifers had been overestimated because other sources of high-chloride water to wells were not recognized. Control of these other sources may require new solutions, such as the plugging, sealing, or destruction of abandoned wells. More accurate understanding of the areal extent of seawater intrusion also may allow increased withdrawals from the upper aquifer system, or the use of management solutions, such as injection barriers, which had previously been rejected as impractical for control of the widespread seawater intrusion believed to exist in the upper aquifer system underlying the Oxnard Plain [Ventura County Public Works Agency, 1995].

Reichard [1995] recently completed a simulation-optimization modeling study of alternative water-resource management strategies in the Oxnard Plain. He showed that a significant reduction in pumping, particularly in the lower aquifer system, and a large quantity of additional water (either recharged to aquifers or substituted for groundwater supply) would be required to control seawater intrusion. Numerous projects to reduce water demand and develop supplemental water sources through the expansion of existing groundwater recharge facilities, expansion of existing water-distribution facilities, use of reclaimed water, and injection of imported or reclaimed water into aquifers underlying the Oxnard Plain are planned by local water agencies [Ventura County Public Works Agency, 1995]. Accurate definitions of the hydrogeologic framework and of the true three-dimensional extent of seawater intrusion are essential for the effective management of groundwater resources of the basin.

Nishikawa [1997] applied the SUTRA variable-density transport code [Voss, 1984a] to this area to help evaluate the physical processes controlling seawater intrusion. The model analysis supports the concept that the submarine canyons represent a main avenue for seawater to enter the aquifer system. One of the conclusions is that seawater intrudes mostly through thin basal layers having relatively high hydraulic conductivity. Nishikawa and Reichard [1996] subsequently used the SUTRA model of this area in a Monte Carlo framework to evaluate the decision rules derived from the simulation-optimization model of Reichard [1995], who did not directly incorporate transport into his analysis. Their results indicate for a 15-year period that chloride concentrations will probably be lower in both aquifer systems under the optimized decision rules than if heads for 1993 (wet-year) conditions were simply maintained for the entire period.

13.4 Alternative Modeling Approaches

The application of a numerical simulation model to an aquifer system subject to seawater intrusion offers an opportunity to quantify the level of understanding of the problem and the system, by integrating the complex theoretical understanding of the physics and hydraulics with the geological and hydrological data and concepts describing the system of interest. However, because the governing system of equations are nonlinear, a comprehensive and detailed model may require more computational resources than are available or practical. Hence, as in many engineering analyses, assumptions and simplifications must be made to render the problem tractable. Judgment and common sense are keys to balancing the trade-off between accuracy and efficiency, and to assure that simplifications are physically reasonable and appropriate, yet preserve the essential features of the system and problem. If this is accomplished, then the resulting model can be a valuable tool to help test and evaluate alternative management options.

The purpose of this section is to review a variety of approaches to modeling seawater intrusion that have been applied in the United States. The content of this section is extracted, in part, from Reilly [1993].

13.4.1 System Conceptualization

In general terms, the object of quantitative analysis of seawater intrusion is to understand and describe the relation between saline and fresh groundwater. The physical system is highly complex, and therefore is rarely treated in terms of fully three-dimensional density-dependent miscible fluid flow in a porous medium. Instead, simplifying assumptions are usually made to (1) reduce the dimensionality of the problem, (2) simplify the boundary conditions, geometry, and heterogeneity of the system, and (or) (3) simplify the physics and governing equations. All of these represent a simplification of the conceptual model in one way or another. Perhaps the most important assumption concerns the tendency of the freshwater and the saltwater to mix.

Under certain conditions, these two miscible fluids can be considered as immiscible and separated by a sharp "interface" or boundary. This assumption of a sharp interface has been used successfully and has had an important effect on the mathematical representation of the physical process. The sharp-interface problem can be formulated in terms of two distinct flow fields—the freshwater flow field and the saltwater flow field. The two systems are coupled through their common boundary, although it is assumed that fluid from one region cannot cross the interface boundary into the other region. The sharp interface approach is generally most useful when the problem at hand involves a regional or large-scale perspective on a system, and it is acceptable to ignore local variations in salinity, say at particular observation or pumping wells.

When analyzing systems under steady-state conditions, where saltwater is assumed to be stationary and in equilibrium with freshwater, the problem can be further simplified. If the saltwater is static, there are no saltwater head gradients. This formulation then only requires flow in the freshwater system to be analyzed, and the sharp interface to move in response to freshwater heads. This analysis simplifies the two-fluid problem into a single-fluid problem with a free-surface boundary condition. This method has been used by Glover [1959], Bennett and Giusti [1971], Fetter [1972], Voss [1984b], Guswa and

LeBlanc [1985], Reilly et al. [1987], and others.

When the saltwater is not static, then the sharp interface problem must be formulated in terms of flow in two separate but adjoining systems having a common free-surface boundary (the saltwater interface). Analysis of the problem using this conceptualization requires the solution of two simultaneous partial differential equations for the flow of freshwater and saltwater. This approach has been used by Bonnet and Sauty [1975], Pinder and Page [1977], Mercer et al. [1980a], Wilson and Sa Da Costa [1982], and Essaid [1990a, b]. All but one of these applications assumed each flow domain was a two-dimensional system. Essaid [1990a, b] developed a quasi-three-dimensional approach whereby three-dimensional systems can be simulated.

A more rigorous conceptualization of saltwater-freshwater systems is that of one miscible fluid having heterogeneous physical properties (fluid density and viscosity), which are controlled by variations in the dissolved-solids concentration and temperature of the fluid. In turn, the concentration and temperature are affected by transport processes (including advection, dispersion, and diffusion). It is common to assume that the dependence of fluid properties on concentration can be adequately described by an appropriate single reference constituent, such as chloride. Under that assumption, the HST3D model [Kipp, 1987] is an example of a code that can solve the coupled groundwater flow, solute transport, and heat transport equations in three dimensions. For many intrusion problems in coastal areas, the total thickness of the aquifer system of interest is small enough that temperature variations and heat transport can be safely ignored. The simulation problem is then reduced to solving two simultaneous non-linear partial differential equations that express the conservation of mass of fluid and conservation of mass of salt. Most applications to date in the United States have utilized such models in only two dimensions. Examples include Souza and Voss [1987], Voss and Souza [1987], and Bush [1988].

In limited cases, this variable-density approach can be further simplified for large regional simulations to assume that the concentration (which defines the density) is known (and varies spatially) but will not change with time during the period of the analysis. Under this assumption, a solution to the transport equation is not required, and a solution of the density-dependent flow equation is sufficient to de-

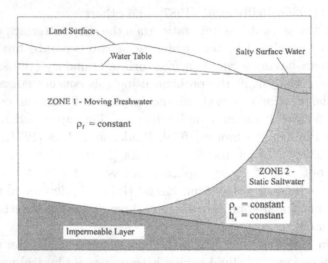

Figure 13.13. A simplified cross section showing sharp-interface approach applied at Cape Cod, MA (from Reilly [1993]).

fine the flow system. This approach has been used by Weiss [1982], Garven and Freeze [1984], Kuiper [1985], and Kontis and Mandle [1988].

13.4.2   Regional Sharp-Interface Conceptualization

13.4.2.1   Cape Cod, Massachusetts

Groundwater is the principal source of freshwater supply for Cape Cod, Massachusetts. A lens-shaped reservoir of fresh groundwater is maintained in dynamic equilibrium by recharge from precipitation and discharge to the ocean and streams. To evaluate the eventual equilibrium impacts of regional groundwater development, quantitative analyses of the area were undertaken by Guswa and LeBlanc [1985].

The system was analyzed as a three-dimensional flow field under steady-state conditions, assuming that a sharp interface separates shallow freshwater flowing over static saltwater, as illustrated in Figure 13.13. The observed interface on Cape Cod was defined as narrow, allowing for this approximation. The data available and used in assessing the reasonableness of the simulations included water budgets, geologic framework information, aquifer-test information, water-table information, and estimates of the location of the saltwater-freshwater transition zone. Additional information that is useful with this approach, but is rarely available, is measurements of saltwater heads.

The advantage of this approach was that sufficient information was available to show and test the reasonableness of the simulations. Because only regional estimates were required, the steady-state, sharp-interface approach was a reasonable approximation. The disadvantage was that the ability of individual wells to supply water with low chloride concentrations could not be evaluated. Overall, the results are useful in developing regional water management plans.

13.4.2.2   Montauk area, Long Island, New York

The Montauk peninsula, at the extreme eastern end of Long Island, New York is a popular resort area that draws tens of thousands of vacationers annually. The sole source of fresh groundwater is a series of Pleistocene glacial deposits that are bounded below and laterally by saltwater. The permanent population is only a few thousand; thus, the seasonal increase in population imposes a large, fluctuating demand on the groundwater system. Prince [1986] developed a simulation model of the system and then estimated its response to the present use and future demand.

The system was analyzed as a two-dimensional flow field assuming a sharp interface separates moving freshwater over moving saltwater, as illustrated in Figure 13.14. The thickness of the saltwater-freshwater transition zone was usually less than 7 m, indicating that a sharp-interface approximation was reasonable.

The advantage and disadvantage of this approach are the same with the Cape Cod case. Overall, the results were useful in determining that the principal aquifer is capable of producing more than 2.3 million liters per day, and for developing regional water-management plans.

13.4.2.3   Soquel-Aptos Basin, Santa Cruz County, California

The physiography of the Soquel-Aptos Basin, Santa Cruz County, California ranges from very steep valley slopes and angular landforms in the Santa Cruz Mountains to nearly flat marine terraced, sea cliffs, and narrow beaches along Monterey Bay. The region is mainly an urban area, and growth is projected with concurrent in-

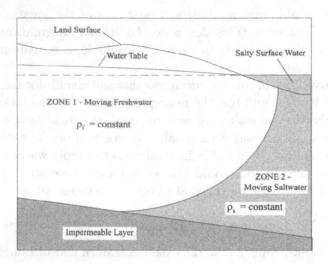

Figure 13.14. A simplified cross section showing sharp interface approach applied at Montauk, NY (from Reilly [1993]).

creases in water demand. The principal hydrologic unit of interest is the Purisima Formation, which is a layered aquifer of variable thickness. Saltwater is not yet present in the aquifer inland of the coast, and the position of the interface offshore is not known. An analysis was undertaken by Essaid [1990b] to quantitatively estimate (1) the amount of freshwater flow through the system, (2) the quantity of freshwater outflow to the sea, (3) the undisturbed position of the saltwater interface offshore, (4) the quantity of discharge to the sea that must be maintained to keep the interface at or near the shore, and (5) the rate at which the interface will move because of onshore development.

The system was analyzed as a three-dimensional flow field under time-varying conditions and assuming a sharp interface with moving freshwater over moving saltwater, as illustrated in Figure 13.15. The three-dimensional approach incorporates the ability of freshwater to discharge from deeper layers into overlying layers containing saltwater. The data available and used in assessing the reasonableness of the simulations were: water budgets, geologic framework information, aquifer-test information, and time-varying water-table information. An estimate of the location of the saltwater-freshwater transition zone was not available. Additional information that is also

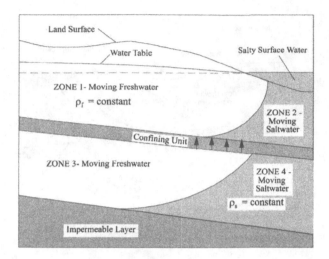

Figure 13.15. A simplified cross section showing sharp interface approach applied at Soquel-Aptos Basin, CA (from Reilly [1993]). Density is constant over time.

useful with this approach, but is rarely available, is measurements of saltwater heads.

The advantage of this method of analysis was that the three-dimensional nature of the flow field and staggered interface location could be estimated. Also, the importance of the transient response of the system was better understood. In fact, it was shown that the present interface is probably still responding to long-term Pleistocene sea-level fluctuations and has not achieved equilibrium with present day sea-level conditions. This is consistent with the results of Meisler et al. [1985], who concluded that in the northern Atlantic Coastal Plain, the transition zone between fresh and salty groundwater is not in equilibrium with present sea level, but probably reflects a position consistent with a long-term average sea level 15 to 30 m below present sea level. Thus, they also concluded that the transition zone is moving slowly landward and upward towards equilibrium with present sea level.

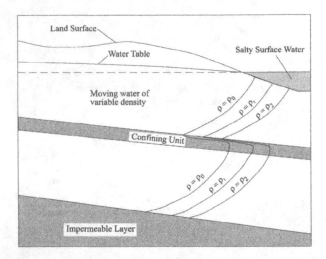

Figure 13.16. A simplified cross section showing density-dependent flow approach (from Reilly [1993]). Application to Gulf Coastal Plain assumed density is constant over time, so only flow equation need be solved. Application to southern Broward County calculated changes in density over time by also solving the transport equation.

13.4.3   Variable-Density Flow with Steady-State Solute Distribution

13.4.3.1   Gulf Coastal Plain

The aquifer systems of the Gulf Coastal Plain of the United States encompass 750,000 km², including the Mississippi embayment and offshore areas beneath the Gulf of Mexico. The gulfward thickening wedge of unconsolidated to semiconsolidated sediments is a complexly interbedded sequence of sand, silt, and clay with minor beds of lignite, gravel, and limestone. The sediments crop out in bands approximately parallel to the present-day coastline of the Gulf of Mexico. Approximately 36 billion liters per day of groundwater was pumped from the system in 1980. The presence of saltwater at very different concentrations in complex areal and vertical distributions necessitated the analysis of the system as a variable-density flow problem. The vast extent of the area necessitated a regional approach, and a simulation was undertaken by Williamson [1987] to describe this complex system.

The system was analyzed as a three-dimensional flow field having a variable-density fluid, as illustrated schematically in Figure 13.16. However, the density distribution was assumed to remain constant in space and time. The three-dimensional approach incorporates the ability of freshwater to discharge from deeper layers into overlying layers containing saltwater. The variable-density approach does not treat the dispersed boundary between the freshwater and saltwater as an interface, and allows for some fluid flow to occur between and within the freshwater and saltwater zones. The major assumption in this approach is that the density (concentration) distribution does not change in time because of the fluid flow. This allows the analysis to be based on only the variable-density flow equation and does not require analysis of salt transport. Williamson [1987] states that this approach is valid for this case because the volume of water simulated as moving into an adjacent model block in a few decades at maximum flow rates will not significantly affect the average density in the large regional model blocks, and therefore would not affect the solution. This method of analysis, however, may not be appropriate to assess long-term steady-state results. The data available and used in assessing the reasonableness of the simulations were: water budgets, geologic framework information, aquifer-test information, and time-varying water-table information. An important data element that is required in this analysis, which is input to the model as known information, is the three-dimensional density distribution for the entire system. These data are very important and are rarely known to the extent necessary. As with the previous examples, additional information that is also useful with this approach, but is rarely available, is measurements of pressures in the saltwater part of the system.

The advantage of this method of analysis is that the importance of the complex variable-density distribution on the three-dimensional flow field can be understood and quantified to some extent, while saving the computational costs of solving the three-dimensional solute-transport equation. This is probably the only approach that will allow such an analysis at the present time. The disadvantage is that any errors in conceptualizing the system and describing the distribution of the densities in the flow field directly affect the results and may induce errors that compensate for other uncertainties. Therefore, it is difficult to assess the reasonableness or accuracy of the analysis. Williamson's [1987] analysis of this system indicated that on a regional basis, the resistance to vertical flow caused by many thin, localized fine-grained beds within the permeable zones can be as important as the resistance caused by regionally mappable confining units.

13.4.4 Variable-Density Flow and Solute Transport

13.4.4.1 Southern Broward County, Florida

As discussed previously, the Biscayne aquifer is an unconfined carbonate aquifer that underlies southernmost Florida. The aquifer is critical in that it is used extensively for public supply. The Biscayne aquifer has been comprehensively studied (see section 13.3.1); the system has been relatively well defined and the extent of the saltwater intrusion monitored since 1939. This comprehensive monitoring indicated that between 1945 and 1993 the saltwater front has moved as much as 0.5 mile (0.8 km) inland in parts of southern Broward County, Florida.

A study was undertaken in the early 1990s to use a variable-density flow and solute-transport simulation approach to determine the relative importance of various hydrologic and water-management factors in determining the previous positions and estimating future positions of the saltwater front. Merritt [1996] used the Subsurface Waste Injection Program (SWIP) [INTERCOMP Resource Development and Engineering, Inc., 1976; INTERA Environmental Consultants Inc., 1979] to simulate the salinity distribution under predevelopment conditions. Merritt also used a sharp interface model, SHARP [Essaid, 1990a], to check the consistency and reasonableness of the simulations made by both methods.

Merritt [1996] used the SWIP model to simulate the system using different system layering and characteristics. This approach is represented schematically in Figure 13.16 for conditions in which hydraulic properties can vary within and between layers, fluid density is variable spatially throughout the system, and temporal changes in density are determined by simulating the transport of salt through the system. In the layered simulations, near-equilibrium positions of the saltwater transition zone occurred in the more permeable zones within one to several years, indicating the dynamic nature of the system in response to development.

Both the sharp interface simulation of the area and the density-dependent simulation produced similar positions of the saltwater-

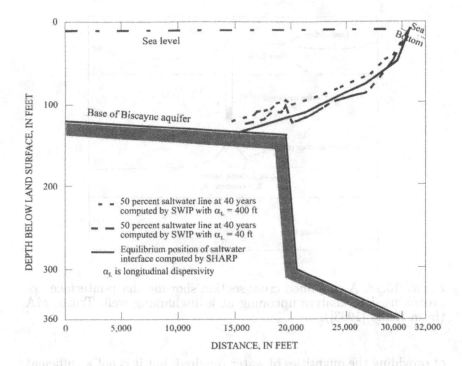

Figure 13.17. Positions of the center of the transition zone in southern Broward County, Florida, simulated by the variable-density model SWIP and the interface simulated by the SHARP model under predevelopment conditions (from Merritt [1996]).

freshwater boundary, as shown in Figure 13.17. Neither simulation, however, reproduced the actual location very well. This leads to the conclusion that even after years of comprehensive monitoring and study, the conceptual model of this aquifer system may not be complete. This result of these simulations is very helpful in that it clearly shows the gaps in our understanding and paves the way for a better understanding and representation of this critical system.

13.4.5   Local Analysis of Upconing Beneath a Discharging Well

Although the other cases focused on regional problems, it is important to note that in order for an aquifer to supply freshwater to wells, it is a necessary condition that the regional system be capable

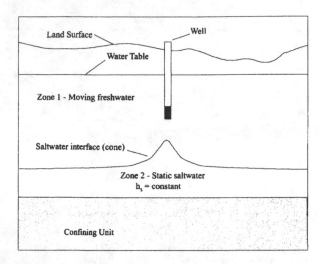

Figure 13.18. A simplified cross section showing sharp-interface approach used to analyze upconing at a discharging well, Truro, MA (from Reilly [1993]).

of providing the quantities of water required, but it is not a sufficient condition. Even though the regional system may be in equilibrium, local saltwater movement near discharging wells can make these wells produce unpotable water. Thus, the regional system may be capable of sustaining the rate of production, but the drawdown in the vicinity of an individual well may cause a local, vertically-upward movement of saltwater into the well (called upconing). In his analysis of Montauk, Long Island, NY, Prince [1986] included some estimate of the ability of individual wells of different designs to supply water. Thus, the water management must be a combination of regional considerations and local considerations.

The methods of analysis are the same as those discussed in the regional studies; however, the scale of interest is different. A well in Truro, Cape Cod, MA, was analyzed using the sharp interface approach [Reilly et al., 1987] and the density-dependent flow and transport approach [Reilly and Goodman, 1987], as illustrated conceptually in Figures 13.18 and 13.19, respectively. The analyses at Truro indicated that although the regional system was in equilibrium and capable of sustaining an estimated 4.2 million liters per day at a particular well field [LeBlanc 1982], the actual well was

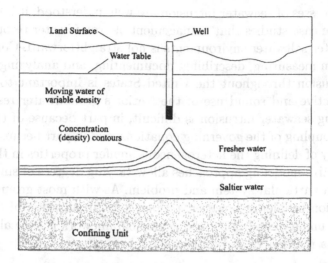

Figure 13.19. A simplified cross section showing density-dependent solute-transport approach used to analyze upconing at a discharging well, Truro, MA (from Reilly [1993]).

not capable of this production because of local effects. A single well withdrawing the total amount can create large drawdowns that can cause the saltwater to move to the well, but still be in equilibrium on a regional basis. In the Truro example, the withdrawals at the well averaged about 1 million liters per day annually and were usually more than 2.5 million liters per day in the summer, which caused the water from the well to show some indications of saltwater contamination. Estimates from the local analysis indicated that the maximum permissible discharge from the single well is probably less than 1.8 million liters per day. Thus, the withdrawal rate should be limited at the one well and additional wells installed if more capacity (up to the regional capacity) is required.

13.5   Closing Remarks

Saltwater intrusion into fresh groundwater systems has been recognized in the United States since 1824. An evolution of technical approaches in the study and management of the movement of fresh and salty groundwater is provided in this chapter by summarizing the long history of case studies available. Although the basic concepts and processes of seawater intrusion are well understood, it is obvious from the case studies that management of groundwater resources in freshwater-saltwater environments is not straightforward. Continued efforts in measuring, describing, documenting, and analyzing seawater intrusion throughout the United States is important to enable the effective and sound use of the Nation's groundwater resources. Modeling seawater intrusion is difficult, in part because of the nonlinear coupling of the governing equations, and in part because of the difficulty of defining the heterogeneous aquifer properties in the field. Alternative modeling approaches allow varying degrees of simplification of a particular system and problem. As with most groundwater simulation problems, however, proper conceptualization of the flow field, boundary conditions, and geologic framework is critical for obtaining a reasonable and reliable model.

Acknowledgments: Our USGS colleagues Bill Back, Roy Sonenshein, John Izbicki, and Tracy Nishikawa generously provided source material and helpful ideas. Special thanks go to Mr. Roy Herndon (Orange County Water District) and Mr. Keith Smith (South Florida Water Management District) and their colleagues for generously providing their time, helpful discussions, and source material. We appreciate the helpful review comments and suggestions on the manuscript from Larry Land, Peter Martin, Roy Sonenshein, Chet Zenone, and Roy Herndon.

Chapter 14

Impact of Sea Level Rise in the Netherlands

G .H. P. Oude Essink

14.1   Introduction

At this moment, a large number of coastal aquifers, especially shallow ones, already experience a severe saltwater intrusion caused by both natural as well as man-induced processes. Coastal aquifers, which are situated within the zone of influence of mean sea level (M.S.L.), will be threatened even more by the rise in global mean sea level. Present estimates of global mean sea level rise, as presented in the Intergovernmental Panel of Climate Change (IPCC) Second Assessment Report, range from 20 to 86 cm from the year 1990 to 2100, with a best estimate of 49 cm, including the cooling effect of aerosols [Warrick et al., 1995]. The extent of the impact of sea level rise on coastal aquifers depends on: (1) the time lag before a new state of dynamic equilibrium of the salinity distribution is reached and (2) the zone of influence of sea level rise in the aquifers.

The rise in sea level will accelerate the salinization process in the aquifers which can reduce the freshwater resources. For aquifers that lie below M.S.L., the present capacity of freshwater discharge may not be sufficient to cope with the excess of saltwater seepage inflow. As a consequence, crops may suffer from salt damage and fertile arable land might eventually become barren land.

Due to sea level rise, the mixing zone between fresh and saline groundwater will be shifted further inland. Groundwater extraction wells, which were originally located in fresh groundwater, will then be located in or nearby brackish or saline groundwater, so that up-

J. Bear et al. (eds.), Seawater Intrusion in Coastal Aquifers, 507–530.

coning can more easily occur. This can be considered as one of the most serious impacts of sea level rise for every coastal aquifer where groundwater is heavily exploited.

In some areas, shoreline retreat due to sea level rise will also affect coastal aquifers by narrowing the width of the sand-dune areas, thus diminishing the area over which natural groundwater recharge occurs. This may lead to a decrease in freshwater resources. In addition, higher salinity due to an increased saltwater intrusion in rivers and estuaries may jeopardize adjacent aquifers through recharge from surface water with a higher solute content. Consequently, the extraction rates of existing pumped wells may have to be reduced or the wells might have to be abandoned.

However, countermeasures can be executed to compensate the impact of sea level rise:

- extracting saline or brackish groundwater, thus reducing the inflow of saline groundwater and decreasing seepage quantity and chloride load in low-lying areas;
- infiltrating fresh surface water, thus pushing saline groundwater away;
- reclaiming land in front of the coast, thus generating new freshwater lenses;
- inundating low-lying areas, thus removing the driving force of the salinization process;
- widening existing sand-dune areas where natural groundwater recharge occurs, thus creating thicker freshwater lenses;
- increasing (artificial) recharge in upland areas, thus reducing the inflow of saline groundwater into the coastal aquifer;
- modifying pumping practice through reduction of withdrawal rates and/or adequate location of extraction wells;
- creating physical barriers, thus blocking the free entrance of saline groundwater and halting the salinization process.

A combination of countermeasures can probably reduce the stress in the coastal aquifers.

Note that, nowadays, dramatic lowering of the piezometric levels due to excessive overpumping already occurs in many coastal aquifers around the world, such as in the regions of Bangkok, Venice and Shanghai. It is obvious that for those coastal aquifers, the impact of a (relatively small) sea level rise on the aquifer system will be of marginal importance compared to the effect of an increase in withdrawal rate.

Only a few quantitative studies of saltwater intrusion in coastal aquifers induced by sea level rise in the world are available: e.g. the North Atlantic Coastal Plain, USA [Meisler et al., 1984]; the Galveston Bay, Texas, USA [Leatherman, 1984]; the Potomac-Raritan-Magothy aquifer system, New Jersey, USA [Lennon et al., 1986; Navoy, 1991]; and the Netherlands [Oude Essink, 1996].

In the Netherlands, coastal aquifers have been under the stress of saltwater intrusion, and will be threatened even more in the future by sea level rise. The salinization process can accelerate which will result in a reduction of freshwater resources. Furthermore, the seepage quantity through the Holocene aquitard may increase in areas situated below M.S.L. The present capacity of the discharge systems (pumping stations and water courses) in several low-lying areas, called polders, near the coast may be insufficient to cope with the excess of seepage water. This seepage will probably have a higher salinity than at present. A substantial increase in volumes of water required for flushing will be needed to counteract the deterioration of the water quality. In addition, crops may suffer from salt damage due to the increased salinity of the soil, which could have serious economic implications.

The possible impacts of sea level rise and human activities (countermeasures) on vulnerable coastal aquifers in the Netherlands for the next millennium are investigated through numerical modeling. A two-dimensional groundwater flow model is applied to assess: (1) the propagation of sea level rise in the coastal aquifers; (2) the changes in the salinity distributions of these aquifers; (3) the changes in the volumes of freshwater lenses in sand-dune areas and (4) the changes in the seepage (both quantity and quality) in low-lying areas.

14.2   The Model

Profiles have been chosen perpendicular to the Dutch coastline, as a two-dimensional model is applied. Each profile characterizes a representative coastal area in the Netherlands with its own specific subsoil conditions. Strictly speaking, the number of conceivable profiles is

infinite. For reasons of simplicity, however, the number of representative profiles has been limited to eight. Figure 14.1 shows the coastal
areas in which these eight representative profiles are located. Note
that, unfortunately, it is not possible to model the islands of Zeeland
by means of a representative profile as they cannot be schematically
represented in two dimensions. For instance, sand-dune areas occur
in the western seaside parts of the islands of Zeeland, which would
complicate a proper selection of the orientation of the profile. In
addition, the geometry of the polders is such that axial-symmetric
distributions of piezometric levels often occur. Moreover, the characteristics of these islands, such as the thickness of the hydrogeologic
system, the boundary conditions and the phreatic groundwater levels, differ too much from each other to join them together in one
representative profile.

Apparently, three-dimensional effects, such as groundwater flow
perpendicular to the profile due to groundwater extraction or low
phreatic groundwater levels in polders, are left out of consideration.
In fact, these effects may be important for the interpretation of the
results on a local scale, but for the applied profiles, in which subsoil parameters are averaged over areas of several hundred square
kilometers, the effects are probably of minor importance.

14.2.1 Subsoil Conditions of the Profiles

Figure 14.2 shows the schematized geometry, the values of subsoil
parameters and the phreatic groundwater levels in the polders of
each of the eight profiles. The data of the hydrogeological scheme
have been borrowed from numerous sources. Throughout the coastal
aquifer, the effective porosity $= 0.35$, the longitudinal dispersivity
$\alpha_L = 0.2$ m and the anisotropy factor $= 0.1$. (Unfortunately, the applied computer code uses only constant values for these three subsoil
parameters.) Sand-dune areas with freshwater lenses occur in the
profiles *Ber*, *Haa*, *Lei* and *Wad*. Note that the phreatic groundwater
levels in the low-lying polders are below N.A.P. in all profiles, except
in the profile *Vla* (N.A.P. stands for Normaal Amsterdams Peil and
is the reference level in the Netherlands. N.A.P. roughly equals Mean
Sea Level.). Obviously, this situation reveals that a permanent salinization process must be taking place in the direction of the low-lying
polders. Figure 14.3 shows the chloride distributions in 1990 in the
eight profiles. As concentration measurements at great depths are

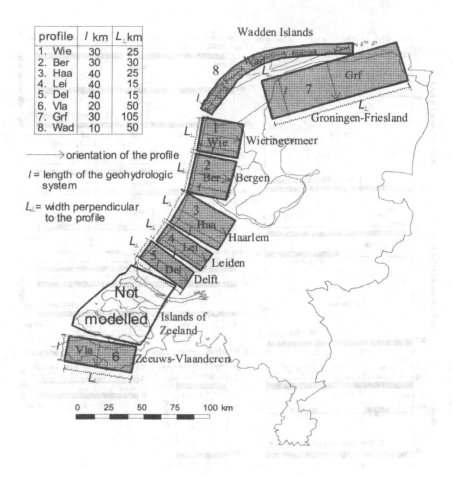

profile	l km	$L_\perp$ km
1. Wie	30	25
2. Ber	30	30
3. Haa	40	25
4. Lei	40	15
5. Del	40	15
6. Vla	20	50
7. Grf	30	105
8. Wad	10	50

——→ orientation of the profile

l = length of the geohydrologic system

$L_\perp$ = width perpendicular to the profile

Figure 14.1. Location of the eight coastal areas in which the representative profiles perpendicular to the Dutch coastline are defined. The names of the profiles are arbitrarily chosen: they refer to a city, region or province positioned in that profile. The islands of Zeeland are not modeled.

generally rare, the chloride distribution in some places might be disputable. Specific characteristics of each profile are briefly discussed below:

1. *Wie* = Wieringermeer: The transmissivity of this hydrogeologic system is huge, as the thickness is enormous, some 260 m, and the hydraulic conductivity high. Moreover, the hydraulic resistance of the Holocene aquitard is high. The Wieringermeer polder (reclaimed in 1930) in the reach 15,000-29,500 m has a

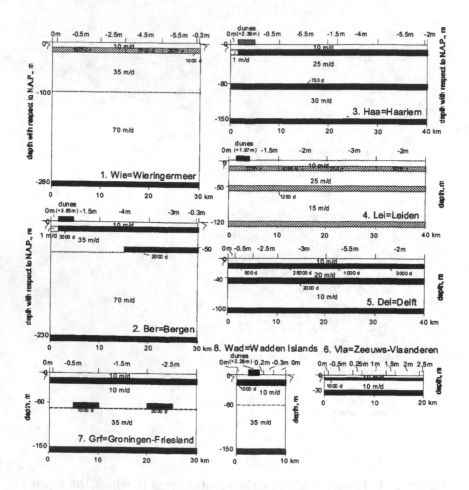

Figure 14.2. Schematized geometry, the subsoil parameters and the phreatic groundwater levels with respect to N.A.P. in the polders of the eight representative profiles. As can be seen, there is a great variety of subsoil conditions along the Dutch coast. The groundwater recharge f in the sand-dune area is 360 mm/yr. The highest maximum phreatic groundwater level in the sand-dune areas of the profiles *Ber*, *Haa*, *Lei* and *Wad* is given in brackets.

very low phreatic groundwater level: between -4.5 and -5.5 m N.A.P. At the inland boundary, the water level in the freshwater lake IJsselmeer is a constant piezometric level boundary. Fresh groundwater, which belongs to the freshwater body of Hoorn, is present in the central part of this profile.

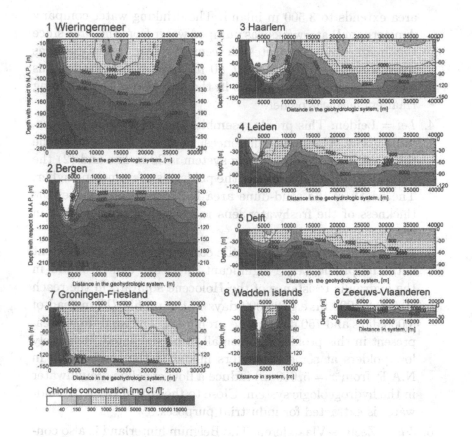

Figure 14.3. Present fresh-saline distributions (in mg Cl⁻/ℓ) in 1990 in the eight representative profiles.

2. *Ber* = Bergen: In the central part of this profile, a low-lying polder (−4 m N.A.P.) occurs, which represents the areas reclaimed in the seventeenth century: the Beemster (1608-1612), the Wormer (1625-1626) and the Schermer (1633-1635). The main difference with the profile *Wie* in terms of chloride distribution is caused by the presence of a sand-dune area of 3000 m width under which a freshwater lens has evolved.

3. *Haa* = Haarlem: The hydraulic conductivity is lower than that of the two previous profiles. Various low-lying polders are located at a considerable distance from the coast, such as the Haarlemmermeer polder (reclaimed during 1840-1852) in the reach 10,000-20,000 m. From the coastline, a broad sand-dune

area extends to 3,500 m inland. The drinking water company Amsterdam Waterworks has pumped water from this area since the middle of the nineteenth century. The freshwater lens reaches to a depth of at least 80 m below N.A.P. Under this lens, saline groundwater flows in the direction of the low-lying Haarlemmermeer polder.

4. *Lei* = Leiden: This profile resembles the profile *Haa* to a certain extent. The main differences, however, are that (1) the thickness of the hydrogeologic system is 30 m less and (2) the phreatic groundwater level of the polders is a few meters higher. The width of the sand-dune area is 2,500 m, and hence, the thickness of the freshwater lens is somewhat smaller than in the profile *Haa*.

5. *Del* = Delft: Though this profile corresponds to a great extent with the profile *Lei*, a significant difference can be found in the hydraulic resistance of the Holocene aquitard in the reach 10,000-20,000 m: e.g. 25,000 days in the profile *Del* instead of 3,000 days and 750 days in the profile *Lei*. The sand-dune area, present in this profile, is too small to be simulated. Relatively low polders at some kilometers inland of the coast (-2.5 m N.A.P. from $x = 5,000$ m) induce a flow of saline groundwater in the hydrogeologic system. Close to the city of Delft, groundwater is extracted for industrial purposes.

6. *Vla* = Zeeuws-Vlaanderen: The Belgium hinterland is also considered in this profile. The phreatic groundwater levels in the polders at some kilometers inland of the coast are above N.A.P. The thickness of the hydrogeologic system is the smallest of all profiles, only some 20 m. In the past, a severe salinization process has caused high chloride concentrations in this hydrogeologic system.

7. *Grf* = Groningen-Friesland: The phreatic groundwater levels are not very low in this profile. The length of the saline groundwater tongue is limited to some kilometers. More inland of the tongue, brackish groundwater is present in the entire hydrogeologic system.

8. *Wad* = Wadden Islands: At both boundaries, the piezometric level of the sea is constant. At the Waddenzee side, polders are present with phreatic groundwater levels around N.A.P. At the North Sea side, the width of the sand-dune area is 2,000 m,

under which a freshwater lens has evolved with a thickness of about 80 m.

14.2.2 Theoretical Background of MOC, Adapted for Density Differences

Originally, MOC (also called USGS 2-D TRANSPORT) [Konikow and Bredehoeft, 1978] was applied as a horizontal two-dimensional groundwater flow code. It appeared that the code could easily be adapted to model flow with non-uniform density distributions. Lebbe [1983] was the first to adapt MOC for simulating density-dependent groundwater flow in vertical profiles (see section 12.11.6). Later Oude Essink [1996] adapted the updated version 3.0 of MOC [1989], named MOC-DENSITY (see also section 12.11), using documentation of Lebbe [1983] and van der Eem [1987].

This paragraph comprises the fundamentals of the numerical algorithm to represent two-dimensional density dependent groundwater flow. The equation of motion for two-dimensional laminar groundwater flow in an anisotropic non-homogeneous porous medium in the principle directions is described by Darcy's law [Bear, 1972]:

$$q_x = -\frac{\kappa_x}{\mu_i}\frac{\partial p}{\partial x}$$

$$q_z = -\frac{\kappa_z}{\mu_i}(\frac{\partial p}{\partial z} + \gamma) \qquad (14.1)$$

where q_x, q_z = Darcian specific discharges in the principal directions, respectively horizontal and vertical (LT^{-1}), p = pressure ($ML^{-1}T^{-2}$), κ_x, κ_z = principal intrinsic permeabilities, respectively horizontal and vertical (L^2), μ_i = dynamic viscosity of water ($ML^{-1}T^{-1}$), $\gamma = \rho_i g$ = specific weight ($ML^{-2}T^{-2}$), ρ_i = density of groundwater (ML^{-3}), and g = gravity acceleration (LT^{-2}).

The relation between the pressure and the so-called fictive freshwater head is as follows (if the atmospheric pressure equals zero):

$$\phi_f = \frac{p}{\rho_f g} + z \qquad (14.2)$$

where ϕ_f = fictive freshwater head (L), $p/\rho_f\, g$ = pressure head, expressed in freshwater (L), ρ_f = density of fresh groundwater (ML^{-3}), and z = elevation with respect to the reference level (L).

The density distribution in the deep hydrogeologic systems is assumed to be non-uniform and varies with depth. As the density ρ_i (as γ in Eq. 14.1) varies with position, effects of density difference have to be considered. Henceforth, the Darcian specific discharge in vertical direction takes into account density differences. Inserting Eq. 14.2 into Eq. 14.1 gives:

$$q_x = -\frac{\kappa_x \rho_f g}{\mu_i} \frac{\partial \phi_f}{\partial x}$$

$$q_z = -\frac{\kappa_z \rho_f g}{\mu_i} \left(\frac{\partial \phi_f}{\partial z} - 1 + \frac{\rho_i}{\rho_f} \right) \tag{14.3}$$

As the density varies with position, the intrinsic permeabilities κ_x, κ_z and the dynamic viscosity μ should be applied instead of the horizontal and vertical hydraulic conductivities k_x or k_z of fresh groundwater $(L T^{-1})$, which are defined as follows:

$$k_x = \frac{\kappa_x \rho_f g}{\mu_f}$$

$$k_z = \frac{\kappa_z \rho_f g}{\mu_f} \tag{14.4}$$

However, small viscosity differences may be disregarded in case density differences are taken into account in groundwater problems in vertical profiles [Bear and Verruijt, 1987]. As such, the factor μ_f/μ_i, which is close to 1, is ignored in the development of the MOC-DENSITY code from this point on, also because of the lack of accuracy by which the horizontal and vertical hydraulic conductivities are determined. Making use of Eq. 14.4 and $\mu_f/\mu_i \approx 1$, Eq. 14.3 becomes:

$$q_x = -k_x \frac{\partial \phi_f}{\partial x}$$

$$q_z = -k_z \left(\frac{\partial \phi_f}{\partial z} + \Upsilon \right) \tag{14.5}$$

where $\Upsilon = (\rho_i - \rho_f)/\rho_f$ is the relative density difference, the so-called buoyancy. As can be seen, the vertical Darcian specific discharge has an extra term in comparison with groundwater flow with a uniform density: this term is also called the vertical density gradient velocity.

The equation of continuity for density dependent groundwater flow in two-dimensions is as follows:

$$-\left[\frac{\partial q_x}{\partial x} + \frac{\partial q_z}{\partial z} \right] = S_s \frac{\partial \phi_f}{\partial t} + W'(x, z, t) \tag{14.6}$$

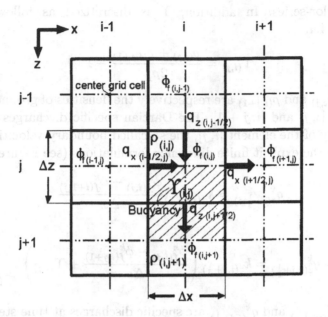

Figure 14.4. Nodal array for development of finite difference expressions. The buoyancy of the vertical velocity $q_{z(i,j+1/2)}$ is determined over the hatched area.

where $W'(x,z,t)$ = source function, which describes the mass flux of the fluid into (negative sign, e.g. well injection) or out of (positive sign, e.g. groundwater extraction) the system (T^{-1}), and S_s = specific storativity (L^{-1}). Here we note that as the MOC-DENSITY code is a code for 2D in vertical profiles, a 'well' in the model actually simulates a drain perpendicular to the profiles in the x–z plane.

Through combining the Eqs. 14.5 and 14.6, and consider a unit thickness of the aquifer in the y-direction, the groundwater flow equation can be defined:

$$\frac{\partial}{\partial x}\left(k_x \frac{\partial \phi_f}{\partial x}\right) + \frac{\partial}{\partial z}\left(k_z \frac{\partial \phi_f}{\partial z}\right) + \frac{\partial k_z \Upsilon}{\partial z} = S_s \frac{\partial \phi_f}{\partial t} + W'(x,z,t) \tag{14.7}$$

where k_x, k_z = respectively horizontal and vertical hydraulic conductivities $(L\,T^{-1})$.

To solve the groundwater flow equation numerically, the derivatives are replaced in the finite difference approximation by values of the difference quotients of the function in separate discrete points, using Taylor-series. In addition, Υ is discretized as follows, see Figure 14.4:

$$\Upsilon_{(i,j)} = \frac{\rho_{(i,j)} + \rho_{(i,j+1)}}{2\rho_f} - 1 \tag{14.8}$$

where $\rho_{(i,j)}$ and $\rho_{(i,j+1)}$ are respectively the densities of groundwater at node $[i,j]$ and $[i,j+1]$. The Darcian specific discharges at the boundary of the element $[i,j]$, the so-called boundary velocities, are given by the explicit finite difference formulations (see Figure 14.4):

$$q^k_{x(i+\frac{1}{2},j)} = k_{x(i+\frac{1}{2},j)} \frac{\phi^k_{f(i,j)} - \phi^k_{f(i+1,j)}}{\Delta x} \tag{14.9}$$

and

$$q^k_{z(i,j+\frac{1}{2})} = k_{z(i,j+\frac{1}{2})} \left(\frac{\phi^k_{f(i,j)} - \phi^k_{f(i,j+1)}}{\Delta z} + \Upsilon_{(i,j)} \right) \tag{14.10}$$

where $q^k_{x(i+\frac{1}{2},j)}$ and $q^k_{z(i,j+\frac{1}{2})}$ are specific discharges at time step k on the boundary between nodes $[i,j]$ and $[i+1,j]$, and between $[i,j]$ and $[i,j+1]$, respectively. $k_{x(i+1/2,j)}$ and $k_{z(i,j+1/2)}$ are weighted harmonic means of the hydraulic conductivities:

$$
\begin{aligned}
k_{x(i+\frac{1}{2},j)} &= \frac{2k_{x(i+1,j)}k_{x(i,j)}}{k_{x(i+1,j)} + k_{x(i,j)}} \\
k_{z(i,j+\frac{1}{2})} &= \frac{2k_{z(i,j+1)}k_{z(i,j)}}{k_{z(i,j+1)} + k_{z(i,j)}}
\end{aligned}
\tag{14.11}
$$

The final finite difference approximation for the groundwater flow equation becomes (see also Pinder and Bredehoeft [1968]):

$$
\begin{aligned}
k_{x(i-\frac{1}{2},j)} & \left[\frac{\phi^k_{f(i-1,j)} - \phi^k_{f(i,j)}}{(\Delta x)^2} \right] + k_{x(i+\frac{1}{2},j)} \left[\frac{\phi^k_{f(i+1,j)} - \phi^k_{f(i,j)}}{(\Delta x)^2} \right] \\
+ k_{z(i,j-\frac{1}{2})} & \left[\frac{\phi^k_{f(i,j-1)} - \phi^k_{f(i,j)}}{(\Delta z)^2} + \frac{\Upsilon_{(i,j-1)}}{\Delta z} \right] \\
+ k_{z(i,j+\frac{1}{2})} & \left[\frac{\phi^k_{f(i,j+1)} - \phi^k_{f(i,j)}}{(\Delta z)^2} - \frac{\Upsilon_{(i,j)}}{\Delta z} \right] \\
= S_s & \left[\frac{\phi^k_{f(i,j)} - \phi^{k-1}_{f(i,j)}}{\Delta t} \right] + W'_{(i,j)}
\end{aligned}
\tag{14.12}
$$

The advection-dispersion equation is still solved by means of the method of characteristics (see Garder, Peaceman and Pozzi [1964]). The basic concept underlying the application of the method of characteristics is to uncouple the advective and the dispersive component of the equation, and to solve them separately [Konikow and Bredehoeft, 1978].

The conversion from the (conservative) chloride concentration to density, that is applied in the MOC-DENSITY model, is as follows:

$$\rho_{(i,j)} = \rho_f \cdot \left(1 + \frac{\rho_s - \rho_f}{\rho_f} \cdot \frac{C_{(i,j)}}{C_s} \right) \qquad (14.13)$$

where $\rho_{(i,j)}$ = density of groundwater in grid cell $[i,j]$ $(M\,L^{-3})$, ρ_f = reference density, usually the density of fresh groundwater (without dissolved solids) at mean subsoil temperature $(M\,L^{-3})$, ρ_s = density of saline groundwater at mean subsoil temperature $(M\,L^{-3})$, $C_{(i,j)}$ = chloride concentration or the so-called chlorinity in grid cell $[i,j]$ (mg Cl^-/ℓ), C_s = reference chloride concentration (mg Cl^-/ℓ). A linear relation is assumed between ρ_s and C_s.

In this case study, the following data are applied for the Dutch situation: $\rho_f = 1000$ kg/m^3; $\rho_s = 1025$ kg/m^3, thus $(\rho_s - \rho_f)/\rho_f = 0.025$; and $C_s = 18,630$ mg Cl^-/ℓ. In Dutch coastal aquifers, the chloride concentration of groundwater normally does not exceed 17,000 mg Cl^-/ℓ, as seawater that intrudes the groundwater flow regime is mixed with water from mainly the river Rijn. The density of that saline groundwater ρ_s equals about 1022.8 kg/m^3.

The following model parameters are applied in the MOC-DENSITY code (see also Oude Essink [1996]). Each rectangular grid cell has a length Δx of 250 m and a height Δz of 10 m (except for the profile Vla, the height Δz is 5 m in that case). Each grid cell contains five particles. The convergence criterion TOL for the iterative calculation of the freshwater head in the groundwater flow equation is $3.048 \cdot 10^{-6}$ m (10^{-5} ft). The time step Δt for recalculating the groundwater flow equation equals 1 year (during this step, the velocity field remains constant). During this computation, the buoyancy is deduced from the computed solute concentration, and a new freshwater head distribution as well as a new velocity field is calculated. The maximum relative distance across one grid cell, in which a particle is allowed to move during one solute time step, is set to 0.9.

14.2.3 Scenarios of Sea Level Rise

Three scenarios with different rates of sea level rise are considered: one with a sea level rise of 0 meter per century (SLR = 0 m/c), one with SLR = +0.6 m/c, and one with SLR = −0.6 m/c (a sea level fall). The total simulation time is one millennium.

The scenario SLR = 0 m/c is applied as the reference case. The present situation concerning boundary and initial conditions, phreatic groundwater levels in the polders and groundwater extraction rates is maintained during this scenario. Accordingly, many processes, such as those of hydrological nature (e.g. less precipitation, more evapotranspiration due to climate change), of morphological nature (e.g. shoreline retreat), or of man-induced nature (e.g. sand-renourishment at the coastal zone to counteract shoreline retreat), that may affect the hydrogeologic system are left out of consideration. The results from this scenario clarify the effects of (past) human activities, such as land reclamation during the past centuries which has created low-lying polders. In addition, the time lag is determined until a state of dynamic equilibrium regarding the salinity distribution will be reached.

Bear in mind that a sea level rise with a rate of e.g. +0.6 m/c means a sea level elevation of 6 m after a simulation time of one millennium. This rise in sea level induces an elevation in phreatic groundwater level in the sand-dune area. In fact, the rise in phreatic groundwater level could be impeded if the position of the land surface is fixed. In order to assure that the phreatic groundwater level can rise freely, sand-renourishment is supposed to nullify the possible impediment.

14.3   Propagation of Sea Level Rise

The characteristic length λ of a formation indicates the zone over which a sea level rise at the seaside boundary influences the hydrogeologic system: the so-called *zone of influence*. For example, in a simple hydrogeological system, which consists of a Pleistocene sandy aquifer overlain by a Holocene (clayey) aquitard, the characteristic length λ equals $\sqrt{kDc}$, where k is the hydraulic conductivity of the semi-confined aquifer (LT^{-1}), D is the saturated thickness of the aquifer (L), and c is the hydraulic resistance of the aquitard (T).

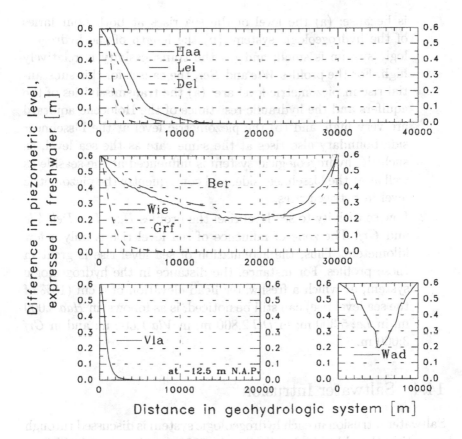

Distance in geohydrologic system [m]

Figure 14.5. The propagation of a sea level rise in the hydrogeologic system: differences in piezometric level, expressed in freshwater, at −25 m N.A.P. (in the profile *Vla* at −12.5 m N.A.P.) in 2090 for the scenario SLR = +0.6 m/c with respect to the scenario SLR = 0 m/c.

In general, the geometry of the representative profiles is complex, the zone of influence is determined by calculating the difference in piezometric level, expressed in freshwater for the scenario SLR = +0.6 m/c in the year 2090 with respect to the scenario SLR = 0 m/c (see Figure 14.5). Two types of sensitivity for sea level rise can be perceived for the eight Dutch profiles:

1. High sensitivity to sea level rise: the profiles *Wad*, *Wie* and *Ber*. The zone of influence of sea level rise is very long. Consequently, piezometric levels in the hydrogeologic system elevate significantly due to sea level rise. For the profile *Wad*, this

is because: (a) the level of the sea rises at both boundaries of the hydrogeologic system; (b) the length of the hydrogeologic system is small; and (c) the transmissivity is relatively high. For the profiles *Wie* and *Ber*, the causes for this substantial rise in piezometric level are: (a) the transmissivities of the aquifers and the hydraulic resistance of the Holocene aquitard are very high; and (b) the piezometric level at the IJsselmeer side boundary also rises at the same rate as the sea level, as such the hydrogeological system is influenced at the seaside as well as at the IJsselmeer side, and consequently the piezometric level in between rises.

2. Low sensitivity to sea level rise: the profiles *Haa*, *Lei*, *Del*, *Vla* and *Grf*. The zone of influence of sea level rise is only a few kilometers. Thus, the attenuation of sea level rise is great in these profiles. For instance, the distance in the hydrogeologic system, at which a freshwater head elevation of 6 cm (10% of the sea level rise) can still be noticed, is as follows: in *Haa* 8,000 m; in *Lei* 5,300 m; in *Del* 2,800 m; in *Vla* 1,600 m; and in *Grf* 3,000 m.

14.4   Saltwater Intrusion

Saltwater intrusion in each hydrogeologic system is discussed through analyzing the chloride distribution in 2990 for the scenario SLR = +0.6 m/c with respect to the scenario SLR = 0 m/c. In addition, the time lag, before a state of dynamic equilibrium in the salinization process is reached, is also given.

14.4.1   Chloride Distribution in 2990 for SLR = +0.6 m/c

Figure 14.6 shows the chloride distributions in 2990 in the eight profiles for the scenario SLR =+0.6 m/c. Through comparison with Figure 14.3 (the situation in 1990), one can deduce the following:

1. *Wie*: The entire hydrogeologic system will be saline in 2990. The two main causes for the severe saltwater intrusion are the high transmissivities and the low-lying Wieringermeer polder in the reach 15,000–29,500 m.

2. *Ber*: A severe saltwater intrusion also occurs in this profile. The freshwater lens remains, though the volume obviously decreases

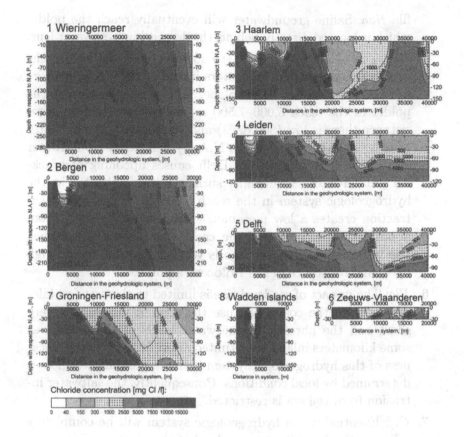

Figure 14.6. Future fresh-saline distributions (in mg Cl$^-$/ℓ) in 2990 in the eight representative profiles for the situation with a sea level rise of +0.6 m/c.

due to sea level rise. The aquifer, which underlies the low-lying polder in the central part of the profile, will contain much more saline groundwater in the future than in 1990.

3. *Haa*: The freshwater lens decreases and the saltwater intrusion in the center of the low-lying polder (e.g. the Haarlemmermeer polder) will be severe.

4. *Lei*: The saltwater intrusion is rather limited for several reasons: (a) the thickness of the hydrogeologic system is only 120 m; (b) as such, the freshwater lens is obstructing the inflow of saline groundwater to the hinterland; and (3) the phreatic groundwater levels of the polders are not as low as in the pro-

file *Haa.* Saline groundwater will eventually reach the polder which is located directly behind the decreasing freshwater lens.

5. *Del:* The results correspond, to some extent, to those in the profile *Lei.* In the first kilometers inland of the coast, a severe saltwater intrusion is taking place. Saline seepage occurs in the polder in the reach 5,000–7,500 m, because the hydraulic resistance there is low (500 days) and the phreatic groundwater level is relatively low (−2.5 m N.A.P.). Industrial groundwater extraction in the vicinity of Delft causes upconing of brackish groundwater. Meanwhile, surface water infiltrates in the hydrogeologic system in the reach 8,000-20,000 m, as the extraction creates a low piezometric level in the aquifer. From about 10,000 m inland of the coast, the impact of sea level rise in terms of saline seepage is very limited due to the high hydraulic resistance of the Holocene aquitard.

6. *Vla:* The impact of sea level rise is limited to the first few kilometers from the coast, because the zone of influence is very short and the phreatic groundwater levels of the polders at some kilometers inland are situated above N.A.P. As the thickness of this hydrogeologic system is small, groundwater flow is determined by local conditions. Consequently, the saltwater intrusion from the sea is restricted.

7. *Grf:* Eventually, the hydrogeologic system will be completely saline in the first kilometers inland of the coast.

8. *Wad:* The hydrogeologic system underneath the polder becomes more saline at great pace. However, the freshwater lens remains, though its volume decreases. Note that the phreatic groundwater level in the sand-dune area is supposed to rise with sea level rise. Sand should be supplied where necessary in order to assure that the phreatic groundwater level can rise freely.

14.4.2   Time Lag to Dynamic Equilibrium

Based on analyses of the changes in volume distribution of fresh, brackish and saline groundwater based on the scenario SLR = 0 m/c for 10,000 years (as the state of dynamic equilibrium will not be reached within one millennium in several profiles, the simulation time has been extended to 10,000 years), it appears that most of the hydrogeologic systems along the Dutch coast at the present have not yet reached a state of dynamic equilibrium as far as the salinity distribution in the subsoil is concerned [Oude Essink, 1996]. The time lag of a hydrogeologic system between the cause of changes and the ultimate effect on the salinization process is mainly determined by three causes: (a) the driving force of the salinization process, that is the difference in piezometric level between mean sea level and the piezometric level inland (the low piezometric level inland is induced by low phreatic groundwater levels in the polders and groundwater extraction); (b) the transmissivity and porosity of the hydrogeologic system, that is the velocity with which groundwater can pass through the system; and (c) the geometry (size) of the hydrogeologic system under consideration. Based on the causes mentioned above, four durations of time lags are classified:

1. A time lag of several centuries: hydrogeologic systems with small geometries, such as the profiles *Vla* and *Wad*.

2. A time lag of a large number of centuries: hydrogeologic systems with high transmissivities and low-lying polder areas, such as the profiles *Wie* and *Ber*.

3. A time lag of a few millennia: hydrogeologic systems with moderate transmissivities and polder areas with high phreatic groundwater levels, such as the profiles *Haa* and *Grf*.

4. A time lag of several millennia: hydrogeologic systems with relatively low transmissivities and polder areas with high phreatic groundwater levels, such as the profiles *Lei* and *Del*.

Considering the absolute change in the volume of saline groundwater, the salinization appears to be greatest in the northern part of the Netherlands in the profiles *Wie*, *Ber* and *Grf*. Saltwater intrusion is still substantial in the profiles *Haa*, *Lei* and *Del*. The salinization is limited in the profiles *Wad* and especially *Vla*, which is obvious as the hydrogeologic system of the profile *Vla* is small.

14.5   Freshwater Lenses in Sand-Dune Areas

Figure 14.7 shows the volume of fresh groundwater in the sand-dune areas as a function of time in the four profiles *Ber*, *Haa*, *Lei* and *Wad*. During the next century, the impact of sea level rise on the fresh groundwater resources along the coast appears to be of minor

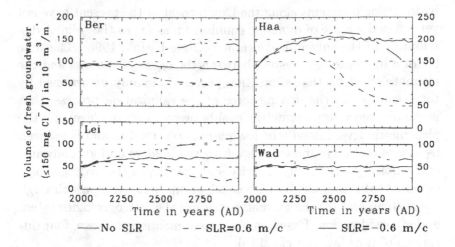

Figure 14.7. Volume of fresh groundwater ($\leq$ 150 mg Cl$^-$/ℓ) in the sand-dune area in 10^3 m^3/m$'$ as a function of time for the three scenarios of sea level rise.

importance, at most a few percents. After a millennium, however, the changes are significant.

Two phenomena occur in the profile *Haa*. Firstly, for the scenario SLR = 0 m/c, the volume of fresh groundwater increases significantly (e.g. +23.4% in 2090 with respect to 1990) because of two reasons: (a) the groundwater extraction rate has been decreased and artificial recharge has started since 1956 and the freshwater lens is still rebounding to a larger volume and (b) the freshwater lens evolves towards the low-lying polder in the reach 10,000-20,000 m, and as a result, the volume increases. Secondly, for the scenario SLR = −0.6 m/c, the volume of the freshwater lens drops below the volume of the scenario SLR = 0 m/c after some seven centuries. The reason for this phenomenon is that, after some centuries, the groundwater flow in the sand-dune area alters in the direction of the sea, as the sea level is falling. The shape of the freshwater lens also changes, which results in a decreasing volume of fresh groundwater. In the profile *Wad*, the volume of fresh groundwater decreases somewhat during the second half of the next millennium for the scenario SLR = −0.6 m/c. The reason is that, from then on, sea level fall causes fresh surface water, which originates from the polder adjacent to the sand-dune area, to infiltrate into the hydrogeologic system. This water, which contains a higher chloride concentration than 150 mg Cl^-/ℓ, supersedes the fresh groundwater with a chloride concentration of 40 mg Cl^-/ℓ that recharges the sand-dune area.

14.6   Seepage in Low-Lying Polders

The impact of sea level rise in a specific polder can be significant in terms of seepage quantity, mean chloride concentration and chloride load if the following situations occur: (a) the polder is situated directly inland of the coast; (b) the phreatic groundwater level in the polder is low, namely several meters below N.A.P.; (c) and/or the zone of influence of sea level rise in the hydrogeologic system is long as a result of high transmissivities of aquifers.

14.6.1   Seepage Quantity

The propagation of the sea level rise (see Figure 14.5) already indicates the possible changes in seepage quantity due to sea level rise. Obviously, if the sea level rise is attenuated considerably in the hydrogeologic system, the changes in seepage quantity are limited. In general, for Dutch polders at a great distance from the coast, the increase in seepage due to sea level rise appears to be of minor importance with respect to the already existing seepage quantities in those polders. For the scenario SLR $= 0$ m/c, the decrease in seepage quantity through the Holocene aquitard due to changes in density distribution (caused by the salinization of the hydrogeologic system) is not insignificant in some polders (up to a few tens of percents), and thus remarkable. The impact of a sea level rise of $+0.6$ m (during the next century) on the change in seepage quantity is considerable (up to several tens of percents) in polders in the first kilometers from the coast, whereas it is marginal (mostly up to only a few percents) in polders located further inland.

14.6.2   Mean Chloride Concentration

In general, it takes several decades or even centuries before brackish and saline groundwater, which just enters the aquifers as seawater or is already present in the deeper aquifers, reach polders that are located several kilometers inland of the coast. It is clear that the nearer the polder is to the seaside boundary, the earlier the impact of sea level rise on the chloride distribution can be noticed. During the next century, the mean chloride concentration of the subsoil will probably increase with several hundreds of mg Cl^-/ℓ in many low-lying polders along the Dutch coast due to the combined effect of a sea level rise of +0.6 m and the non-equilibrium state of the present salinization process as a result of the delayed effect of previous human activities. After several centuries, the impact of sea level rise on the mean chloride concentration becomes substantial in almost all polders, and accordingly, a considerable increase in the salinity of seepage can be expected in the distant future.

14.6.3 Chloride Load

During the next century, the increase in chloride load is drastic in nearly all polders in the low-lying regions of the Netherlands. Even those polders that lie beyond the direct zone of influence of sea level rise are affected. It appears that the changes in chloride load are much greater than the changes in seepage quantity. As an indicative estimate: the chloride load through the Holocene aquitard in the low-lying coastal regions of the Netherlands will be doubled in 2090, three fifths of which is caused by a sea level rise of +0.6 m and two fifths by the delayed effect of previous human activities (namely the reclamation of lakes or parts of the sea) on the future salinization process.

14.7 Compensating Measures

When no sea level rise occurs, effective human interventions are likely to counteract the long-term salinization process in most of the profiles. Such interventions could be: (a) reclaiming land off the coast, thus evolving new freshwater lenses if areas with groundwater recharge are created; (b) extracting (saline) groundwater, thus decreasing the seepage quantity and the chloride load in the polders; (c) inundating low-lying polders, thus removing the driving force of the salinization process; (d) widening existing sand-dune areas, thus generating thicker freshwater lenses; and (e) creating physical barriers, thus blocking the free entrance of saline groundwater and halting the salinization process. Note that some of these countermeasures may not be economically feasible. Nevertheless, they provide insight into the range of conceivable human interventions. Initially, the deep-well infiltration of surface water increases the seepage quantity in the polders, and thus increases the chloride load as well. However, after some decades to centuries, the chloride load will drop substantially when the salinization process towards the low-lying hinterland is blocked by the infiltrated surface water.

When the sea level rises +0.6 m/c, even the most effective countermeasures cannot stop the salinization of the subsoil [Oude Essink, 1996]. Therefore, it is likely that human interventions can only retard the impact of sea level rise on the hydrogeologic systems in the long-term at the expense of major investments. Whether these investments are considered feasible or not depends on the prevailing economic, environmental and political circumstances at the time when a (political) decision is taken. If taken, they should be analyzed and, if necessary, adapted and optimized throughout the years based on changes in the salinization of the subsoil.

Note that countermeasures should be taken in time, since the time lag is considerable (several decades to centuries) before these measures can effectively change the salinity distribution of the subsoil.

14.8 Conclusions and Recommendations

14.8.1 Conclusions

Most of the hydrogeologic systems of the eight studied profiles along the Dutch coast have not yet reached a state of dynamic equilibrium as far as the salinity distribution in the subsoil is concerned. This present salinization process is already severe and is generated by previous human activities such as the reclamation of lakes or parts of the sea during the past centuries. It is to be expected that sea level rise significantly accelerates the salinization. Eventually, many hydrogeologic systems will become completely saline due to: (1) the future sea level rise; and (2) the present difference between polder level and sea level.

When the sea level rises +0.6 m during the next century, the volumes of the freshwater lenses in the four profiles with existing sand-dune areas hardly decrease. However, a few centuries later, the decrease in volume may reach several tens of percents. When the sea level rises +0.6 m during the next century, the chloride load increases substantially in nearly all polders in the low-lying regions of the Netherlands. To some extent, feasible countermeasures may retard, but not stop, the salinization process.

In conclusion, it is clear that in the future, the Dutch have to cope with much more saline groundwater in their coastal hydrogeologic systems than at present.

14.8.2   Recommendations

Due to high groundwater extraction rates and due to the diversity and complexity of the Dutch polder landscape, groundwater flow is in fact three-dimensional. In order to simulate the impact of sea level rise on coastal aquifers in a quantitative way and to analyze the possible human interventions which could counteract the ongoing salinization of the Dutch subsoil, it is recommended to conduct three-dimensional numerical modeling. For this purpose, the three-dimensional computer code MOC3D [Konikow et al., 1996], which is interconnected with MODFLOW, is adapted for density dependent modeling [Oude Essink, 1998] (see also Chapter 12). However, three-dimensional modeling of saltwater intrusion in large-scale coastal aquifers may face major restrictions as discussed in Oude Essink and Boekelman [1996].

Chapter 15

Movement of Brackish Groundwater Near a Deep-Well Infiltration System in the Netherlands

A. Stakelbeek

15.1 Introduction

The dune area along the Northsea coast in the western part of the Netherlands is one of the most valuable areas in the country. To keep the groundwater, for ecological reasons, at an adequately high level and to avoid regional rising of brackish groundwater, groundwater abstraction has been reduced in the last 30 years. However, in the Netherlands soil passage is preferred as disinfection method to produce drinking water. Therefore, the reduced groundwater abstraction is compensated by artificial recharge and abstraction of pre-treated surface water, which has become the predominant production method in the western part of the Netherlands. Particularly the method of deep-well infiltration is winning ground because of small space demands and relatively low hydrological or ecological impacts.

PWN Water Supply Company North-Holland, is operating a deep-well infiltration system since the beginning of 1990 (Figure 15.1). Its capacity is 5 Mm^3/year. The infiltration of pre-treated surface water takes place by means of 20 infiltration wells into an sandy aquifer 50 to 100 meters below sea level. The simultaneous abstraction is carried out through 12 abstraction wells in the same aquifer where the infiltration wells are situated. To the lower side the aquifer is bounded by a thin clay layer, under which the groundwater is brackish. The greatest difficulty when operating a deep-well infiltration system close to a formation with brackish groundwater is to avoid

J. Bear et al. (eds.), Seawater Intrusion in Coastal Aquifers, 531–541.

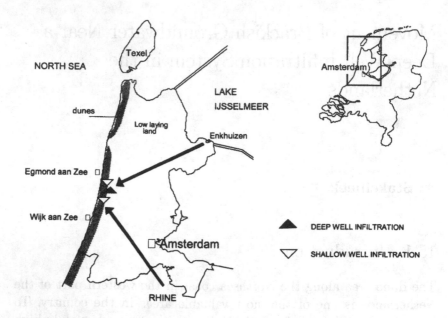

Figure 15.1. Location of the deep-well infiltration project and transportation of pre-treated surface water.

brackish water to be drawn into the aquifer used for infiltration and abstraction.

15.2   Design of the Deep-Well Infiltration System

The deep-well infiltration project DWAT consists of 20 infiltration, 12 abstraction and 10 observation wells. The infiltration and simultaneous abstraction happens in a sandy aquifer between 50 and 100 meter below sea level. The aquifer is bounded at the lower side by a thin clay layer. Below this layer the groundwater is brackish (Figure 15.2).

In order to achieve a maximum spread of retention times, and to reduce as much as possible the quality fluctuations of the infiltrated water, the infiltration and abstraction wells are laid out in a rhombus-like arrangement, taking into account the existing road and path infrastructure (Figure 15.3).

Because the lower side of the aquifer is bounded by only a thin clayey layer, there is a risk of local upconing of brackish groundwater,

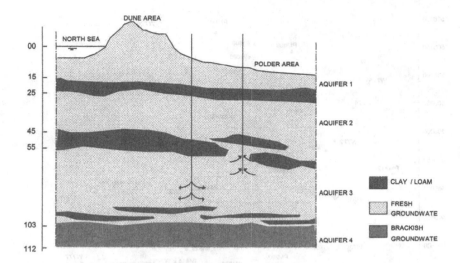

Figure 15.2. Hydrogeological situation near the deep-well infiltration plant.

especially near the abstraction wells. To prevent brackish groundwater from upconing near a deep-well infiltration system which consists of infiltration wells and abstraction wells, there are two points worthy paying attention. The first is the position of the abstraction wells in relation to the infiltration wells. The second is the amount of over-infiltration.

To calculate the necessary amount of over-infiltration, model calculations of the movement of the brackish groundwater are worked out. For this deep-well project calculations were carried out with a 3-dimensional sharp interface, transient finite-element model. The computer code of the calculation model is called SALINA (IWACO).

In the SALINA computer code the pressure in both the fresh and the salt part of the aquifer is expressed in the freshwater head at the bottom of the salt aquifer (ϕ) (Figure 15.4)

$$p(z) = (\rho_f \, \phi - \rho_s \, z) \, g \qquad (15.1)$$

with p = pressure (Pa), ρ_f = density of freshwater (kg/m^3), ρ_s = density of saltwater (kg/m^3), ϕ = freshwater head at the bottom of the salt part of the aquifer (m), and g = gravity constant (m/s^2). The flow equation can be expressed by the gradient of pressure:

$$\mathbf{q} = K \, \nabla p \qquad (15.2)$$

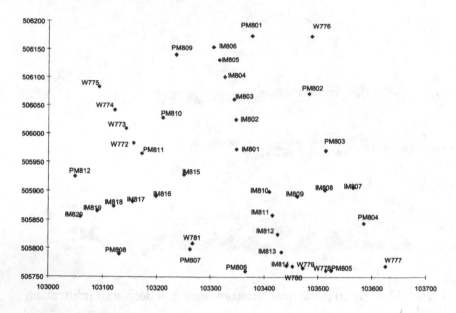

Figure 15.3. Well-field of deep-well infiltration project DWAT. (IM = infiltration well, PM = abstraction well, W = observation well).

with K = hydraulic conductivity (m/day), and $\mathbf{q}$ = specific flux (m/day). The continuity in the fresh part of the aquifer is a function of flow, storage (S), position of the interface (h) and abstraction N (positive or negative):

$$\nabla \mathbf{q}_f (H - h) = S\frac{\partial h}{\partial t} + N \tag{15.3}$$

and in the salt part of the aquifer:

$$\nabla \mathbf{q}_s\, h = -S\frac{\partial h}{\partial t} + N \tag{15.4}$$

Combining equations of flow, Eqs. 15.1 and 15.2 and continuity, Eqs. 15.3 and 15.4, for both the fresh and the salt part of the aquifer, gives an equation of the form:

$$S\frac{\partial h}{\partial t} = \text{function}(\phi, h) \tag{15.5}$$

for both the fresh and the salt part. Combining these equations for the fresh and the salt part of the aquifers gives 2 equations and 2 variables, ϕ and h, to be solved.

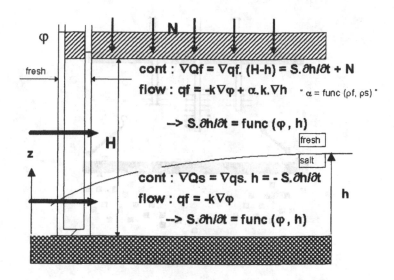

Figure 15.4. Parameters used in the computer code SALINA.

With the SALINA-model the lowering of the brackish zone, here defined as the 10,000 mg/l zone, was calculated [Stakelbeek, 1992]. These calculations showed that in order to prevent upconing of brackish groundwater in the well field, an over-infiltration of 10% of the infiltrated volume was needed near the abstraction wells on the edges of the well field.

The maximum lowering of the 10,000 mg/l interface was found to 5 m and occurred at the center of the well field. The lowering diminishes on the sides of the well field and decreased to zero at the edges of the well field near the abstraction wells.

The calculations with a 3-dimensional steady state sharp interface model also showed that a rhombus-like arrangement of infiltration and abstraction wells with the infiltration wells in the middle and the abstraction wells on the sides of the well field is the best possible layout to prevent upconing of brackish groundwater. This arrangement required the lowest amount of over-infiltration to prevent upconing of brackish groundwater.

However the sharp interface approach did not give any information about the movement of brackish groundwater with low salinity. Due to dispersion an upconing of brackish water with low salinity could be expected at the edges of the well field near the abstraction wells

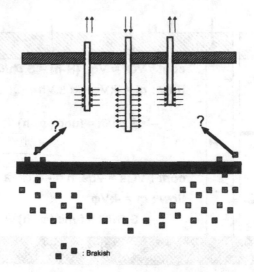

Figure 15.5. Upconing of brackish groundwater with low salinity, due to dispersion, at the edges of the well field near the abstraction wells.

(Figure 15.5). Therefore close monitoring of the position of brackish groundwater during operation is needed.

15.3 Measurements of Movement of Brackish Groundwater During Operation

During operation the position of the brackish zone was permanently measured by means of electrode-cables installed below the filters of the abstraction wells and in the observation wells (Figure 15.6). The position of the brackish zone is determined by measuring the electric conductivity of the groundwater between electrode-pairs on several depths on the electrode cables. The distance between the electrode-pairs is only 4 m, so it was possible to measure slight movements of the brackish zone quite accurately.

Before the infiltration started in 1990 the measured salt concentration increased as the depth increased and the "interface" with different salt-concentrations were positioned more or less horizontal. In the well field the maximum salinity of the infiltration/abstraction aquifer (50–100 m msl, Figure 15.2) was 200 mg/l. Because of the (over-)infiltration of freshwater this balance was disturbed and the

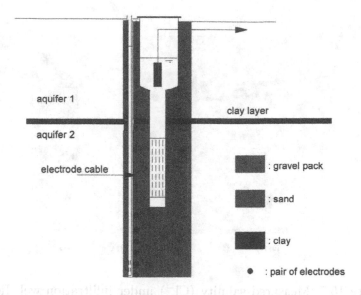

Figure 15.6. Electrode cable below the filter of an abstraction well.

position of the brackish and saltwater was remodeled.

The measurements during the experiments between 1990 and 1995 show that near the infiltration wells in the center of the well field brackish groundwater is pushed downwards and that a storage of freshwater has been built up. In Figure 15.7 the measured salinity under infiltration well IM801 in the center of the well field is shown. As Figure 15.7 shows, the salinity at depths of 115 m msl and less decreases in time. The salinity at greater depths, for example 119 m msl, stays at the same level, so in the center of the well field the thickness of the brackish zone decreased.

As Figure 15.8 shows, the salinity at the edges of the well field (observation well W774, Figure 15.3) does not decrease at that depth.

In Figure 15.9 the measured position in of the brackish zone in a cross section of the well field after 3 years simultaneous infiltration/abstraction, with 10% over-infiltration, is given. As Figure 15.9 shows, during the period of over-infiltration a storage of freshwater has been built up in the center of the well field, near the infiltration wells. In the period 1990–1995 several experiments were conducted to investigate the use of the storage which has been built up due to the over-infiltration. This storage can be used in times of high demand

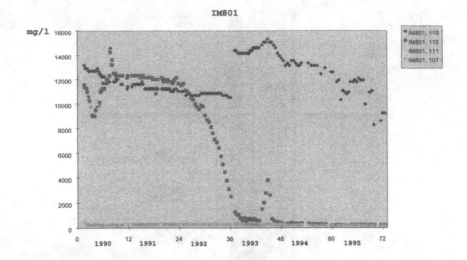

Figure 15.7. Measured salinity (Cl⁻) under infiltration well IM801 in the center of the well field during operation of the deep-well infiltration system with 10% over-infiltration.

or when the quality of the surface water is not good enough for infiltration. These experiments showed that the over-infiltrated volume, needed to prevent brackish groundwater from upconing during the operation, cannot be recovered fully during calamity circumstances when the infiltration will be interrupted and the abstraction is continued. Of course the loss depends on the geological conditions and on the position of the wells. Because of density differences, the brackish zone will rise to its natural position during the abstraction. The process is not fully reversible. If the infiltration/abstraction, and so the over-infiltration, will be continued for a long period, the position of the brackish zone comes into equilibrium with the groundwater flow. From that moment on the position of the brackish zone will not change anymore. The infiltrated freshwater will drain horizontally away above the saltwater layer, and all the over-infiltration will be lost that way.

Experiments have been carried out to investigate the period the abstraction can be continued without infiltration. Due to dispersion this period will be much shorter than is indicated by the calculation results with the 10,000 mg/l zone. Regarding to the DWAT deep-well infiltration project, the conclusion is that the period the abstraction

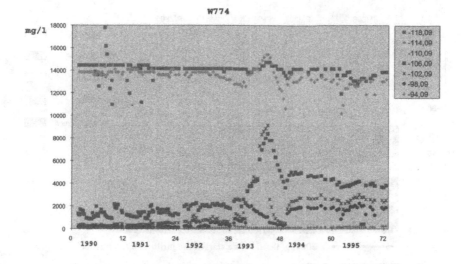

Figure 15.8. Measured salinity (Cl⁻) near observation well W774 at the edge of the well field (Figure 15.2).

can be continued without infiltration depends on the length of the preceding period in which over-infiltration had been performed. After 5 years simultaneous infiltration and abstraction with 10% over-infiltration, experiments made clear that is was possible to continue the abstraction, without infiltration, for 3 months. In these months no serious problems concerning upconing of brackish groundwater occurred. Despite the fact that 20–40% of the earlier over-infiltrated volume could not be recovered, the deep-well infiltration system is considered to have an important value for the drinking water supply in case of calamities.

Another important point of concern is the increasing dispersion zone between fresh and salt groundwater. The measurements show that dispersion especially at the sides of the well field cause upconing of brackish water with low salinity (Figure 15.9). For the production of drinking water these small quantities of brackish water might cause unacceptable difficulties. A sharp interface approach neglects this dispersion which leads to a serious underestimation of the risk of short-term upconing of brackish water with low salt concentrations. If a 3-dimensional density depended groundwater model with dispersion is not available, for short time problems a convection-dispersion model without density depended flow might give a better approach

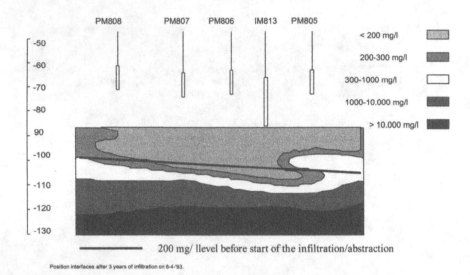

Figure 15.9. Measured postion of the brackish zone (Cl⁻) in a cross-section over the well field (Figure 15.2) after 3 years silmultaneous infiltration and abstraction with 10% overinfiltration.

than a sharp interface approach.

15.4 Conclusions and Recommendations

First, it is clear that, if brackish groundwater occurs near a deep-well infiltration system, a certain amount of over-infiltration is mostly necessary in order to prevent brackish groundwater from upconing. A large quantity of this over-infiltration drains away horizontally above the saltwater layer and will be lost. Only a part can be recovered during calamities when the infiltration is interrupted, or in situations with over abstraction instead of over-infiltration. The loss depends on local hydrogeological conditions, and on how long over-infiltration had been performed previously after which the over-abstraction occurs. At a certain moment there will be an equilibrium between the infiltration/abstraction and the position of the brackish water. From that moment the over-infiltrated water will drain horizontally away above the brackish zone. In spite of the loss of the over-infiltrated water, a deep-well infiltration system, even when its near brackish groundwater, could have a value for the drinking water supply during calamities, in which infiltration is not possible. For the DWAT project as described above, the period of abstraction without infiltration can be as long as several months if the period of the proceeding over-infiltration amounts to at least 5 years.

Secondly, model-calculations are a useful tool to design and to optimize the operation of a deep-well plant. However, nothing is more unexpected than the things that are invisible such as natural processes at a great depth. Therefore control, for example by measurements by electrode-cables, in addition to calculations, is absolutely necessary.

Finally, sharp interface calculations are only useful to predict the movement of brackish groundwater with a relatively high density. However, the movement of brackish groundwater with a lower density might cause more problems for drinking water production systems. To predict the movement of this low density brackish water, a 3-dimensional density diffusion/dispersion solute transport model is needed. For 3-dimensional problems and for early time steps a decoupled convection-diffusion/dispersion model for which the density effect is not part of the problem gives a better estimate than a sharp interface approach [Esch and Stakelbeek, 1996].

Chapter 16

A Semi-Empirical Approach to Intrusion Monitoring in Israeli Coastal Aquifer

A. J. Melloul & D. G. Zeitoun

16.1 Introduction

The basic elements of groundwater quality monitoring programs in general, and saltwater intrusion control in particular, involve the definition of monitoring policy and objectives. The implementation of these objectives include design of facilities and instruments, field surveys, sample collection, data analysis, evaluation of the information, characterization of possible scenarios of saltwater intrusion and operational decisions. The strategy of designing a monitoring program depends mainly upon the natural and political-administrative boundaries of the area of concern, the quantity and quality of existing records regarding aquifer geology, the groundwater regime, other sources and types of contaminants, and available financial resources [WMO, 1988; Melloul and Goldenberg, 1994].

Coastal aquifers located beneath highly populated areas are often operated under stress management with an overdraft of pumpage, which may lead to saltwater intrusion. Such over-exploitation often results in saltwater encroachment into the coastal aquifer which can intrude hundreds of meters inland from the seashore, as has occurred in many coastal aquifers in the world [Melloul and Gilad, 1993; Suleiman, 1995; Siraz, 1998; Dempser, 1998]. In such areas, the main objective of the monitoring of saltwater intrusion is to locate the depth of the saltwater/freshwater interface in order to estimate the toe of saltwater intrusion, the heterogeneity of the aquifer lithology, the existence of varied saline sources, and the radical changes

J. Bear et al. (eds.), Seawater Intrusion in Coastal Aquifers, 543–558.
© 1999 Kluwer Academic Publishers.

in groundwater quality over space and time [Melloul Bachmat 1975; Vengosh et al., 1991a; Siraz, 1998].

This monitoring is also essential to control the degree of groundwater deterioration in such aquifers, and determine a sustainable level of management of the aquifer [US EPA, 1990; Goldenberg and Melloul, 1994; Appleyard, 1995; Smithers and Walkers, 1995].

The main issues of monitoring methodology in general are network design and operation. The physical design of the network begins with determining the number and locations of sampling sites [WMO, 1988]. A sampling plan must be established to produce optimal sample collection (where, how, frequency, etc.), as set forth in Loaiciga [1989]. Ongoing observation well monitoring is the norm. Monitoring of saltwater intrusion has to consider the means of groundwater extraction. Additionally, the techno-geo-hydrological planning of facilities, such as well drilling and adequate screen location in each observation well, as well as choosing the most effective geophysical methods for the purpose are the necessary conditions for establishing an efficient monitoring network.

Most sub-aquifers can have a wide range of physical parameters, matrix materials, and groundwater qualities (fresh, brackish, seawater, and brines) which change over time and space, and which can influence the resulting physical parameters of the aquifer. For instance, hydraulic conductivity can thus decrease due to water-rock interactions, or increase due to rapid groundwater movement through preferred flow pathways [Pope et al., 1978; Hirasaki, 1982; Wong, 1988; Faust et al., 1989; Dagan, 1990; Warick, et el., 1991; Goldenberg, et el., 1993]. Thus, the heterogeneity of such aquifers, the groundwater quality varieties, and the various and numerous change which occur between the different solids, liquids and gas phases within the aquifer media [Goldenberg, et el., 1986] make the monitoring of seawater a complicated task.

These difficulties can be augmented by the fact that the observation wells in this saline environment can rapidly deteriorate. Therefore, in many countries saltwater intrusion is based on limited or deteriorated network. This means that saltwater monitoring network must be continuously augmented and improved by new wells. However, because the drilling of new wells is expensive and the budget for that is not always available, the use of unconventional methods must be considered to add data. Conventional methods consist of in-situ measurement of the groundwater salinity profile by means of observation wells. Indirect methods may utilize geo-electromagnetic TDEM method that can give the depth to saline water body in the aquifer (see Chapter 2).

The objectives of this study is to show that a combination of salt-water monitoring network and the development of a simple procedure can answer questions related to the magnitude and the dynamics of saltwater intrusion into the aquifer. Moreover, this methodology can pinpoint abnormal levels of saltwater intrusion, indicate the directives to take for preventing further deterioration in aquifer water quality. It leads to a better understanding of the magnitude of the expected uncertainties [US EPA, 1990]. This methodology, suitable for aquifers with scarce data, is applied to the Israel Coastal Aquifer, which suffers from a lack of data.

16.2   Hydrogeological Background and Saltwater Monitoring of Study Area

16.2.1   Hydrogeological Background

The area of investigation in the Coastal Aquifer of Israel extends from the towns of Binyamina and Hadera in the north to Ashqelon in the south, parallel to the coast for about 120 km, and between 0.1 to 2 km inland to the mountain aquifer border in the east, as presented in Figure 16.1. The Coastal Aquifer is composed of Pleistocene sandstone, calcareous sandstone, silt, and intercalation of clays and loam, which can appear as lenses (Figure 16.2). The basis of the aquifer is built upon sea clays and shales (Saqiya) of Neogene age. Most clay lenses or layers present along the coast farther out towards the east. Within approximately 5 km from the sea, these clay layers divide the aquifer into three major sub-aquifers A, B and C [Tolmach, 1979; Rosensaft et al., 1995]. Each of these sub-aquifers constitutes mostly a separate hydrologic unit, resulting in several water producing zones. In some areas, sub-aquifer B can in turn be subdivided into B1, B2, and sometimes B3, making the geometry complicated.

When dealing with saltwater intrusion into coastal aquifers in general and with the Israel coastal aquifer in particular, variations in lithological aquifer media must be considered. Such lithology can consist not only of uniform rock but also lithologic combinations

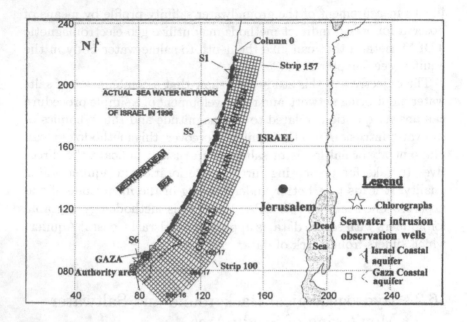

Figure 16.1. Location map of seawater intrusion network of the Coastal Aquifer of Israel and Gaza Strip.

such as sandstone with calcareous cement, clayey or silty sandstone, sandstone, etc. [Melloul, 1988; Rosensaft, et el., 1995]. Hydraulic conductivity (K), and storativity (S) will therefore vary over space. Aquiclude layers such as silt, or clays and loam intercalations can be either short or continuous layers. Depending upon their format of development, they may change the conceptual hydrogeological model of the aquifer. Where clay layers are more developed and thick, they can delineate separate hydrologic units or sub-aquifer production zones, in which different types of water may be stored [Fink, 1970; Melloul and Bachmat, 1975; Vengosh et al., 1991a].

16.2.2 Aquifer Abstraction and Saltwater Intrusion

The Coastal Aquifer is a major component of Israeli water resource system, and is utilized both as a long-term reservoir and as a main source of drinking and agricultural irrigation water. Its phreatic portion is located beneath a highly populated area and it supplies significant quantities of water for agriculture, domestic, and industrial

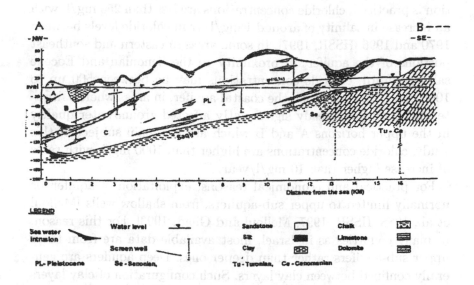

Figure 16.2. A typical hydrogeological cross section along a strip in the Coastal Aquifer of Israel.

purposes [Collin, 1995]. The exploitation of the coastal aquifer in Israel is based upon around 3,000 wells. Pumpage in the region over recent years has been between 350 to 400 MCM/yr, whilst recommended pumpage, as noted in HSSR [1987] has been around 280 MCM/yr. The aquifer is naturally recharged by precipitation as well as being artificially recharged in some areas from the National Water Carrier and water percolating from agricultural, industrial, and domestic sources. The recent rainy year 1991/92 added a significantly quantity of water to the system (rainfall exceeded the normal by 200%). Management decisions to encourage use of groundwater for drinking purposes rather than irrigation, and activities which lower evaporation and leakage from the system have been implemented during the recent years. This has contributed to a more positive water balance. Recharge has increased to between 100 to 150 MCM/yr (input from the Dan metropolitan sewage treatment plant has added about 80 to 100 MCM/yr), dropping the overdraft to zero and leading in some years to a positive quantitative hydrological balance. However, salinity still threatens the aquifer from all directions. In the internal areas of the coastal aquifer, where the major abstraction is practiced, chloride concentrations are less than 250 mg/l, with an increase in salinity of around 1 mg/l/yr in chloride levels between 1970 and 1996 [HSSR, 1997]. In some areas in eastern and southeast portions of the aquifer, in proximity to the Senonian and Eocene salty aquitard, chloride concentrations may be between 500 up to 1000 mg/l. However along the coastal aquifer, in areas where saltwater intrusion has already significantly affected groundwater quality in the upper horizons A and B which are the main subject of this study, chloride concentrations are higher than 1000 mg/l, with rates of increase higher than 10 mg/l/year.

For practical and economical reasons, exploitation of aquifers is normally limited to upper sub-aquifers, from shallow wells [Melloul et al., 1988; HSSR, 1997; Melloul and Gilad, 1993]. For this reason, in many countries as in Israel, most available data are from these upper sub-aquifers rather than deeper ones. Deep aquifers are generally confined between clay layers. Such configuration of clay layers toward the sea can become thicker and the lower sub-aquifers can sometimes become disconnected from seawater and are saturated with local brines [Fink, 1970; Melloul and Bachmat, 1975; Kolton, 1988]. Therefore, saltwater intrusion is mostly found in the upper aquifers where the abstraction is significant, and the connection to the sea is more evident. This situation not only fits the Coastal Aquifer of Israel, but can be expected in other countries as well.

Another concern is the sharp increase in salinity in the deep sub-aquifers from fossil saline sources, when the abstraction from those sub-aquifers is not well controlled. Therefore, it is very important to monitor these sub-aquifers for the sudden salinity increases for forecasting purposes [Suleiman, 1995; Melloul and Atsmon, 1992, 1996; Sariz, 1998]. But such monitoring which are based only on observation well data is relatively high cost.

16.2.3 Changes in Salinity in Intruded Areas

A concern of saltwater intrusion is the rapid change in groundwater salinity in wells located in various areas near the seashore [Suleiman, 1995; Dempster, 1998; Siraz, 1998]. In Israel, this can be illustrated by a north-south transect of the Coastal Aquifer for wells denoted as S1 and S5 located in the north and the central part of the Coastal Aquifer, and S6 located around 100 km south in the Gaza Strip area.

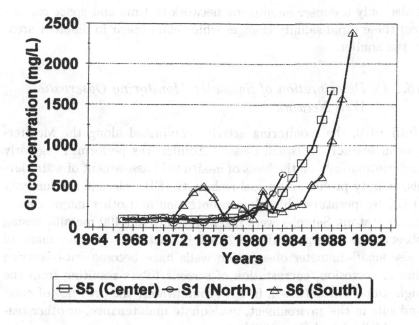

Figure 16.3. Characteristic chlorographs, noted as S1, S5 and S6 in the text, illustrating the process of saltwater intrusion along the Israel and the Gaza Strip Coastal Aquifer.

All these wells are reported to be between 500 to 1500 m from the seashore, in areas where saltwater intrusion is expected (Figure 16.1). Their chlorographs behavior indicates similarities despite the varied distances between these wells (Figure 16.3). These chlorographs show a historical record of small changes in fluctuation in low chloride content, followed by a sudden and sharp increase in salinity [Melloul and Collin 1994; Melloul and Atsmon, 1996]. These chlorographs indicates that initially, Cl values were at low levels and fluctuated from 50 to 100% around a mean value. After some time, Cl values began to increase continuously and rapidly, varying in one to three years from 200 to around 2500 mg/l, at a rate of around 30 mg/l of chlorides per year.

This last stage happened in a relatively short time. The intensity and magnitude of such a salinization process is influenced by heterogeneous lithological aquifer matrix, salinity sources, and the management of groundwater in the area in question [Melloul and Goldenberg 1998; Collin, 1995; Melloul and Atsmon, 1992, 1996].

Thus, only a denser monitoring network in time and space can detect these rapid salinity changes which can appear in different areas of the aquifer.

16.2.4   Deterioration of Saltwater Monitoring Observation Well Network

Until 1970, the monitoring activity conducted along the Mediterranean seashore of Israeli Coastal Aquifer was performed regularly and continuously on the basis of in-situ field assessment of water levels, salinity profiles, electrical resistivity (ER), electric conductivity (EC), temperature, chloride concentration and other major chemical indicators. Samples were taken from around 400 metallic casing observation wells generally of 2 inch diameter. Over time, many of these small-diameter observation wells have become out-of-service due to corrosion, encrustation of screens/filters resulting from the high saline concentration in the groundwater, the presence of clays and silts in the environment, inadequate maintenance, or other reasons [Melloul and Dax, 1990].

In 1990, the saltwater intrusion monitoring network consisted of only about 200 observation wells of small diameter, sampled at the end of the summer once a year. Only in a limited set of about 30 observation wells, where open filters had already been reached by seawater, sampling was conducted twice a year, at the end of summer and spring. An example is shown for one of these specific wells, Afridar 4, in Figure 16.4. Data like this are becoming more and more scarce. Therefore, it becomes urgent to improve the existing saltwater intrusion network, without waiting for budget to drill new wells.

16.3   Semi-Empirical Procedure to Assess Saltwater Intrusion

16.3.1   Previous Approaches to Assess Saltwater Intrusion

During the previous years, the monitoring activity conducted along the Mediterranean seashore of the Israeli Coastal Aquifer was performed regularly by a denser observation wells network. As a result,

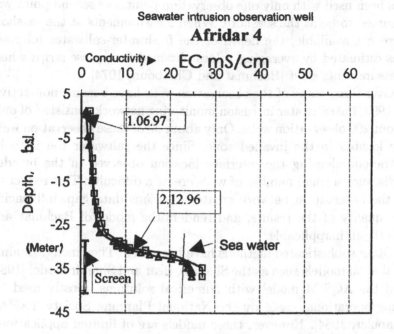

Figure 16.4. An observation well for saltwater intrusion, with salt-water interface within the open screen.

in 1970 an "acceptable" clear image of the saltwater-freshwater interface along the sea shore was available, mostly for the upper layers (A and B). One assessment was based upon the Bachmat and Chetboun [1974] model. This model assuming the saltwater interface as a parabolic line based on a simpler steady state model, and determined from at least two available points. In this model, the toe of saltwater into the aquifer being mathematically linked with water levels, would be given by the point of intersection of the parabolic freshwater/saltwater interface with the base of sub-aquifer A, B and/or AB. One of the difficulties encountered in this model is the presence of several sub aquifers which are separated from each other in some locations by impervious layers, while in other locations by semi pervious ones. Other difficulties are due to the approximate nature of the mathematical model and errors for interpolation in time and space, and the high uncertainty related to the values of parameters like hydraulic conductivity (K), storativity (S), effective porosity (n), etc., because of the scarcity of the observations wells. When this model has been used with only one observation point, the second point was assumed to be at the seashore. When measurements at the seashore were not available, the depth of the freshwater-saltwater interface was estimated by averaging values obtained from close strips where measurements exist [Bachmat and Chetboun, 1974].

Over time, many of the observation wells have become non-active. In 1990, the saltwater intrusion monitoring network consisted of only about 200 observation wells. Only about 30 of these observation wells are located in the invaded zone. Since the saltwater toe is to be extrapolated using the interface location observed in the invaded wells, such a small number of wells create a difficulty. This reduction of the observation network creates a serious data gap, influencing the quality of the results, and renders the model of Bachmat and Chetboun inapplicable.

Other sophisticated methods have been used. They include mainly analytical models such as the Shapiro, Bear and Shamir model [1983] and the AQSIM model with numerical solutions, mostly used for some operational cases by the National Planning Society TAHAL [Kapuler, 1984]. However, these models are of limited applications. They need a large number of physical parameters such as K, S, n, that are not always available (see section 16.2.2).

Therefore, to overcome the lack of available data, a semi-empirical methods has been constructed to be used mainly in the upper layers of the aquifer. Its central objective is to assess saltwater intrusion in spite of the above mentioned difficulties. The solution we propose is to combine the analysis of TDEM results and the observed wells data with a simple procedure to estimate saltwater intrusion.

16.3.2   The New Methodology

In this procedure only parameters as depth to saline water, and the geometry of the sub aquifers A, B, and/or AB (unified sub-aquifers) are used. The freshwater/saltwater interface is sharp and is estimated in the field as around 17500 mg/l of total dissolved liquids (TDS). This value corresponds to half of the Mediterranean seawater (around 35,000 mg/l TDS). The procedure consists of:

1. Division of the aquifer into strips all along the coastal aquifer. Each strip is characterized by a representative hydrogeological cross section.

2. Estimation of the freshwater/saltwater interface and the position of the toe for each strip using a semi-empirical procedure. This procedure is based on a simplified mathematical model of the saltwater encroachment and a least squares estimation.

16.3.2.1   Simplified Mathematical Model

Each strip is modeled as an aquifer of thickness D, bounded by an impervious horizontal bottom at $z = 0$ and an impervious top layer at $z = D$, while x is an inland horizontal coordinate (Figure 7.1). The hydraulic conductivities may be written in terms of the intrinsic permeability k, the dynamic viscosity μ and the fluid density ρ as follows

$$K_f = \frac{k\,\rho_f}{\mu_f}; \quad K_s = \frac{k\,\rho_s}{\mu_s} \qquad (16.1)$$

We consider the simplest case of saltwater at rest, thus a steady interface. In this case the discharge of freshwater in the strip j (Q_j) is a constant and the position of the interface is given by $x = \Xi(z)$. Integration of the continuity equation leads to the general solution (see Chapter 7):

$$\frac{d\Xi}{dz} = -\frac{K''}{Q_j}(D - \xi) \qquad (16.2)$$

where $\xi = \xi(x)$ is the interface elevation and

$$K'' = \frac{(\rho_s - \rho_f)}{\rho_s}K_f \qquad (16.3)$$

Consistent with Dupuit assumption we neglect the seepage face and assume that the interface originates at $x = 0$, $z = D$, i.e. the sea serves as a sink (Figure 7.1), and the boundary condition is $\Xi(D) = 0$. This approximation becomes accurate for a shallow interface, i.e. for $\Xi(0) \gg D$ (see, Bear and Dagan [1964a]).

With the change of variable $\zeta = D - \xi$ (Figure 7.1) and after integration of the last equation with $\Xi(0) = 0$, the exact solution may be written as follows:

$$\Xi(\zeta) = \frac{K''}{Q}\frac{\zeta^2}{2} \qquad (16.4)$$

This equation is precisely the Dupuit parabola obtained for a homogeneous medium, provided the constant hydraulic conductivity is taken equal with K_f.

16.3.2.2   Least Squares Procedure

Results provided by geo-electromagnetic method (TDEM) are related to the depth of saline water obtained by resistivity values within a range of 1.3 to 2.9 ohms. This range corresponds to typical values for saltwater body. In spite of the fact that some of TDEM results have high discrepancies with observation well data, a significant number of TDEM measurements show relative small discrepancies with the observation well. Therefore, we used the TDEM results in conjunction with observation wells data in order to enrich the saltwater monitoring data.

In Table 16.1 we summarize the number of observation points of the saltwater intrusion. For a given strip j, the total number of observation points is denoted by ns_j. Each observation point in the strip may be described by the distance from the sea Ξ_{ij} and the depth of the interface at this distance ζ_{ij}, for $i = 1, \cdots, ns_j$.

In the least squares procedure, we assume that the total observation error is the sum of the theoretical estimation and the random error. The random error is assumed to be a linear function of the position of the interface. Then, for a given strip j, the total observation error ϵ_{ij} for the measurement (Ξ_{ij}, ζ_{ij}) may be expressed as:

$$\epsilon_{ij} = \Xi_{ij} - \frac{K''}{Q_j} \frac{\zeta_{ij}^2}{2} - a_j \zeta_{ij} - b_j \qquad (16.5)$$

In Eq. 16.5, the parameters a_j and b_j are determined by the minimization of the function $\sum_{i=1}^{ns_j} \epsilon_{ij}^2$ with respect to a_j and b_j. The parameters K''/Q_j are determined using an estimation of Q_j from water level measurement near the coast and the Geographic Information System developed by Zeitoun and Soyeux [1998]. In this estimation the hydraulic conductivity of the freshwater is not needed.

Hydrologic		Dist to sea	TDEM measurement				Well measurement	
			1990		1995		1990	1995
Strp	Col	(m)	Resis (Ω)	Dpth (m)	Resis (Ω)	Dpth (m)	Dpth (m)	Dpth (m)
101	0	100	1.5	28			26	
106	0	650	1.5	59			68	62
107	0	180	1.5	27	1.5	32		33
108	0	70	1.4	18			21	23
108	0	250	1.6	41	1.1	47	53	52
109	0	70	1.5		1.4	26		
109	0	100	1.5	22			55	
109	0	510	2.5	56	2.5	59	60	61
109	0	590	2.3	51			57	
111	0	110	1.5	21	1.5	25	27	
111	0	470	1.4	56	1.3	55	54	51
111	0	480	1.5	72			61	60
112	0	180			1.6	34	51	
115	0	140	1.4	26	1.7	26	27	
119	0	150	1.3	24			31	
121	0	130	1.3	22			17	
121	0	200	1.8	24			77	
122	0	200	2.1	25			52	48
133	0	400	2.1	67			64	66
136	0	200	1.6	23	1.8	25	27	
136	0	580	2.0	94			91	
138	0	700	1.4	52	1.7	50	55	46
139	0	420	1.9	47			57	49
139	0	860	1.5	52			40	58
141	0	700	1.5	40	1.5	45	42	45
143	0	100	1.3	13			28	
147	0	800	2.7	44			89	110
148	1	800			1.4	53	56	
148	1	800	1.4	50			51	
151	0	300	1.5	28			58	
151	0	300			1.4	31	58	
151	0	500	1.3	32			32	

Table 16.1. Well and TDEM depths (above sea level) measurements in the Coastal Aquifer of Israel. (After: Goldman, et al. [1991, 1995])

Hydrologic		Dist to sea	TDEM measurement				Well measurement	
			1990		1995		1990	1995
Strp	Col	(m)	Resis (Ω)	Dpth (m)	Resis (Ω)	Dpth (m)	Dpth (m)	Dpth (m)
152	0	550	1.8	32			34	35
153	0	500	2.9	91	2.2	74	86	
154	0	100	1.5	36	1.5	37	36	
155	0	400	2.4	43			32	35

(continued)

16.4   Actual Implementation to Israel Coastal Aquifer

The procedure described above has been implemented to this aquifer to assess saltwater intrusion for different years as follows. First, the Coastal Plain Aquifer is subdivided into 57 hydrologic strips of 2 km width each from hydrologic strip 100 to 157 as shown in Figure 16.1. Then for each strip, whenever possible, the recent assessment of saltwater intrusion based observation well data is assigned together with the TDEM results relating to the saltwater depth along the upper sub-aquifer AB. The depths are estimated from TDEM resistivity values varying within a range of 1.3 to 2.9 ohms. When the discrepancies between the TDEM and borehole data are less than 5 m, results are then considered as "acceptable". An example is given in Figure 16.5. The delineation of freshwater/saltwater depth then follow the procedure explained in section 16.3 to estimate within 20% of uncertainty. Saltwater intrusion along the upper hydrologic horizons AB of the Coastal Aquifer since 1990 is shown in Figure 16.6.

The results given in Figure 16.6 indicate that the probable maximum saline water intrusion into the aquifer has reached a distance of between 0.1 to 2 km along the seashore. The farthest intrusion appears to be in the Dan Metropolitan and Netanya regions. Taking into consideration uncertainty, comparison of this data with that of 1960 shows a maximum intrusion rate of around 30 m/year, mainly in the northern part of the aquifer. In 1995, when observation wells and TDEM data are combined in Figure 16.6, we note some sign of

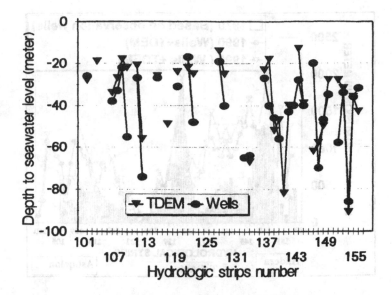

Figure 16.5. Comparison of saltwater/ freshwater interface depth based on observation wells and TDEM measurements.

stagnation or even retreat of saltwater intrusion into the Tel Aviv area and in Emek Efer, and Hadera areas. This may be explained by a reduction in pumpage from wells in this area, as well as by artificial and natural freshwater recharge during and after the rainy winter of 1991/92. However, this interpretation of interface retreat may be questioned when considering the laboratory work of Golden-berg et al. [1986], which suggests that saltwater intrusion·is basically an irreversible process. For this and other reasons, more research are required for better interpretation of the results and the management of the aquifer.

16.5   Conclusions

Complexity of groundwater monitoring networks for seawater in-trusion stems from the lithological changes which characterize the aquifer media and radical groundwater quality changes near the coast. Thus, only a denser monitoring network can readily supply good quality data to overcome the problem of heterogeneity and radical groundwater quality alteration. Only such monitoring can

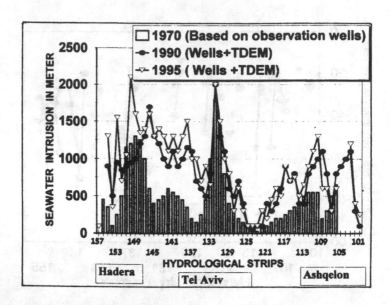

Figure 16.6. Saltwater intrusion (in meter) into the Coastal Aquifer of Israel in 1960, 1990, and 1995.

supply adequate data to calibrate models, and check their feasibility and assumptions. This is required in order to make models more operational, and mitigate the fact that validation of any model remains a challenging task [Konikow, and Bredehoeft, 1992]. Israeli experience has been presented as an example to indicate how budgetary problems for drilling more observation wells can be offset by combining conventional means with other methods, such as the TDEM geoelectromagnetic method, to obtain sufficient data to delineate seawater intrusion location, and utilize simple linear models as an initial step to assess more accurately the approximate location and dynamism of seawater intrusion.

Acknowledgments: The authors are greatly indebted to Dr. Martin Collin for his help in editing this paper.

Chapter 17

Nile Delta Aquifer in Egypt

M. Sherif

17.1 Introduction

Egypt lies between latitudes 22° and 32° North, and longitudes 25° and 35° East. The North boundary of Egypt is the Mediterranean Sea, and the East is the Red Sea. The South and West are political boundaries with Sudan and Libya, respectively. Land area of Egypt is about one million km^2, 94% of which is desert. The population of the country is currently estimated as 62 million.

There are three dependent sources of water in Egypt: Nile water, groundwater, and drainage water. Rainfall is rare and does not contribute to the water budget of the country. The Nile water is almost the only renewable source of water. The Egyptian share from the Nile water, according to the international agreements, is limited to 55.5 billion m^3 per year since 1959. The per capita share from the renewable water is thus about 900 m^3/year.

In 1992, 4.5 billion m^3 of drainage water were reused, while 11.7 billion m^3 of low quality drainage water were expelled to the Mediterranean Sea and northern lakes. Additional reuse of agriculture and industrial drainage water is not recommended because of its low quality and high contents of chemicals and pesticides. Unwise reuse of drainage water may cause serious environmental impacts. Currently 5.3 billion m^3/year of groundwater are pumped from different aquifers, 85% of which originates from the Nile water and the rest is fossil water.

The Nile Delta is one of the most fertile areas in Egypt. However, less than two-thirds of its area is under cultivation. The uncultivated

J. Bear et al. (eds.), Seawater Intrusion in Coastal Aquifers, 559–590.
© 1999 *Kluwer Academic Publishers.*

lands are found along the sea to the north and the Delta fringes close to the desert. Irrigation is mostly practiced by surface water through an intensive network of canals branching out from the Nile River. The growth of population in the Nile Delta and hence, the increase in human, agricultural, and industrial activities has imposed an increasing demand for freshwater. This increase in demand is covered by intensive pumping of fresh groundwater, causing subsequent lowering of the piezometric heads and upsetting the dynamic balance between fresh and saline water bodies in the Nile Delta aquifer. Like any coastal aquifer, an extensive seawater body has intruded the aquifer forming the major constraint against its exploitation.

Many studies were conducted to simulate the seawater intrusion in the Nile Delta aquifer using various numerical techniques. Most of which were based on the sharp interface approach, while few accounted for the dispersion zone and density variation. Examples of the previous investigations include among others, Wilson et al. [1979], Amer et al. [1979], Amer and Farid [1981], Farid [1980; 1985], Sherif [1987], Sherif et al. [1988; 1990a; 1990b], Darwish [1994], Amer and Sherif [1995; 1996], and Sherif and Singh [1997]. Field investigations and experiences revealed the existence of a considerable dispersion zone between the aquifer freshwater and the intruded seawater. The sharp interface assumption is therefore not justified in such an aquifer. The material presented in this chapter is therefore based on the variable density approach.

17.2   Nile Delta Aquifer

The Nile Delta region lies within the temperature zone which is a part of the great Desert belt. It occupies a portion of the arid belt of the Southern Mediterranean region. The desert fringes on both sides of the Delta cause a rise in temperature and affect the changes in daily temperature. The average temperatures in January and July at Cairo are 12°C and 31°C, respectively. Minimum and maximum temperatures at Cairo are 3°C and 48°C, respectively. Rainfall over the Nile Delta is rare and occurs in Winter. The maximum annual rainfall (180 mm) is encountered along the shore line of the Mediterranean Sea. This amount decreases very rapidly and reaches 26 mm at Cairo.

One of the groundwater reservoirs in Egypt lies beneath the Nile

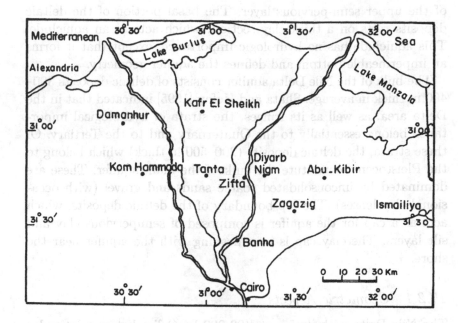

Figure 17.1. The Nile Delta aquifer.

Delta. It is among the largest underground water reservoirs in the world, with a total capacity of 500 billion m³ of water. The Nile Delta aquifer is over six million acres in extent, with eastern boundaries near the Suez Canal and western boundaries well into the desert, as shown in Figure 17.1. It fills a vast underground bowl situated between Cairo and the Mediterranean Sea. If it were not for the presence of seawater along the bottom of this bowl, the Nile Delta aquifer could have been easily exploited to the best interests of Egypt.

Seepage out of the Nile River, canals and drains and surplus of irrigation water are the main sources of aquifer recharge. It may also be recharged, nominally, by northward flow from the Nile Valley aquifer. On the other hand, the aquifer loses some of its water to the Mediterranean Sea and to the drainage system in the northern part of the Delta.

The Nile Delta aquifer forms an immense, and complex groundwater system. It is a leaky Pleistocene aquifer overlain by a semipervious Holocene aquitard (clay cap) and underlain by an impermeable Miocene aquiclude. The aquifer status (phreatic, confined or leaky) is defined according the thickness and vertical permeability of the upper semi-pervious layer. The basal portion of the deltaic deposits rests on a thick clay section, which acts as an aquiclude. This aquiclude has no hydrologic importance, except that it forms an impermeable bottom and defines the aquifer geometry.

The bulk of the Nile Delta aquifer consists of deltaic deposits 300–400 m thick in average. Shata and Hefny [1995] indicated that in the Delta area, as well as its fringes, the strata of hydrological importance belong essentially to the Quaternary and to the Tertiary. Of these strata, the deltaic deposits (200–500 m thick) which belong to the Pleistocene, constitute the bulk of the main aquifer. These are dominated by unconsolidated coarse sands and gravel (with occasional clay lenses). The top boundary of the deltaic deposits, which acts as a cap for the aquifer is composed of semipervious clay and silt layers. The clay cap is intermeshing with the aquifer near the shore.

17.2.1 Geometric Aspects

The Nile Delta with its fringes (22,000 km^2) lies between latitudes 30°25′ and 31°30′ North, and longitudes 29°50′ and 32°15′ East, Figure 17.1. The Nile Delta aquifer is generally considered a leaky aquifer, however, phreatic conditions are encountered in some areas east and west of the Delta fringes, where the upper clay layer vanishes. The depth of the aquifer at the Mediterranean Sea varies from one point to the other and may reach 1000 m in some locations. Since the bottom (impermeable) layer approaches the upper clay layer near Cairo at El-Manawat (Figure 17.2), the aquifer is considered partly isolated from the Nile Valley aquifer to the south.

Horizontal and vertical dimensions of the Nile Delta aquifer were defined using deep oil borings and data from test bore holes [Farid 1980]. The boundary between the Pleistocene aquifer and the Pliocene and/or Miocene aquiclude was defined. Eight longitudinal and lateral cross sections were drawn. Figures 17.2 and 17.3 present two vertical cross sections in the middle and east of the Nile Delta. Figure 17.4 presents the contour lines of the aquifer thickness [Farid, 1980]. The thickness of the Pleistocene aquifer increases toward the Mediterranean Sea, and decreases southward, almost vanishing near El Manawat. South of Tanta in the transversal direction, the bottom boundary forms a concave shape. The aquifer thickness decreases eastward and southward with a maximum depth in the middle of

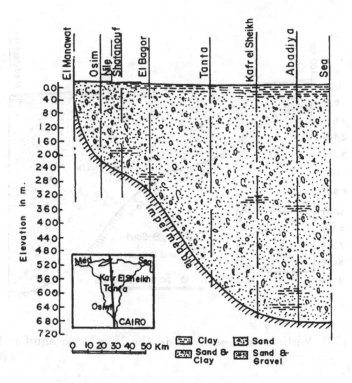

Figure 17.2. Vertical cross-sections in the Nile Delta aquifer: Middle Delta.

the Delta. North of Tanta, the maximum thickness of the aquifer shifts toward the east.

17.2.2   Clay Cap

The clay cap takes different profiles along the shore of the Mediterranean Sea. Generally, it is divided into two layers. The first layer is the upper clay cap which acts as an aquitard, about 20 m thick. The second layer is the lower clayey sand layer, a few meters in thickness, with higher permeability than the former. The thicknesses of the upper clay cap (aquitard) and the clayey sand layer are well defined at many locations in the Delta. Their thicknesses are mostly less than 20 m and 15 m, respectively. Only along the Mediterranean shore, the thickness of the clay cap may reach 70 m. No clay layer exists at the Delta fringes. Figure 17.5 presents the contour map for

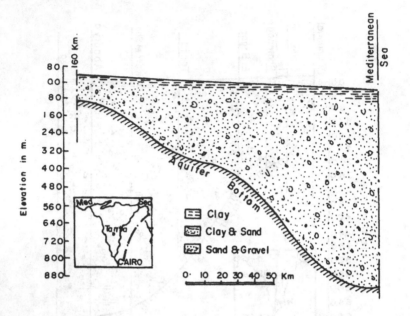

Figure 17.3. Vertical cross-sections in the Nile Delta aquifer: East Delta.

the combined layers (semipervious layer) in the Nile Delta revealing their sedimentary pattern.

The vertical movement of water through the upper semipervious layer affects, to a great extent, the water balance of the system. Downward movement of water occurs in two different stages. The first is the downward infiltration of irrigation water from the ground surface to the subsoil water table through the unsaturated zone of the clay cap. The velocity of this movement is defined by the downward infiltration velocity. The second is the movement of the subsoil water from the water table to the groundwater in the aquifer through the saturated zone of the clay cap. The velocity of the flow in this zone is defined by seepage velocity and can be evaluated from Darcy's law. Values of the vertical hydraulic conductivity, K_{cv}, for the clay layer in the Nile Delta area are given in Figure 17.6. It has an average value of to 2.5 mm/day.

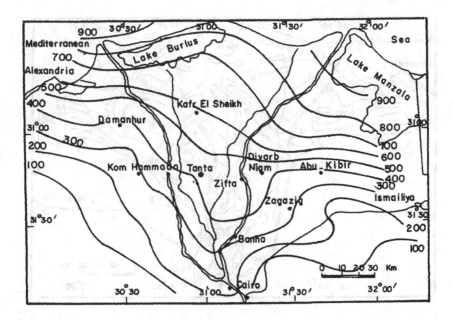

Figure 17.4. Contour lines for the aquifer thickness (Farid, 1980).

17.2.3 Hydraulic Parameters

Field experiments were carried out to determine the hydraulic parameters for the Nile Delta aquifer. A hydraulic conductivity of 100 m/day and a storativity of about 10^{-4} to 10^{-3} are considered to represent the regional values for the Nile Delta aquifer. Farid [1980] reported different values for hydraulic conductivity and storativity at various locations. The hydraulic conductivity of the aquifer decreases toward the south and west. An effective porosity of 0.3 is considered to represent the aquifer medium.

The dispersivity of the medium in the Nile Delta aquifer has not yet been measured. Dispersivity has traditionally been considered as a characteristic single-valued property of the entire medium [Bear and Bachmat, 1967; Bear, 1979; Bear and Verruijt, 1987]. Studies have suggested that dispersivity is not a constant but rather depends on the mean travel distance and/or the scale of the system.

For an isotropic medium, the number of non zero components of the dispersivity tensor is 21. All are related to two parameters only, the longitudinal dispersivity, α_L, and the lateral dispersivity,

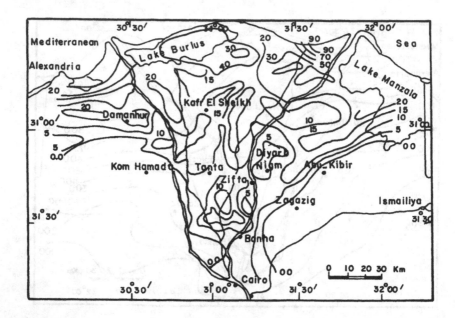

Figure 17.5. Contour lines for the thickness of the upper semipervious layer (Farid, 1980).

α_T. Laboratory experiments showed that α_L is of the order of magnitude of the average sand grain size. Transversal dispersivity α_T is estimated as 10 to 20 times smaller than α_L. For longitudinal dispersivity, values between 0.1 and 500 m can be found in the literature. Based on similar studies and on Sherif et al. [1988], longitudinal dispersivity, α_L, and lateral dispersivity, α_T, for the Nile Delta aquifer are set equal to 100 m and 10 m, respectively.

The free water table is measured at various locations, however, when missing, it is generally assumed to be 1 m below the ground surface. The piezometric head is monitored periodically and is given in Figure 17.7.

17.2.4 Pumping Activities

Pumping activities from different governorates are monitored by the Groundwater Research Institute (GWRI) in Egypt. According to the well inventory in 1992 the total pumping from the Nile Delta aquifer was estimated as 1.92 billion m^3/year (Table 17.1). The current pumping for various uses in the Nile Delta is estimated at 2.3

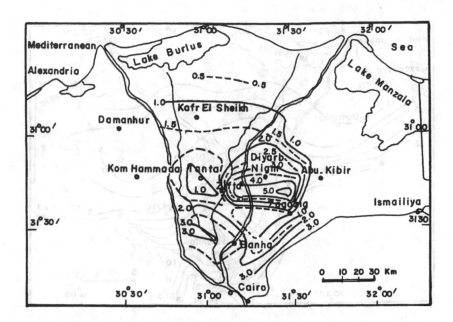

Figure 17.6. Vertical hydraulic conductivity of the upper semipervious layer.

Governorate	Number of wells	Q (10^3 m^3/year)
Alexandria	18	813
El Beheira	40	109,640
El Gharbia	3,391	792,851
Dakahlia	2	626
Sharkia	1,953	404,615
Ismaailia	396	62,148
Menoufia	1,719	549,457
Total	7,492	1,920,150

Table 17.1. Groundwater extraction from different governorates [GWRI, 1992].

billion m^3/year.

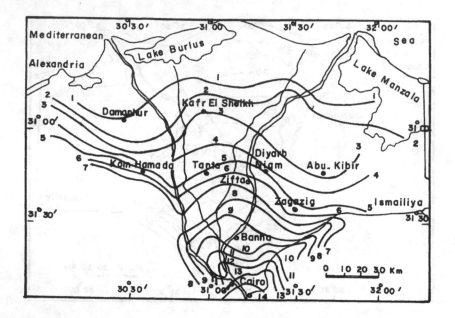

Figure 17.7. Piezometric head in the Nile Delta aquifer (GWRI, 1992).

17.3   Origin of Brackish Water

It is important to know whether the saline water in the Nile Delta aquifer, with specific reference to northern parts, is attributed to sea-water intrusion or is simply a paleo-formation water. To that end, a hydrochemical study, [Farid, 1980] was conducted to evaluate the Chloride-Bicarbonate $Cl/(CO_3 + HCO_3)$ ratio for the groundwater in different locations of the Nile Delta aquifer. Revelle [1963] con-cluded that this ratio, in milligram equivalents per liter, could be used as a criterion for the recognition and evaluation of seawater intrusion in coastal aquifers.

Based on the Revelle ratio Farid [1980] indicated that the brackish water in the Nile Delta is of seawater origin. He also showed that the saline water wedge had retreated seaward by about 2.5 km during the years 1968–1978. This movement was attributed to the continuous rise in piezometric heads during this specific period of time.

During the years 1958–1962, a number of wells were drilled by a Yugoslavian company in the Nile Delta aquifer [Sherif, 1987]. Short

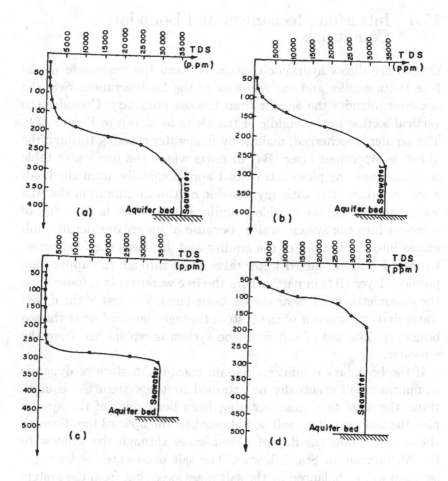

Figure 17.8. Variation of TDS with depth at different locations in the Nile Delta aquifer: (a) at Abu-Kibir, (b) at Diyarb Nigm, (c) at Zifta, (d) at Tanta.

and long normal resistivity curves, self potential and two microsonde curves were recorded in some bore holes. Water salinity in the permeable layers was calculated using the true resistivity and self potential methods. The salinity distribution with depth at different locations indicated a typical case of a seawater invasion. Examples for the measured salinity with depth at different locations are provided in Figure 17.8.

17.4   Intrusion Mechanism and Boundary Conditions

Due to the direct hydraulic contact between the freshwater in the Nile Delta aquifer and the seawater of the Mediterranean Sea, the seawater intrudes the aquifer from the sea boundary. Consider the vertical section in the middle of the Delta as shown in Figure 17.9. The aquifer is recharged, mainly, by freshwater entering through the upper semipervious layer (B4) in parts where the free water table is higher than the piezometric head and, nominally, from the landward boundary (B1) with any possible northward flux from the Nile Valley aquifer. At the seaside boundary (B3), there is an influx of seawater into the system which, because of its greater density, migrates into the bottom of the aquifer and displaces the freshwater. Upward leakage of mixed water takes place through the upper semi pervious layer (B4) in parts where the free water table is lower than the piezometric head (near the sea boundary). The rest of the mixed water finds its way out of the system through the window at the sea boundary. The loss of salt from the system is replenished from the seawater.

If the boundary conditions remain constant, a state of dynamic equilibrium will eventually be attained by the system. At equilibrium, the total fluid mass entering from both sides of the system plus the leakage influx, will be balanced by the upward flux through the aquitard plus the flux of mixed water through the window to the Mediterranean Sea. Likewise, The salt mass entering from the seaward will be balanced by the salt mass swept out from the system with the mixed water.

The lateral boundary (B1) at the landward side, Figure 17.9, should be located at a point where either the concentration is constant and equals to the freshwater concentration, C_f, or where the concentration gradient across the boundary is negligible, i.e., is equal to zero. The former condition of prescribed concentration is preferable because it accelerates the convergence. The pressure at this boundary is hydrostatic and the freshwater flux through it can be calculated from Darcy equation. The bottom of the aquifer (B2) is impermeable, i.e., the normal flux through the bed for both fluid and salt ions is equal to zero.

The seaward boundary (B3), where the Nile Delta aquifer is ex-

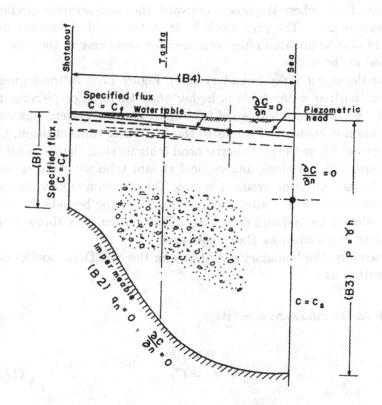

Figure 17.9. Intrusion mechanism and boundary conditions in the Nile Delta aquifer.

posed to the Mediterranean Sea, can be treated in different ways. It has been customary [Huyakorn et al., 1987; Sherif et al., 1988; Galeati et al., 1992] to deal with the seaside boundary as prescribed concentration boundary when the flow across it is directed inward, and as zero dispersive flux boundary when the flow direction is outward. Since conditions along this boundary vary with time, it may not be adequately prescribed on the basis of preliminary simulations as done by Huyakorn et al. [1987]. We follow the approach utilized by Sherif et al. [1988] and Galeati et al. [1992]. During each iteration the directions of the velocity vectors at all nodal points located on B3 are checked, and the appropriate boundary conditions are assigned accordingly. The boundary conditions at B3 are updated after each iteration. Over the segment of inward flow, the concentration is prescribed and equals to seawater concentration, C_s. Over the window

(Figure 17.9), where the flow is outward, the concentration gradient is equal to zero. The pressure is hydrostatic at this boundary and should also be updated after each iteration according to the concentration of the nodes.

For the upper leaky boundary (B4), Figure 17.9, in the segment where the free water table is higher than the aquifer piezometric head, i.e., there is a leakage of freshwater into the system, the concentration is known and equals to the freshwater concentration, C_f. Otherwise, where the piezometric head is higher than the water table, the concentration along any vertical stream tube will be constant, i.e., the concentration gradient is zero. The direction of flow through B4 should be checked after each iteration and the boundary conditions should be updated accordingly. The vertical flux through the aquitard is governed by Darcy equation.

Therefore, the boundary conditions for the Nile Delta aquifer can be written as:

(a) At the landward side (B1):

$$\psi = \psi_L$$
$$C = C_f \tag{17.1}$$

where ψ_L is the piezometric head at the land side.

(b) At the bottom boundary (B2), impermeable:

$$q_n = 0$$
$$\frac{\partial C}{\partial n} = 0 \tag{17.2}$$

(c) At the seaside boundary (B3):

$$\psi = \psi_s$$
$$\frac{\partial C}{\partial n} = 0 \quad \text{over the window}$$
$$C = C_s \quad \text{below the window} \tag{17.3}$$

where ψ_s is the equivalent freshwater hydraulic head at seaside.

(d) At the upper boundary (B4):

$$q_z = \frac{K'_z}{b}(\psi_t - \psi) + \frac{aK'_z}{\rho_f}(C - C_f)$$

$$C = C_f \qquad \text{for downward flux}$$

$$\frac{\partial C}{\partial n} = 0 \qquad \text{for upward flux} \qquad (17.4)$$

In the above K'_z is the vertical hydraulic conductivity for the aquitard, b is the aquitard thickness, ψ_t is the free water table elevation, ρ_f is the freshwater density and a is a constant given as

$$a = \frac{\rho_s - \rho_f}{C_s - C_f} \qquad (17.5)$$

where ρ_s is the seawater density.

17.5 Modeling Seawater Intrusion in Nile Delta Aquifer

The seawater intrusion phenomenon in the Nile Delta aquifer is a fully three dimensional process. Water and salt ions can move in any direction according to the hydraulic and concentration gradients. However, three dimensional simulation of seawater intrusion in the Nile Delta aquifer is unfeasible, not only because of its huge volume and requirements of fine discretization, but also because of the lack of the required data.

17.5.1 Vertical Simulation

To investigate the seawater intrusion in the Nile Delta aquifer Sherif et al. [1988] considered the vertical section in the middle of the Delta, as shown in Figure 17.2. A two-dimensional Finite Element model for Dispersion in coastal aquifers (2D-FED) was employed to predict the concentration distribution and the flow pattern in the aquifer. The description of the model and its complete mathematical development are found in Sherif et al. [1988].

Based on some field data, it was assumed that seawater will not migrate inland up to Shatanuf, which is 150 km from the Mediterranean Sea. Under this assumption a study domain of 150 km in length was considered; the depth of the domain varies from 680 m at the seaside to 240 m at the land side. The thickness of the upper semipervious (clay) layer is taken as 40 m from the sea boundary to Om-Sin, which is about 17 km inland. From Om-Sin to Kafr-Elsheikh, 52 km from the sea boundary, it is varied from 40 m to 15 m after which the thickness of the semipervious layer is constant. The bottom boundary of the Nile Delta aquifer is impermeable.

A calibrated value for the hydraulic conductivities K_{xx} and K_{zz}, of 100 m/day was considered. The vertical hydraulic conductivity for the upper semipervious layer, K_{zz}, was set equal to 0.005 m/day. The piezometric head at the land boundary was 14 m above sea level. At the sea boundary the piezometric head was 0.6 m. The free water table was measured in some stations, and between these stations it was assumed linear. Based on similar studies, the longitudinal and transversal dispersivities, α_L and α_T, were set equal to 100m and 10m, respectively.

The domain was subdivided into five subdomains, each was then divided into a number of triangular elements. Smaller elements were placed in the regions where the concentration gradient is relatively high. An intensive grid was also employed near the shore boundary to account for the expected change in the direction of the velocity vectors due to the cyclic flow. The domain was finally represented by a nonuniform grid with 4020 nodes and 7600 triangular elements. The convergence criterion was set equal to 10^{-5}.

Due to deep and wide opening of the Nile Delta aquifer at the Mediterranean Sea, the seawater intruded the aquifer under a high potential head, even higher than that at the land side. Mixed water found its way back again to the sea through the window. At the last 22km from the sea boundary, there was some upward flux of mixed water through the upper semipervious layer. This result is consistent with observations in the northern part of the Delta. The drainage system of the cultivated lands captures considerable amounts of saline water directed from the aquifer to the upper semipervious layer.

Equiconcentration line 35 (35,000 ppm), intruded into the Nile Delta aquifer to a distance of 63 km from the sea side measured along the bottom boundary, as shown in Figure 17.10a. Equiconcentration line 1.0 intruded to a distance of about 108 km measured along the same boundary. The width of the dispersion zone, measured between equiconcentration lines 1 and 35, is about 45 km, as shown in Figure

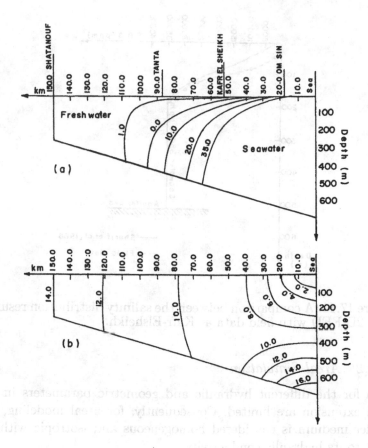

Figure 17.10. Vertical simulation for seawater intrusion in the Nile Delta aquifer: (a) Equiconcentration lines, (b) Equipotential lines.

17.10a. Figure 17.11 presents a comparison of the results obtained by the 2D-FED model with measured data for the salinity distribution with depth at Kafr-Elsheikh.

It can be concluded from the shape of the equipotential lines (Figure 17.10b) that the depth of the window at the seaside is about 350m. Strong cyclic flow at the sea boundary was detected. The seawater intrudes into the aquifer from the lower part of the seaside under a high potential head and either rotates back to the sea through the window or finds its way to the agriculture drainage system in the upper semipervious layer.

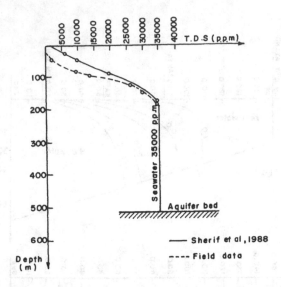

Figure 17.11. A comparison between the salinity distribution resulted from 2D-FED with field data at Kafr-Elsheikh.

17.5.2   Areal Simulation

Data for the different hydraulic and geometric parameters in the areal extension are limited. Consequently, for areal modeling, the aquifer medium is considered homogeneous and isotropic with respect to its hydraulic conductivity.

Sherif and Singh [1997] investigated the seawater in the Nile Delta aquifer in the aerial view using SUTRA [Voss, 1984a]. The study domain was discretized into 998 quadratieral elements with 1084 nodal points, as shown in Figure 17.12. This intensive grid was provided to account for any variation in geometric and hydraulic parameters. This discretization ensures the numerical stability and avoids improper dimensions of the elements throughout the domain. The cylinderness ratio of the various element was kept less than 1:8.

The depth of the aquifer and the thickness of the upper clay layer at the various elements were adapted from Figures 17.4 and 17.5, respectively. Values of horizontal and transverse hydraulic conductivities (K_{xx} and K_{yy}) were assigned as 100 m/day. Longitudinal and lateral dispersivities were taken equal to 100 m and 10 m, respectively. Pumping activities at each governorate were evaluated and

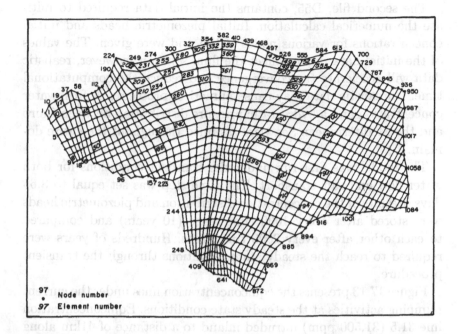

Figure 17.12. The grid system for horizontal simulation.

the share of each node of the finite element grid was calculated.

The free water table in the upper clay layer was set 1m below the ground level and the piezometric head at the different nodes was adapted from Figure 17.7. A specified head was considered below the two branches of the Nile river. The pressure at the Mediterranean Sea was set equal to the atmospheric pressure. At the southern boundaries the pressure was defined. Open boundaries were considered along both Nubaria and Ismaillia canals. Concentration along the Mediterranean Sea was set equal to seawater concentration, C_s.

All data were prepared and presented in two files, D5 and D55, as required by SUTRA. The first file, D5, includes geometric, hydraulic and transport parameters. The default values for various parameters were defined first and the multiplication factors were then introduced for each element. Nodal coordinates of the grid system in the horizontal plane were defined. The convergence criterion for both the concentration and the pressure was set equal to 10^{-3}. The boundary conditions and the pumping activities at the various nodes were introduced as well in D5.

The second file, D55, contains the initial data required to initiate the numerical calculation. Initial piezometric heads and initial concentrations for various nodes of the grid were given. The values of the initial data do not affect the final solution, however, realistic data will accelerate the convergence and reduce the computational time. Unrealistic data may cause numerical problems. A freshwater concentration and a hydrostatic pressure distribution between Cairo and the Mediterranean Sea were assumed throughout the entire domain.

The model was first run under the transient conditions for both water flow and salt transport. The time step was set equal to 3.65 days. Results of the concentration distribution and piezometric heads were stored after every 1000 time steps (10 years) and compared to each other after every 2000 time steps. Hundreds of years were required to reach the steady state conditions through the transient procedure.

Figure 17.13 presents the equiconcentration lines under the current pumping activities at the steady state conditions. Equiconcentration line 31.5 (31,500 ppm) intruded inland to a distance of 41km along latitude 31°00' while equiconcentration line 17.5 intruded to a distance of 61.5 km along the same latitude measured from the sea side boundary. Equiconcentration line 3.5, representing 0.1 of the maximum concentration, intruded inland to a distance of 84 km measured from the same boundary along the same latitude.

It should be noted that the verification of the resulted equiconcentration lines is not justified in this case. The salinity of any point in the aerial view may vary dramatically in the vertical direction from freshwater concentration to seawater concentration as indicated in Figure 17.8. The concentration of groundwater at a specific point in the plane is not indicative unless the depth of that point is defined. This variation can not be included in the areal simulation. Nevertheless, the resulted concentration distribution (Figure 17.13) is quite consistent with the few observations recorded from some production wells in the region. For example, the salinity of the groundwater at Kafr El-Sheikh, 52 km from the shore line, is equal to the seawater salinity at a depth of less than 100 m. On the other hand, the groundwater at Tanta, 92 km from the shore line, is mostly fresh up to a depth of about 250 m. To the south of Tanta, the groundwater has a salinity of less than 1000 ppm throughout the entire depth of

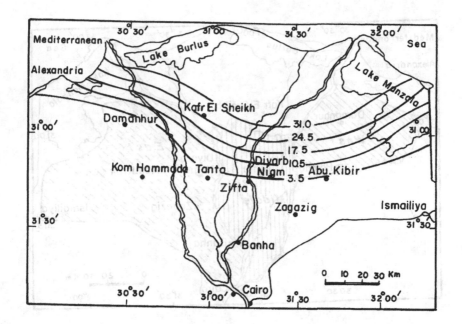

Figure 17.13. Equiconcentration lines in the areal view.

the aquifer.

17.6   Effect of Pumping

To examine the effect of pumping on seawater intrusion and define the best locations for additional pumping, six different scenarios were selected. The resulted equiconcentration lines were then compared with those encountered under the current pumping (basic run), shown in Figure 17.13. The simulation was performed via SUTRA and the results were compared in the aerial view.

The area of the Nile Delta was divided into three main zones for pumping activities; the middle zone, the eastern zone, and the western zone, as shown in Figure 17.14. The following scenarios were thus considered:

Scenario 1 The current pumping was redistributed in the eastern and western parts of the Delta without any pumping from the middle Delta. The 2.3 billion m^3/year of groundwater pumping were distributed among the nodal points in the eastern and

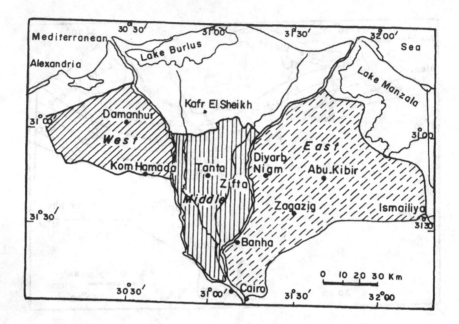

Figure 17.14. Zones of pumping in the Nile Delta.

western Delta only.

Scenario 2 The 2.3 billion m^3/year were pumped from the middle Delta only without any pumping in the eastern and western parts.

Scenario 3 In this scenario, the current pumping was doubled, i.e., 4.6 billion m^3/year were pumped from the same governorates. The share of each node, as calculated in the basic run is thus doubled.

Scenario 4 In this scenario, 4.6 billion m^3/year were pumped from the eastern and western sides of the Delta. No pumping took place in the middle Delta.

Scenario 5 The distribution of the current pumping (2.3 billion m^3/year) was maintained, while additional pumping of 1.5 billion m^3/year in the eastern Delta and 0.8 billion m^3/year in the western Delta were considered. The total pumping was thus to 4.6 billion m^3/year.

Scenario 6 In this scenario, an additional pumping of 2.3 billion m^3/year was considered in the middle Delta only, while the

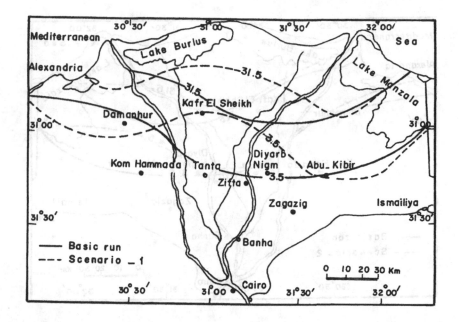

Figure 17.15. Comparison between equiconcentration lines (basic run and Scenario 1).

distribution of the current pumping was maintained.

Figures 17.15 through 17.20 compare equiconcentration lines 31.5 and 3.5 in the various scenarios with those in the basic run (current conditions). Scenarios 1 and 2, consider the current pumping but from different areas. Figures 17.15 and 17.16 reveal that both scenarios 1 and 2 are actually better than the current policy for groundwater pumping regarding seawater intrusion. Although the same amounts of water were pumped, yet less intrusion was encountered. Scenario-1 reduces the intrusion in the middle Delta considerably with slight inland intrusion in the east and west. Scenario-2 reduces the intrusion throughout the Delta with specific reference to the eastern part of the Delta.

Scenarios 3, 4, 5 and 6 represent policies for pumping 4.6 billion m^3/year from the Nile Delta aquifer. Figures 17.17 through 17.20 reveal that scenarios 4 and 6 cause less impact than scenarios 3 and 5. In scenario 4 (Figure 17.18), although the pumping from the entire Delta was doubled yet less intrusion was found in the middle Delta. On the other hand, equiconcentration line 3.5 advanced inland by

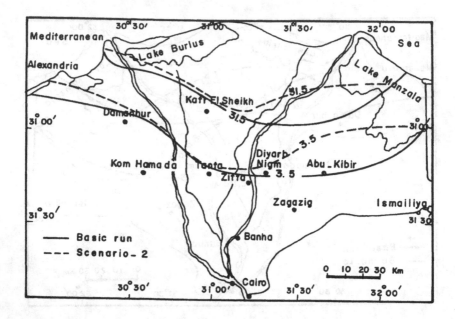

Figure 17.16. Comparison between equiconcentration lines (basic run and Scenario 2).

a distance of about 21 km and 7.5 km in the western and eastern parts, respectively. Under scenario 6, where an additional pumping of 2.3 billion m^3/year was considered from the middle Delta only, equiconcentration line 3.5 advanced inland by a limited distance in the middle and western Delta and retreated slightly in the eastern Delta as shown in Figure 17.20. Scenario 6 has the least impact under the condition of doubling the current pumping.

The rate of intrusion migration or retardation under any of the proposed scenarios is relatively small, in the range of 10 to 40 m/year. A movement of 1km of any equiconcentration line would require a time between 25 and 100 year, according to the proposed scenario. The response of the system to the different scenarios is quite tardy due to the huge capacity of the Nile Delta aquifer and to the slow nature of the dispersion process. The dead storage (nonrenewable water) in the aquifer is estimated in the order of 500 billion m^3 [Farid, 1980]. Hundreds of years were required to achieve the steady state conditions under any scenario.

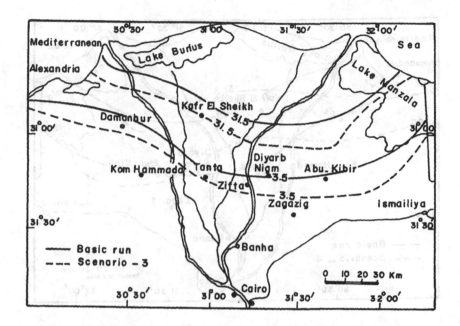

Figure 17.17. Comparison between equiconcentration lines (basic run and Scenario 3).

17.7   Effect of Climate Change

17.7.1   Introduction

Climate is a dynamic regime subject to natural variations at various time scales from years to millennia. Climate change is defined as a significant change in climate variables, from one period to another, and can be caused both by natural conditions and by human activities [Refsgaard at al., 1989].

The steady increase in concentrations of greenhouse gases in the earth's atmosphere is expected, within the next century, to produce a significant global warming [Buddemeneir, 1988]. Concentration of carbon dioxide in the atmosphere has increased by about 25 percent since 1860. This increase was mainly due to fossil fuel consumption and industrial and agricultural activities. The concentration of carbon dioxide has increased significantly over the past quarter century. On the other hand, measurements show that the average temperature of the earth has increased by 0.5 to 0.7 degree since the turn of

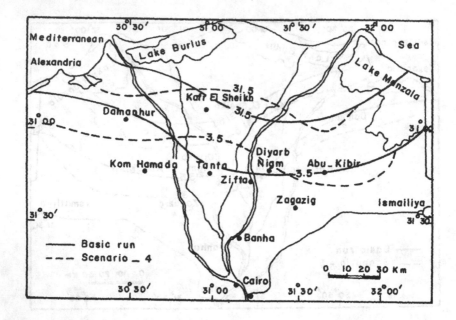

Figure 17.18. Comparison between equiconcentration lines (basic run and Scenario 4).

this century.

Despite the many uncertainties involved in global warming calculations, and regardless of the ongoing efforts to control emissions of greenhouse gases, it is expected that, under any circumstances, considerable increase in the atmospheric concentration of greenhouse gases will occur in the next century [Mimikou, 1995]. The rise of the global-mean temperature is expected to be accelerated because of the growing concentration of other greenhouse gases (CH_4, N_2O, and CFCs) and is likely to occur even before the doubling of atmospheric concentration of CO_2 [Burger, 1989]. Studies indicated that, by the middle of the next century, the increase in the atmospheric concentrations of greenhouse gases will cause an increase in the earth's average temperature by 1.5 to 4.5° Celsius.

Greenhouse gases allow solar radiation to pass through the atmosphere and hit the surface of the earth. On the other hand, greenhouse gases intercept and store the infrared radiation emitted from the surface of the earth back to the space. This process warms up the atmosphere and gives us our present climate conditions. If there were

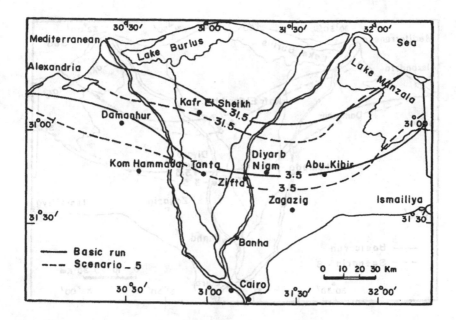

Figure 17.19. Comparison between equiconcentration lines (basic run and Scenario 5).

no greenhouse gases in the atmosphere, the average temperature of the earth would be 30° colder. Likewise, any increase in the concentration of the greenhouse gases will be associated by a corresponding increase in the global temperature.

17.7.2 Climate Change and Seawater Intrusion

Arid regions may be subject to more dryer weather under the expected global warming. Therefore, less groundwater will be available and more demand for groundwater will be imposed. A study on the effects of climate variability and change of groundwater in Europe estimated the expected increase or decrease in groundwater recharge according to the regional distribution of rainfall, which was evaluated by Global Circulation Models "GCMs" [Thomsen, 1989]. Another study in Australia revealed that a ±20% change in rainfall would result in a ±30% change in recharge beneath grasslands, while beneath pine plantation the corresponding change in groundwater is four times higher [Sharma, 1989].

The expected increase in the global temperature will warm to some

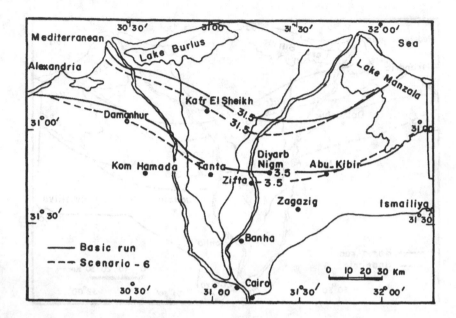

Figure 17.20. Comparison between equiconcentration lines (basic run and Scenario 6).

extent, the land surface, oceans, and seas. The atmospheric pressure will also be affected. Any variation in the atmospheric pressure will cause inverse variation in sea levels. With a decrease in atmospheric pressure of one millibar, sea levels will rise by ten millimeters. A series of depressions in atmospheric pressure can cause a considerable rise of the water levels in shallow ocean basins.

On the other hand, sea level would also rise for two reasons: first, warmer oceans would expand; and second, melting ice sheets and glaciers would add to the total mass of water in the oceans [Theon, 1993]. Data obtained from tidal gauges indicates that sea level has risen by 10 to 20 cm over the past century. It is estimated that about 25 percent of which has resulted from the thermal expansion of the oceans. The rest must be the result of melting ice sheets and glaciers on the land surface. The predicted increase in the sea level is in the range of 0.5 to 1.5 m over the next 50 to 150 years [Buddemeneir, 1988].

A projected global sea level rise of 1.1±0.6 m by the year 2080 AD is ascribed to a combination of thermal expansion of ocean water and

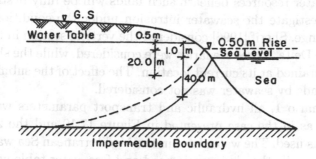

Figure 17.21. Sharp interface and sea level rise.

melting of glaciers and ice sheets. According to Ghyben-Herzberg relation, a one meter height of free water table above the mean sea level ensures 40 meter of freshwater below the sea level. Likewise, a 50 cm rise in the sea level causes 20 meter reduction in the freshwater thickness, as shown in Figure 17.21.

Rise of seawater will impose an additional hydraulic pressure at the sea side boundary. The intrusion process is dependent on many hydraulic, geometric and transport parameters. Each case has its own conditions and no general formula can be generalized for all coastal aquifers. Numerical models can be used to assess the expected intrusion under the condition of climate change and sea level rise.

17.7.3   Climate Change and Seawater Intrusion in the Nile Delta Aquifer

The Nile Delta aquifer in Egypt is naturally bounded from the north by the Mediterranean Sea. Measurements indicate that the water level in the Mediterranean Sea has risen by an average value of 2.5 mm/year during the last 70 years. Meanwhile, the coastal region of the Delta is subject to land subsidence with an average value of about 4.7 mm/year. Therefore, the relative difference between the seawater level and the ground surface in the northern region of the Nile Delta is approximately 70 cm/century [Alnaggar et al., 1995]. It is also known that the ground level of coastal region is varied between −0.5 m to +2.0 m relative the mean sea level [Alnaggar et al., 1995]. Investigations showed that, by the end of the next century, about 4000 km^2 of areas of low altitude will be submerged by seawater, if global warming trend and sea level rise continue. The whole groundwater resources beneath such lands will be fully destroyed.

To investigate the seawater intrusion under the condition of climate change, Sherif [1996] considered the vertical section in the middle of the Delta. Three scenarios were considered, while the shore line was maintained at its current location. The effect of the submergence of low lands by seawater was not considered.

In scenario 1, all hydraulic and transport parameters were kept constant as in the case presented in Figure 17.10 and the 2D-FED model was used. The water level in the Mediterranean Sea was raised by 0.2 m, while the piezometric head and free water table were kept as before. Under this condition, equiconcentration line 35.0 advanced slightly while equiconcentration line 5.0 advanced by a distance of 2km, measured along the bottom boundary. Equiconcentration line 1.0 advanced inland by a distance of 2.5 km, as shown in Figure 17.22a.

In scenario 2, the seawater level was increased by 0.5 m and other parameters were kept at their basic values. Equiconcentration line 35.0 migrated further inland by 1.5 km as compared to the initial case. Equiconcentration line 5.0 moved inland to a distance of 96km measured along the bottom boundary as shown in Figure 17.22b, indicating an inland movement of 4.5 km as compared to the initial case. Equiconcentration line 1.0 advanced inland by a distance of about 9km.

Global warming may also impose additional demands for groundwater resources to substitute for any shortage in the surface water. To investigate the effect of the additional pumping from the Nile Delta aquifer on seawater intrusion, the piezometric head at the land side was lowered by 0.5 m (due to additional pumping) in scenario 3, while other parameters were kept unchanged. Under this scenario, Equiconcentration line 35.0 advanced inland to a distance of 70 km measured along the bottom boundary, while equiconcentration line 5.0 intruded to a distance of 103 km, measured along the same boundary. The width of the dispersion zone has increased considerably, as shown in Figure 17.22c. Any additional pumping from the Nile Delta aquifer will cause a significant increase in the seawater intrusion.

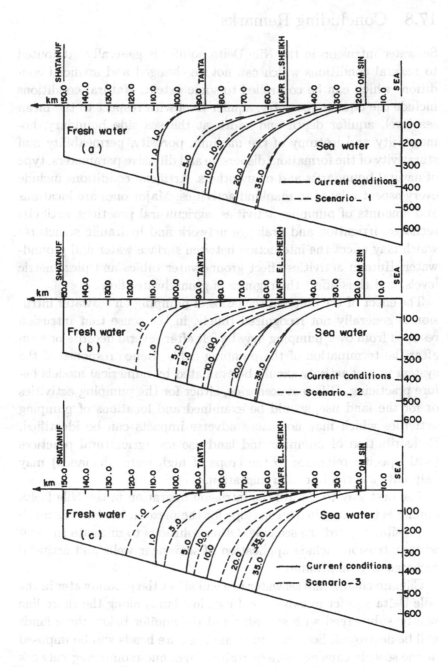

Figure 17.22. Effect of sea level rise and additional pumping: Equiconcentration lines under (a) Scenario 1, (b) Scenario-2 (c) Scenario-3.

17.8 Concluding Remarks

Seawater intrusion in the Nile Delta aquifer is generally attributed to natural conditions which can not be changed and artificial conditions which can be controlled to some extent. Natural conditions include the aquifer geometry, aquifer level with respect to the mean sea level, aquifer depth and width at the sea side boundary, homogeneity and isotropy of the medium, porosity, permeability and storativity of the formation, dispersive and diffusive parameters, type of natural boundaries and many others. Artificial conditions include every aspect related to mankind activities. Major ones are locations and amounts of pumping activities, agricultural practices, artificial recharge, irrigation and drainage network and hydraulic structures which may affect the interaction between surface water and groundwater. Human activities affect groundwater tables and piezometric levels, and thus affect the regimes of groundwater flow.

The effect of the different scenarios of pumping on seawater intrusion is generally not recognized timely, in the sense that intrusion resulted from over pumping may be felt after several decades or even after the termination of the pumping fields. Hence response of the system to any activity should be evaluated by numerical models before practicing. Different scenarios, either for the pumping activities or for the land use, should be examined and locations of pumping activities which may not cause adverse impacts can be identified. Redistribution of pumping and land use for agricultural practices (with specific reference to the crops of high water demands) may help mitigate the intrusion migration.

The best way to control the intrusion migration in the Nile Delta aquifer is to maintain a steep slope for the piezometric head as much as possible toward the sea side. Other solutions to mitigate the seawater intrusion include application of scavenger wells and artificial recharge of groundwater.

Climate change and seawater rise will affect the groundwater in the Nile Delta aquifer in two ways. First, low lands along the shore line will be submerged with seawater and the aquifer below these lands will be destroyed. Second, additional pressure heads will be imposed at the seaside causing more intrusion. Dynamic monitoring network for piezometric head and salinity is needed for better assessment and mitigation of seawater intrusion in the Nile Delta aquifer.

Bibliography

[1] Abriola, L.M., Dekker, T.J. and Rathfelder, K.M., "Recent advances in the modelling of organic liquid contaminant migration and persistence in aquifer systems," In: *Hydrogeologic Investigation, Evaluation, and Ground Water Modeling*, eds. Y. Eckstein and A. Zaporozec, Water Environment Federation, Alexander, Virginia, 303-316, 1993.

[2] Achmad, G. and Wilson, J.M., *Hydrogeologic framework and the distribution and movement of brackish water in the Ocean City-Manokin Aquifer System at Ocean City, Maryland*, Report of Investigations, Maryland Geological Survey, No. 57, 125 p., 1993

[3] Alnaggar, D., Soliman, S. and Labib, G., "Exploratory study concerning the consequences of the Egyptian north coast and the Mediterranean sea inter-relation," *Proc. Annual Conf. National Water Research Center*, Ministry of Public Works & Water Resources, Egypt, 1995.

[4] Amer, A. and Farid, M.S., "About sea water intrusion phenomenon in the Nile Delta aquifer," *Proc. Int. Workshop on Management of the Nile Delta Groundwater Aquifer*, CU/Mit, Cairo, 1981.

[5] Amer, A., Hefny, K. and Farid, M.S., "Hydrological aspects of the Nile Delta aquifer," *Proc. Conf. Water Resources in Egypt*, Cairo University, Cairo, 1979.

[6] Amer, A. and Sherif, M.M., "Behavior of seawater intrusion in the Nile Delta under different conditions," Working Paper Series No. 32-1, SRP, NWRC-MPWWR, Cairo, Egypt, 1995.

[7] Amer, A. and Sherif, M.M., "An integrated study for seawater intrusion in the Nile Delta aquifer," Working Paper for SRP, NWRC-MPWWR, Cairo, Egypt, 1996.

[8] Anderson, W.A., *MARQLOOPS-Marquardt Inversion of Loop-Loop Frequency Soundings*, U.S. Geol. Survey Open File Rept. 79-240, 75 p., 1979.

[9] Andersson, J.E., Ekman, L., Nordqvist, R. and Winberg, A., "Hydraulic testing and modelling of a low-angle fracture zone at Finnsjön, Sweden," J. Hydrology, **126**, 45-77, 1991.

[10] Andrews, J.E. and Bainbridge, C., "Submarine canyons off eastern Oahu," *Pacific Science*, **26**(1), 108-113, 1972.

[11] Anthony, S.S., "Electromagnetic methods for mapping freshwater lenses on Micronesian atoll islands," J. Hydrology, **137**, 99-111, 1992.

[12] Appleyard, S., "The impact of urban development on recharge and groundwater quality in a coastal aquifer near Perth, western Australia," *Hydrogeology J.*, **3**, 65-75, 1995.

[13] Appelo, C.A.J. and Geirnart, W., "Processes accompanying the intrusion of salt water," In: *Hydrogeology of Salt Water Intrusion—A Selection of SWIM Papers*, **II**, 291-304, 1991.

[14] Appelo, C.A.J. and Postma, D., *Geochemistry, Groundwater and Pollution*, Balkema, Rotterdam, 536p., 1993.

[15] Appelo, C.A.J. and Willemsen, A., "Geochemical calculations and observations on salt water intrusions, a combined geochemical/mixing cell model," *J. Hydrology*, **94**, 313-330, 1987.

[16] ASCE, *Groundwater Basin Management, Manuals of Engineering Practice*, **40**, New York, 1961.

[17] Ayers, J. and Vacher, H.L., "Hydrogeology of an atoll island; a conceptual model from a detailed study of a Micronesian example," *Ground Water*, **24**(2), 185-198, 1986.

[18] Ayers, J.F., "Shallow seismic refraction used to map the hydrostratigraphy of Nukuoro Atoll," *J. Hydrology*, **113**, 123-133, 1990.

[19] Aziz, K. and Settari, A., *Petroleum Reservoir Simulation*, Applied Science Publishers, London, UK, 1979.

[20] Bachmat, Y. and Chetboun, G., *Seawater Encroachment in the Coastal Plain of Israel During the Period 1958-1971*, Israeli Hydrological Service Rep., 2/74, 52p, 1974.

[21] Back, W., *Hydrochemical Facies and Groundwater Flow Patterns in the Northern Part of Atlantic Coastal Plain*, U.S. Geol. Survey Prof. Paper 498-A, 1-42, 1966.

[22] Back, W., "Geologic significance of the groundwater mixing zone associated with seawater intrusion," In: *5th Int. Symp. Water-Rock Interaction*, Reykjavik, Iceland, 25-28, 1986.

[23] Back, W. and Freeze, R.A. [eds.], *Chemical Hydrogeology*, Benchmark Papers in Geology, **73**, Hutchinson Ross Publ. Co., Stroudsburg, PA, 1983.

[24] Back, W., Hanshaw, B.B., Herman, J.S. and Van Driel, J.N., "Differential dissolution of a Pleistocene reef in the ground-water mixing zone of coastal Yucatan, Mexico," *Geology*, **14**, 137-140, 1986.

[25] Back, W., Hanshaw, B.B., Pyle, T.E., Plummer, L.N. and Weidie, A.E., "Geochemical significance of groundwater discharge and carbonate solution to the formation of Caleta Xel Ha, Quintana Roo, Mexico," *Water Resour. Res.*, **15**, 1521-1535, 1979.

[26] Back, W., Hanshaw, B.B. and Van Driel, J.N., "Role of groundwater in shaping the eastern coastline of the Yucatan Peninsula, Mexico," *In: Groundwater as a Geomorphoic Agent*, (ed.) R.G. LaFleur, The "Binghamton" Symp. Geomorphol., **13**, 281-293, 1984.

[27] Back, W., Herman, J.S. and Paloc, H., (eds), *Hydrogeology of Selected Karst Regions, IAH*, **13**, Verlag Heinz Heise GmbH, Hannover, 1992.

[28] Badiozamani, K., "The Dorag dolomitization model, application to the Middle Ordovician of Wisconsin," *J. Sed. Petrology*, **43**(4), 965-984, 1973.

[29] Badon-Ghyben, W., "Nota in verband met de voorgenomen putboring nabij Amsterdam (Notes on the probable results of well drilling near Amsterdam)," *Tijdschrift van het Koninklijk Instituut van Ingenieurs*, The Hague, 1888/9, 8-22, 1888.

[30] Bajracharya, K. and Barry, D.A., "Non-equilibrium solute transport parameters and their physical significance: numerical and experimental results," J. Contaminant Hydrology, 24(3-4), 185-204, 1997.

[31] Baptista, A.M., Adams, E.E. and Stolzenback, K.D., "The 2-D unsteady transport equation solved by the combined use of the finite element method and the method of characteristics," In: Proc. 5th Int. Conf. on Finite Elements in Water Resources, 353–362, Springer-Verlag, 1984.

[32] Barker, R.D., "Investigation of groundwater salinity by resistivity methods," In: Geotechnical and Environmental Geophysics, v. II, n. 5, Soc. Explor. Geophysicists, Inv. in Geophysics, (ed.) S. Ward, 201-212, 1990.

[33] Bear, J., Dynamics of fluids in Porous Media, American Elsevier, New York, 764 p., 1972.

[34] Bear, J., Hydraulics of Groundwater, McGraw-Hill, New York, 567 p., 1979.

[35] Bear, J. and Bachmat, Y., "A generalized theory on hydrodynamic dispersion in porous media," IASH Symp. Artificial Recharge and Management of Aquifers, IASH, 1967.

[36] Bear, J. and Bachmat, Y., Introduction to Modeling of Transport Phenomena in Porous Media, Kluwer, Dordrecht, 1990.

[37] Bear, J. and Dagan, G., "Some exact solutions of interface problems by means of the hodograph method," J. Geophys. Res., 69, 1563-1572, 1964a.

[38] Bear, J. and Dagan, G, "Moving interface in coastal aquifers," J. Hydraul. Div., Am. Soc. Civ. Eng., 90, 193-216.1964b.

[39] Bear, J., Sorek, S. and Borisov, V., "On the Eulerian–Lagrangian Formulation of Balance Equations in Porous Media," Numer. Meth. Partial Diff. Eqn., 13(5), 505-530, 1997.

[40] Bear, J. and Verruijt, A., Modeling Groundwater Flow and Pollution, D. Reidel Publ. Co., Dordrecht, the Netherlands, 414 p., 1987.

[41] Beck, A.E., Physical Principles of Exploration Methods, Wuerz Publ., Winnipeg, 292 p., 1991.

[42] Beekman, H.E. and C.A.J., Appelo, "Ion chromatography of fresh- and salt-water displacement: laboratory experiments and multicomponent transport modelling," J. Contaminant Hydrology, 7, 21-37, 1990.

[43] Bekins, B.A., McCaffrey, A.M. and Dreiss, S.J., "Episodic and constant flow models for the origin of low-chloride waters in a modern accretionary complex," Water Resour. Res., 31(12), 3205-3215, 1995.

[44] Bennett, G.D. and Giusti, E.V., "Coastal ground-water flow near Ponce, Puerto Rico," U.S. Geol. Survey Prof. Paper 750-D, D206-D211, 1971.

[45] Bergamaschi, L., Gallo, C., Manzini, G., Paniconi, C. and Putti, M., "A mixed finite element/TVD finite volume scheme for saturated flow and transport in groundwater," Proc. 9th Int. Conf. on Finite Elements in Fluids: New Trends and Applications, Part II, 1223–1232, Padua, Italy, Università di Padova, 1995.

[46] Bloyd, R. M., Approximate Ground-Water-Level Contours, April 1981, for the Soquel-Aptos Area, Santa Cruz County, California, U.S. Geol. Survey Open File Rept. 81-680, 1981.

[47] Bodine, M.W. and Jones, B.F., THE SALT NORM: A Quantitative Chemical-Mineralogical Characterization of Natural Waters, U.S. Geol. Survey Water-Res. Inv. Rept. 86-4068, 130 p., 1986.

[48] Bond, L. D. and Bredehoeft, J. D., "Origins of seawater intrusion in a coastal aquifer-A case study of the Pajaro Valley," *J. Hydrol.*, **92**, 363-388, 1987.

[49] Bonnet, M. and Sauty, J.P., "Un modele simplifie pour la simulation des nappes avec intrusion saline," In: *Application of Mathematical Models in Hydrology and Water Resources Systems*. Proc. Bratislava Symp., Int. Assoc. Sci. Hydrology, Publ. 115, 45-56, 1975.

[50] Brogan, S.D., "Aquifer remediation in the presence of rate-limited sorption," M.S. Thesis, Stanford University, 1991.

[51] Brogan, S.D. and Gailey, R.M., "A method for estimating field-scale mass-transfer rate parameters and assessing aquifer cleanup times," *Ground Water*, **33**(6), 997-1009, 1995.

[52] Brown, J.S., *Study of Coastal Ground Water with Special Reference to Connecticut*, U.S. Geol. Survey Water-Supply Paper 537, 1925.

[53] Brezzi, F. and Fortin, M., *Mixed and Hybrid Finite Element Methods*, Springer-Verlag, Berlin, 1991.

[54] Buddemeneir, R.W., "The impacts of climate change on the Sacramento San Joaquin Delta," Testimony before the Water and Power Resources Subcommittee, U.S. House of Representatives, San Francisco, CA., 1988.

[55] Bundschuh, J., "Modelling annual variations of spring and groundwater temperatures associated with shallow aquifer systems," *J. Hydrology*, **142**(1-4), 427-444, 1993.

[56] Burger, A. and Dubertret, L., (eds), *Hydrogeology of Karstic Terrains*, IAH, Paris, 1975.

[57] Burger, A. and Dubertret, "Hydrogeology of karstic terrains, case histories," *IAH*, 1, Verlag Heinz Heise GmbH, Hannover, 1984.

[58] Burger, A., "Expected climate change," *Report on Session III in Carbon Dioxide and Other Greenhouse Gases: Climate and Associated Impacts*, edited by Fentechi, R. and Ghazi, A., Kluwer Acad. Publ., Dordrecht, 1989.

[59] Burger, H.R., *Exploration Geophysics of the Shallow Subsurface*, Prentice-Hall, New York, 489 p., 1992.

[60] Bush, P.W., *Simulation of Saltwater Movement in the Floridan Aquifer System, Hilton Head Island, South Carolina*, U.S. Geol. Survey Water-Supply Paper 2331, 1988.

[61] Bush, P.W. and Johnston, R.H., *Ground-Water Hydraulics, Regional Flow, and Ground-Water Development of the Floridan Aquifer System in Florida and in Parts of Georgia, South Carolina, and Alabama*, U.S. Geol. Survey Prof. Paper 1403-C, 1988.

[62] California Dept. Water Resources, *Santa Ana Gap Salinity Barrier*, Orange County, Calif. Dept. Water Resources Bulletin no. 147-1, 1966.

[63] Cant, R.V. and Weech, P.S., "A review of the factors affecting the development of Ghyben-Herzberg lenses in the Bahamas," *J. Hydrology*, **84**, 333-343, 1986.

[64] Carbognin, L., "Land subsidence: a worldwide environmental hazard," *Nature and Resources*, **XXI** (1), 2-12, 1985.

[65] Carlston, C.W., "An early American statement of the Badon Ghyben-Herzberg principle of static fresh-water-salt-water balance," *Am. J. Sci.*, **261**, 88-91, 1963.

[66] Carpenter, A.B., "Origin and chemical evolution of brines in sedimentary basins," Oklahoma Geol. Surv. Circular 79, 60-77, 1978.

[67] Carpenter, A.B., "The chemistry of dolomite formation, 1. The stability of dolomite," In: *Concepts and Models of Dolomitization*, (eds.) D.H. Zenger, J.B. Dunhan and R.L. Ethington, SEPM Special Publ., **28**, Tulsa, OK, 1980.

[68] Carrier, G.F., "The mixing of ground water and sea water in permeable subsoils," *J. Fluid Mech.*, **4**, 479-488, 1958.

[69] Cedeno, W. and Vemuri, V.R., "Genetic algorithms in aquifer management," J. Network & Computer Applications, **19**(2), 171-187, 1996.

[70] Chan, N., "Robustness of the multiple realization method for stochastic hydraulic aquifer management," *Water Resour. Res.*, **29**(9), 3159-3167, 1993.

[71] Chan, N., "Partial infeasibility method for chance-constrained aquifer management," *J. Water Resour. Planning & Management, ASCE*, **120**(1), 70-89, 1994.

[72] Chapelle, F.H., Landmeyer, J.E. and Bradley, P.M., *Assessment of intrinsic bioremediation of jet fuel contamination in a shallow Aquifer, Beaufort, South Carolina*, U.S. Geol. Survey Water-Res. Inv. Rept. 95-4262, 30 p., 1996.

[73] Chavent, G. and Roberts, J.E., "A unified physical presentation of mixed, mixed-hybrid finite elements and standard finite difference approximations for the determination of velocities in waterflow problems," *Adv. Water Resour*, **14**(6), 329–348, 1991.

[74] Cheng, H.P., Yeh, G.T., Xu, J., Li, M.H. and Carsel, R. "A study of incorporating the multigrid method into the three-dimensional finite element discretization: A modular setting and application," *Int. J. Numer. Meth. Engng.*, **41**, 499-526, 1998.

[75] Cherkauer, D.S. and McKereghan, P.F., "Ground-water discharge to lakes: focusing in embayments," *Ground Water*, **29**(1), 72-80, 1991.

[76] Cherkauer, D.S., Mckereghan P.F. and Schalch, L.H., "Delivery of chloride and nitrate by ground-water to the Great-Lakes—Case study for the Door Peninsula Wisconsin," *Ground Water*, **30**(6), 885-894, 1992.

[77] Chidley, T.R.E. and Lloyd, J.W., "A mathematical study of fresh-water lenses," *Ground Water*, **15**(3), 215-222, 1972.

[78] CHO-TNO (Commissie voor Hydrologisch Onderzoek TNO), "Research on possible changes in the distribution of saline seepage in the Netherlands," *J Verslagen en Mededelingen*, no. 26, TNO, The Hague J, 1980.

[79] Collin, M., *POLLSITE Land-Use Register*, Israel Hydrological Service Rep., 2/95, Jerusalem, 35p, 1995.

[80] Collins, M.A. and Gelhar, L.W., "Seawater intrusion in layered aquifers," *Water Resour. Res.*, **7**(4), 971–979, 1971.

[81] Cooley, R.L., "Some new procedures for numerical solution of variably saturated flow problems," *Water Resour. Res.*, **19**(5), 1271–1285, 1983.

[82] Cooper, H.H. Jr., Kohout, F.A., Henry, H.R. and Glover, R.E., *Sea Water in Coastal Aquifers*, U.S. Geol. Survey Water-Supply Paper 1613-C, 1964.

[83] Cordes, C. and Kinzelbach, W., "Continuous groundwater velocity-fields and path lines in linear, bilinear, and trilinear finite-elements," *Water Resour. Res.*, **28**(11), 2903-2911, 1992.

[84] Cordes, C. and Putti, M., "Triangular mixed finite elements versus finite differences and finite volumes in groundwater flow modeling," In: *Computational Methods in Subsurface Flow and Transport Problems, Vol. 1*, 61–68, Comp. Mech. Publ., 1996.

[85] Correns, C.W., "Zur Geochemie der Diagenese-Geochim. et Cosmochim," *Acta*, **1**, 49-54, 1950.

[86] Countryman, R.A., II, "Geophysical delineation of the saltwater interface in the lower Suwanee River Basin, FL," M.S. thesis, Univ. South Florida, 538 p., 1996.

[87] Cox, R.A. and Nishikawa, T., "A new total variation diminishing scheme for the solution of advective-dominant solute transport," *Water Resour. Res.*, **27**(10), 2645-2654, 1991.

[88] Curtis, B.F., *Geometry of Sandstone Bodies*, Oklahoma, USA, 1961.

[89] Custodio, E., "Salt-fresh water interrelationship under natural conditions," Ch. 3, "Effect of human activities on salt-fresh water relationships in coastal aquifers," Ch. 4, In: *Groundwater Problems in Coastal Areas*, UNESCO-IHP, 14-112, 1987.

[90] Custodio, E., "Studying, monitoring and controlling seawater intrusion in coastal aquifers," In: *Guidelines for Study, Monitoring and Control*, FAO Water Reports No. 11, 7-23, 1997.

[91] Custodio, E., Bayo, A., Pascual M. and Boch, X., "Results from studies in several karst formations in southern Catalonia (Spain)," In: *Hydrogeological Processes in Karst Terrains*, Proc. Int. Symp. and Field Seminar, Antalya, Turkey, (eds.) G. Gunay, I. Johnson & W. Back, IAHS Publ. No. 207, 295-326, 1993.

[92] Custodio, E., Bruggeman, G.A. and Cotecchia, V., "Groundwater problems in coastal areas," *Studies and Reports in Hydrology*, **35**, UNESCO, Paris, 1987.

[93] Custodio, E. and Llamas, M.R., *Hidrologia Subterranea*, 2 vols., Ed. Omega, Barcelona, 1976.

[94] Dagan, G., "A note on higher-order corrections of the head covariances in steady aquifer flow," *Water Resour. Res.*, **21**, 573-578, 1985.

[95] Dagan, G., *Flow and Transport in Porous Formations*, Springer-Verlag, 465 p., 1989.

[96] Dagan, G., "Transport in heterogeneous porous formation: spatial moments, ergodicity and effective dispersion," *Water Resour. Res.*, **24**(6), 1281-1290, 1990.

[97] Dagan, G., "A discussion of 'A density-dependent flow and transport analysis of the effects of groundwater development in a freshwater lens of limited areal extent: the Geneva area (Florida, U.S.A.) case study,' by Panday et al. (1993)," *J. Contaminant Hydrology*, **18**, 332-334, 1995.

[98] Dagan, G. and Bear, J., "Solving the problem of local interface upconing in a coastal aquifer by the method of small perturbation," *J. Hyd. Res.*, **6**, 15-44, 1968.

[99] Dagan, G. and Zeitoun, D.G., "Seawater-freshwater interface in a stratified aquifer of random permeability distribution," *J. Contaminant Hydrology*, **29**, 185-203, 1998a.

[100] Dagan, G. and Zeitoun, D.G., "Free-surface flow toward a well and interface upconning in stratified aquifers of random conductivity," *Water Resour. Res.*, (submitted) 1998b.

[101] Dale, R.H. and Takasaki, K.J., *Probable effects of increasing pumpage from the Schofield ground-water body, Island of Oahu, Hawaii*, U.S. Geol. Survey Water-Res. Inv. Rept. 76-47, 45 p., 1976.

[102] Dam, J.C. van, "Partial depletion of saline groundwater by seepage," *J. Hydrology*, **29** (3/4), 315- 339, 1976a.

[103] Dam, J.C. van, "Possibilities and limitations of the resistivity method of geo-electrical prospecting in the solution of geo-hydrological problems," *Geoexploration*, **14**, 179-193, 1976b.

[104] Dam, J.C. van, "Characterization of the interaction between groundwater and surface water," Salinity keynote paper for the Budapest symposium of IAHS, 1986.

[105] Dam, J.C. van, "Problems associated with saltwater intrusion into coastal aquifers and some solutions," *Proc. ILT Seminar on Problems of Lowland Development*, Institute of Lowland Technology, Saga University, Japan, 1992.

[106] Dam, J.C. van, "Impact of sea level rise on saltwater intrusion in estuaries and aquifers," Keynote lecture delivered at the International Workshop Seachange '93, Noordwijkerhout, 19-23 April 1993, Tidal Waters Division (presently RIKZ) of the Directorate General for Public Works and Watermanagement (Rijkswaterstaat), The Hague, the Netherlands, 1993.

[107] Dam, J.C. van, "Groundwater hydrology in lowland," In: *Lowlands, Development and Management*, N. Miura, M.R. Madhav and K. Koga, (eds), Balkema, Rotterdam/Brookfield, 1994.

[108] Dam, J.C. van, (ed), *Saltwater Intrusion in Coastal Aquifers, Guidelines for Studying, Monitoring and Controlling*, Water Reports **11**, FAO, Rome, 1997.

[109] Dam, J.C. van and Meulencamp, J.J., "Some results of the geoelectrical resistivity method in groundwater investigations in the Netherlands," *Geophysical Prospecting*, **15**, 92-115, 1967.

[110] Daniels, J.J. and Keys, W.J., "Geophysical well logging for evaluating hazardous waste sites," In: *Geotechnical and Environmental Geophysics*, v. I, Soc. Explor. Geophysicists, Inv. in Geophysics, n. 5, (ed.) S. Ward, 263-286, 1990.

[111] Darwish, M.M., "Effect of probable hydrological changes on the Nile Delta aquifer system," Ph. D. Thesis, Faculty of Engineering, Cairo University, Cairo, 1994.

[112] Davis, J.L. and Annan, A.P., "Ground-penetrating radar for high resolution mapping of soil and rock stratigraphy," *Geophysical Prospecting*, **37**(5), 531-551, 1989.

[113] Davis, S.N., Wittemore, D.O. and Febryka-Martin, J., "Uses of chloride/bromide ratios in studies of potable water," *Ground Water*, 338-350, 1998.

[114] De Breuck, W., (ed.), *Hydrogeology of Salt Water Intrusion—A Selection of SWIM Papers*, International Contributions to *Hydrogeology*, IAH, **II**, Verlag Heinz Heise GmbH, Hannover, 422 p., 1991.

[115] De Breuck, W. and G., De Moor, "The evaluation of coastal aquifer of Belgium," In: *Hydrogeology of Salt Water Intrusion—A Selection of SWIM Papers*, **II**, 35-48, 1991.

[116] Dempser, D.J., "Management practices for sustainable groundwater extraction in coastal aquifer," IAH, IGC *Groundwater Sustainable Solutions*, Eds. T.R. Weaver and C.R. Lawrence, Melbourne Univ., Australia, 31-38, 1998.

[117] Dennis, J.E. and Moré, J.J., "Quasi-Newton methods, motivation and theory," *SIAM Review*, **19**(1), 46–89, 1977.

[118] Dennis, J.E. and Schnabel, R.B., *Numerical Methods for Unconstrained Optimization and Nonlinear Equations*, Prentice-Hall, 1983.

[119] Desai, B.I., Gupta, S.K., Shaw, M.V. and Sharma, S.C., "Hydrochemical evidence of sea water intrusion along the Mangrol-Chorwad coast of Saurashtra," *Gujarat. Hydrological Sciences*, **24**, 71-82, 1979.

[120] Diersch, H.J., "Finite element modeling of recirculating density driven saltwater intrusion processes in groundwater," *Adv. Water Resources*, **11**, 25–43, 1988.

[121] Dobrin, M. and Savit, C., *Introduction to Geophysical Prospecting*, McGraw-Hill, New York, 867 p., 1988.

[122] Dodds, A.R. and Dragan, I., "Integrated geophysical methods used for groundwater studies in the Murray Basin, South Austrailia," In: *Geotechnical and Environmental Geophysics*, v. II, Soc. Explor. Geophysicists, Inv. in Geophysics, n. 5, (ed.) S. Ward, 303-310, 1990.

[123] Domagalski, J.L., Eugster, H.P. and Jones, B.F., "Trace metal geochemistry of Walker, Mono and Great Salt Lakes," In: *Fluid-Mineral Interactions—A Tribute to H.P.Eugster*, (eds.) R.J. Spencer & I-Ming Chou, Sp. Publ. Geochemical Society, **2**, 315-353, 1990.

[124] Douglas Jr., J., Ewing, R.E. and Wheeler, M.F., "Approximation of the pressure by a mixed method in the simulation of miscible displacement," *R.A.I.R.O. Anal. Numer.*, **17**, 17–33, 1983.

[125] Du Commun, J., "On the cause of freshwater springs, fountains, etc.," *American Journal of Science and Arts*, **14**, 174-175, 1828.

[126] Duffy, C.J. and Al-Hassan, S., "Groundwater circulation in a closed desert basin: Topographic scaling and climatic forcing," *Water Resour. Res.*, **24**(10), 1675-1688, 1988.

[127] Duffy, C.J. and Lee, D.H., "Base-flow response from non-point source contamination—Simulated spatial variability in source, structure, and initial condition," *Water Resour. Res.*, **28**(3), 905-914, 1992.

[128] Durlofsky, L.J., "A triangle based mixed finite element–finite volume technique for modeling two phase flow through porous media," *J. Comput. Phys.*, **105**, 252–266, 1993.

[129] Eem, J.P., van der, "Adaption Konikow-Bredehoeft for density differences," (in Dutch). *Internal notition KIWA*, Aug. 1987.

[130] Emekli, N., Karahanoglu, N., Yazicigil, H. and Doyuran, V., "Numerical-simulation of saltwater intrusion in a groundwater basin," *Water Environment Research*, **68**(5), 855-866, 1996.

[131] Esch, J.M. van and Stakelbeek, A., "Density dependent groundwater flow near deepwell infiltration systems," Modelling Aspects, 14th Salt Water Intrusion Meeting, Malmö, Sweden, 1996.

[132] Essaid, H. I., "A comparison of the coupled fresh water-salt water flow and the Ghyben- Herzberg sharp interface approaches to modeling of transient behavior in coastal aquifer systems," *J. Hydrol.*, **86**, 169-193, 1986.

[133] Essaid, H.I., *The Computer Model SHARP, a Quasi-Three-Dimensional Finite-Difference Model to Simulate Freshwater and Saltwater Flow in Layered Coastal Aquifer Systems*, U.S. Geol. Survey Water-Res. Inv. Rept. 90-4130, 1990a.

[134] Essaid, H.I., "A multilayered sharp interface model of coupled freshwater and saltwater flow in coastal systems: model development and application," *Water Resour. Res.*, **26**(7), 1431-1454, 1990b.

[135] Essaid, H.I., *Simulation of Freshwater and Saltwater Flow in the Coastal Aquifer System of the Purisima Formation in the Soquel - Aptos Basin, Santa Cruz County, California*, U.S. Geol. Survey Water-Res. Inv. Rept. 91-4148, 1992.

[136] Eyre, P.R., "Simulation of ground-water flow in Southeastern Oahu, Hawaii" *Ground Water*, **23**, 325-330, 1985.

[137] Farid, M.S., "Nile Delta ground water study," MS. Thesis, Faculty of Engineering, Cairo University, Cairo, 1980.

[138] Farid, M.S., "Management of groundwater system in the Nile Delta," Ph.D. Thesis, Faculty of Engineering, Cairo University, Cairo, 1985.

[139] Faust, C.R., Guswa, J.H. and Mercer, J.W., "Simulation of three dimension flow of immiscible fluids within and below the saturated zone," *Water Resour. Res.*, **26**(6), 2449-2464, 1989.

[140] Fetter, C.W., "Position of the saline water interface beneath oceanic islands," *Water Resour. Res.*, **8**, 1307-1314, 1972.

[141] Fidelibus, M.D. and Tulipano, L., "Mixing phenomena owing to sea water intrusion for the interpretation of chemical and isotopic data of discharge waters in the Apulian coastal carbonate aquifer (southern Italy)," Proc. 9th Salt Water Intrusion Meeting, Delft, 591-600, 1986.

[142] Fink, M., *The Hydrogeology of the Gaza Strip*, (in Hebrew), TAHAL Rep., 3, 70 p., 1970.

[143] Fitterman, D.V., "Geophysical mapping of the freshwater/saltwater interface in Everglades National Park, Florida," U.S. Geol. Survey Fact Sheet, Reston, VA, 2 p., 1996.

[144] Fitterman, D.V. and Stewart, M., "Transient electromagnetic sounding for groundwater," *Geophysics*, **51**, 995-1005, 1986.

[145] Fletcher, R., *Practical Methods of Optimization, Vol. 1: Unconstrained Optimization*, John Wiley & Sons, 1980.

[146] Forsyth, P.A., "A control volume finite element approach to NAPL groundwater contamination," *SIAM J. Sci. Stat. Comput*, **12**(5), 1029–1057, 1991.

[147] Fraser, D.C., "Resistivity mapping with an airborne multicoil electromagnetic system," *Geophysics*, **43**, 144-172, 1978.

[148] Fretwell, J.D. and Stewart, M., "Resistivity study of a coastal karst terrain," *Ground Water*, **19**(4), 219-223, 1981.

[149] Freund, R.W., "A transpose-free quasi-minimal residual algorithm for non-Hermitian linear systems," *SIAM J. Sci. Comput.*, **14**(2), 470–482, 1993.

[150] Freund, R.W., Golub, G.H. and Nachtigal, N.M., "Iterative solution of linear systems," *Acta Numerica*, 57–100, 1992.

[151] Frind, E.O., "Simulation of long-term transient density-dependent transport in groundwater," *Adv. Water Resour.*, **5**, 73–88, 1982a.

[152] Frind, E. O., "Seawater intrusion in continuous coastal aquifer aquitard system," *Adv. Water Resour.*, **5**, 89-97, 1982b.

[153] Gailey, R.M. and Gorelick, S.M., "Design of optimal, reliable plume capture schemes—Application to the Gloucester Landfill groundwater contamination problem," *Ground Water*, **31**(1), 107-114, 1993.

[154] Galeati, G. and Gambolati, G., "On boundary conditions and point sources in the finite element integration of the transport equation," *Water Resour. Res.*, **25**(5), 847–856, 1989.

[155] Galeati, G., Gambolati, G. and Neuman, S.P., "Coupled and partially coupled Eulerian-Lagrangian model of freshwater-saltwater mixing," *Water Resour. Res.*, **28**(1), 149-165, 1992.

[156] Gambolati, G., "Equation for one-dimensional vertical flow of groundwater, 1, The rigorous theory," *Water Resour. Res.*, **9**(4), 1022–1028, 1973.

[157] Gambolati, G. and Perdon, A., "The conjugate gradients in subsurface flow and land subsidence modelling," In: *Fundamentals of Transport Phenomena in Porous Media*, 953–984, Martinus Nijhoff, Dordrecht, Holland, 1984.

[158] Gambolati, G., Paniconi, C. and Putti, M., "Numerical modeling of contaminant transport in groundwater," *Migration and Fate of Pollutants in Soils and Subsoils*, vol. 32 of *NATO ASI Series G: Ecological Sciences*, 381–410, Springer-Verlag, Berlin, 1993.

[159] Gambolati, G., Pini, G., Putti, M. and Paniconi, C., "Finite element modeling of the transport of reactive contaminants in variably saturated soils with LEA and non-LEA sorption," *Environmental Modeling, Vol. II: Computer Methods and Software for Simulating Environmental Pollution and its Adverse Effects*, Chap. 7, 173–212, Comp. Mech. Publ., 1994a.

[160] Gambolati, G., Putti, M. and Rangogni, R., "Saltwater contamination of a coastal Italian aquifer by a coupled finite element model of flow and transport," *Excerpta*, **7**, 145–186, 1994b.

[161] Gambolati, G., Putti, M. and Paniconi, C., "Projection methods for the finite element solution of the dual-porosity model in variably saturated porous media," *Advances in Groundwater Pollution Control and Remediation*, Vol. 9 of *NATO ASI Series 2: Environment*, 97–125, Kluwer Academic, 1996.

[162] Gangopadhyay, S. and Das Gupta, A., "Simulation of salt-water encroachment in a multi-layer groundwater system, Bangkok, Thailand," *Hydrogeology J.*, **3**(4), 74-88, 1995.

[163] Garder, A.O., Jr., Peaceman D.W. and Pozzi, A.L., Jr. "Numerical calculation of multidimensional miscible displacement by the method of characteristics," *Soc. Pet. Eng. J.*, **4**(1), 26-36, 1964.

[164] Garven, G. and Freeze, R.A., "Theoretical analysis of the role of groundwater flow in the genesis of stratabound ore deposits," *Am. J. Sci.*, **284**, 1085-1112, 1984.

[165] Geonics Ltd., PCLOOP, Geonics Ltd., Mississauga, Ontario, Canada, 1985.

[166] Ghassemi, F., Chen, T.H., Jakeman, A.J. and Jacobson, G., "Two and three-dimensional simulation of seawater intrusion: performances of the 'SUTRA' and 'HST3D' models," *AGSO J. Australian Geology & Geophysics*, **14**(2/3), 219-226, 1993.

[167] Ghassemi, F., Jakeman, A.J. and Jacobson, G., "Simulation of sea water intrusion," *Mathematics & Computers in Simulation*, **32**, 71-76, 1990a.

[168] Ghassemi, F., Jakeman, A.J. and Jacobson, G., "Mathematical modelling of sea water intrusion, Nauru Island," *Hydrological Processes*, **4**(3), 269-281, 1990b.

[169] Gleick, P.H., (ed.), *Water in Crisis—A Guide to the World's Fresh Water Resources*, Oxford Univ. Press, 1993.

[170] Glover, R.E., "The pattern of fresh-water flow in a coastal aquifer," *J. Geophys. Res.*, **64**(4), 457-59, 1959.

[171] Glover, R.E., "The pattern of fresh-water flow in a coastal aquifer," In: *Sea Water in Coastal Aquifers*, H.H. Cooper, F.A. Kohout, H.R. Henry and R.E. Glover, eds., U.S. Geol. Survey Water-Supply Paper 1613-C, C32-C35, 1964.

[172] Goddard, M., "Geophysical delineation of the saltwater interface near Weeki Wachee Spring, FL," M.S. thesis, Univ. South Florida, Tampa, FL, 1997.

[173] Goldenberg, L.C., Mandel, S. and Magaritz, M., "Fluctuating non-homogeneous changes of hydraulic conductivity in porous media," *Q. J. Eng. Geology*, **19**, 183-190, 1986.

[174] Goldenberg, L.C., Hutcheon, I., Waldlaw, N. and Melloul, A.J., "Rearrangement of fine particles in porous media causing reduction of permeability and formation of preferred pathways of flow: experimental findings and conceptual model," *Transport in Porous Media*, **13**, 221-237, 1993.

[175] Goldenberg, L.C. and Melloul, A.J., "Hydrological and chemical management in the rehabilitation of an aquifer," *J. Env. Management*, **42**, 247-260, 1994.

[176] Goldman, M., Gilad, D., Ronen, A. and Melloul, A., "Mapping of sea water intrusion into the coastal aquifer of Israel by the time domain electromagnetic method," In: *Geoexploration*, **28**, Elsevier, 153-174, 1991.

[177] Goldstein, N.E., Benson, S.M. and Alumbaugh, D., "Saline groundwater plume mapping with electromagnetics," In: *Geotechnical and Environmental Geophysics*, v. II, Soc. Explor. Geophysicists, Inv. in Geophysics, n. 5, (ed.) S. Ward, 17-26, 1990.

[178] Goode, D.J., "Direct simulation of groundwater age," *Water Resour. Res.*, **32**(2), 289-296, 1996.

[179] Gorhan, H.L., "The determination of the saline/fresh water interface by resistivity soundings," *Assoc. Engineering Geologists Bull.*, **13**, 163-168, 1976.

[180] Greene, H. G., *Geology of the Monterey Bay Region*, U.S. Geol. Survey Open File Rept. 77-718, 1977.

[181] Gray, W.G. and Pinder, G.F., "An analysis of the numerical solution of the transport equation," *Water Resour. Res.*, **12**, 1976.

[182] Gureghian, A.B., "TRIPM, a two-dimensional finite element model for the simultaneous transport of water and reacting solutes through saturated and unsaturated porous media," *Technical Report ONWI-465*, Off. of Nuclear Water Isolation, Columbus, Ohio, 1983.

[183] Guswa, J.H. and LeBlanc, D.R., *Digital Models of Ground-Water Flow in the Cape Cod Aquifer System, Massachusetts*, U.S. Geol. Survey Water-Supply Paper 2209, 1985.

[184] Gutjahr, A.L. and Gelhar, L.W., "Stochastic models of subsurface flow: Infinite versus finite domains and stationarity," *Water Resour. Res.*, **17**, 337-350, 1981.

[185] Hagemeyer, T. and Stewart, M., "Resistivity investigations of salt-water intrusion near a major sea-level canal," In: *Geotechnical and Environmental Geophysics*, v. II, Soc. Explor. Geophysicists, Inv. in Geophysics, n. 5, (ed.) S. Ward, 67-78, 1990.

[186] Haggerty, R., "Design of multiple contaminant remediation in the presence of rate limitations," M.S. Thesis, Stanford University, 1992.

[187] Haggerty, R., Schroth, M.H. and Istok, J.D., "Simplified method of 'Push-Pull' test data analysis for determining in situ reaction rate coefficients," *Ground Water*, **36**(2), 314-324, 1998.

[188] Haggerty, R. and Gorelick, S.M., "Design of multiple contaminant remediation: sensitivity to rate-limited mass transfer", *Water Resour. Res.*, **30**(2), 435-446, 1994.

[189] Hahn, J., "Aspects of groundwater salinization in the Wittmund (East Freisland) coastal area," In: *Hydrogeology of Salt Water Intrusion—A Selection of SWIM Papers*, **II**, 251-270, 1991.

[190] Hallaji, K. and Yazicigil, H., "Optimal management of a coastal aquifer in Southern Turkey," *J. Water Resources Planning and Management, ASCE*, **122**(4), 233-244, 1996.

[191] Halley, R.B., Vacher, H.L. and Shinn, E.A., "Geology and hydrogeology of the Florida Keys," Chap. 5 in *Geology and Hydrogeology of Carbonate Islands*, (eds.) H.L. Vacher and E.A. Shinn, Developments in Sedimentology, v. 54, Elsevier, New York, 217-248, 1997.

[192] Hanor, J.S., "Precipitation of beachrock cements—mixing of marine and meteoric waters vs CO_2 degassing," *J. Sed. Petrology*, **48**(2), 489-501, 1978.

[193] Hanor, J.S., *Aquifers as Processing Plants for the Modification of Injected Water*, Louisiana Water Res. Inst., **11**, 50 p., 1980.

[194] Hanson, C.F., *Water resources of Big Pine Key, Monroe County, Florida*, U.S. Geol. Survey Open File Rept. 80-447, 36 p., 1980.

[195] Hanshaw, B.B., Back, W. and Deike, R.G., "A geochemical hypothesis for dolomitization by groundwater. In: A paleoaquifer and its relation to economic mineral deposits," *Economic Geology and Bull.Soc. Economic Geologists*, **66**(5), 710-724, 1971.

[196] Hardie, L.A., "Dolomitization, a critical view of some current views," *J. Sed. Petrology*, **57**(1), 166-183, 1987.

[197] Harris, W.H., "Stratification of salt water on barrier islands as a result of differences in sediment permeability," *Water Resour. Res.*, **3**(1), 89-97, 1967.

[198] Harvey, C.F. and Gorelick, S.M., "Mapping hydraulic conductivity—sequential conditioning with measurements of solute arrival time, hydraulic head and local conductivity," *Water Resour. Res.*, **31**(7), 1615-1626, 1995.

[199] Harvey, C.F.; Haggerty R. and Gorelick, S.M., "Aquifer remediation—a method for estimating mass-transfer rate coefficients and an evaluation of pulsed pumping," *Water Resour. Res.*, **30**(7), 1979-1991, 1994.

[200] Heathcote, J.A.; Jones M.A. and Herbert A.W., "Modelling groundwater-flow in the Sellafield Area," *Quart. J. Eng. Geol.*, **29**, s59-s81, 1996.

[201] Helton, J.C., "Risk, uncertainty in risk, and the EPA release limits for radioactive-waste disposal," *Nuclear Technology*, **101**(1), 18-39, 1993.

[202] Hem, J.D., *Study and Interpretation of the Chemical Characteristics of Natural Water*, U.S. Geol. Survey Water-Supply Paper 2254, 1985.

[203] Henry, H.R, "Salt intrusion into freshwater aquifers," *J. Geophys. Res.*, **64**, 1911-1919.1959.

[204] Henry, H.R., "Effects of dispersion on salt encroachment in coastal aquifers," U.S. Geol. Survey Water-Supply Paper 1613-C, C71-C84, 1964.

[205] Herbert, Jr., Roger B., "Evaluating the effectiveness of a mine tailing cover," *Nordic Hydrology*, **23**, 193-208, 1992.

[206] Herzberg, A., "Die Wasserversorgung einiger Nordseebder (The water supply of parts of the North Sea coast in Germany)," *Z. Gasbeleucht. Wasserversorg.*, **44**, 815-819, and **45**, 842- 844, 1901.

[207] Hestenes, M.R. and Stiefel, E., "Methods of conjugate gradients for solving linear systems," *J. Res. Nat. Bur. Standards*, **49**(6), 409–436, 1952.

[208] Hickey, J. J., *Hydrogeologic Study of the Soquel-Aptos Area, Santa Cruz County, California*, U.S. Geol. Survey Open File Rept., 1968.

[209] Hill, M. C., "A comparison of coupled freshwater-saltwater sharp interface and convective-dispersive models of saltwater intrusion in a layered aquifer system," In: *Proc. VII Int. Conf. on Computational Methods in Water Resources, Vol. 1*, Elsevier, 211-216, 1988.

[210] Hill, P.J. and Jacobsen, G., "Structure and evolution of Nauru Island, central Pacific Ocean," *Aust. J. Earth Sci.*, **36**, 365-381, 1989.

[211] Hirasaki, G.J., "Ion exchange with clays in the presence of surfactant," *Soc. Pet. Eng. J.*, **22**, 181-192, 1982.

[212] Hoekstra, P. and Blohm, M.W., "Case histories of time-domain electromagnetic soundings in environmental geophysics," In: *Geotechnical and Environmental Geophysics*, v. II, Soc. Explor. Geophysicists, Inv. in Geophysics, n. 5, (ed.) S. Ward, 1-16, 1990.

[213] Hubbert, M.K., "The theory of ground-water motion," *J. Geol.*, **48**, 785-944, 1940.

[214] Humphreys, G.L., Linford, J.G. and West, S.M., "Application of geophysics to the reclamation of saline farmland in western Australia," In: *Geotechnical and Environmental Geophysics*, v. II, Soc. Explor. Geophysicists, Inv. in Geophysics, n. 5, (ed.) S. Ward, 175-186, 1990.

[215] Hunt, C.D., Ewart, C.J. and Voss, C.I., "Region 2—Hawaiian Islands," In: Back, W., Rosenshein, J.S. and Seaber, P.R., eds., *Hydrogeology: The Geology of North America, Vol. O-2*, Geological Society of America, Boulder, Colorado, 255-262, 1988.

[216] Hunter, J.A., Pullan, S.E., Burns, R.A., Gagne, R.M. and Good, R.S., "Shallow seismic reflection mapping of the overburden-bedrock interface with the engineering seismograph-some simple techniques," *Geophysics*, **49**, 1381-1385, 1984.

[217] Huyakorn, P.S., Andersen, P.F., Mercer, J.W. and White, H.O., "Salt intrusion in aquifers: development and testing of a three dimensional finite element model," *Water Resour. Res.*, **23**, 293-319, 1987.

[218] Huyakorn, P.S., Mercer, J.W. and Ward, D.S., "Finite element matrix and mass balance computational schemes for transport in variably saturated porous media," *Water Resour. Res.*, **21**(3), 346-358, 1985.

[219] Huyakorn, P.S. and Pinder, G.F., *Computational Methods in Subsurface Flow*, Academic Press, 1983.

[220] Huyakorn, P.S., Springer, E.P., Guvanasen, V. and Wadsworth, T.D., " A three-dimensional finite-element model for simulating water flow in variably saturated porous media," *Water Resour. Res.*, **22**(13), 1790-1808, 1986.

[221] Huyakorn, P. S. and Taylor, C., "Finite element models for coupled groundwater and convective dispersion," *Proc. 1st Int. Conf. Finite Elements in Water Resources*, 1.131-1.151, Pentech Press. London, 1976.

[222] Huyakorn, P.S., Thomas, S.D. and Thompson, B.M., "Techniques for making finite elements competitive in modeling flow in variably saturated porous media," *Water Resour. Res.*, **20**(8), 1099-1115, 1984.

[223] Hyndman D.W., Harris J.M. and Gorelick S.M., "Coupled seismic and tracer test inversion for aquifer property characterization," *Water Resour. Res.*, **30**(7), 1965-1977, 1994.

[224] Ingebritsen, S.E. and Sanford, W.E., *Groundwater in Geologic Processes*, Cambridge Univ. Press, 341 p., 1998.

[225] INTERA Environmental Consultants, Inc., *Revision of the Documentation for a Model Calculating Effects of Liquid Waste Disposal in Deep Saline Aquifers*, U.S. Geol. Survey Water-Res. Inv. Rept. 79-96, 79 p., 1979.

[226] INTERCOMP Resource Development and Engineering, Inc., *A Model for Calculating Effects of Liquid Waste Disposal in Deep Saline Aquifers*, U.S. Geol. Survey Water-Res. Inv. Rept. 76-61, 263 p., 1976.

[227] Interpex Ltd., EMIX, Golden, Colorado, 1988.

[228] Interpex Ltd., RESIX, Golden, Colorado, 1995.

[229] Interpex Ltd., TEMIX, Golden, Colorado, 1996.

[230] Israel Hydrological Service, *Israel Hydrological Service Situation Rep.*, (in Hebrew), Hydrological Service Situation Rep., 72 p., Jerusalem, 1987.

[231] Israel Hydrological Service, *Development of Groundwater Resources in Israel up to Autumn 1996*, (in Hebrew), Hydrological Service Situation Rep., 0793-1093, 242 p., Jerusalem, 1997.

[232] Izbicki, J.A., "Chloride sources in a California coastal aquifer," In: Peters, H. [ed.], *Ground Water in the Pacific Rim Countries*, ASCE, Irrigation Div. Proc., 71-77, 1991.

[233] Izbicki, J.A., "Seawater intrusion in a coastal California aquifer," U.S. Geol. Survey Fact Sheet 125-96, 1996.

[234] Jackson, P.D., Taylor-Smith, D. and Stanford, P.N., "Resistivity-porosity-particle shape relationships for marine sands," *Geophysics*, **43**, 1250-1268, 1978.

[235] Jinno, K., Momii, K. and van Dam, J.C., "Activities of SWIM in the topic of groundwater salinization," (in Japanese), *J. Groundwater Hydrology*, **32**(1d), 1990.

[236] Johansson, S., "Localization and quantification of water leakage in aging embankment dams by regular temperature measurements," Commission Internationale Des Grands Barrages, Dix-septieme Congres, des Grands Barrages, Vienna, 991-1005, 1991.

[237] Johnson, V.M. and Rogers, L.L., "Location analysis in ground-water remediation using neural networks," *Ground Water*, **33**(5), 749-758, 1995.

[238] Jol, H.M., Smith, D.G. and Meyers, R.A., "Digital ground penetrating radar (GPR) a new geophysical tool for coastal barrier research (examples from the Atlantic, Gulf, and Pacific coasts, U.S.A.)," *J. Coastal Res.*, **12**(4), 960-968, 1996.

[239] Jones, B.F., "Clay mineral diagenesis in Lacustrine sediment," In: *Studies in Diagenesis*, (ed.) F.A. Mumpton, U.S. Geol.Survey Bull. 1578, 291-300, 1986.

[240] Jones, G., Whitaker, F., Smart, P. and Sanford, W., "Dolomitization of carbonate platforms by saline ground water: Coupled numerical modelling of thermal and reflux circulation mechanisms," In: Geofluids II: 2nd Int. Conf. Fluid Evolution, Migration and Interaction in Sedimentary Basins and OrganicBbelts. Extended Abstracts Volume, (eds.) J. Jendry, P. Carey, J. Parnell, A. Ruffel and R. Worden, 378-381, 1997.

[241] Jones, L.E., *A Real-time Aquifer Management Tool*, Georgia Institute of Technology, 90 p., 1990.

[242] Kakinuma, T., Kishi, Y. and Inouchi, K., "The behavior of groundwater with dispersion in coastal aquifers," *J. Hydrol.*, **98**, 225–248, 1988.

[243] Kapuler, R., "Groundwater quality in the Coastal Aquifer. A predicting model for seawater intrusion," TAHAL Report, no. 3, 12 pp, 1984.

[244] Kauahikaua, J., *Description of a Fresh-Water Lens at Laura Atoll, Majuro Atoll, Republic of the Marshall Islands, Using Electromagnetic Profiling*, U.S. Geol. Survey Open File Rept. 87-582, 32 p., 1987.

[245] Kauahikaua, J., *An Evaluation of Electric Geophysical Techniques for gGround Water Exploration in Truk, Federated States of Micronesia*, U.S. Geol. Survey Open File Rept. 87-146, 76 p., 1986.

[246] Kauffman, S.J., Herman, J.S. and Jones, B.F., "Lithological and hydrological influences on groundwater composition in a heterogeneous carbonate-clay aquifer system," *Geological Society of America Bulletin*, in press, 1998.

[247] Kearey, P. and Brooks, M., *An Introduction to Geophysical Exploration*, Blackwell Scientific Publications. London, 254 p., 1991.

[248] Keller, G.V. and F.C. Frischnecht, 1966. Electrical Methods in Geophysical Prospecting. Pergamon, New York, 519 p.

[249] Kershaw, D.S., "The incomplete Cholesky-conjugate gradient method for the iterative solution of systems of linear equations," *J. Comput. Phys.*, **26**, 43–65, 1978.

[250] Keulegan, H.D., "An example report on model laws for density currents," National Bureau of Standards Report, Dept. of Commer., Washington D.C., 1954.

[251] Kimrey, J.O., *Ground-Water Supply for the Dare Beaches Sanitary District, N.C.*, Dept. of Water Resources, Rept of Inv., **3**, 20 p., 1961.

[252] Kinzelbach, W., *Groundwater Modeling*, Elsevier, 1986.

[253] Kipp, K.L., Jr., *HST3D: A Computer Code for Simulation of Heat and Solute Transport in Three-Dimensional Ground-Water Flow Systems*, U.S. Geol. Survey Water-Res. Inv. Rept. 86-4095, 1987.

[254] Kipp, K.L., Jr. *Guide to the Revised Heat and Solute Transport Simulator: HST3D*, U.S. Geol. Survey Water-Res. Inv. Rept. 97-4157, 149p., 1997.

[255] Kitanidis, P.K., *Introduction to Geostatistics*, Cambridge Univ. Press, 249 p., 1997.

[256] Klein, H. and Waller, B.G., *Synopsis of Saltwater Intrusion in Dade County, Florida, through 1984*, U.S. Geol. Survey Water-Res. Inv. Rept. 85-4101, 1985.

[257] Klinge, H., Vogel, P. and Schelkes, K., "Chemical composition and origin of saline formation waters from the Konrad Mine, Germany," Proc. 7th Int. Symp. Water-Rock Interaction-WRI 7, 1117-1120, 1992.

[258] Kohout, F.A., "Cyclic flow of salt water in the Biscayne aquifer of southeastern Florida," *J. Geophys. Res.*, **65**, 2133-2141, 1960.

[259] Kohout, F.A., "A hypothesis concerning cyclic flow of salt water related to geothermal heating in the Floridan aquifer," *Trans. New York Acad. Sci., Ser. II*, **28**(2), 249-271, 1965.

[260] Kolton, Y., *The Connection Between Groundwater Aquifer and Sea Water Along the Coastal Plain Aquifer of Israel*, (in Hebrew), TAHAL Rep., 01/88/31, 61 p., 1988.

[261] Konikow, L.F. and Bredehoeft, J.D. "Computer model of two-dimensional solute transport and dispersion in ground water," *USGS Techniques of Water-Resources Investigations*, Book 7, Chap. C2, 90 p., 1978.

[262] Konikow, L.F. and Bredehoeft, J.D., "Ground-water models cannot be validated," *Adv. Water Resour.*, **15**, 75-83, 1992.

[263] Konikow, L.F., Goode, D.J. and Hornberger, G.Z., *A three-dimensional method-of-characteristics solute-transport model (MOC3D)*, U.S. Geol. Survey Water-Res. Inv. Rept. 96-4267, 87 p., 1996.

[264] Konikow, L.F., Sanford, W.E. and Campbell, P.J., "Constant-concentration boundary condition: Lessons from the HYDROCOIN variable-density groundwater benchmark problem," *Water Resour. Res.*, **33**(10), 2253-2261, 1997.

[265] Kontis, A.L. and Mandle, R.J., *Modification of a Three-Dimensional Ground-Water Flow Model to Account for Variable Density Water Density and Effects of Multiaquifer Wells*, U.S. Geol. Survey Water-Res. Inv. Rept. 87-4265, 1988.

[266] Koszalka, E.J., *Delineation of Saltwater Intrusion in the Biscayne Aquifer, Eastern Broward County, Florida, 1990*, U.S. Geol. Survey Water-Res. Inv. Rept. 93-4164, 1995.

[267] Krieger, R.A., Hatchett, J.L. and Poole, J.F., *Preliminary Survey of the Saline-Water Resources of the United States*, U.S. Geol. Survey Water-Supply Paper, 1374, 1957.

[268] Kuiper, L.K., *Documentation of a Numerical Code for the Simulation of Variable Density Ground-Water Flow in Three Dimensions*, U.S. Geol. Survey Water-Res. Inv. Rept. 84-4302, 1985.

[269] Lahm, T.D., Bair, E.S. and Vander Kwaak, J., "Role of salinity-derived variable-density flow in the displacement of brine from a shallow, regionally extensive aquifer," *Water Resour. Res.*, **34**(6), 1469-1480.

[270] LaMoreaux, P.E., (editor in chief), *Hydrology of Limestone Terranes*, IAH, International Contributions to *Hydrogeology*, **2**, Verlag Heinz Heise GmbH, Hannover, 1986.

[271] LaMoreaux, P.E., (editor in chief), *Hydrology of Limestone Terranes*, Annotated bibliography of carbonate rocks, Volume Four, IAH, International Contributions to *Hydrogeology*, **10**, Verlag Heinz Heise GmbH, Hannover, 1989.

[272] Lanczos, C., "Solution of systems of linear equations by minimized iterations," *J. Res. Nat. Bur. Standard*, **49**, 33-53, 1952.

[273] Land, L.S., "Holocene meteoric dolomitization of Pleistocene limestones, North Jamaica," *Sedimentology*, **20**, 411-424, 1973.

[274] Land, L.S., "The origin of massive dolomite," *J. Geol.*, **33**, 112-125, 1985.

[275] Lankston, R.W., "High resolution refraction seismic data acquisition and interpretation," In: *Geotechnical and Environmental Geophysics*, v. I, Soc. Explor. Geophysicists, Inv. in Geophysics, n. 5, (ed.) S. Ward, 45-74, 1990.

[276] Leatherman, S.P., "Coastal geomorphic responses to sea level rise: Galveston Bay, Texas," In: *Greenhouse Effect and Sea Level Rise: A Challenge for this Generation*, eds. M.C. Barth and J.G. Titus, Van Nostrand Reinhold Co., New York, 151-178, 1984.

[277] Lebbe, L.C., "The subterranean flow of fresh and salt water underneath the western Belgian beach," *Sveriges geologiska unders'kning*, Rapporter och meddelanden, **27**, 193-218, 1981.

[278] Lebbe, L.C. "Mathematical model of the evolution of the fresh-water lens under the dunes and beach with semi-diurnal tides." *Proc. 8th Salt Water Intrusion Meeting*, Bari, Italy. Geologia Applicata e Idrogeologia, Vol. XVIII, Parte II: 211-226, 1983.

[279] LeBlanc, D.R., *Potential Hydrologic Impacts of Ground-Water Withdrawal from Cape Cod National Seashore, Truro, Massachusetts*, U.S. Geol. Survey Open-File Rept. 82-438, 1982.

[280] Lee, C.-H. and Cheng, R.T., "On seawater encroachment in coastal aquifers," *Water Resour. Res.*, **10**, 1039-1043, 1974.

[281] Lennon, G.P., Wisniewski, G.M. and Yoshioka, G.A., " Impact of increased river salinity on New Jersey aquifers," In: *Greenhouse Effect, Sea Level Rise, and Salinity in the Delaware Estuary*, eds. C.H.J. Hull and J.G. Titus, Environmental Protection Agency and Delaware River Basin Commission, Washington DC, 40-54, 1986.

[282] Lessoff S.C. and Konikow, L.F., "Ambiguity in measuring matrix diffusion with single-well injection recovery tracer tests," *Ground Water*, **35**(1), 166-176, 1997.

[283] Levy, Y., "The origin and evolution of brines in coastal sabkhas, northern Sinai," *J. Sed. Petrol.*, **47**, 451-462, 1977.

[284] Lewis, F.M., Voss, C.I. And Rubin, J., *Numerical simulation of advective-dispersive multisolute transport with sorption, ion exchange and equilibrium chemistry*, U.S. Geol. Survey Water-Res Inv. Rept. 86-4022, 165 p., 1986.

[285] Lewis, F.M., Voss, C.I. And Rubin, J., "Solute transport with equilibrium aqueous complexation and either sorption or ion exchange," *J. Hydrology*, **90**, 81-115, 1987.

[286] Little, B.G., Cant, R., Buckley, D.K., Jefferies, A., Stark, J. and Young, R.N., *Land Resources of the Commonwealth of the Bahamas*, v. 6A and 6B, Great Exuma, Little Exuma, and Long Island, Land Resources Division, Ministry of Overseas Development, Suribiton, Surrey, 1976.

[287] Liu, H.H. and Dane, J.H., "An interpolation-corrected modified method of characteristics to solve advection-dispersion equations," *Adv. Water Resour.*, **19**(6), 359-368, 1996.

[288] Liu, P.L.-F., Cheng, A.H.-D., Liggett, J.A. and Lee, J.H., "Boundary integral equation solutions of moving interface between two fluids in porous media," *Water Resour. Res.*, **17**, 1445-1452, 1981.

[289] Loaiciga, H.A., "An optimization approach for groundwater quality monitoring network design," *Water Resour Res.*, **25**(8), 1771-1782, 1989.

[290] Loke, M.H. and Barker, R.D., "Rapid least-squares inversion of apparent resistivity pseudo-sections by a quasi-Newton method," *Geophysical Prospecting*, **44**(1), 131-152, 1996a.

[291] Loke, M.H. and Barker, R.D., "Practical techniques for 3D resistivity surveys and data inversion," *Geophysical Prospecting*, **44**(3), 499-523, 1996b.

[292] Luhdorff and Scalmanini Consulting Engineers, *Groundwater Resources and Management Report, Soquel Creek Water District, 1983*, Woodland, Calif., 1984.

[293] Lusczynski, N.J., "Head and flow of groundwater of variable density," *J. Geophy. Res.*, **66**, 4247-4256, 1961.

[294] Magaritz, M. and Luzier, J.E., "Water-rock interactions and seawater-freshwater mixing effects in the coastal dunes aquifer, Coos Bay, Oregon," *Geochim. Cosmochim. Acta*, **49**, 2515-2525, 1985.

[295] Maimone, M., Keil, D., Lahti, R. and Hoekstra, P., "Geophysical surveys for mapping boundaries of fresh water and salty waters in southern Nassau County, Long Island, NY," Proc. 3rd National Outdoor Action Conf., National Water Well Assoc., Orlando, FL, 1989.

[296] Makinde-Odusola, B.A. and Mariño, M.A., "Optimal control of groundwater by the feedback method of control," *Water Resour. Res.*, **25**(6), 1341-1352, 1989.

[297] Mandel, S., "The mechanism of sea water intrusion into calcareous aquifers," Int. Assoc. Hydrol. Sci. Publ. No. 64, 127-131, 1964.

[298] Mandel, S., "The design and instrumentation of hydrogeological observation network," Int. Assoc. Hydrol. Sci. Publ. No. 68, 413-424, 1965.

[299] Manzano, M., Custodio, E. and Jones, B.F., "Progress in the understanding of groundwater flow through the aquitard of the Llobregat Delta, Barcelona, Spain," In: *Livro de homenagem a Carlos Romariz.* A.C.A. de Matos (prefacer), Univ. Lisboa. Secc. Geol. Econ. Apl., Lisbon, Portugal, 115-126, 1990.

[300] McBride, E.F., "Secondary porosity importance in sandstone reservoirs in Texas," *Gulf Coast Assoc. Geol. Soc. Trans.*, **27**, 121-122, 1977.

[301] McCreanor, P.T. and Reinhart, D.R., "Hydrodynamic modelling of leachate recirculating landfills," *Water Science & Technology*, **34**(7-8), 463-470, 1996.

[302] McKenzie, D., *Water-Resources Potential of the Freshwater Lens at Key West, Florida*, U.S. Geol. Survey Water-Res. Inv. Rept. 90-4115, 24 p., 1990.

[303] McNeill, J.D., "Use of electromagnetic methods for groundwater studies," In: *Geotechnical and Environmental Geophysics*, v. I, Soc. Explor. Geophysicists, Inv. in Geophysics, n. 5, (ed.) S. Ward, 191-218, 1990.

[304] McNeill, J.D., "Electromagnetic terrain conductivity measusrement at low induction numbers," Geonics Ltd., Technical Note TN-6, Mississanga, Ontario, Canada, 15 p., 1980a.

[305] McNeill, J.D., "Interpretation procedures for the EM-34," Geonics Ltd., Technical Note TN-8, Mississauga, Ontario, Canada, 16 p., 1980b.

[306] Meijerink, J.A. and van der Vorst, H.A., "An iterative solution method for linear systems of which the coefficient matrix is a symmetric M-matrix," *Math. Comp.*, **31**, 148–162, 1977.

[307] Meisler, H., Leahy, P.P. and Knobel, L.L., *Effect of Eustatic Sea-Level Changes on Saltwater-Freshwater in the Northern Atlantic Coastal Plain*, U.S. Geol. Survey Water-Supply Paper 2255, 1984.

[308] Meissner, U., "A mixed finite element model for use in potential flow problems," *Int. J. Numer. Methods Eng.*, **6**, 467–473, 1972.

[309] Melloul, A.J., *Hydrogeological Atlas of Israel Coastal Plain Aquifer Geometry and Physical Properties*, (in Hebrew), Hydrologic Service Rep., 8/88, Jerusalem, 30 p., 1988.

[310] Melloul, A.J., Aberbach, S. and Kahanovitch, Y., *Renovation of Sea Water Intrusion Network in the Coastal Plain Aquifer of Israel*, (in Hebrew), TAHAL Rep., 01/88/13, 20 p., 1988.

[311] Melloul, A.J. and Atsmon, B., "Some typical features of salt encroachment and pollution in groundwater basins of Israel," In: *Environmental Quality and Ecosystem Stability*, **V/A**, 352-359, 1992.

[312] Melloul, A.J. and Atsmon, B., "A graphic expression of salinization and pollution of groundwater—The case of Israel's groundwater," In: *Environmental Geology*, **28**(3), 1-11, 1996.

[313] Melloul, A.J. and Bachmat, Y., *Assessment of the Hydrological Situation in the Gaza Strip as a Basis for the Management of the Coastal Plain Aquifer in that Region*, (in Hebrew), Hydrological Service Rep., 3, 83 p., Jerusalem, 1975.

[314] Melloul, A.J. and Bibas, M., *Regional and Local Hydrological Situation in the Coastal Plain Aquifer and Water Quantities According to Quality Standards (Chlorides and Nitrates) from 1987/88 and Anticipated Through 1992*, (In Hebrew), Hydrological Service, 1990/3, 38 p., Jerusalem, 1990.

[315] Melloul, A.J. and Collin, M., "The hydrological malaise of the Gaza Strip," *Israel Earth Sciences J.*, **43**, 105-116, 1994.

[316] Melloul, A.J. and Dax, A., " Monitoring the deterioration of small-diameter observation wells," *Water Resources Management*, **4**, 135-153, 1990.

[317] Melloul, A.J. and Gilad, D., "Movement of sea water intrusion in the Coastal aquifer of Israel during the years 1983-1990," (in Hebrew), *Water and Irrigation J.*, **318**, 46-48, 1993.

[318] Melloul, A.J. and Goldenberg, L.C., "Monitoring of Groundwater contaminants," In: *Groundwater Contamination and Control*, Ed. U. Zoller, Marcel-Dekker, 529-545, 1994.

[319] Mercado, A., "The use of hydrogeochemical patterns in carbonate sand and sandstone aquifers to identify intrusion and flushing of old saline water," *Ground Water*, **23**, 635-644, 1985.

[320] Mercer, J.W., Larson, S.P. and Faust, C.R., *Finite-Difference Model to Simulate the Areal Flow of Saltwater and Freshwater Separated by an Interface*, U.S. Geol. Survey Open-File Rept. 80-407, 1980a.

[321] Mercer J.W., Larson, S.P. and Faust, C.R., "Simulation of salt water interface motion," *Ground Water*, **18**, 374-385, 1980b.

[322] Merritt, M.L., *Assessment of Saltwater Intrusion in Southern Coastal Broward County, Florida*, U.S. Geol. Survey Water-Res. Inv. Rept. 96-4221, 1996.

[323] Merritt, M. L., *Tests of Subsurface Storage of Freshwater at Hialeah, Dade County, Florida, and Numerical Simulation of the Salinity of Recovered Water*, U.S. Geol. Survey Water-Supply Paper 2431, 1997.

[324] Meyer, F.W., *Hydrogeology, Ground-Water Movement, and Subsurface Storage in the Floridan Aquifer System in Southern Florida*, U.S. Geol. Survey Prof. Paper 1403-G, 1989.

[325] Mijatovic, B.F., "Hydrogeology of the Dinaric karst," *IAH*, **4**, Verlag Heinz Heise GmbH, Hannover, 1984.

[326] Milanovic, P.T., *Karst Hydrogeology*, Water Resources Publications, Littleton, Colorado, 1981.

[327] Miller, J.A., *Ground Water Atlas of the United States: Segment 6, Alabama, Florida, Georgia, South Carolina*, U.S. Geol. Survey Hyd. Inv. Atlas 730-G, 1990.

[328] Miller, P.T., McGeary, S. and J.A. Madsen, J.A., "High resolution seismic reflection images of New Jersey coastal aquifers," *J. Environmental Engineering Geophysics*, **1**(1), 55-66, 1996.

[329] Mimikou, M.A., "Climate change," Chap. 3 in *Environmental Hydrology*, ed. Singh, V.P., Kluwer Acad. Publ., Dordrecht, 1995.

[330] Mink, J.F., *State of the groundwater resources of southern Oahu*, Board of Water Supply, City and County of Honolulu, 83 p., 1980.

[331] MOC, "Computer Model of Two Dimensional Solute Transport and Dispersion in Ground Water," International Ground Water Modeling Center, Delft, ver 3.0, Nov., 1989.

[332] Morell, I., Medina, J., Pulido, A. and Fenandez-Rubio, R., "The use of bromide and strontium as indicators of marine intrusion in the aquifer of Oropesa-Torreblanca," *Proc. 14th Salt Water Intrusion Meeting*, Malmo, Sweden, 629-640, 1996.

[333] Morse, J.W. and Mackenzie, F.T., "Geochemistry of sedimentary carbonates," *Developments in Sedimentology*, **48**, Elsevier, 707 p., 1990.

[334] Mualem Y. and Bear, J., "The shape of the interface in steady flow in a stratified aquifer," *Water Resour. Res.*, **10**, 1207-1215, 1974.

[335] Muskat, M., *The Flow of Homogeneous Fluids through Porous Media*, McGraw-Hill, 1937.

[336] Naji, A., Cheng, A.H.-D. and Ouazar, D., "Analytical stochastic solutions of saltwater/freshwater interface in coastal aquifers," to appear in *Stochastic Hydrology & Hydraulics*, 1998.

[337] Narayan, K.A. and Armstrong, D., "Simulation of groundwater interception at Lake Ranfurly, Victoria, incorporating variable-density flow and solute transport," *J. Hydrology*, **165**(1-4), 161-184, 1995.

[338] National Research Council, *Valuing Ground Water: Economic Concepts and Approaches*, National Academy Press, Washington, DC. 1997.

[339] Navoy, A.S., "Aquifer-estuary interaction and vulnerability of groundwater supplies to sea level rise-driven saltwater intrusion," Ph.D. thesis, Pennsylvania State University, U.S.A., 225 p., 1991.

[340] Neuman, S.P., "A Eulerian-Lagrangian numerical scheme for the dispersion-convection equation using conjugate space-time grids," *J. Comp. Phys.*, **41**(2), 270-294, 1981.

[341] Neuman, S.P., "Adaptive Eulerian-Lagrangian finite element method for advection—dispersion," *Int. J. Num. Meth. Eng.*, **20**, 321-337, 1984.

[342] Neuman, S.P. and Sorek, S., "Eulerian-Lagrangian methods for advection—dispersion," *Proc. 4th Int. Conf. Finite Element Water Resour.*, Germany, 14.41-14.68, 1982.

[343] Newport, B.D., *Salt water intrusion in the United States*, Report No.600-8-77-011, U.S. EPA, Washington, D.C, 1977.

[344] Nikolaevskij, V.N., *Mechanics of Porous and Fractured Media*, World Scientific, Singapore, 1990.

[345] Nishikawa, T., "Testing alternative conceptual models of seawater intrusion in a coastal aquifer using computer simulation, southern California, USA," *Hydrogeology J.*, **5**(3), 60-74, 1997.

[346] Nishikawa, T. and Reichard, E.G., "Evaluating strategies to manage seawater intrusion," In: Bathala, C.T. [ed.], *Proc. North American Water & Envir. Cong.*, Anaheim, CA, ASCE, New York, 1996.

[347] Nordqvist, R. and Voss, C.I., "A simulation-based approach for designing effective field-sampling programs to evaluate contamination risk," *Hydrogeology J.*, **4**(4), 23-39, 1996.

[348] Oberdorfer, J.A., Hogan, P.J. and Buddemeier, R.W., "Atoll island hydrogeology: flow and freshwater occurrence in a tidally dominated system," *J. Hydrology*, **120**, 327-340, 1990.

[349] Oki, D., Souza, W.R., Bohlke, E. and Bauer, G., "Numerical analysis of the hydrogeologic controls in a layered coastal aquifer system, Oahu, Hawaii," *Hydrogeology J.*, **6**(2), 243-26, 1983.

[350] Oldenburg, C.M. and Pruess, K., "Dispersive transport dynamics in a strongly coupled groundwater-brine flow system," *Water Resour. Res.*, **31**(2), 289-302, 1995.

[351] Orange County Water District, *Reflections on 60 Years*, OCWD 1993 Annual Report, Fountain Valley, CA, 1993.

[352] Orange County Water District, *The Groundwater Management Plan*, OCWD, Fountain Valley, CA, 1994.

[353] Oude Essink, G.H.P. "Impact of sea level rise on groundwater flow regimes. A sensitivity analysis for the Netherlands," Ph.D. thesis, Delft University of Technology, 411 p., 1996.

[354] Oude Essink, G.H.P., "Simulation of 3D density-dependent groundwater flow: the adapted MOC3D code," (in Dutch), Stromingen, in press, 1998.

[355] Oude Essink, G.H.P. and Boekelman, R.H. "Problems with large-scale modelling of saltwater intrusion in 3D." *Proc.* 14th *Salt Water Intrusion Meeting*, Malmö, Sweden, 288-299, 1996.

[356] Palmer, D., *The Generalized Reciprocal Method of Seismic Refraction Interpretation*, Society Explor. Geophysicists, Tulsa, OK, 104 p., 1980.

[357] Paniconi, C. and Putti, M., "A comparison of Picard and Newton iteration in the numerical solution of multidimensional variably saturated flow problems," *Water Resour. Res.*, **30**(12), 3357-3374, 1994.

[358] Paniconi, C. and Putti, M., "Newton-type linearization and line search methods for unsaturated flow models," *Advances in Groundwater Pollution Control and Remediation*, Vol. 9 of NATO ASI Series 2: Environment, 155-172, Kluwer Academic,, 1996.

[359] Papadrakakis, M., "Solving large-scale nonlinear problems in solid and structural mechanics," *Solving Large-Scale Problems in Mechanics*, 183-223, John Wiley & Sons, 1993.

[360] Parasnis, D.J., *Principles of Applied Geophysics*, Chapman and Hall, New York, 402 p., 1986.

[361] Parker, G.G., "Salt-water encroachment in southern Florida," *Am. Water Works Assoc. J.*, **37**(6), 526-542, 1945.

[362] Parker, G.G., Ferguson, G.E. and Love, S.K., *Interim Report on the Investigations of Water Resources in Southeastern Florida with Special Reference to the Miami Area in Dade County*, Florida Geol. Survey Rept. of Invest. No. 4, 1944.

[363] Parker, G.G., Ferguson, G.E., Love, S.K., et al., *Water Resources of Southeastern Florida*, U.S. Geol. Survey Water-Supply Paper 1255, 1955.

[364] Parkhurst, D.L., Thorstenson, D.C. and Plummer, L.N., *PHREEQE—A Computer Program for Geochemical Calculations*, U.S. Geol. Survey Water Resources Inv. Rept. 80-96, 195 p., 1980, revised and reprinted 1990.

[365] Peyret, R. and Taylor, T.D., *Computational Methods for Fluid Flow*, Springer-Verlag, New York, NY, 1983.

[366] Piggott, A.R., Bobba A.G. and Novakowski, K.S., "Regression and inverse analyses in regional groundwater modelling," *J. Water Resources Planning & Management, ASCE*, **122**(1), 1-10, 1996.

[367] Piggott, A.R., Bobba, A.G. and Xiang, J.N., "Inverse analysis implementation of the SUTRA groundwater model," *Ground Water*, **32**(5), 829-836, 1994.

[368] Pinder, G.F. and Bredehoeft, J.D. "Application of the Digital Computer for Aquifer Evaluation," *Water Resour. Res.*, **4**(5), 1069-1093, 1968.

[369] Pinder, G.F. and Cooper, H.H. Jr., "A numerical technique of calculating the transient position of the saltwater front," *Water Resour. Res.*, **3**, 875-881, 1970.

[370] Pinder, G.F. and Page, R.H. "Finite element simulation of salt water intrusion on the South Fork of Long Island," In: *Finite Elements in Water Resources*, Proc. 1st Int. Conf. Finite Elements in Water Resources, Pentech, London, 2.51–2.69, 1977.

[371] Pini, G. and Putti, M., "Krylov methods in the finite element solution of groundwater transport problems," *Computational Methods in Water Resources X*, Vol. 1, 1431–1438, Kluwer Academic, 1994.

[372] Planert, M. and Williams, J.S., *Ground Water Atlas of the United States: Segment 1, California, Nevada*, U.S. Geol. Survey Hyd. Inv. Atlas 730-B, 1995.

[373] Ploethner, D., Zomenis, S., Avraamides, C. and Charalambides, A., "Hydrochemical studies for the determination of salt water intrusion processes in the Larnaca coastal plain, Cyprus," *Proc. 9th Salt Water Intrusion Meeting*, Delft, The Netherland, 233-243, 1986.

[374] Plummer, L.N., Parkhurst, D.L. and Kosiur, D.R., *MIX-2, A Computer Program for Modeling Chemical Reactions in Natural Waters*, U.S. Geol. Survey Water-Res. Inv. Rept. 61-75, 68 p., 1975.

[375] Plummer, L.N., "Stable isotope enrichment in paleowaters of the south east Atlantic Coastal Plain, U.S.," *Science*, **262**, 2016-2020, 1993.

[376] Poland, J. F., *Hydrology of the Long Beach-Santa Ana Area, California*, U.S. Geol. Survey Water-Supply Paper 1471, 1959.

[377] Poland, J. F., Piper, A.M., et al., *Ground-Water Geology of the Coastal Zone, Long Beach-Santa Ana Area, California*, U.S. Geol. Survey Water-Supply Paper 1109, 1956.

[378] Pope, G.A., Lake, L.W. and Helfferich, F.G., "Cation exchange in chemical flooding," *Soc. Pet. Eng. J.*, **18**, 418-434, 1978.

[379] Prakash, C., "Examination of the upwind formulation in the control volume finite element method for fluid flow and heat transfer," *Numer. Heat Transfer*, **11**, 401–416, 1987.

[380] Press, W.H., Flannery, B.P., Teukolsky, S.A. and Vetterling, W.T., *Numerical Recipes: The Art of Scientific Computing*, Cambridge University Press, 1989.

[381] Price, R.N. and Herman, J.S., "Geochemical investigation of salt-water intrusion into a coastal carbonate aquifer: Mallorca, Spain," *Geol. Soc. Amer. Bull.*, **103**, 1270-1279, 1991.

[382] Price, J.S. and Woo, M.-K., "Studies of a subarctic coastal marsh. III. Modelling the subsurface water fluxes and chloride distribution," *J. Hydrology*, **120**, 1-13, 1990.

[383] Prince, K.R., *Ground-Water Assessment of the Montauk Area, Long Island, New York*, U.S. Geol. Survey Water-Res. Inv. Rept. 85-4013, 1986.

[384] Provost, A.M., Voss, C.I. and Neuzil, C.E., *Glaciation and regional groundwater flow in the Fennoscandian shield*, Swedish Nuclear Power Inspectorate SKI Report 96:11, Stockholm, Sweden, 82 p., 1998.

[385] Pullan, S.E. and MacAulay, H.A., "An in-hole shotgun source for engineering seismic surveys," *Geophysics*, **52**, 985-996, 1987.

[386] Putti, M. and Cordes, C., "Finite element approximation of the diffusion operator on tetrahedra," *SIAM J. Sci. Comput.*, (to appear), 1998.

[387] Putti, M. and Paniconi, C., "Finite element modeling of saltwater intrusion problems with an application to an Italian aquifer," *Advanced Methods for Groundwater Pollution Control*, Vol. 364 of *CISM (Int. Centre for Mechanical Sciences) Courses and Lectures*, 65–84, Springer-Verlag, 1995a.

[388] Putti, M. and Paniconi, C., "Picard and Newton linearization for the coupled model of saltwater intrusion in aquifers," *Adv. Water Resour.*, 18(3), 159–170, 1995b.

[389] Putti, M., Yeh, W. W.-G. and Mulder, W.A., "A triangular finite volume approach with high-resolution upwind terms for the solution of groundwater transport equations," *Water Resour. Res.*, 26(12), 2865–2880, 1990.

[390] Ranganathan, V., "Basin dewatering near salt domes and formation of brine plumes," *J. Geophys. Res.-Solid Earth*, 97(B4), 4667-4683, 1992.

[391] Ranganathan, V. and Hanor, J.S., "Density-driven groundwater flow near salt domes," *Chemical Geology*, 74, 173-188, 1988.

[392] Ranganathan, V. and Hanor, J.S., "Perched brine plumes above salt domes and dewatering of geopressured sediments," *J. Hydrology*, 110, 63-86, 1989.

[393] Refsgaard, J.C., Alley, W.M. and Vuglinsky, V.S, "Methodology for distinguishing between man's influence and climate effects on the hydrological cycle," IHP-III Project 6.3, UNESCO, Paris, 1989.

[394] Reichard, E.G., "Groundwater-surface water management with stochastic surface water supplies: A simulation optimization approach," *Water Resour. Res.*, 31(11), 2845-2865, 1995.

[395] Reilly, T.E., "Simulation of dispersion in layered coastal aquifer systems," *J. Hydrology*, 114, 211-228, 1990.

[396] Reilly, T.E., "Analysis of ground-water systems in freshwater-saltwater environments," In: *Regional Ground-Water Quality*, W.M. Alley, ed., Van Nostrand Reinhold, New York, 1993.

[397] Reilly, T.E., Frimpter, M.H., LeBlanc, D.R. and Goodman, A.S., "Analysis of steady-state saltwater upconing with application at Truro Well Field, Cape Cod, Massachusetts," *Ground Water*, 25(2), 194-206, 1987.

[398] Reilly, T.E. and Gibs, J., "Effects of physical and chemical heterogeneity on water-quality samples obtained from wells," *Ground Water*, 31(5), 805-813, 1993.

[399] Reilly, T. E. and Goodman, A.S., "Quantitative analysis of saltwater–freshwater relationships in ground-water systems— a historical perspective," *J. Hydrol.*, 80, 125-160, 1985.

[400] Reilly, T.E. and Goodman, A.S., "Analysis of saltwater upconing beneath a pumping well," *J. Hydrology*, 89, 169-204, 1987.

[401] Reilly, T.E. and LeBlanc, D., "Experimental evaluation of factors affecting temporal variability of water samples obtained from long-screened wells," *Ground Water*, 36(4), 566-571, 1986.

[402] Revelle, R., "Criteria for recognition of seawater in groundwater," *Trans. Amer. Geophy. Union*, 22, 1963.

[403] Rizzo, D.M. and Doughterty, D.E., "Design optimization for multiple management period groundwater remediation," *Water Resour. Res.*, 32(8), 2549-2561, 1996.

[404] Roe, P.L., "Characteristic-based schemes for the Euler equation," *Annu. Rev. Fluid Mech.*, **18**, 337–365, 1986.

[405] Rogers, D.B. and Dreiss S.J., "Saline groundwater in Mono Basin, California—1. Distribution," *Water Resour. Res.*, **31**(12), 3131-3150, 1995a.

[406] Rogers, D.B. and Dreiss S.J., "Saline Groundwater in Mono Basin, California—2. Long-term control of lake salinity by groundwater," *Water Resour. Res.*, **31**(12), 3151-3169, 1995b.

[407] Rogers, L.L., "History matching to determine the retardation of PCE in ground-water," *Ground Water*, **30**(1), 50-60, 1992.

[408] Rogers, L.L. and Dowla, F.U., "Optimization of groundwater remediation using artificial neural networks with parallel solute transport modelling," *Water Resour. Res.*, **30**(2), 457-481, 1994.

[409] Rogers, L.L., Dowla, F.U. and Johnson, V.M., "Optimal field-scale groundwater remediation using neural networks and the genetic algorithm," *Environ. Sci. Tech.*, **29**(5), 1145-1155, 1995.

[410] Rosensaft, M., Eker, A., Levitah, D., Shimron, I., Shimron, A., Rosenfeld, A., Sneh, A. and Bein, A., *The Three-Dimensional Configuration and Lithological Composition of the Coastal Aquifer and the Unsaturated Zone, Israel*, Geological Survey of Israel Rep. GSI/15/95, 25 p., 1995.

[411] Rumer R.R. and Shiau, J.C., "Salt water interface in a layered coastal aquifer," *Water Resour. Res.*, 4, 1235-1247, 1968.

[412] Saad, Y., "Krylov subspace methods for solving large unsymmetric linear systems," *Math. Comp.*, **37**(155), 105–126, 1981.

[413] Saad, Y., "Krylov subspace methods: Theory, algorithms, and applications," *Computing Methods in Applied Sciences and Engineering*, 24–41, SIAM, Philadelphia, 1990.

[414] Saad, Y., "ILUT: A dual threshold incomplete ILU factorization," *Num. Lin. Alg. Appl.*, **1**, 387–402, 1994.

[415] Saad, Y. and Schultz, M.H., "Conjugate gradient-like algorithms for solving nonsymmetric linear systems," *Math. Comp.*, **44**(170), 417–424, 1985.

[416] Saad, Y. and Schultz, M.H., "GMRES: A generalized minimum residual algorithm for solving nonsymmetric linear systems," *SIAM J. Sci. Stat. Comput.*, **7**(3), 856–869, 1986.

[417] Sacks,L.A., Herman, J.S., Konikow, L.F. and Vela, A., "Seasonal dynamics of groundwater-lake interaction at Donana National Park, Spain," *J. Hydrol.*, **136**, 123-154, 1992.

[418] Saiers, J.E., Hornberger, G.M., Harvey, C., "Colloidal silica transport through structured, heterogeneous porous media," *J. Hydrology*, **163**(3-4), 271-288, 1994.

[419] Sahoo, D., Smith, J.A., Imbrigiotta, T.E. and McLellan, H.M., "Surfactant-enhanced remediation of a trichloroethene-contaminated aquifer, 2. Transport of TCE," *Environ. Sci. Tech.*, **32**, 1686-1693, 1998.

[420] Samarsky, A. A. and Nikolaev, E. S., *Methods of Solution of Grid Equations*, (in Russian), Nauka, Moscow, 1978.

[421] Sanford, W.E. and Konikow, L.F., *A two constituent solute transport model for groundwater having variable density*, U.S. Geol. Survey Water-Resour. Inv. Rept. 85-4279, 1985.

[422] Sanford, W.E. and Konikow, L.F., "Porosity development in coastal carbonate aquifers—Reply," *Geology*, **17**(10), 962-963, 1989.

[423] Sanford, W., Whitaker, F., Smart, P., "Geothermal circulation of sea water in carbonate platforms: I. Theoretical and sensitivity analysis," *American J. Science* (in press), 1998.

[424] Savenije, H.H.G., "Rapid assessment technique for salinity intrusion in alluvial estuaries," Ph.D. Thesis, Delft University of Technology, Delft, the Netherlands, 1992.

[425] Sayles, F.L. and Mangelsdorf, P.C., "The equilibrium of clay minerals with sea water; exchange reactions," *Geochim. et Cosmochim. Acta*, **41**, 951-960, 1977.

[426] Shachnai, E. and Zeitoun, D.G., "Monitoring and management of the Coastal Aquifer for 1995," TAHAL 5718-96.026, 1996.

[427] Schelkes, K. and Vogel, P., "Paleohydrological information as an important tool for groundwater modelling of the Gorleben Site," In: *Paleohydrogeological Methods and Their Applications*, Proc. Nuclear Energy Agency Workshop, Nov. 1992, OECD Publ., Paris, 237-250, 1992.

[428] Schelkes, K., Vogel, P., Klinge, H. and Knoop, R-M., "Modelling of variable-density groundwater flow with respect to planned radioactive waste-disposal sites in West Germany—Validation activities and first results," In: *GEOVAL-1990*, Symp. Validation of Geosphere Flow and Transport Models, Proceedings, Stockholm, May 1990, OECD Publ., Paris, 328-335, 1991.

[429] Schincariol, R.A. and Schwartz, F.W., "On the generation of instabilities in variable-density flow," *Water Resour. Res.*, **30**(4), 913-927, 1994.

[430] Schoeller, H., *Geochemie des eaux souterraines.Application aux eaux des gisements de petrole*, Soc. Ed. Technip, Paris, 213 p., 1956.

[431] Schmorak, S. and Mercado, A., "Upconing of fresh water–sea water interface below pumping wells, field study," *Water Resour. Res.*, **5**(6), 1290–1311, 1969.

[432] Schwertfeger, B.C., "On the occurrence of submarine freshwater discharges," *Proc. 6th Salt Water Intrusion Meeting*, in Hanover, 1979, Geologisch Jahrbuch, Reihe C, Heft 29, Bundesanstalt fr Geowissenschaften und Rohstoffe und den Geologischen Landesmtern in der Bundesrepublik Deutschland, Hannover, 1981.

[433] Screaton, E.J., Wuthrich, D.R. and Dreiss, S.J., "Permeabilities, fluid pressures, and flow rates in the Barbados Ridge Complex," *J. Geophys. Res.*, **95**(B6), 8997-9007, 1990.

[434] Screaton, E.J. and Ge, S., "An assessment of along-strike fluid and heat transport within the Barbados Ridge accretionary complex: Results of preliminary modeling," *Geophys. Res. Let.*, **24**(23), 3085-3088, 1997.

[435] Ségol, G., *Classic Groundwater Simulations: Proving and Improving Numerical Models*, Prentice Hall, 531 p., 1993.

[436] Ségol, G. and Pinder, G.F., "Transient simulation of salt water intrusion in south eastern Florida," *Water Resour. Res.*, **12**, 65-70, 1976.

[437] Ségol, G., Pinder, G.F. and Gray, W.G., "A Galerkin finite element technique for calculating the transient position of the saltwater front," *Water Resour. Res.*, **11**(2), 343–347, 1975.

[438] Senger, R.K., "Paleohydrology of variable-density groundwater -flow systems in mature sedimentary basins—Example of the Palo Duro Basin, Texas, USA," *J. Hydrology*, **151**(2-4), 109-145, 1993.

[439] Senger, R.K. and Fogg, G.E., "Stream functions and equivalent freshwater heads for modelling regional flow of variable-density groundwater—1. Review of theory and verification," *Water Resour. Res.*, **26**(9), 2089-2096, 1990.

[440] Sengpiel, K.P., "Groundwater prospecting by multifrequency airborne EM techniques," In: *Airborne Resistivity Mapping*, (ed.) G.J. Palacky, Geol. Surv. Canada, Paper 86-22, 128-131, 1986.

[441] Sengpiel, K.P., "Resistivity/depth mapping with airborne electromagnetic survey data," *Geophysics*, **48**, 181-196, 1983.

[442] Shapiro, A.M., Bear, J. and Shamir, U., "Development of numerical model for predicting the movement of the regional interface in the coastal aquifer of Israel," Technion IIT Report, Haifa, 1983.

[443] Sharma, M.L., "Impact of climate change on groundwater recharg," *Conf. on Climate and Water*, Publ. of Academy of Finland, **1**, 511-520, 1989.

[444] Shata, A. and Hefny, K., "Strategies for planning and management of groundwater in the Nile Valley and Nile Delta in Egypt," Working Paper Series No. 31-1, Strategic Research Program (SRP), NWRC-MPWWR, Cairo, 1995.

[445] Sherif, M.M., "A Two-dimensional finite element model for dispersion (2D-FED) in coastal aquifers," Ph.D. Thesis, Faculty of Engineering, Cairo University, Giza, Egypt, 1987.

[446] Sherif, M.M., "Climate changes and groundwater," *Proc. 2nd Int. Conf. Civil Eng. and Computer Application, Research and Practice*, Bahrain, 1996.

[447] Sherif, M.M. and Singh, V.P., "Saltwater intrusion," *Hydrology of Disasters*, Chap. 10, 269–316, Kluwer Academic, 1996.

[448] Sherif M.M. and Singh, V.P., "Groundwater development and sustainability in the Nile Delta aquifer," Final report submitted to Binational Fulbright Commission, Egypt, 1997.

[449] Sherif, M.M., Singh, V.P. and Amer, A.M., "A two-dimensional finite element model for dispersion (2D-FED) in coastal aquifer," *J. Hydrology*, **103**, 11-36, 1988.

[450] Sherif, M.M., Singh, V.P. and Amer, A.M., "A note on saltwater intrusion in coastal aquifers," *Water Resources Management*, **4**, 123-134, 1990a.

[451] Sherif, M.M., Singh, V.P. and Amer, A.M, "A sensitivity analysis of (2D FED), a model for sea water encroachment in leaky coastal aquifers," *J. Hydrol.*, **103**, 11-36, 1990b.

[452] Sheriff, S.D., "Spreadsheet modeling of electrical sounding experiments," *Ground Water*, **30**(6), 971-974, 1992.

[453] Siegel, D.J. and Mandle, R.J., "Isotopic evidence for glacial meltwater recharge to the Cambrian-Ordovician aquifer, north-central U.S.," *Quaternary Res.*, **22**(3), 328-335, 1984.

[454] Simmons, C.T., "Density-induced groundwater flow and solute transport beneath saline disposal basins," Doctoral Dissertation, Flinders University of South Australia, 239 p., 1997.

[455] Simmons, C.T. and Narayan, K.A., "Mixed convection processes below a saline disposal basin," *J. Hydrology*, **194**, 263-285, 1997.

[456] Simmons, C.T. and Narayan, K.A., "Modelling density-dependent flow and solute transport at the Lake Tutchewop saline disposal complex, Victoria," *J. Hydrology*, **206**, 219-23, 1998.

[457] Siraz, L., "Salinisation mechanism in the coastal groundwaters of Palghar Taluka, Maharashtra, India," IAH, IGC *Groundwater Sustainable Solutions*, Eds. T.R. Weaver and C.R. Lawrence, Melbourne Univ., Australia, 347-352, 1998.

[458] Slaine, D.D., Pehme, P.E., Hunter, J.A., Pullan, S.E. and Greenhouse, J.P., "Mapping overburden stratigraphy at a proposed hazardous waste facility using shallow seismic reflection methods," In: *Geotechnical and Environmental Geophysics*, v. II, Soc. Explor. Geophysicists, Inv. in Geophysics, n. 5, (ed.) S. Ward, 273-280, 1990.

[459] Smith, B.S., *Saltwater movement in the Upper Floridan Aquifer beneath Port Royal Sound, South Carolina*, U.S. Geol. Survey Water-Supply Paper 2421, 40 p., 1994.

[460] Smith, J.A., Sahoo, D., McLellan, H.M. and Imbrigiotta, T.E., "Surfactant-enhanced remediation of a trichloroethene-contaminated aquifer, 1. Transport of Triton X-100," *Environ. Sci. Tech.*, **31**, 3565-3572, 1997.

[461] Smith, R.T. and Ritzi, R.W., "Designing a nitrate monitoring program in a heterogeneous, carbonate aquifer," *Ground Water*, **31**(4), 576-584, 1993.

[462] Smith, Z.A., *Groundwater in the West*, Academic Press, San Diego, CA, 1989.

[463] Smithers, H. and Walker, S., "The practical application of modelling to the sustainable management of water resources in northern England," In: *Modelling and Management of Sustainable Basin-scale Water Resources Systems*, Proc. of a Boulder Symp., IAHS Publ. no. 231, 15-20, 1995.

[464] Solley, W.B., Pierce, R.R. and Perlman, H.A., *Estimated use of water in the United States in 1990*, U.S. Geol. Survey Circular 1081, 1993.

[465] Sonenshein, R.S., *Delineation and Extent of Saltwater Intrusion in the Biscayne Aquifer, Eastern Dade County, Florida*, U.S. Geol. Survey Water-Res. Inv. Rept. 96-4285, 1997.

[466] Sonenshein, R.S. and Koszalka, E.J., *Trends in Water-Table Altitude (1984-93) and Saltwater Intrusion (1974-93) in the Biscayne Aquifer, Dade County, Florida*, U.S. Geol. Survey Open-File Rept. 95-705, 1996.

[467] Sorek, S., "Eulerian-Lagrangian formulation for flow in soil," *Adv. Water Resour.*, **8**, 118-120, 1985a.

[468] Sorek, S., "Adaptive Eulerian-Lagrangian method for transport problems in soils," *Scientific Basis for Water Water Resources Management*, IASH Publication, **153**, 393-403, 1985b.

[469] Sorek, S., "Eulerian-Lagrangian method for solving transport in aquifers," *Adv. Water Resour.*, **11**(2), 67-73, 1988.

[470] Sorek, S. and Braester, C., "Eulerian-Lagrangian formulation of the equations for groundwater denitrification using bacterial activity," *Adv. Water Resour.*, **11**(4), 162-169, 1988.

[471] Souza, W.R., *Documentation of a graphical-display program for the saturated-unsaturated transport (SUTRA) finite-element simulation model*, U.S. Geol. Survey Water-Res. Inv. Rept. 87-4245, 122 p., 1987.

[472] Souza, W.R. and Voss, C.I., "Analysis of an anisotropic coastal aquifer system using variable-density flow and solute transport simulation," *J. Hydrology*, **92**, 17-41, 1987.

[473] Souza, W.R. and Voss, C.I., "Assessment of potable ground water in a fresh-water lens using variable-density flow and solute transport simulation," In: Proc. National Water Well Assoc. (NWWA) Conf. on Solving Ground Water Problems with Models, Indianapolis, Indiana, 1023-1043, 1989.

[474] Spear, W.E., *Long Island Sources of Additional Supply of Water for the City of New York*, New York Board of Water Supply, 1912.

[475] Stakelbeek, A., "Movement of brackish groundwater near a deep-well infiltration system," 12th Salt Water Intrusion Meeting, Barcelona, Spain, 1992.

[476] Starinsky, A., "Relationship between Ca-chloride brines and sedimentary rocks in Israel," Ph.D. dissertation., Hebrew University, Jerusalem, (in Hebrew), 1974.

[477] Starinsky, A., Bielsky, M., Lazar, B., Steinitz, G. and Raab, M., "Strontium isotope evidence on the history of oilfield brines, Mediterranean Coastal Plain, Israel," *Geochim. Cosmochim. Acta*, **47**, 687-695, 1983.

[478] Starinsky, A., Vengosh, A., Shapiro, B. and Darbysheir, P., "Geochemical and isotopic natural tracers as indicators for ground water salinization in Yarqon-Taninim aquifer," Research Report to the Water Comission, Israel, (in Hebrew), 1995.

[479] Steeples, D.W. and Miller, R.D., "Seismic reflection methods applied to engineering, environmental and groundwater problems," In: *Geotechnical and Environmental Geophysics*, v. I, Soc. Explor. Geophysicists, Inv. in Geophysics, n. 5, (ed.) S. Ward, 1-30, 1990.

[480] Stewart, M., "Rapid reconnaissance mapping of fresh-water lenses on small oceanic islands," In: *Geotechnical and Environmental Geophysics*, v. II, Soc. Explor. Geophysicists, Inv. in Geophysics, n. 5, (ed.) S. Ward, ed., 57-66, 1990.

[481] Stewart, M., "Evaluation of electromagnetic methods for rapid mapping of saltwater interfaces in coastal aquifers," *Ground Water*, **20**, 538-545, 1982.

[482] Stewart, M. and Bretnall, R.E. Jr., "Interpretation of VLF resistivity data for ground-water contamination studies," *Ground Water Monitoring Review*, **6**(1), 71-75, 1986.

[483] Stewart, M. and Gay, M., "Evaluation of transient electromagnetic soundings for deep detection of conductive fluids," *Ground Water*, **24**, 351-356, 1986.

[484] Stewart, M. and Hermeston, S., "Monitoring saltwater interfaces in PVC-cased boreholes using induction logs," Southwest Florida Water Management District, Project Report, Brooksville, FL, 43 p., 1990.

[485] Stewart, M., Layton, M. and Lizanec, T., "Application of resistivity surveys to regional hydrogeologic reconnaissance," *Ground Water*, **21**(1), 42-48, 1983.

[486] Stone, H., "Iterative solution of implicit approximations of multidimensional partial differential equations," *SIAM J. Appl. Math.*, **5**, 530-559, 1968.

[487] Strack, O.D.L., "A single-potential solution for regional interface problems in coastal aquifers," *Water Resour. Res.*, **12**, 1165-1174, 1976.

[488] Strack, O.D.L., "A Dupuit-Forchheimer model for three-dimensional flow with variable-density," *Water Resour. Res.*, **12**, 3007-3017, 1995.

[489] Street, G.J. and Engel, R., "Geophysical surveys of dryland salinity," In: *Geotechnical and Environmental Geophysics*, v. II, Soc. Explor. Geophysicists, Inv. in Geophysics, n. 5, (ed.) S. Ward, 187-200, 1990.

[490] Sukhija, B.S., Varma, V.N., Nagabhushanam, P. and Reddy, D.V., "Differentiation of paleomarine and modern intruded salinities in coastal groundwaters (of Karaikal and Tanjavur, India) based on inorganic chemistry, organic biomarker fingerprints and radiocarbon dating," *J. Hydrology*, **174**, 173-201, 1996.

[491] Suleiman S, El-Baruni, "Deterioration of quality of groundwater from Suani Wellfield, Tripoli, Libya 1976-93," *Hydrogeol. J.*, **3**(2), 58-64, 1995.

[492] Swartz, J.H., "Resistivity studies of some salt-water boundaries in the Hawaiian Islands," Am. Geophys. Union, Trans., **18**, 387-393, 1937

[493] Swartz, J.H., "Part II–Geophysical investigations in the Hawaiian Islands," Am. Geophys. Union, Trans., **20**, 292-298, 1939.

[494] SWIM (Salt Water Intrusion Meetings), Proceedings:

Delft (9th), 1986:
Boekelman, R.H., van Dam, J.C., Evertman, M., Hoorn, W.H.C., (editors), Proceedings of the 9th Salt Water Intrusion Meeting Delft, 12-16 May, 1986, Water Management Group, Department of Civil Engineering, Delft University of Technology, Delft, the Netherlands.

Ghent (10th), 1988:
Breuck, W. de and Walschot, L., (editors), Ghent, (Belgium), 16-20 May 1988, Natuurwetenschappelijk Tijdschrift, Ghent, Belgium

Gdansk (11th), 1990:
Kozerski, B. and Sadurski, A., (editors), Proceedings of the 11th Salt Water Intrusion Meeting, Gdansk, 14-17 May, 1990 Technical University of Gdansk, Department of Hydrogeology and Water Supply, Gdansk, Poland.

Barcelona (12th), 1992:
Custodio, E. and Galofre, A., (editors), Study and modelling of saltwater intrusion into aquifers, CIMNE, Barcelona, Spain.

Cagliari (13th), 1994:
Proceedings of the 13th Salt Water Intrusion Meeting, University of Cagliari, Italy.

Malmö (14th), 1996:
Leander, B., et al., 14th Salt Water Intrusion Meeting, SWIM 96, Rapporter och meddelanden, no. 87, Geological Survey of Sweden, Uppsala, Sweden.

[495] Syriopoulou, D. and Koussis, A.D., "Two-dimensional modelling of advection-dominated solute transport in groundwater by the matched artificial dispersivity method," *Water Resour. Res.*, **27**(5), 865-872, 1991.

[496] Taigbenu, A.E., Liggett, J.A. and Cheng, A.H.-D., "Boundary integral solution to seawater intrusion into coastal aquifers," *Water Resour. Res.*, **20**, 1150–1158, 1984.

[497] Takasaki, K.J. and Mink, J.F., *Water Resources of Southeastern Oahu, Hawaii*, U.S. Geol. Survey Water-Res. Inv. Rept. 82-628, 1982.

[498] Takasaki, K.J. and Mink, J.F., *Evaluation of major dike-impounded ground-water reservoirs, Island of Oahu*, U.S. Geol. Survey Water-Supply Paper 2217, 77 p., 1985.

[499] Task Committee on Saltwater Intrusion, "Saltwater intrusion in the United States," *J. Hyd. Div.*, Proc. ASCE, **95**(HY5), 1969.

[500] Telford, W.M., Geldart, L.P. and Sheriff, R.E., *Applied Geophysics*, Cambridge Univ. Press, Cambridge, 770 p., 1990.

[501] Theon, J.S., "Global warming and environmental changes on the surface of the earth," Memorial Seminar for the Silver Jubilee on the Occasion of the XXV IAHR Biennial Congress, Tokyo, Japan, 1993.

[502] Thomsen, R., "The effect of climate variability and change on groundwater in Europe," *Conf. Climate and Water*, Publ. of the Academy of Finland, **1**, 486-500, 1989.

[503] Todd, D.K., "Sea water intrusion in coastal aquifers," *Am. Geophys. Union Trans.*, **34**, 749-754, 1953.

[504] Todd, D.K., "Sources of saline intrusion in the 400-foot aquifer, Castroville area, California,".Report for Monterey County Flood Control & Water Conservation District, Salinas, California, 41 p., 1989.

[505] Tolmach, Y., "Coastal aquifer, Areas of Tel-Aviv through Hadera," (in Hebrew), *Hydrogeological Atlas of Israel*, **3–6**, 70 p., 1979.

[506] Toth, J., "A theoretical analysis of groundwater flow in small drainage basins," *J. Geophys. Res.*, **68**, 4795-4812, 1963.

[507] Underwood, M.R., Peterson, F.L. and Voss, C.I., "Groundwater lens dynamics of Atoll Islands," *Water Resour. Res.*, **28**(11), 2889-2902, 1992.

[508] Unger, A. J.A., Forsyth, P.A. and Sudicky, E.A., "Variable spatial and temporal weighting schemes for use in multi-phase compositional problems," *Adv. Water Resources*, **19**(1), 1–27, 1996.

[509] US EPA, *Progress in Groundwater Protection and Restoration.*, 440/6-90-001, 40 p., 1990.

[510] Vacher, H.L., "Introduction: Varieties of carbonate islands and a historical perspective," Chap. 1 in *Geology and Hydrogeology of Carbonate Islands*, (eds.) H.L. Vacher and E.A. Shinn, Developments in Sedimentology, **54**, Elsevier, New York, 1-34, 1997.

[511] Vacher, H.L., Wightman, M.J. and Stewart, M., "Hydrology of meteoric diagenesis: Effect of Pleistocene stratigraphy on the Ghyben-Herzberg lens of Big Pine Key, FL," In: *Quaternary Coasts of the United States: Marine and Lacustrine Systems*, (eds.) Fletcher and Wehmiller, SEPM Spec. Publ., 48, 213-219, 1992.

[512] Vacher, H.L., Wallis, T.N. and Stewart, M., "Hydrogeology of a freshwater lens beneath a Holocene strandplain: Great Exuma, Bahamas," *J. Hydrology*, **125**, 93-109, 1991.

[513] van der Vorst, H., "Bi-CGSTAB: A fast and smoothly converging variant of BI-CG for the solution of nonsymmetric linear systems," *SIAM J. Sci. Stat. Comput.*, **13**, 631–644, 1992.

[514] van Genuchten, M.T. and Nielsen, D.R., "On describing and predicting the hydraulic properties of unsaturated soils," *Ann. Geophys.*, **3**(5), 615–628, 1985.

[515] van Leer, B., "Towards the ultimate conservative difference scheme, IV, A new approach to numerical convection," *J. Comp. Phys.*, **23**, 263–275, 1977.

[516] Vengosh, A., Gill, J., Reyes, A. and Thoresberg, K., "A multi-isotope investigation of the origin of ground water salinity in Salinas Valley, California," American Geophsical Union, San Francisco, California, 1997.

[517] Vengosh, A., Heumann, K.G., Juraske, S. and Kasher, R., "Boron isotope application for tracing sources of contamination in groundwater," *Environm. Sci. & Tech.*, **28**, 1968-1974, 1994.

[518] Vengosh, A. and Pankratov, I., "Chloride/bromide and chloride/fluoride ratios of domestic sewage effluents and associated contaminated ground water," *Ground Water*, 1998.

[519] Vengosh, A., Spivack, A.J., Artzi, Y. and Ayalon, A., "Boron, Strontium and Oxygen Isotopic constraints on the origin of brackish ground water From the Mediterranean Coast, Israel," American Geophysical Union, Fall Meeting, San Francisco, 1996.

[520] Vengosh, A., Starinsky, A., Melloul, A., *Salinization of the Coastal Aquifer Water by Ca-Chloride Solutions at the Interface Zone Along the Coastal Plain of Israel*, (in Hebrew), Geological Survey of Israel Rep. 27/91, 21 p., 1991a.

[521] Vengosh, A., Starinsky, A., Melloul, A., Fink, M. and Erlich, S., "Salinization of the coastal aquifer water by Ca-chloride solutions at the interface zone, along the Coastal Plain of Israel," Hydrological Service Report, Hydro/20/1991, (in Hebrew), 1991b.

[522] Vengosh, A. and Rosenthal, A., "Saline groundwater in Israel: Its bearing on the water crisis in the country," *J. Hydrology*, **156**, 389-430, 1994.

[523] Ventura County Public Works Agency, *Annual Report–Fox Canyon Groundwater Management Agency*, Ventura, CA., 1995.

[524] Visher, F.N. and Mink, J.F., *Ground-Water Resources in Southern Oahu, Hawaii*, U.S. Geol. Survey Water-Supply Paper 1778, 39 p., 1964.

[525] Vogel, P., Schelkes, K. and Giesel, W., " Modelling of variable-density flow in an aquifer crossing a salt dome—First Results," Proc. 12th Saltwater Intrusion Meeting, Barcelona, Nov. 1992, CIMNE, 359-369, 1993.

[526] Volker, R. and Rushton, K., "An assessment of the importance of some parameters for sea-water intrusion in aquifers and a comparison of dispersive and sharp-interface modelling approaches," *J. Hydrol.*, **56**, 239-250, 1982.

[527] Voss, C.I., *SUTRA—Saturated Unsaturated Transport—A Finite-Element Simulation Model for Saturated-Unsaturated, Fluid-Density-Dependent Ground-Water Flow with Energy Transport or Chemically-Reactive Single-Species Solute Transport*, U.S. Geol. Survey Water-Res. Inv. Rept. 84-4369, 409 p., 1984a.

[528] Voss, C.I., *AQUIFEM-SALT: A Finite-Element Model for Aquifers Containing a Seawater Interface*, U.S. Geol. Survey Water-Res. Inv. Rept. 84-4263, 1984b.

[529] Voss, C.I. and Andersson, J., "Regional flow in the Baltic Shield during holocene coastal regression, *Ground Water*, **31**, 6, 989-1006, 1993.

[530] Voss, C.I., Boldt, D. and Shapiro, A.M., *A graphical-user interface for the U.S. Geological Survey's SUTRA code using ArgusONE (for simulation of saturated-unsaturated ground-water flow with solute or energy transport)*, U.S. Geol. Survey Open-File Rept. 97-421, 106 p., 1997.

[531] Voss, C.I. and Souza,W.R., "Variable density flow and solute transport simulation of regional aquifers containing a narrow freshwater-saltwater transition zone," *Water Resour. Res.*, **23**(10), 1851-1866, 1987.

[532] Voss, C.I. and Souza, W.R., *Dynamics of a regional freshwater-saltwater transition zone in an anisotropic coastal aquifer system*, U.S. Geol. Survey Open-File Rept. 98-398, 88 p., 1998.

[533] Voss, C.I. and Wood, W.W., "Synthesis of geochemical, isotopic and ground-water modeling analysis to explain regional flow in a coastal aquifer of southern Oahu, Hawaii," In: *Mathematical Models and their Applications to Isotope Studies in Groundwater Hydrology*, International Atomic Energy Agency (IAEA) Vienna, Austria, IAEA-TECDOC-777, 147-178, 1994.

[534] Wagner, B.J., "Simultaneous parameter estimation and contaminant source characterization for coupled groundwater-flow and contaminant transport modelling," *J. Hydrology*, **135**(1-4), 275-303, 1992.

[535] Wagner, B.J., "Sampling design methods for groundwater modelling under uncertainty," *Water Resour. Res.*, **31**(10), 2581-2591, 1995.

[536] Wagner, B.J. and Gorelick, S.M., "Optimal groundwater quality management under parameter uncertainty," *Water Resour. Res.*, **23**(7), 1162-1174, 1987.

[537] Wagner, B.J. and Gorelick, S.M., "Reliable aquifer remediation in the presence of spatially variable hydraulic conductivity: from data to design," *Water Resour. Res.*, **25**(10), 2211-2225, 1989.

[538] Wait, R.L. and Callahan, J.T., "Relations of fresh and salty groundwater along the southeastern U.S. Atlantic Coast," *Ground Water*, **1**, 3-17, 1965.

[539] Ward, W.C. and Halley, R.B., "Dolomitization in a mixing zone of near-seawater composition, late Pleistocene, northeastern Yucatan Peninsula," *J. Sed. Petrology*, **55**(3), 407-420, 1985.

[540] Ward, S.H., *Geotechnical and Environmental Geophysics*, Soc. Exploration Geophysicists, Inv. in Geophysics, n. 5, vols. I, II, III., 1990a.

[541] Ward, S.H., "Resistivity and induced polarization methods," In: *Geotechnical and Environmental Geophysics*, v. I, Soc. Explor. Geophysicists, Inv. in Geophysics, n. 5, (ed.) S. Ward, 147-190, 1990b.

[542] Warick, A.W., Islas, A. and Lommen, D.O., "An analytical solution to Richards' equation for time-varying in filtration," *Water Resour. Res.*, **27**(5), 763-766, 1991.

[543] Warrick, R.A., Oerlemans, J., Woodworth, P.L., Meier, M.F. and le Provost, C., "Changes in sea level," In: *Climate Change 1995: The Science of Climate Change, Contribution of Working Group I to the Second Assessment Report of the Intergovernmental Panel on Climate Change*, eds. J.T. Houghton, L.G. Meira Filho and B.A. Callander, Cambridge University Press, Cambridge, 359-405, 1995.

[544] Weinstein, H., Stone, H. and Kwan, T., "Iterative procedure for solution of systems of parabolic and elliptic equations in three dimensions," *Ind. Eng. Chem. Fundam.*, **8**, 281-287, 1969.

[545] Weinstein, H., Stone, H. and Kwan, T., "Simultaneous solution of multi-phase reservoir flow equations," *Soc. Pet. Eng. J.*, **10**, 99-110, 1970.

[546] Weiss, E., *A Model for the Simulation of Flow of Variable-Density Ground Water in Three Dimensions under Steady-State Conditions*, U.S. Geol. Survey Open-File Rept. 82-352, 1982.

[547] Wentworth, C. K., *Geology and Ground-Water Resources of the Kalihi District*, Board of Water Supply, City and County of Honolulu, Hawaii, 1941.

[548] Wesner, G.M. and Herndon, R.L., *Orange County Water District Water Reclamation and Seawater Intrusion Barrier Project*, OCWD Engineering Report, 1990.

[549] Whitaker, F.F. and Smart, P.L., "Bacterially-Mediated Oxidation of Organic matter: A Major Control on Groundwater Geochemistry and Porosity Generation in Oceanic Carbonate Terrains, Breakthroughs in Karst Geomicrobiology and Redox Geochemistry," Abstracts and Field-Trip Guide for the symposium held Feb. 16-19, 1994, Colorado Springs, CO, Special Publication 1, 72-74, 1994.

[550] White, P.A., "Electrode arrays for measuring groundwater flow direction and velocity," *Geophysics*, **59**(2), 192-201, 1994.

[551] Wicks, C.M. and Herman, J.S., "Regional hydrogeochemistry of a modern coastal mixing zone," *Water Resour. Res.*, **32**(2), 401-407, 1996.

[552] Wicks, C.M. and Herman, J.S., Randazzo, A.F. and Jee, J.L., "Water-rock interactions in a modern coastal mixing zone," *Geol. Soc. Am. Bull.*, **107**(9), 1023-1032, 1995.

[553] Wicks, C.M. and Troester, J.W., "Hydrologic and hydrochemical analysis of the Coastal Plain aquifer in Isla de Mona, Puerto Rico," 30th Annual Meeting of the Geol. Soc. Am., north-central section, **28**(6), 70, 1997.

[554] Wightman, M.J., "Geophysical analysis and Dupuit-Ghyben-Herzberg modeling of fresh-water lenses on Big Pine Key, Florida," M.S. thesis, Univ. South Florida, Tampa, FL, 122 p., 1990.

[555] Wigley, T.M.L. and Plummer, L.N, "Mixing of carbonate waters," *Geochem. Cosmochem Acta*, **40**(9), 989-995, 1976.

[556] Williams, M.D. and Ranganathan, V., "Ephemeral thermal and solute plumes formed by upwelling groundwaters near salt domes," *J. Geophys. Res.-Solid Earth*, **99**(B8), 15667-15681, 1994.

[557] Williamson, A.K., "Preliminary simulation of ground-water flow in the Gulf Coast Aquifer Systems, South-Central United States," In: *Regional Aquifer Systems of the United States: Aquifers of the Atlantic and Gulf Coastal Plain*, ed. John Vecchioli and A.I. Johnson, Am. Water Resour. Assoc. Monograph Ser. No. 9, Bethesda, MD, 119-137, 1987.

[558] Wilson, J.L. and Sa da Costa, A., "Finite element simulation of a saltwater/freshwater interface with indirect toe tracking," *Water Resour. Res.*, **18**(4), 1069-1080, 1982.

[559] Wilson, J.L., Townley, H. and Sa da Costa, A., "Mathematical development and verification of a finite element aquifer flow model AQUIFEM-1," Technology Adaptation Program Report no. 79-2, MIT, Cambridge, 1979.

[560] Wirojanagud, P. and Charbeneau, R.J., "Salt water upconing in unconfined aquifers," *J. Hydraul. Eng. Am. Soc. Civ. Eng.*, **111**, 417–434, 1985.

[561] W.M.O., *Manual on Water-Quality Monitoring*, Operational Hydrology Rep. 27, 197 p., 1988.

[562] Wolf, A.V., Brown, M.G. and Prentiss, P.G., *Concentrative Properties of Aqueous Solutions Conversion Tables*, CRC Handbook of Chemistry and Physics, Weast, R. C. and Astle, M. J. (Eds.), CRC Press, 1980.

[563] Wong, P., "The statistical physics of sedimentary rock," *Physics Today*, 22-30, 1988.

[564] Xue, Y. and Xie, C., "A three-dimensional miscible transport model for seawater intrusion in China," *Water Resour. Res.*, 4, 903-912, 1995.

[565] Yechieli, Y., Ronen, D. and Vengosh, A., "Preliminary 14C study of groundwater at the fresh-saline water interface of the Mediterranean coastal plain aquifer in Israel," *Proc. 14th Salt Water Intrusion Meeting*, Malmo, Sweden, 84-90, 1996.

[566] Zeitoun, D.G. and Soyeux, E. "A Geographic Information System for the hydrological control of the coastal aquifer of Israel," submitted to J. Water Resources Planning and Management, ASCE, 1998.

[567] Zeitoun, D.G., Zelinger, A. and Chachnai, E., "Design of the coastal collectors of Rubin North," TAHAL 04/94/18, 1994.

[568] Zienkiewicz, O.C., *The Finite Element Method*, McGraw-Hill, 1986.

[569] Zohdy, A.A.R., "A new method for the automatic interpretation of Schlumberger and Wenner sounding curves," *Geophysics*, **54**(2), 245-253, 1989.

[570] Zohdy, A.A.R., "Automatic interpretation of Schlumberger sounding curves, using modified Dar Zarrouk functions," U.S. Geol. Survey Bull. 1313-E, 1975.

[571] Zohdy, A.A.R., "A computer program for the automatic interpretation of Schlumberger sounding curves over horizontally stratified media," Nat. Tech. Infor. Serv., PB-232-703, 1973.

[572] Zohdy, A.A.R. and Bisdorf, R.J., *Programs for the Automatic Processing and Interpretation of Schlumberger Sounding Curves*, U.S. Geol. Survey Open-File Rept. 89-137 A & B, 1989.

Theory and Applications of Transport in Porous Media

Series Editor:
Jacob Bear, *Technion – Israel Institute of Technology, Haifa, Israel*

1. H.I. Ene and D. Polisševski: *Thermal Flow in Porous Media*. 1987
    ISBN 90-277-2225-0

2. J. Bear and A. Verruijt: *Modeling Groundwater Flow and Pollution*. With Computer
    Programs for Sample Cases. 1987   ISBN 1-55608-014-X; Pb 1-55608-015-8

3. G.I. Barenblatt, V.M. Entov and V.M. Ryzhik: *Theory of Fluid Flows Through Natural
    Rocks*. 1990   ISBN 0-7923-0167-6

4. J. Bear and Y. Bachmat: *Introduction to Modeling of Transport Phenomena in Porous
    Media*. 1990   ISBN 0-7923-0557-4; Pb (1991) 0-7923-1106-X

5. J. Bear and J-M. Buchlin (eds.): *Modelling and Applications of Transport Phenomena
    in Porous Media*. 1991   ISBN 0-7923-1443-3

6. Ne-Zheng Sun: *Inverse Problems in Groundwater Modeling*. 1994
    ISBN 0-7923-2987-2

7. A. Verruijt: *Computational Geomechanics*. 1995   ISBN 0-7923-3407-8

8. V.N. Nikolaevskiy: *Geomechanics and Fluidodynamics*. With Applications to Reser-
    voir Engineering. 1996   ISBN 0-7923-3793-X

9. V.I. Selyakov and V.V. Kadet: *Percolation Models for Transport in Porous Media*.
    With Applications to Reservoir Engineering. 1996   ISBN 0-7923-4322-0

10. J.H. Cushman: *The Physics of Fluids in Hierarchical Porous Media: Angstroms to
    Miles*. 1997   ISBN 0-7923-4742-0

11. J.M. Crolet and M. El Hatri (eds.): *Recent Advances in Problems of Flow and
    Transport in Porous Media*. 1998   ISBN 0-7923-4938-5

12. K.C. Khilar and H.S. Fogler: *Migration of Fines in Porous Media*. 1998
    ISBN 0-7923-5284-X

13. S. Whitaker: *The Method of Volume Averaging*. 1999   ISBN 0-7923-5486-9

14. J. Bear, A.H.-D. Cheng, S. Sorek, D. Ouazar and I. Herrera (eds.): *Seawater Intrusion
    in Coastal Aquifers – Concepts, Methods and Practices*. 1999
    ISBN 0-7923-5573-3

Kluwer Academic Publishers – Dordrecht / Boston / London